T0323572

Introduction to Plasma Physics

INTRODUCTION TO PLASMA PHYSICS

Robert J Goldston
and
Paul H Rutherford

Plasma Physics Laboratory
Princeton University

Taylor & Francis
Taylor & Francis Group
New York   London

Published in 2000 by
Taylor & Francis Group
270 Madison Avenue
New York, NY 10016

© 1995 by Taylor & Francis Group, LLC

No claim to original U.S. Government works
Printed in the United States of America on acid-free paper
10 9 8 7 6 5 4 3 2

International Standard Book Number-10: 0-7503-0183-X (Softcover)
International Standard Book Number-13: 978-0-7503-0183-1 (Softcover)
Library of Congress catalog number- 95-37117

The software that originally came with this book is now available for download on the Web site.

Library of Congress Cataloging-in-Publication Data

Catalog record is available from the Library of Congress

Taylor & Francis Group
is the Academic Division of Informa plc.

**Visit the Taylor & Francis Web site at
http://www.taylorandfrancis.com**

The two diskettes that originally came this book are now available for download on the Web site.

Dedicated to

Ruth Berger Goldston

and

Audrey Rutherford

Contents

Preface

Plasmas occur pervasively in nature: indeed, most of the known matter in the Universe is in the ionized state, and many naturally occurring plasmas, such as the surface regions of the Sun, interstellar gas clouds and the Earth's magnetosphere, exhibit distinctively plasma-dynamical phenomena arising from the effects of electric and magnetic forces. The science of plasma physics was developed both to provide an understanding of these naturally occurring plasmas and in furtherance of the quest for controlled nuclear fusion. Plasma science has now been used in a number of other practical applications, such as the etching of advanced semiconductor chips and the development of compact x-ray lasers. Many of the conceptual tools developed in the course of fundamental research on the plasma state, such as the theory of Hamiltonian chaos, have found wide application outside the plasma field.

Research on controlled thermonuclear fusion has long been a world-wide enterprise. Major experimental facilities in Europe, Japan and the United States, as well as smaller facilities elsewhere including Russia, are making remarkable progress toward the realization of fusion conditions in a confined plasma. The use, for the first time, of a deuterium–tritium plasma in the tokamak experimental fusion device at the Princeton Plasma Physics Laboratory has recently produced slightly in excess of ten megawatts of fusion power, albeit for less than a second. In 1992, an agreement was signed by the European Union, Japan, the Russian Federation and the United States of America to undertake jointly the engineering design of an experimental reactor to demonstrate the practical feasibility of fusion power.

This book is based on a one-semester course offered at Princeton University to advanced undergraduates majoring in physics, astrophysics or engineering physics. If the more advanced material, identified by an asterisk after the Chapter heading or Section heading, is included then the book would also be suitable as an introductory text for graduate students entering the field of plasma physics.

We have attempted to cover all of the basic concepts of plasma physics with reasonable rigor but without striving for complete generality—especially where this would result in excessive algebraic complexity. Although single-particle, fluid and kinetic approaches are introduced independently, we emphasize the interconnections between different descriptions of plasma behavior; particular phenomena which illustrate these interconnections are highlighted. Indeed, a unifying theme of our book is the attempt at a deeper understanding of the underlying physics through the presentation of multiple perspectives on the same physical effects. Although there is some discussion of weakly ionized gases, such as are used in plasma etching or occur naturally in the Earth's ionosphere, our emphasis is on fully ionized plasmas, such as those encountered in many astrophysical settings and employed in research on controlled thermonuclear fusion, the field in which both of us work. The physical issues we address are, however, applicable to a wide range of plasma phenomena. We have included problems for the student, which range in difficulty from fairly straightforward to quite challenging; most of the problems have been used as homework in our course.

Standard international (SI) units are employed throughout the book, except that temperatures appearing in formulae are in units of energy (i.e. joules) to avoid repeated writing of Boltzmann's constant; for practical applications, temperatures are generally stated in electron-volts (eV). Appendices A and C allow the reader to convert from SI units to other units in common use.

The student should be well-prepared in electromagnetic theory, including Maxwell's equations, which are provided in SI units in Appendix B. The student should also have some knowledge of thermodynamics and statistical mechanics, including the Maxwell–Boltzmann distribution. Preparation in mathematics must have included vectors and vector calculus, including the Gauss and Stokes theorems, some familiarity with tensors or at least the underlying linear algebra, and complex analysis including contour integration. Appendix D contains all of the vector formulae that are used, while Appendix E gives expressions for the relevant differential operators in various coordinate systems. Higher transcendental functions, such as Bessel functions, are avoided. Suggestions for further reading are given in Appendix F.

In addition to the regular problems, which are to be found in all chapters, we have provided a disk containing two graphics programs, which allow the student to experiment visually with mathematical models of quite complex plasma phenomena and which form the basis for some homework problems and for optional semester-long student projects. These programs are provided in both Macintosh[1] and IBM PC-compatible format. In the first of these two computer programs, the reader is introduced to the relatively advanced topic of area-preserving maps and Hamiltonian chaos; these topics, which form another of the underlying themes of the book, reappear later in our discussions both of the magnetic islands caused by resistive tearing modes and of the nonlinear

---

[1] Macintosh is a registered trademark of Apple Computer, Inc.

phase of electron plasma waves.

We are deeply indebted to Janet Hergenhan, who prepared the manuscript in LATEX format, patiently resetting draft after draft as we reworked our arguments and clarified our presentations. We would also like to thank Greg Czechowicz, who has drawn many of the figures, John Wright, who produced the IBM-PC versions of our programs, and Keith Voss, who served for three years as our 'grader', working all of the problems used in the course and offering numerous excellent suggestions on the course material.

We are grateful to Maureen Clarke and, more recently, James Revill of Institute of Physics Publishing, who have suffered patiently through our many delays in producing a completed manuscript.

Our own research in plasma physics and controlled fusion has been supported by the United States Department of Energy, Contract No. DE-AC02-76-CHO-3073.

Robert J Goldston
Paul H Rutherford
Princeton, 1995

Introduction

After an initial Chapter, which introduces plasmas, both in the laboratory and in nature, and derives the defining characteristics of the plasma state, this book is divided into six 'Units'. In Unit 1, the plasma is considered as an assemblage of charged particles, each moving independently in prescribed electromagnetic fields. After deriving all of the main features of the particle orbits, the topic of 'adiabatic' invariants is introduced, as well as the conditions for 'non-adiabaticity', illustrating the latter by means of the modern dynamical concepts of mappings and the onset of stochasticity. In Unit 2, the fluid model of a plasma is introduced, in which the electromagnetic fields are required to be self-consistent with the currents and charges in the plasma. Particular attention is given to demonstrating the equivalence of the particle and fluid approaches. In Unit 3, after an initial Chapter which describes the most important atomic processes that occur in a plasma, the effects of Coulomb collisions are treated in some detail. In Unit 4, the topic of small-amplitude waves is covered in both the 'cold' and 'warm' plasma approximations. The treatment of waves in the low-frequency branch of the spectrum leads naturally, in Unit 5, to an analysis of three of the most important instabilities in non-spatially-uniform configurations: the Rayleigh–Taylor (flute), resistive tearing, and drift-wave instabilities. In Unit 6, the kinetic treatment of 'hot' plasma phenomena is introduced, from which the Landau treatment of wave–particle interactions and associated instabilities is derived; this is then extended to the non-uniform plasma in the drift-kinetic approximation.

Chapter 1

Introduction to plasmas

1.1 WHAT IS A PLASMA?

First and foremost, a plasma is an ionized gas. When a solid is heated sufficiently that the thermal motion of the atoms breaks the crystal lattice structure apart, usually a liquid is formed. When a liquid is heated enough that atoms vaporize off the surface faster than they recondense, a gas is formed. When a gas is heated enough that the atoms collide with each other and knock their electrons off in the process, a plasma is formed: the so-called 'fourth state of matter'. Exactly when the transition between a 'very weakly ionized gas' and a 'plasma' occurs is largely a matter of nomenclature. The important point is that an ionized gas has unique properties. In most materials the dynamics of motion are determined by forces between near-neighbor regions of the material. In a plasma, charge separation between ions and electrons gives rise to electric fields, and charged-particle flows give rise to currents and magnetic fields. These fields result in 'action at a distance', and a range of phenomena of startling complexity, of considerable practical utility and sometimes of great beauty.

Irving Langmuir, the Nobel laureate who pioneered the scientific study of ionized gases, gave this new state of matter the name 'plasma'. In greek $\pi\lambda\alpha\sigma\mu\alpha$ means 'moldable substance', or 'jelly', and indeed the mercury arc plasmas with which he worked tended to diffuse throughout their glass vacuum chambers, filling them like jelly in a mold[1].

---

[1] We also like to imagine that Langmuir listened to the blues. Maybe he was thinking of the song 'Must be Jelly 'cause Jam don't Shake Like That', recorded by J Chalmers MacGregor and Sonny Skylar. This song was popular in the late 1920s, when Langmuir, Tonks and Mott-Smith were studying oscillations in plasmas.

1.2 HOW ARE PLASMAS MADE?

A plasma is not usually made simply by heating up a container of gas. The problem is that for the most part a container cannot be as hot as a plasma needs to be in order to be ionized—or the container itself would vaporize and become plasma as well.

Typically, in the laboratory, a small amount of gas is heated and ionized by driving an electric current through it, or by shining radio waves into it. Either the thermal capacity of the container is used to keep it from getting hot enough to melt—let alone ionize—during a short heating pulse, or the container is actively cooled (for example with water) for longer-pulse operation. Generally, these means of plasma formation give energy to free electrons in the plasma directly, and then electron–atom collisions liberate more electrons, and the process cascades until the desired degree of ionization is achieved. Sometimes the electrons end up quite a bit hotter than the ions, since the electrons carry the electrical current or absorb the radio waves.

1.3 WHAT ARE PLASMAS USED FOR?

There are all sorts of uses for plasmas. To give one example, if we want to make a short-wavelength laser we need to generate a population inversion in highly excited atomic states. Generally, gas lasers are 'pumped' into their lasing states by driving an electric current through the gas, and using electron–atom collisions to excite the atoms. X-ray lasers depend on collisional excitation of more energetic states of partially ionized atoms in a plasma. Sometimes a magnetic field is used to hold the plasma together long enough to create the highly ionized states.

A whole field of 'plasma chemistry' exists where the chemical processes that can be accessed through highly excited atomic states are exploited. Plasma etching and deposition in semiconductor technology is a very important related enterprise. Plasmas used for these purposes are sometimes called 'process plasmas'.

Perhaps the most exciting application of plasmas such as the ones we will be studying is the production of power from thermonuclear fusion. A deuterium ion and a tritium ion which collide with energy in the range of tens of keV have a significant probability of fusing, and producing an alpha particle (helium nucleus) and a neutron, with 17.6 MeV of excess energy (alpha particle ~ 3.5 MeV, neutron ~ 14.1 MeV). A promising way to access this energy is to produce a plasma with a density in the range 10^{20} m^{-3} and average particle energies of tens of keV. The characteristic time for the thermal energy contained within such a plasma to escape to the surrounding material surfaces must exceed about five seconds, in order that the power produced in alpha particles can sustain the temperature of the plasma. This is not a simple requirement to meet, since electrons within a fusion plasma travel at velocities of $\sim 10^8 \, \text{m s}^{-1}$, while a fusion device must have a characteristic size of $\sim 2 \, \text{m}$, in order to be an economic power source. We will learn how magnetic fields are used to contain a hot plasma.

The goal of producing a plentiful and environmentally benign energy source is still decades away, but at the present writing fusion power levels of 2–10 MW have been produced in deuterium–tritium plasmas with temperatures of 20–40 keV and energy confinement times of 0.25–1 s. This compares with power levels in the 10 mW range that were produced in deuterium plasmas with temperatures of $\sim 1 \, \text{keV}$ and energy confinement times of $\sim 5 \, \text{ms}$ in the early 1970s. It is the quest for a limitless energy source from controlled thermonuclear fusion which has been the strongest impetus driving the development of the physics of hot plasmas.

1.4 ELECTRON CURRENT FLOW IN A VACUUM TUBE

Let us look more closely now at how a plasma is made with a dc electric current. Consider a vacuum tube (not filled with gas), with a simple planar electrode structure, as shown in Figure 1.1. Imagine that the cathode is sufficiently heated that copious electrons are boiling off of its surface, and (in the absence of an applied electric field) returning again. Now imagine we apply a potential to draw some of the electrons to the anode. First, let us look at the equation of motion for the electrons:

$$m_e \frac{d\mathbf{v}_e}{dt} = -e\mathbf{E} = e\nabla\phi \tag{1.1}$$

where m_e is the electron mass (9.1×10^{-31} kg), $\mathbf{v}_e$ is the vector electron velocity (m s^{-1}), e is the unit charge (1.6×10^{-19} C), $\mathbf{E}$ is the vector electric field (V m^{-1}), and ϕ is the electrical potential (V). To derive energy conservation, we take the dot product of both sides with $\mathbf{v}_e$:

$$m_e \mathbf{v}_e \cdot \frac{d\mathbf{v}_e}{dt} = \tfrac{1}{2} m_e \frac{dv_e^2}{dt} = e\mathbf{v}_e \cdot \nabla\phi. \tag{1.2}$$

The total (or convective) derivative, moving with the particle, is defined by

$$\frac{d}{dt} \equiv \frac{\partial}{\partial t} + \mathbf{v}_e \cdot \nabla. \tag{1.3}$$

Thus the total (convective) time derivative of the electric potential, ϕ, moving with the electron, can be viewed as being made up of a part having to do with the potential changing in time at a fixed location (the partial derivative, $\partial/\partial t$), plus a part having to do with the changing location at which we must evaluate ϕ. Since in this case we are considering a *steady-state* electric field, the partial (non-convective) time derivatives are zero. Thus we have

$$\frac{d}{dt}\left(\frac{m_e v_e^2}{2}\right) = \frac{d}{dt}(e\phi) \qquad (1.4)$$

or, moving along the trajectory of an electron,

$$\frac{m_e v_e^2}{2} - e\phi = \text{constant}. \qquad (1.5)$$

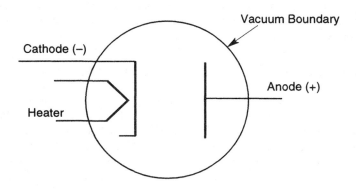

Figure 1.1. Vacuum-tube geometry for a hot-cathode Child–Langmuir calculation.

Equation (1.5) gives us some important information about the electron velocity in the inter-electrode space of our vacuum tube. If for simplicity we assign $\phi = 0$ to the cathode (since the offset to ϕ can be chosen arbitrarily), and negligibly small energy to the random 'boiling' energy of the electrons near the cathode, then the constant on the right-hand side of equation (1.5) can be taken to be zero, and

$$v_e \approx \left(\frac{2e\phi}{m_e}\right)^{1/2}. \qquad (1.6)$$

Note that, in this case, v_e is not a random thermal velocity, but rather a directed flow of the electrons—the individual velocities of the electrons and the average velocity of the electron 'fluid' are the same. As a consequence of this 'fluid' velocity of the electrons, there is a net current density $\mathbf{j}$ (amperes/meter^2) $\equiv -n_e e \mathbf{v}_e$ flowing between the two electrodes, where n_e is the number density of electrons—the electron 'count' per cubic meter. In order to understand this current, it is helpful to think of a differential cube, as shown in Figure 1.2, with edges of length dl, volume $(dl)^3$, and total electron count in the cube of

$n_e(dl)^3$. Imagine that the electron velocity is directed so that the contents are flowing out of one face of the cube (see Figure 1.2). If the fluid is moving at v_e (meters/second), the cube of electrons is emptied out across that face in time dl/v_e seconds. Thus, $en_e(dl)^3$ units of charge cross $(dl)^2$ square meters of surface in dl/v_e seconds—the current density is thus $en_e(dl)^3/[(dl/v_e)(dl)^2] = n_e e v_e$ (coulombs/second · meter^2, i.e. amperes/meter^2), as we stated above.

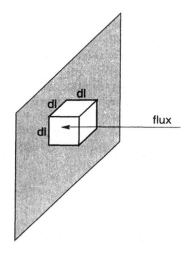

Figure 1.2. Geometry for interpreting $\mathbf{j} = -n_e e \mathbf{v}_e$.

If we now consider the integral of this particle current over the surface area of a given volume, we have the total flow of particles out of the volume per second, and so the time derivative of the total number of particles in a given volume of our vacuum tube is given by

$$\frac{\partial N_e}{\partial t} = -\int n_e \mathbf{v}_e \cdot d\mathbf{S} = 0 \qquad (1.7)$$

where N_e is the total number of particles in a volume, and $d\mathbf{S}$ is an element of area of its surface. Here we assume that there are no sources or sinks of electrons within the volume; by setting the result to zero we are positing a steady-state condition. By Gauss's theorem, this can be expressed in differential notation as

$$\frac{\partial n_e}{\partial t} = -\nabla \cdot (n_e \mathbf{v}_e) = 0. \qquad (1.8)$$

Poisson's equation is of course

$$\nabla \cdot (\epsilon_0 \nabla \phi) = en_e \qquad (1.9)$$

where ϵ_0, the permittivity of free space, is $8.85 \times 10^{-12}\,\mathrm{C\,V^{-1}\,m^{-1}}$.

The complete set of equations we need to solve in order to understand the current flow in our evacuated tube is then made up of equations (1.6), (1.8), and (1.9). Before we go on to solve these equations, we can immediately see a useful overall scaling relation. If we imagine taking any valid solution of this set of equations, and scaling ϕ by a factor α everywhere, then equation (1.9) tells us that n_e must scale by the same factor α. Equation (1.6) says that v_e must scale everywhere by $\alpha^{1/2}$. Equation (1.8) is also satisfied by this result, since $n_e v_e$ is scaled everywhere equally by $\alpha^{3/2}$. In the conditions we have been describing, with plenty of electrons boiling off the cathode (so there is no limit to the source of electrons at the boundary of our problem), the total current in the tube scales as $\phi^{3/2}$. This is called the Child–Langmuir law.

The condition we are considering is called space-charge-limited current flow. If too few electrons are available from the cathode, the current can fall below the Child–Langmuir law. It is then called emission-limited current flow. For the specific case of planar electrodes, with a gap smaller than the typical electrode dimensions, we can approximate the situation using one-dimensional versions of equations (1.8) and (1.9):

$$-n_e e v_e = j = \text{constant} \tag{1.10}$$

and

$$\frac{d}{dx}\left(\epsilon_0 \frac{d\phi}{dx}\right) = e n_e. \tag{1.11}$$

Substituting equation (1.6), we have

$$\epsilon_0 \frac{d^2\phi}{dx^2} = e n_e = -j/v_e = -j\left(\frac{m_e}{2e\phi}\right)^{1/2} \tag{1.12}$$

We can find a solution to this nonlinear equation simply by assuming that $\phi \propto x^\beta$, where β is some constant power. Looking at the powers of x that occur on each side, we come to the conclusion that

$$\beta - 2 = -\beta/2 \quad \text{or} \quad \beta = 4/3. \tag{1.13}$$

So now we can assume that $\phi = Ax^{4/3}$ which, when substituted into equation (1.12), gives

$$\epsilon_0 A (4/3)(1/3) = -j\left(\frac{m_e}{2eA}\right)^{1/2} \tag{1.14}$$

or

$$\phi(x) = \left(\frac{-9j}{4\epsilon_0}\right)^{2/3}\left(\frac{m_e}{2e}\right)^{1/3} x^{4/3}. \tag{1.15}$$

This solution is appropriate for our conditions, where we have taken the potential to be zero at the cathode, and since so many electrons are 'boiling' around the cathode, we have assumed that negligible electric field strength is required to extract electrons from this region. Thus we have chosen the solution that has $d\phi/dx = 0$ where $\phi = 0$, i.e. at $x = 0$. Let us now make the last step of deriving the current–voltage characteristics of our vacuum tube. At $x = L$ (where L is the inter-electrode spacing), let the potential be V volts. Then we can solve equation (1.15) for the current density:

$$ j = -\frac{4\epsilon_0}{9L^2} \left(\frac{2e}{m_e} \right)^{1/2} V^{3/2}. \tag{1.16} $$

Finally, let us evaluate the performance of a specific configuration. Let us take a fairly large tube: an inter-electrode spacing of 0.01 m, and an electrode area of $0.05\,\text{m} \times 0.20\,\text{m} = 0.01\,\text{m}^2$. For a voltage drop of 50 V, we get a current drain of $8.3\,\text{A}\,\text{m}^{-2}$, or only 83 mA—we need much larger electric fields to draw significant power in a vacuum tube. The cloud of electrons at a density of about $2 \times 10^{13}\,\text{m}^{-3}$ impedes the flow of current rather effectively. For perspective, note that a tungsten cathode of this area can provide an emission current of hundreds of amperes.

1.5 THE ARC DISCHARGE

We have now in our vacuum tube a population of electrons with energies up to 50 eV. Let us imagine introducing gas at a pressure of $\sim 1\,\text{Pa}$ (about 10^{-5} of an atmosphere). The electrons emitted from the cathode will collide with the gas molecules, transferring momentum and energy efficiently to the bound electrons within these gas molecules. Since typical binding energies of outer-shell electrons are in the few eV range, these collisions have a good probability of ionizing the gas, resulting in more free electrons. The 'secondary' electrons created in this way are then heated by collisions with the incoming primary electrons from the hot cathode, and cause further ionizations themselves. Eventually the ions and electrons come into thermal equilibrium with each other at temperatures corresponding to particle energies in the range of 2 eV, in the plasma generated in such an 'arc' discharge. Since most of the electrons are now thermalized—not monoenergetic as in the Child–Langmuir problem—they have a range of velocities. The energy of some of the secondary electrons, as well as that of the primaries, is high enough to continue to cause ionization. This continual ionization process balances the loss of ions which drift out of the plasma and recombine with electrons at the cathode or on the walls of the discharge chamber, and the system comes into steady state. Ion and electron densities in the range of $10^{18}\,\text{m}^{-3}$ are easily obtained in such a system.

Matters have changed dramatically from the original Child–Langmuir problem. The electron density has risen by five orders of magnitude, but nonetheless the space-charge effect impeding the flow of the electron current is greatly reduced. The presence of the plasma, which is an excellent conductor of electricity, greatly reduces the potential gradient in most of the inter-electrode space. Only in the region close to the cathode are the neutralizing ions absent—because there they are rapidly drawn into the cathode by its negative potential. Almost all of the potential drop occurs then across this narrow 'sheath' in front of the cathode. If we return to equation (1.16), we see that the current extracted from the cathode must then increase by about the ratio $(L/\lambda_s)^2$, where λ_s is the width of the cathode sheath.

The current–voltage characteristic of an arc plasma is very different from the Child–Langmuir relation: indeed in a certain sense its resistance is negative. The external circuit driving the arc must include a resistive element as well as a voltage source. If the resistance of this element is reduced, allowing *more current* to flow through the arc, the plasma density increases due to the increased input power, the cathode sheath narrows due to the higher plasma density, *and the voltage drop across the arc falls!* Of course even though the voltage decreases with rising current, the input power, IV, increases. This nonetheless strange situation pertains up to the point where the full electron emission from the cathode is drawn into the arc. The voltage drop at this point might be 10–20 V in our case, the current hundreds of amperes, and the input power would be thousands of watts. If the current is raised further the arc makes the transition from space-charge-limited to emission-limited, and the voltage across the arc rises with rising current, since a higher voltage is needed to pull ions into the cathode.

Thus, as we can see, by introducing gas—and therefore plasma—into the problem, we have created a very different situation. From an engineering point of view, we now have to consider how to handle kilowatts of heat outflow from a small volume. From a physics point of view, it is interesting now to try to understand the behavior of the new state of matter we have just created.

Of course we do not always have to make a plasma in order to study one. The Sun is a plasma; so are the Van Allen radiation belts surrounding the Earth. The solar wind is a streaming plasma that fills the solar system. These plasmas in our solar system provide many unsolved mysteries. How is the Sun's magnetic field generated, and why does it flip every eleven years? How is the solar corona heated to temperatures greater than the surface temperature of the Sun? What causes the magnetic storms that result in a rain of energetic particles into the Earth's atmosphere, and disturbances in the Earth's magnetic field? Outside of the solar system there are also many plasma-related topics. What is the role of magnetic fields in galactic dynamics? The signals from pulsars are thought to be synchrotron radiation from rotating, highly magnetized neutron stars. What can we learn from these signals about the atmospheres of neutron stars and about the interstellar medium? All of these are very active areas of research.

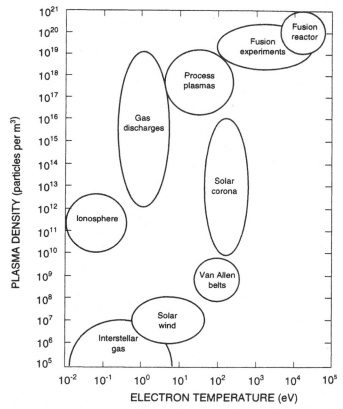

Figure 1.3. Typical parameters of naturally occurring and laboratory plasmas.

Some typical parameters of naturally occurring and laboratory plasmas are given in Table 1.1. Their density and temperature parameter regimes are illustrated in Figure 1.3. We see that the plasma state spans enormous ranges in scale-length, density of particles and temperature.

1.6 THERMAL DISTRIBUTION OF VELOCITIES IN A PLASMA

If we have a plasma in some form of near-equilibrium, i.e. where the particles collide with each other frequently compared to the characteristic time-scale over which energy and particles are replaced, it is reasonable to expect the laws of equilibrium statistical mechanics to give a good approximation to the distribution of velocities of the particles. We will assume for the time being that the distribution with respect to space is uniform.

Table 1.1. Typical parameters of naturally occurring and laboratory plasmas.

	Length scale (m)	Particle density (m^{-3})	Electron temperature (eV)	Magnetic field (T)
Interstellar gas	10^{16}	10^6	1	10^{-10}
Solar wind	10^{10}	10^7	10	10^{-8}
Van Allen belts	10^6	10^9	10^2	10^{-6}
Earth's ionosphere	10^5	10^{11}	10^{-1}	3×10^{-5}
Solar corona	10^8	10^{13}	10^2	10^{-9}
Gas discharges	10^{-2}	10^{18}	2	—
Process plasmas	10^{-1}	10^{18}	10^2	10^{-1}
Fusion experiment	1	10^{19}–10^{20}	10^3–10^4	5
Fusion reactor	2	10^{20}	10^4	5

Consider any one specific particle, labeled 'r', in the plasma as a distinguishable microsystem. We will ignore quantum-mechanical effects that make distinguishability invalid, and consider only particles that behave classically.

Problem 1.1: What are some plasma parameters (electron temperatures and densities) where quantum-mechanical effects might be important?

We now ask the question: what is the probability P_r of finding our specific particle in any *one* particular state of energy W_r? The particle has to have gained this energy W_r from its interaction with the others, so the remaining thermal 'bath' of particles must have energy $W_{tot} - W_r$, where W_{tot} is the total thermal energy in the plasma. If the particles have collided with each other enough, we can expect the fundamental theorem of statistical mechanics to hold. This theorem amounts to saying that we know as little as could possibly be known about any given thermal system: all possible accessible microstates of the total system are populated with equal probability. Thus in order to determine the probability P_r of any given state of our specific particle, we need only evaluate the number of microstates accessible to the 'bath' with energy $W_{tot} - W_r$. Let us define Ω as the number of microstates accessible to the bath with total energy W. Then, for any thermal system statistical mechanics *defines* its temperature,

T, by the relation

$$\frac{1}{T} \equiv \frac{k\,\mathrm{d}\ln\Omega}{\mathrm{d}W} \equiv \frac{\mathrm{d}S}{\mathrm{d}W} \tag{1.17}$$

where k is the Boltzmann constant, and the entropy, S, of the system is defined by $S \equiv k\ln\Omega$. Since the energy of our specific particle is small compared to the energy of the bath, we can approximate the number of microstates available to the system by

$$\ln\,\Omega|_{W_{\text{tot}}-W_r} \approx \ln\,\Omega|_{W_{\text{tot}}} - W_r/kT. \tag{1.18}$$

Taking the exponential of both sides, we obtain

$$\Omega|_{W_{\text{tot}}-W_r} \approx \Omega\,|_{W_{\text{tot}}}\exp(-W_r/kT) \tag{1.19}$$

which is just the result we are seeking. The relative probability P_r of the particle having energy W_r is given by the famous 'Boltzmann factor', $\exp(-W_r/kT)$, since Ω evaluated at W_{tot} is not a function of W_r.

If we ignore, for the time being, any potential energy associated with the position of the particle, we have the result that the relative probability that the velocity of our particle lies in some range of velocities $\mathrm{d}v_x\mathrm{d}v_y\mathrm{d}v_z$ centered around velocity (v_x, v_y, v_z) is given by

$$\exp\left(\frac{-m(v_x^2 + v_y^2 + v_z^2)}{2kT}\right)\mathrm{d}v_x\mathrm{d}v_y\mathrm{d}v_z \tag{1.20}$$

where m is the mass of the particle. Since there was nothing special about our particular particle (which was chosen arbitrarily from the bath), this same relative probability distribution is appropriate for all the particles in the bath. It is convenient to define a 'phase-space density of particles', $f(\mathbf{x}, \mathbf{v})$, which gives the number of particles per unit of $\mathrm{d}x\mathrm{d}y\mathrm{d}z\mathrm{d}v_x\mathrm{d}v_y\mathrm{d}v_z$, the volume element of six-dimensional phase space. The three-dimensional integral of f over all velocities, $\mathbf{v}$, gives the number density of particles per unit volume of ordinary physical space, which we denote n. The units of f are given by

$$[f] = \mathrm{m}^{-3}(\mathrm{m\,s}^{-1})^{-3} = \mathrm{s}^3\,\mathrm{m}^{-6}. \tag{1.21}$$

For a Maxwell–Boltzmann distribution, f is simply the Boltzmann factor with an appropriate normalization. If we carry through the necessary integral over all $\mathbf{v}$ to ensure that

$$\int f\mathrm{d}v_x\mathrm{d}v_y\mathrm{d}v_z = n \tag{1.22}$$

thereby obtaining the correct normalizing factor, the result is that the Maxwell–Boltzmann (or Maxwellian) distribution function is given by

$$f_{\mathrm{M}} = \frac{n}{(\sqrt{2\pi}\,v_{\mathrm{t}})^3}\exp(-v^2/2v_{\mathrm{t}}^2) \tag{1.23}$$

where the thermal velocity, v_t, is given by

$$v_t \equiv (kT/m)^{1/2}. \qquad (1.24)$$

Equation (1.24) is the last time that we will show the Boltzmann constant, k. Henceforth we will drop k, writing for example simply $v_t = (T/m)^{1/2}$. The Boltzmann constant k has the role of converting temperature from degrees Kelvin to units of energy (see equation (1.17)). In plasma physics, we generally find it more convenient to express temperature directly in energy units. In practical applications, we tend to discuss the temperature in units of electron-volts (eV), the kinetic energy an electron gains in free-fall down a potential of 1 V, but the equations we write, such as $v_t = (T/m)^{1/2}$ above, are in SI units for velocity and mass, so T is expressed in joules. Since when a charge of one coulomb falls down a potential of one volt, the kinetic energy gain is by definition one joule, the energy in an electron-volt, expressed in joules, is numerically equal to the electron charge expressed in coulombs. Rather than refer to a plasma as having temperature 11 600 K, we say its temperature is 1 eV, and evaluate T in SI units as 1.60×10^{-19} J (see Appendix C). Often, however, we will encounter the expressions (T/e) or (W/e) in plasma physics equations. When evaluating such expressions, it is even more convenient to insert the temperature, T, or particle energy, W, in units of eV, for the whole expression. An eV divided by e is a V—a perfectly good unit in SI! In other words, the expression (W/e) for a 10 keV particle becomes in SI 10^4 V. Remember, however, that the average kinetic energy of a particle in a Maxwellian distribution is $\langle W \rangle = (3/2)kT$— or, in our nomenclature, $\langle W \rangle = (3/2)T$. This is because the distribution contains three degrees of freedom per particle, corresponding to the three velocity components (v_x, v_y, v_z). From statistical mechanics we know that the typical energy associated with each degree of freedom is $T/2$.

One important use of the velocity-space distribution function f is to find the value of some quantity averaged over the distribution. For any quantity X, the local velocity-space average of X, which we denote $\langle X \rangle_v$ is given by

$$\langle X \rangle_v = \frac{\int f X \, d^3v}{\int f \, d^3v} = \frac{\int f X \, d^3v}{n}. \qquad (1.25)$$

In particular, if we take $X = W \equiv mv^2/2$, we find, for a Maxwellian distribution, that $\langle W \rangle_v = (3/2)T$, as we discussed above. If we are interested in the average energy of motion that a particle has in any one direction, say the z direction, $W_z \equiv mv_z^2/2$, we find $\langle W_z \rangle_v = T/2$ for a Maxwellian distribution function. The average of v_z^2 is simply T/m, or v_t^2 as defined by equation (1.24). Thus the quantity v_t, as we have defined it, is the 'root-mean-square' of the velocities in any one direction. (Beware that some researchers use an alternative definition, namely $v_t \equiv (2T/m)^{1/2}$.)

In some cases, a plasma has an anisotropic distribution function, which can be approximated as a 'bi-Maxwellian' with a different temperature along the magnetic field than across the field. This can happen in the laboratory or in natural plasmas due to forms of heating that add perpendicular or parallel energy preferentially to the particles, or loss processes that take out one or the other form of energy rapidly compared to collisions. In this case, taking the direction of the magnetic field to be the z direction, we have

$$f = \frac{n}{(\sqrt{2\pi}\,v_{t\parallel})(\sqrt{2\pi}\,v_{t\perp})^2} \exp\left(-\frac{v_z^2}{2v_{t\parallel}^2} - \frac{v_x^2 + v_y^2}{2v_{t\perp}^2}\right) \tag{1.26}$$

where

$$v_{t\perp} \equiv (T_\perp/m)^{1/2} \qquad v_{t\parallel} \equiv (T_\parallel/m)^{1/2} \tag{1.27}$$

and $\langle W_z \rangle_v = \langle W_\parallel \rangle_v = m\langle v_\parallel^2 \rangle_v/2 = T_\parallel/2$, because the parallel direction represents one degree of freedom. Similarly, defining $v_\perp^2 = v_x^2 + v_y^2$, $\langle W_x \rangle_v = \langle W_y \rangle_v = m\langle v_\perp^2 \rangle_v/4 = T_\perp/2$, so $\langle W_\perp \rangle_v = \langle W_x \rangle_v + \langle W_y \rangle_v = T_\perp$, because the perpendicular direction represents two degrees of freedom. In an isotropic plasma, with $T_\parallel = T_\perp = T$, $\langle W_\perp \rangle_v = 2\langle W_\parallel \rangle_v$.

Problem 1.2: Sketch a three-dimensional plot of an anisotropic distribution function f, with $T_\parallel = 2T_\perp$. Show that $\int f \, d^3v = n$ for f given by equation (1.26).

1.7 DEBYE SHIELDING

We have now done some very basic statistical mechanics to understand the Maxwell–Boltzmann distribution function of a plasma. Maxwell–Boltzmann statistics arise repeatedly in plasma physics, and the next example is fundamental to the very definition of a plasma. Consider a charge artificially immersed in a plasma which is in thermodynamic equilibrium. The equilibrium state implies that the plasma must be changing very slowly compared to the particle collision time, and that there is no significant temperature variation over distances comparable to a collision mean-free path. For present purposes, we will assume that the plasma is 'isothermal'—at a constant temperature, independent of position. Once again, consider the particle distribution function to be a heat 'bath' at a given temperature. And again consider a single specific particle, but now allow the particle to have both kinetic and potential energy:

$$W_r = mv^2/2 + q\phi \tag{1.28}$$

where q is the charge of the particle ($-e$ for an electron, $+Ze$ for an ion of charge Z), and so the Boltzmann factor becomes

$$\exp[-(mv^2/2 + q\phi)/T]. \tag{1.29}$$

The relative probability of a given energy of the particle now depends on position implicitly, through ϕ. The point worth noting is that this same Boltzmann factor (with a constant normalization in front—independent of position) gives the relative probability and therefore the relative particle distribution function over the whole volume in thermal equilibrium. If we integrate the distribution function over velocity space to obtain a relative local particle density, we find that the spatial dependence that remains comes only from the Boltzmann factor:

$$n \propto \exp(-q\phi/T). \tag{1.30}$$

This means physically that electrons will tend to gather near a positive charge in a plasma, and therefore they will tend to shield out the electric field from the charge, preventing the field from penetrating into the plasma. By the same token, ions will have the opposite tendency, to 'shy away from' a positive charge, and gather near a negative one.

A fundamental property of a plasma is the distance over which the field from such a charge is shielded out. Indeed, it is considered one of two formal defining characteristics of a plasma that this shielding length (called the Debye length, λ_D, which was first calculated in the theory of electrolytes by Debye and Hückel in 1923) be much smaller than the plasma size. The second defining characteristic of a plasma is that there should be many particles within a Debye sphere, which has volume $(4/3)\pi\lambda_D^3$, with the consequence that the statistical treatment of Debye shielding is valid.

It is fairly easy to calculate the Debye length for an idealized system. Let us suppose that we have immersed a planar charge in a plasma. Assume the plasma ions have charge Ze, and far from the electrode the ion and electron densities are $n_e = Zn_i \equiv n_{e\infty}$. This boundary condition at infinity is required in order to provide charge neutrality over the bulk of the plasma, so as to keep the electric field, $\mathbf{E}$, from building up indefinitely. Let us also choose to set $\phi = 0$ at infinity for simplicity. Given our assumptions at infinity, from the Boltzmann factor we know that

$$n_e(x) = n_{e\infty}\exp(e\phi/T_e)$$
$$Zn_i(x) = n_{e\infty}\exp(-eZ_i\phi/T_i). \tag{1.31}$$

We are allowing $T_e \neq T_i$, for generality, but both T_i and T_e are spatially homogeneous, i.e. the electrons are in thermal equilibrium among themselves, and the ions are in thermal equilibrium among themselves, but the ions and electrons are not necessarily in thermal equilibrium with each other. At first sight this may seem unphysical, but it happens often in plasmas because electron–electron energy transfer by collisions and ion–ion energy transfer by collisions are both faster than collisional electron–ion energy transfer, due to the large mass discrepancy. We will study this in Unit 3. For the time being, it might be helpful to think about the example of collisional equilibration in a system of ping-pong balls and bumper-cars. At first the ping-pong balls and bumper-cars will each, separately, come to thermal equilibrium, because their self-collisions are efficient at transferring energy as well as momentum. It will take longer for the balls and cars to come into thermal equilibrium with each other, because the transfer of energy in their collisions is weak.

The Poisson equation for our one-dimensional planar geometry is

$$\epsilon_0 \frac{d^2\phi}{dx^2} = e(n_e - Zn_i) = en_{e\infty}[\exp(e\phi/T_e) - \exp(-eZ\phi/T_i)] \tag{1.32}$$

where ϵ_0 is again the permittivity of free space. It is difficult to solve this equation in the region near the planar charge, where $e\phi/T$ may be large, but we can obtain a qualitative sense of the solution by assuming that $e\phi/T$ is small, and expanding the exponential in $e\phi/T$. Equation (1.32) then becomes

$$\epsilon_0 \frac{d^2\phi}{dx^2} \approx en_{e\infty}(e\phi/T_e + eZ\phi/T_i) \tag{1.33}$$

i.e.

$$\frac{d^2\phi}{dx^2} \approx \frac{e^2 n_{e\infty}(1 + ZT_e/T_i)}{\epsilon_0 T_e}\phi \tag{1.34}$$

which can be solved to obtain the characteristic exponential decay length which we are seeking:

$$\phi \propto \exp(-x/\lambda_D) \tag{1.35}$$

where

$$\lambda_D \equiv \left(\frac{\epsilon_0 T_e}{n_e e^2(1 + ZT_e/T_i)}\right)^{1/2} \tag{1.36}$$

Often the ion term is not included in the definition of the Debye length, giving $\lambda_D \equiv (\epsilon_0 T_e/n_e e^2)^{1/2}$. For typical laboratory plasmas, the Debye length is indeed small. For a 3 eV electric arc discharge at a density of 10^{19} m^{-3}, we find that $\lambda_D \approx 3 \times 10^{-6}$ m. The number of particles in the Debye sphere for this case is about one thousand, making our statistical treatment reasonably valid.

Problem 1.3: Derive the equivalent of equation (1.34) in spherical coordinates (i.e. for the case of a point charge immersed in a plasma). Show that the solution is $\phi \propto \exp(-r/\lambda_D)/r$.

Problem 1.4: The typical distance between two electrons in a plasma is of order $n_e^{-1/3}$. Show that the potential energy associated with bringing two electrons this close together is much less than their typical kinetic energy, so long as $n_e \lambda_D^3 \gg 1$.

1.8 MATERIAL PROBES IN A PLASMA

In our discussion of Debye shielding, we considered the response of an equilibrium plasma to a localized charge. We did not, however, consider the possibility of collisions between plasma particles and whatever was carrying the charge. The situation is very different in the case of a real material probe inserted into a plasma. Such a probe intercepts particle trajectories, resulting in violation of the assumption of equilibrium in its near vicinity. If the probe is biased negative with respect to the plasma, with potential $\phi \ll -T_e/e$, few electron trajectories are intercepted, since most electrons cannot reach the probe, so the electrons will be close to equilibrium and maintain $n_e \sim n_{e\infty}\exp(e\phi/T)$. A sheath region will develop around the probe, whose width scales with the Debye length, as in the case we just considered, because the electron population will be exponentially depleted close to the negatively biased probe. Ions, however, will be accelerated across the sheath, and into the material electrode. In the case of cold ions, $T_i \ll T_e$, the calculation of the ion density reduces to the ion analog of the Child–Langmuir calculation we performed at the beginning of this Chapter. While the electron density falls exponentially in the vicinity of a negatively biased material probe, the ion density is depressed as well, but more weakly, as $\phi^{-1/2}$ (see equation (1.12)). The ion density, in this case, is *not* enhanced by the negative bias, due to the depleting collisions with the probe surface. The ion current density to a negatively biased probe in a $Z = 1$ plasma is given approximately by $j_i \sim n_{i\infty}eC_s$, where C_s is the so-called 'ion sound speed' $C_s \equiv [(T_e + T_i)/m_i]^{1/2}$, which shows up in situations like this where both ion and electron temperatures contribute to ion motion, and $n_{i\infty}$ is the ion density far from the probe. (We will encounter C_s again when we study ion acoustic waves in Unit 4.) This ion current is called the 'ion saturation current', $j_{sat,i}$, because the ion current saturates at this value as the probe bias is driven further negative. The sheath width grows as the potential becomes more negative, in just such a way as to keep the ion Child–Langmuir current constant at $j_{sat,i}$.

Problem 1.5: Perform an ion Child–Langmuir calculation to model the plasma sheath at a material probe. Assume an inter-electrode spacing of $\lambda_D \equiv (\epsilon_0 T_e/n_e e^2)^{1/2}$ to model the sheath width, and a potential drop of $e\phi = -T_e$. Take $T_i = 0$. You may assume that the electron density is negligible in the sheath region, to make the *ion* Child–Langmuir calculation valid. Determine the ion current density, j_i, across this model sheath.

The electron current to a material probe depends exponentially on the probe potential, since the electron density at the probe face varies exponentially with $e\phi/T$, and the particle flux from a Maxwell–Boltzmann electron distribution into a material wall is given by Γ [particles s^{-1} m^{-2}] $= n_e(8T_e/\pi m_e)^{1/2} \sim n_e v_{t,e}$. A potential of $e\phi \sim 3.3T_e$ is required to reduce the electron current to the probe to equal the ion current, in a hydrogen plasma. This is called the 'floating' potential, because the potential of a probe that is not allowed to draw any net current will 'float' to this value. Such a strong potential is required, of course, because $v_{t,e} \sim C_s(m_i/m_e)^{1/2}$, so the electron current in the absence of negative probe bias is much larger in absolute magnitude than $j_{sat,i}$.

UNIT 1

SINGLE-PARTICLE MOTION

In this Unit we will investigate charged-particle motion in magnetic, electric and even gravitational fields. Natural and laboratory-generated plasmas are frequently immersed in strong externally-generated magnetic fields, because these fields confine charged-particle orbits (and therefore plasmas), at least in the direction perpendicular to the magnetic field. Magnetic and electric fields are also generated by currents and charge accumulations within plasmas, and so an understanding of charged-particle motion in these fields underlies the understanding of the dynamics of plasma motion.

We will begin by studying particle motion in uniform static fields in Chapter 2. Then we will include spatial gradients in Chapter 3. In Chapter 4 we will include time-dependent phenomena, and discuss invariants of particle motion. In Chapter 5 we will introduce the modern nonlinear theory of chaos in particle orbits, using the concept of Hamiltonian maps.

Chapter 2

Particle drifts in uniform fields

Many plasmas are immersed in externally imposed magnetic and/or electric fields. All plasmas have the potential to generate their own electromagnetic fields as well. Thus, as a first step towards understanding plasma dynamics, in this Chapter we begin by considering the behavior of charged particles in uniform fields, thus constructing the most fundamental aspects of a magnetized plasma. We also carefully introduce some of the mathematical formalisms that we will use throughout the book.

2.1 GYRO-MOTION

In the presence of a uniform magnetic field, the equation of motion of a charged particle is given by

$$m\dot{\mathbf{v}} = q\mathbf{v} \times \mathbf{B} \tag{2.1}$$

where q is the (signed) charge of the particle. Taking $\hat{\mathbf{z}}$ to be the direction of $\mathbf{B}$ (i.e. $\mathbf{B} = B\hat{\mathbf{z}}$ or we sometimes say $\hat{\mathbf{b}} \equiv \mathbf{B}/B$ which, in this case, is the same as $\hat{\mathbf{z}}$), we have

$$\dot{v}_x = qv_y B/m \tag{2.2}$$
$$\dot{v}_y = -qv_x B/m \tag{2.3}$$
$$\dot{v}_z = 0. \tag{2.4}$$

For a specific trajectory, we also need initial conditions at $t = 0$: these we take to be $x = x_i$, $y = y_i$, $z = z_i$, $v_x = v_{xi}$, $v_y = v_{yi}$, $v_z = v_{zi}$. If we take the time derivative of both sides of equation (2.2), we can use equation (2.3) to substitute for $\dot{v}_y$, and obtain

$$\frac{\mathrm{d}^2 v_x}{\mathrm{d}t^2} = -\left(\frac{qB}{m}\right)^2 v_x. \tag{2.5}$$

21

If we define $\omega_c \equiv |q|B/m$, it is clear that the solution of this equation is

$$v_x = A\cos(\omega_c t) + B\sin(\omega_c t) \qquad (2.6)$$

where A and B are integration constants. Evidently ω_c, called the 'cyclotron frequency' (also sometimes called the 'Larmor frequency' or the 'gyro-frequency'), is going to prove to be a very important quantity in a magnetized plasma. It is convenient to use complex-variable notation, and rewrite equation (2.6) as

$$
\begin{aligned}
v_x &= \text{Re}[A\exp(i\omega_c t)] - \text{Re}[B i\exp(i\omega_c t)] \\
&= \text{Re}\left[(A - iB)\exp(i\omega_c t)\right] = \text{Re}\left\{\left[v_\perp \exp(i\delta)\right]\exp(i\omega_c t)\right\} \\
&= \text{Re}\left[v_\perp \exp(i\omega_c t + i\delta)\right]
\end{aligned} \qquad (2.7)
$$

where Re indicates the real part of the subsequent expression, $v_\perp$ is an absolute speed perpendicular to $\mathbf{B}$, and δ is a phase angle. The quantities $v_\perp$ and δ have become our new integration constants. (We will now drop the Re in this notation, since it is clear that we are dealing with real quantities.) In this formulation, $v_\perp$ and δ are chosen to match the initial velocity conditions. Equation (2.2) gives

$$v_y = i(|q|/q)v_\perp \exp(i\omega_c t + i\delta) = \pm i v_\perp \exp(i\omega_c t + i\delta) \qquad (2.8)$$

where $\pm$ evidently indicates the sign of q. From the initial conditions, we now can say that $v_\perp = (v_{xi}^2 + v_{yi}^2)^{1/2}$ and $\delta = \mp\tan^{-1}(v_{yi}/v_{xi})$, where the upper sign is for positive q. Note that v_x and v_y are 90° out of phase, so we have circular motion in the plane perpendicular to $\mathbf{B}$. Equation (2.4) indicates that v_z is a constant, and so the motion constitutes a helix along $\mathbf{B}$. If we integrate equations (2.4), (2.7) and (2.8) in time, we obtain

$$
\begin{aligned}
x &= x_i - i(v_\perp/\omega_c)[\exp(i\omega_c t + i\delta) - \exp(i\delta)] \\
y &= y_i \pm (v_\perp/\omega_c)[\exp(i\omega_c t + i\delta) - \exp(i\delta)] \\
z &= z_i + v_{zi}t
\end{aligned} \qquad (2.9)
$$

where the integration constants have been chosen to match the initial position conditions.

Clearly, then, another fundamental quantity in a magnetized plasma is the length $r_L \equiv (v_\perp/\omega_c)$, called the 'Larmor radius' or 'gyro-radius'. This is the radius of the helix described by the particle as it travels along the magnetic field line. Figure 2.1 shows an electron and a proton gyro-orbit, drawn more or less to scale, for equal particle energies $W = mv_\perp^2/2$. The ratio of the two gyro-radii is the square-root of the ratio of the proton mass to the electron mass, $\sqrt{1837} \approx 43$. Note that $v_\perp$ is proportional to $(W/m)^{1/2}$, and ω_c is proportional to $1/m$, so r_L is proportional to $(mW)^{1/2}$.

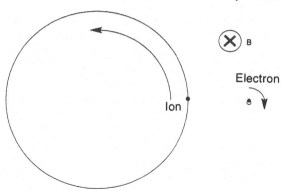

Figure 2.1. Ion and electron gyro-motion in a magnetic field. For fixed energy, the ion's gyro-orbit is much larger than the electron's. 'X' indicates that the magnetic field faces into the page.

The centers of the gyro-orbits are referred to as their 'guiding centers', or 'gyro-centers', and give a measure of a particle's average location during a gyro-orbit. Averaging equation (2.9) over a gyro-period, the guiding-center position for the particular initial values considered here is seen to be given by

$$x_{gc} = x_i + i(v_\perp/\omega_c)\exp(i\delta) \qquad y_{gc} = y_i \mp (v_\perp/\omega_c)\exp(i\delta) \qquad (2.10)$$

so that the particle's position described in terms of its guiding-center position is given by

$$x = x_{gc} - i(v_\perp/\omega_c)\exp(i\omega_c t + i\delta)$$
$$y = y_{gc} \pm (v_\perp/\omega_c)\exp(i\omega_c t + i\delta) \qquad (2.11)$$
$$z = z_{gc} = z_i + v_{zi}t.$$

Thus we can think of particle gyro-centers as sliding along magnetic field lines, like beads on a wire. Note that electrons and ions rotate around the field lines in opposite directions, with the upper sign giving the phase for positively charged particles. If you point your two thumbs along the direction of the magnetic field, the fingers of your left hand curl in the direction of rotation of positively charged ions, while those of your right hand do the same for electrons. These directions of rotation are both such that the tiny perturbation of the magnetic field inside the particle orbits, due to the current represented by the particle motion, acts to *reduce* the ambient magnetic field. High-pressure plasmas reduce the externally imposed magnetic field through the superposition of this 'diamagnetic' effect from a high density of energetic particles.

The ion and electron Larmor radii and gyro-frequencies provide fundamental space-scales and time-scales in a magnetized plasma. Phenomena which occur on space-scales much smaller than the gyro-radius, or on time-scales much faster than a gyro-period, are often insensitive to the presence of the magnetic field, and can be described using equations appropriate for an unmagnetized plasma. In the opposite limit of large space-scales and long time-scales, gyro-motion is crucial to plasma behavior, and generates some surprising phenomena—somewhat akin to the behavior of a gyroscope which responds to any attempt to change the orientation of its axis of rotation by moving at 90° to the applied torque. Some plasma phenomena, especially in the Earth's magnetosphere, can occur at *intermediate* space-scales and time-scales, such that the electrons can be considered magnetized, while the ions are essentially unmagnetized. In our discussion of particle motion, however, we will generally consider space-scales much greater than a gyro-radius, and time-scales much longer than a gyro-period of either species, unless we specifically state otherwise.

Problem 2.1: Look through articles in *Physical Review Letters, Plasma Physics, Physics of Fluids* B (recently renamed *Physics of Plasmas*) or in other journals over recent years and find at least one article each about laboratory, solar or terrestrial, and astrophysical plasmas immersed in magnetic fields. Give the reference and a few-sentence description of each article. For the plasmas you find described, evaluate the ion and electron gyro-radii and the Debye radius (ignoring ion shielding), insofar as the authors give you the required information. Compare these to the system sizes. Calculate how many particles are within a Debye sphere for each case. Evaluate the ion and electron cyclotron frequencies and compare to the evolution time-scale of the overall plasma. Which of these systems are really plasmas? Which of these are magnetized versus unmagnetized plasmas?

2.2   UNIFORM E FIELD AND UNIFORM B FIELD: E × B DRIFT

Starting from the configuration we have just discussed, with $\mathbf{B} = B\hat{\mathbf{z}}$, let us add a uniform electric field $\mathbf{E}$. We will assume that both the electric and the magnetic field are time-independent. The non-relativistic equation of motion becomes

$$m\dot{\mathbf{v}} = q(\mathbf{E} + \mathbf{v} \times \mathbf{B}). \tag{2.12}$$

Now we will employ a mathematical transformation, which we will justify later, in order to solve this equation expeditiously. Let us define a velocity $\mathbf{u}$ by

$$\mathbf{u} \equiv \mathbf{v} - (\mathbf{E} \times \mathbf{B})/B^2. \tag{2.13}$$

In other words, $\mathbf{u}$ is the particle velocity that we would see in a frame moving at velocity $(\mathbf{E} \times \mathbf{B})/B^2$. Since $\mathbf{E}$ and $\mathbf{B}$ are time-independent, we have $\dot{\mathbf{v}} = \dot{\mathbf{u}}$

and so, substituting for **v** in terms of **u**, equation (2.12) for **u̇** becomes

$$m\dot{\mathbf{u}} = q[\mathbf{E} + \mathbf{u} \times \mathbf{B} + (\mathbf{E} \times \mathbf{B}) \times \mathbf{B}/B^2]. \tag{2.14}$$

Now, we use the vector identity

$$(\mathbf{A} \times \mathbf{B}) \times \mathbf{C} = (\mathbf{A} \cdot \mathbf{C})\mathbf{B} - (\mathbf{B} \cdot \mathbf{C})\mathbf{A} \tag{2.15}$$

(see Appendix D) to obtain

$$m\dot{\mathbf{u}} = q[\mathbf{E} + \mathbf{u} \times \mathbf{B} + (\mathbf{E} \cdot \mathbf{B})\mathbf{B}/B^2 - \mathbf{E}]$$
$$= q[\hat{\mathbf{b}}(\mathbf{E} \cdot \hat{\mathbf{b}}) + \mathbf{u} \times \mathbf{B}]. \tag{2.16}$$

To obtain the equation for the velocity *parallel* to **B**, we take the dot-product of equation (2.16) with $\hat{\mathbf{b}}$, giving

$$m\dot{u}_\parallel = qE_\parallel \tag{2.17}$$

where we are defining

$$u_\parallel \equiv \mathbf{u} \cdot \hat{\mathbf{b}} \qquad E_\parallel \equiv \mathbf{E} \cdot \hat{\mathbf{b}} \qquad v_\parallel \equiv \mathbf{v} \cdot \hat{\mathbf{b}}. \tag{2.18}$$

From equation (2.13) we see that $u_\parallel = v_\parallel$, and so the solution for $v_\parallel$ is just free-fall in the electric field:

$$v_\parallel = (qE_\parallel/m)t + v_{\parallel i}. \tag{2.19}$$

To obtain the equation for the velocity *perpendicular* to **B**, we multiply both sides of equation (2.17) by $\hat{\mathbf{b}}$, and subtract from equation (2.16). We obtain

$$m\dot{\mathbf{u}}_\perp = q\mathbf{u}_\perp \times \mathbf{B} \tag{2.20}$$

where $\mathbf{u}_\perp \equiv \mathbf{u} - u_\parallel\hat{\mathbf{b}}$, $\mathbf{E}_\perp \equiv \mathbf{E} - E_\parallel\hat{\mathbf{b}}$ and $\mathbf{v}_\perp \equiv \mathbf{v} - v_\parallel\hat{\mathbf{b}}$.

Thus, in the direction perpendicular to $\hat{\mathbf{b}}$, we have precisely the same equation for **u** as we had for **v** in the absence of an electric field, i.e. equation (2.11). We have found that the solution of this equation implies that the guiding center does not move at all perpendicular to **B**, and we know that it slides along **B** with velocity $u_\parallel = v_\parallel$ as given by equation (2.19). Thus, in the frame moving at speed $(\mathbf{E} \times \mathbf{B})/B^2$, the only guiding-center velocity we see is parallel to **B**, so in the laboratory frame we see a guiding-center velocity

$$\mathbf{v}_{\text{gc}} = v_\parallel\hat{\mathbf{b}} + (\mathbf{E} \times \mathbf{B})/B^2 \equiv v_\parallel\hat{\mathbf{b}} + \mathbf{v}_E. \tag{2.21}$$

The velocity $\mathbf{v}_E \equiv \mathbf{E} \times \mathbf{B}/B^2$ is called the 'E × B drift'. It is particularly easy to evaluate this drift in SI units: **E** is in units of volts/meter, **B** is evaluated in units of teslas and v_E results in meters/second. Notice that v_E is independent of q, m, $v_\parallel$, and $v_\perp$. This means that the whole plasma drifts together across the electric and magnetic fields with the same velocity.

What we have actually done here is performed a simplified Lorentz transformation, using the **B** field to eliminate the **E** field in the moving frame, and so simplified the equation of motion. Of course the Lorentz transformation works the same for all particles, so the whole plasma v_E-drifts together, relative to what it would have done without the **E** field. Since we have chosen to use a non-relativistic equation of motion, our Lorentz transformation is particularly simple. The approximation we have used is equivalent to assuming that $\gamma \equiv [1 - (v/c)^2]^{-1/2} \approx 1$, or $(v/c)^2 \ll 1$.

For a more physical picture of the origin of the $\mathbf{E} \times \mathbf{B}$ drift without resorting to the Lorentz transformation, consider how the particles are accelerated by the electric field during part of their gyro-orbits, and are decelerated during the other part. The result of these accelerations and decelerations is that the radii of curvature of the gyro-orbits will be slightly larger on the side where the particles have greater kinetic energy than on the side where the particles have less kinetic energy, due to having climbed a potential hill. This gives rise to a drift perpendicular to **E**, as illustrated in Figure 2.2.

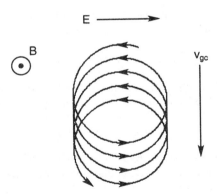

Figure 2.2. Electron $\mathbf{E} \times \mathbf{B}$ drift motion. The half-orbit on the left-hand side is larger than that on the right, because the electron has gained energy from the electric field. The dot indicates that the magnetic field faces out of the page.

Incidentally, in our derivation of the $\mathbf{E} \times \mathbf{B}$ drift, we did not have to assume anything about the relative size of v and $|v_E|$. Indeed, the whole guiding center formalism can be developed for the case where $|v_E|$ is of order v (at the expense of a greater complexity of terms), but we will hereafter assume $|v_E| \ll v$ in our derivations.

2.3 GRAVITATIONAL DRIFT

In the presence of any other simple force on the charged particles in a plasma, we can apply directly the results we have derived for the electric force. In particular, if we imagine a plasma in the Earth's magnetic field, we might wonder what effect the Earth's gravity would have on it. We can simply replace the electric force $q\mathbf{E}$ with a general force, $\mathbf{F}$, in both the equation of motion and in its solution (e.g. in the definition of $\mathbf{u}$). This gives a guiding-center drift

$$\mathbf{v}_F = (\mathbf{F} \times \mathbf{B})/qB^2 \qquad (2.22)$$

or, in the case of gravity, where $\mathbf{F} = m\mathbf{g}$,

$$\mathbf{v}_g = m(\mathbf{g} \times \mathbf{B})/qB^2 \qquad (2.23)$$

which is usually called the 'gravitational drift'.

Note that $\mathbf{v}_g$, unlike $\mathbf{v}_E$, depends on m and q. The presence of gravity gives rise to a net current in a plasma; the ions drift one way and the electrons the other—the ions, which are much heavier, drift much faster. In a finite plasma, this current therefore gives rise to charge separation. Generally speaking, the actual gravitational drift $\mathbf{v}_g$ is very small, and we introduce it mainly for later application of the idea of a 'general force' drift to the case of centrifugal force.

It is interesting to ask why it is that a plasma 'cloud' above the Earth does not seem to fall down in the gravitational field. In fact, the gravitational drift is horizontal, not vertical! (Galileo, for one, might have found this disturbing.) The qualitative answer is that the ion and electron drifts are in opposite directions, and so if the plasma is finite in the horizontal direction, perpendicular to $\mathbf{B}$ and $\mathbf{g}$, charge separation occurs, an electric field builds up (in the horizontal direction and perpendicular to $\mathbf{B}$), and the plasma does indeed drift downwards, after all, due to the $\mathbf{v}_E$ drift. To analyze this situation quantitatively—and to determine whether the plasma falls with acceleration g—we must first understand how a plasma responds to a time-varying electric field, $\dot{\mathbf{E}}$. We will return to this topic in Chapter 4.

Problem 2.2: The ionosphere is composed mostly of a proton–electron plasma immersed in the Earth's magnetic field of about 3×10^{-5} T. How fast is the gravitational drift for each species?

Chapter 3

Particle drifts in non-uniform magnetic fields

In the previous Chapter, we studied particle drifts in uniform fields and developed the fundamental concepts of Larmor radius, gyro-frequency, and gyro-center motion. Now we consider magnetic field gradients both perpendicular and parallel to **B**, and curved magnetic fields. We will find gyro-center drifts across the magnetic field, and acceleration (or deceleration) along **B**. We will develop the concept of 'ordering' the drifts in the ratio of Larmor radius to gradient scale-length. To zeroth order, particles slide along **B** as before (but $v_\parallel$ will now vary), and to first order they drift across **B**, but they still precisely conserve the sum of potential and kinetic energy at each order.

3.1 ∇B DRIFT

We now proceed to examine particle guiding-center drifts in inhomogeneous magnetic fields. We will assume in all of these studies that the gyro-radius, r_L, is much less than the typical scale-length of variation of the magnetic field. Thus

$$\frac{r_L}{B} |\nabla B| \ll 1. \tag{3.1}$$

For example, if B has a sinusoidal variation, $B \propto \exp(ikx)$, or an exponential variation, $B \propto \exp(kx)$, this is equivalent to saying $kr_L \ll 1$, where $1/k$ is a characteristic gradient scale-length for the problem. In this situation, the quantity kr_L becomes a useful 'expansion parameter' for studying the equations of motion by the method of asymptotic expansion.

In our asymptotic expansion procedure, we will assume that the particle velocities can be expressed as a sum of terms

$$\mathbf{v} = \mathbf{v}_0 + \mathbf{v}_1 + \mathbf{v}_2 + \dots \tag{3.2}$$

29

where the leading term is the particle's parallel velocity, $v_\parallel \hat{\mathbf{b}}$, plus its gyro-motion perpendicular to $\mathbf{B}$, and each successive term in the series is assumed to be smaller than the previous one, by approximately kr_L. We will be interested here in calculating the evolution of $\mathbf{v}_0$ and of $\mathbf{v}_1$, and in fact at first order we will only need the guiding-center motion averaged over many gyro-periods. Substituting our form for $\mathbf{v}$ into the equation of motion, we will find that we have terms in the equation *of each order*: $(kr_L)^0$, $(kr_L)^1$, $(kr_L)^2$, etc. If we solve for $\mathbf{v}_0$, $\mathbf{v}_1$, $\mathbf{v}_2$, etc., so as to make the terms in the equation of each order balance separately, we will have an asymptotic series solution for $\mathbf{v}$. This approach is justified by noting that, in the limit $kr_L \to 0$, terms of higher order in kr_L can never be used to balance terms of lower order, because for small enough kr_L, the higher-order terms must be negligible in comparison with the lower-order ones.

We begin by considering the case where we have a perpendicular (i.e. perpendicular to $\mathbf{B}$) gradient in the field strength, B. For simplicity let us assume that $\mathbf{B}$ is in the z direction, and varies only with y. (To generate this field, we need distributed volume currents, since $\nabla \times \mathbf{B} \neq 0$. Such currents are common in plasmas, but do not affect directly our analysis of particle drifts. Of more importance is the fact that our model field does not violate $\nabla \cdot \mathbf{B} = 0$.) We write

$$\mathbf{B} = B_{gc,i}\hat{\mathbf{z}} + (y - y_{gc,i})\frac{\mathrm{d}B}{\mathrm{d}y}\hat{\mathbf{z}} \qquad (3.3)$$

where $y_{gc,i}$ is the initial y position of the particle guiding-center, and $B_{gc,i}$ is the value of B at $y_{gc,i}$. We assume for the validity of our asymptotic expansion procedure that $r_L(\mathrm{d}B/\mathrm{d}y) \ll B$. The equations of motion in the perpendicular (x and y) directions are

$$\begin{aligned} m\dot{v}_x &= qv_y[B_{gc,i} + (y - y_{gc,i})(\mathrm{d}B/\mathrm{d}y)] \\ m\dot{v}_y &= -qv_x[B_{gc,i} + (y - y_{gc,i})(\mathrm{d}B/\mathrm{d}y)]. \end{aligned} \qquad (3.4)$$

Substituting the series expansion for $\mathbf{v}$, we obtain

$$\begin{aligned} m\dot{v}_{x0} + m\dot{v}_{x1} &= q(v_{y0} + v_{y1})[B_{gc,i} + (y_0 - y_{gc,i})(\mathrm{d}B/\mathrm{d}y)] \\ m\dot{v}_{y0} + m\dot{v}_{y1} &= -q(v_{x0} + v_{x1})[B_{gc,i} + (y_0 - y_{gc,i})(\mathrm{d}B/\mathrm{d}y)]. \end{aligned} \qquad (3.5)$$

We have ignored some of the terms that are second order in kr_L, but we have kept all terms that might prove to be of lower order.

In thinking carefully about this procedure, we encounter one of the interesting subtleties of using asymptotic expansions. We will *assume* that $(y - y_{gc,i})(\mathrm{d}B/\mathrm{d}y)$ is smaller than $B_{gc,i}$ by one order in kr_L. This requires that $(y - y_{gc,i})$ always be of order r_L for our series expansion to be correct. However that means that $y(t)$, which we do not yet know, must not grow without bound, because in that case the quantity $(y - y_{gc,i})$ would not remain of order r_L, as our ordering assumes. In particular $y_1(t)$ must not grow without bound, so we must watch out for such 'secularities' in y. In the case at hand this turns out not to be a problem, as we will see; our solution will maintain $(y - y_{gc,i})$ of order r_L—so, *a posteriori*, our assumption will be proven correct. In more complex situations, special techniques may be needed to eliminate secularities, but a valid solution can often still be obtained via this asymptotic expansion procedure.

So let us proceed with our order-by-order solution of equation (3.5). The zeroth-order terms in equation (3.5) constitute simply the equations of motion in a homogeneous magnetic field, which we gave first in equations (2.2) and (2.3), and whose solution is given in equations (2.7), (2.8) and (2.11). Our procedure calls for us to assume that the zeroth-order terms balance, implying that the zeroth-order velocities and positions must be given by our previous solution. Next we gather together all the first-order terms (terms of order kr_L compared to the largest ones) to generate a first-order equation that we must solve:

$$m\dot{v}_{x1} = qv_{y1}B_{gc,i} + qv_{y0}(y_0 - y_{gc,i})(dB/dy)$$
$$m\dot{v}_{y1} = -qv_{x1}B_{gc,i} - qv_{x0}(y_0 - y_{gc,i})(dB/dy).$$
(3.6)

To make further progress, we will now time-average both of these first-order equations over many gyro-periods, since we are only interested in the gyro-averaged particle motion, sometimes called the 'guiding-center drift'. We use the notation $\langle \ \rangle$ here to indicate a time average. The left-hand side of both equations can be set to zero, because all that survives the gyro-averaging process are the time derivatives of $m\langle v_{x1}\rangle$ and $m\langle v_{y1}\rangle$ *due to changes that are slow compared to a gyro-period*, with the result that these terms are now very small compared to the first terms on the right-hand side. We say that the gyro-averaging process 'annihilates' these terms on the left-hand side. In effect it *raises* them by one order, since only time derivatives slow compared to a gyro-period survive the averaging. However for present purposes, the resulting second-order time derivatives can be neglected. Next we note that $\langle v_{y0}(y_0 - y_{gc,i})\rangle = 0$, since equations (2.8) and (2.11) show that v_{y0} and $y_0 - y_{gc,i}$ are 90° out-of-phase, and of course, $\langle v_{y0}y_{gc,i}\rangle = 0$.

Problem 3.1: Prove that $\langle v_{y0}(y_0 - y_{gc,i})\rangle = 0$ for all δ.

Thus $\langle v_{y1}\rangle = 0$, and so the particles do not steadily drift off in the y direction—justifying our expansion procedure (which required that $y - y_{gc,i}$ not grow without bound) *a posteriori*. The particles do, however, steadily drift off in the x direction, since

$$\langle v_{x1} \rangle = -\frac{\langle v_{x0}(y_0 - y_{gc,i}) \rangle}{B_{gc,i}} \frac{\mathrm{d}B}{\mathrm{d}y}. \tag{3.7}$$

Referring to equation (2.11) and taking $\delta = 0$, we arrive at

$$\langle v_{x0}(y_0 - y_{gc,i}) \rangle = \pm \langle \mathrm{Re}[v_\perp \exp(i\omega_c t)] \ \mathrm{Re}[(v_\perp/\omega_c)\exp(i\omega_c t)] \rangle$$
$$= \pm v_\perp^2/(2\omega_c) \tag{3.8}$$

where the $\pm$ sign goes with the charge of the particle. Note that $\langle v_{x1} \rangle$ *does not even have a slow time derivative*, so our assumption that $\langle m\dot{v}_{x1} \rangle$ was negligible is also consistent with our solution. Note also that $B_{gc} = B_{gc,i}$, because the particle is drifting in a direction in which B is constant.

Problem 3.2: Evaluate $\langle v_{x0}(y_0 - y_{gc,i}) \rangle$ for arbitrary δ.

Recognizing that the choices for **B** to be in the z direction and for ∇B to be in the y direction were arbitrary, we have for *perpendicular* gradients of B, a guiding-center drift given by

$$\mathbf{v}_{\mathrm{grad}} = \pm \frac{v_\perp^2}{2\omega_c} \frac{\mathbf{B} \times \nabla B}{B^2} = \frac{W_\perp}{q} \frac{\mathbf{B} \times \nabla B}{B^3} \tag{3.9}$$

where $\mathbf{v}_{\mathrm{grad}}$ is the gyro-averaged drift of the guiding-center, due to a perpendicular gradient in B. We call this the '∇B drift'. In SI units, with energies in eV, equation (3.9) is particularly simple to evaluate: for a $1\,000\,\mathrm{eV}$ particle (and all its energy in $W_\perp$), in a 1 tesla magnetic field, with a gradient scale-length of 1 meter, the ∇B drift velocity is simply 10^3 meters/second.

Note that the ∇B drift, like the gravitational drift, depends on the sign of the charge of the particle, and so it gives rise to a net current, which in turn leads to charge separation in a finite plasma and, consequently, a volumetric electric field. Interestingly, at fixed energy the ∇B drift is independent of particle mass. Notice that if $v_\perp$ is of order $v_\parallel$, this first-order gyro-averaged drift is indeed a factor kr_L smaller than the parallel velocity of the particle along a field line, $v_\parallel \hat{\mathbf{b}}$, which is the only zeroth-order motion that would survive gyro-averaging. This is consistent with our ordering procedure.

Problem 3.3: Assume $e\phi$ is of order W, a particle's kinetic energy, and that the gradient scale-length of the electric potential is roughly the same size, $1/k$, as the scale-length of variation of B. Show that v_E is the same order in kr_L as v_{grad}.

There is a simple physical picture for the ∇B drift, which follows from the fact that the local radius-of-curvature of the gyro-orbit is smaller on the larger-magnetic-field side of the orbit, and correspondingly larger on the smaller-magnetic-field side. If we construct a continuous trajectory from smaller orbits on one side, and larger orbits on the other, we find a net drift perpendicular to both $\mathbf{B}$ and ∇B, as illustrated in Figure 3.1.

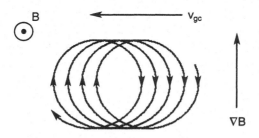

Figure 3.1. Ion ∇B drift motion. The combined effect of smaller gyro-orbits on the high-field side and larger gyro-orbits on the low-field side produces a net leftward drift of the guiding center. The dot indicates that the magnetic field faces out of the page.

3.2 CURVATURE DRIFT

In the previous Section, we made the assumption that there was a gradient in the magnetic field strength, B, but that the vector $\mathbf{B}$ was purely in the z direction, i.e. the magnetic field lines were straight. As we saw, this required volume currents, but these did not affect our analysis. Now we will make another special, but useful, simplifying assumption: that the field lines are locally curved with radius-of-curvature $\mathbf{R}_c$, but that the field strength B is locally constant. A magnetic field with these properties can also be achieved with volume currents. Imagine a current-carrying cylinder with $j_z \propto r^{-1}$ where j_z is the current density in the z direction. The total current I in the z direction within any radius r then increases linearly with r, i.e. $I \propto r$, so that from the usual formula $B \propto I/r$, the θ-directed magnetic field is independent of r. Again, these volume currents are an artifact employed to produce the assumed magnetic field; they do not enter into the analysis of particle drifts.

Now we will solve for the guiding-center drift in a locally cylindrical coordinate system (r, θ, z) matched to the local curvature of the magnetic field lines, such that $\hat{\boldsymbol{\theta}} = \hat{\mathbf{b}}$. To zeroth order in kr_{L}, particles move along the θ-directed field lines with parallel velocity $v_{\parallel}\hat{\mathbf{b}}$, and spiral around the field lines with speed $v_{\perp}$. To solve for the first-order motion, we transform to the rotating frame that is moving with the zeroth-order particle motion in the θ direction. In this frame, the usual equations of motion hold, except for a centrifugal 'pseudo-force' in the radial direction, namely

$$\mathbf{F}_{\mathrm{cf}} = \frac{mv_{\parallel}^2}{R_{\mathrm{c}}}\hat{\mathbf{r}} = mv_{\parallel}^2\frac{\mathbf{R}_{\mathrm{c}}}{R_{\mathrm{c}}^2} \tag{3.10}$$

where we have defined a radius-of-curvature vector $\mathbf{R}_{\mathrm{c}}$ which is drawn from the local center-of-curvature to the field-line, as shown in Figure 3.2. (A Coriolis pseudo-force could also arise from drift motion in this rotating frame, but it will turn out that the drift motion is parallel to the axis of rotation, so the Coriolis force is zero in this particular case.)

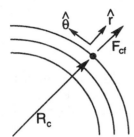

Figure 3.2. Geometry for calculating the curvature drift. The radius-of-curvature vector is drawn from the local center-of-curvature to the field line.

Using equation (2.22), we can then directly deduce

$$\mathbf{v}_{\mathrm{curv}} = \frac{mv_{\parallel}^2}{qB^2}\frac{\mathbf{R}_{\mathrm{c}} \times \mathbf{B}}{R_{\mathrm{c}}^2} = \frac{2W_{\parallel}}{qB^2}\frac{\mathbf{R}_{\mathrm{c}} \times \mathbf{B}}{R_{\mathrm{c}}^2} \tag{3.11}$$

where $W_{\parallel}$ is the particle's parallel energy. The vector radius-of-curvature, $\mathbf{R}_{\mathrm{c}}$, may not be a familiar way to describe local magnetic field geometry. In fact, however, any curved magnetic field can be characterized locally by a radius-of-curvature $\mathbf{R}_{\mathrm{c}}$, meaning that $\mathrm{d}\hat{\mathbf{b}}/\mathrm{d}s$ (where s measures length along the field line) $= -\mathbf{R}_{\mathrm{c}}/R_{\mathrm{c}}^2$. This is easily verified for the locally cylindrical geometry we have assumed, where the equivalent statement is just $(1/r)\mathrm{d}\hat{\boldsymbol{\theta}}/\mathrm{d}\theta = -\hat{\mathbf{r}}/r$. Since the $\mathrm{d}/\mathrm{d}s$ operator is just the derivative along the direction $\hat{\mathbf{b}}$, the radius-of-curvature can be re-expressed

$$\mathbf{R}_{\mathrm{c}}/R_{\mathrm{c}}^2 = -(\hat{\mathbf{b}} \cdot \boldsymbol{\nabla})\hat{\mathbf{b}} \tag{3.12}$$

giving a more common expression for the 'curvature drift'

$$\mathbf{v}_{\mathrm{curv}} = \left(\frac{2W_{\parallel}}{qB^2}\right)\mathbf{B} \times [(\hat{\mathbf{b}} \cdot \boldsymbol{\nabla})\hat{\mathbf{b}}]. \tag{3.13}$$

In many cases, especially when the plasma pressure is low and volume currents are low, the magnetic field in the plasma is approximately curl-free. In such cases, the magnetic field necessarily has *both* a gradient and curvature. For these so-called 'vacuum fields' (with no volume currents), the curvature drift can be put in a simpler form, closely related to that of the ∇B drift. Referring to Figure 3.2, it is clear that if a vacuum field is characterized over a local region by this geometry, then the magnetic field strength must fall off in the perpendicular direction with

$$(\nabla B)_\perp = -B\mathbf{R}_c/R_c^2 = (\mathbf{B} \cdot \nabla)\hat{\mathbf{b}} \tag{3.14}$$

in order for the field to have zero curl in all directions perpendicular to $\mathbf{B}$. (This result is established more formally in Problem 3.9.) Thus we can rewrite the curvature drift for vacuum fields as

$$\mathbf{v}_{\text{curv}} = \pm \frac{v_\parallel^2}{\omega_c} \frac{\mathbf{B} \times \nabla B}{B^2} = \frac{2W_\parallel}{q} \frac{\mathbf{B} \times \nabla B}{B^3} \tag{3.15}$$

which is identical in form to the ∇B drift given in equation (3.9), except that $W_\perp$ has been replaced by $2W_\parallel$. Note that the $\pm$ sign again goes with the sign of the charge. In an anisotropic Maxwellian plasma, $\langle W_\parallel \rangle = T_\parallel/2$ and $\langle W_\perp \rangle = T_\perp$, where the average here is taken over the velocity distribution; so the average of the combined ∇B and curvature guiding-center drifts for the particles in such a plasma, in a vacuum magnetic field, is

$$\langle \mathbf{v}_{\text{curv}} + \mathbf{v}_{\text{grad}} \rangle = \frac{T_\parallel + T_\perp}{q} \frac{\mathbf{B} \times \nabla B}{B^3}. \tag{3.16}$$

For an isotropic plasma, $T_\parallel + T_\perp = 2T$.

We have derived the $\mathbf{E} \times \mathbf{B}$ drift, the ∇B drift, and the curvature drift each in a rather specialized geometry. However, these drifts do not interfere with each other. Imagine adding a magnetic field gradient perpendicular to $\mathbf{B}$ or an electric field perpendicular to $\mathbf{B}$ to the present proof. They would give rise to the same cross-field drift we calculated before: the larger and smaller sides of the gyro-orbits would be formed, and net drifts would result just as before. Parallel gradients in $\hat{\mathbf{b}}$ (with gradient scale lengths $\gg r_L$) which gave rise to the curvature drift do not affect the other drifts, since the parallel motion played no role in those derivations. It is interesting, however, that the presence of these other drifts *would* give rise to a Coriolis force in the present calculation—but only in the direction parallel to $\mathbf{B}$. We will return to this issue when we discuss conservation of energy and magnetic moment to first-order in kr_L.

Problem 3.4: An anisotropic proton–electron plasma is immersed in the magnetic field from an infinite wire carrying current $I_z = 10^6$ A. This plasma has uniform density $n = 10^{19}\,\mathrm{m}^{-3}$, $T_{\perp e} = T_{\perp i} = 2\,\mathrm{keV}$ and $T_{\parallel e} = T_{\parallel i} = 5\,\mathrm{keV}$. At radius R away from the wire, what are the average ion and electron ∇B and curvature drift velocities? What is the total (ion + electron) guiding center current density, $\mathbf{j} = \sum n q \mathbf{v}$, in the plasma (where the summation is over species), and in which direction does this current flow? (Ignore the magnetic field due to the current in the plasma.)

3.3   STATIC B FIELD; CONSERVATION OF MAGNETIC MOMENT AT ZEROTH ORDER

So far we have considered gradients of B which were perpendicular to $\mathbf{B}$, and they gave rise to drifts which were also perpendicular to $\mathbf{B}$. We have considered the equation of motion parallel to $\mathbf{B}$ only in the sense of noting that the solutions of the guiding-center motion have represented free acceleration or deceleration along $\hat{\mathbf{b}}$ by a parallel electric field. Now we will consider the case of gradients of B along the direction of $\mathbf{B}$, which result in significant modifications of the equation for the parallel velocity.

Consider a static magnetic field which is pointed primarily in the z direction, and whose magnitude rises with $|z|$. To satisfy $\nabla \cdot \mathbf{B} = 0$, the field lines must converge away from $z = 0$. This could be arranged, for example, by having a solenoidal current winding, with a higher density of turns near the ends, as illustrated schematically in Figure 3.3.

As we have seen, a particle gyrating around a magnetic field line in

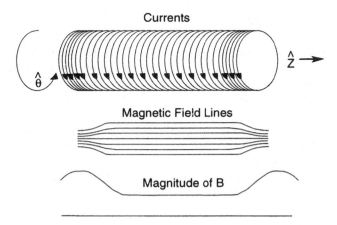

Figure 3.3. Currents in a solenoidal winding and the resulting 'mirror' magnetic fields inside the solenoid, shown schematically.

this system will drift across the non-uniform magnetic field dominantly in the θ direction, due to the ∇B and curvature drifts. However, we are concerned here with particle motion *along* the magnetic field, and—as it will turn out— changes in the mix of parallel and perpendicular velocity which result. Unlike the ∇B and curvature drift velocities, these velocity changes will not be of order kr_{L} compared to v_0—they will be of order unity. The case illustrated in Figure 3.3 is axisymmetric around the z axis, i.e. $\partial/\partial\theta = 0$, and B_θ is zero, so the only component of **B** besides B_z is B_r. This symmetry is not important to our analysis—the only geometrical property of the field that we will use is that the characteristic scale-length of variation of **B** is long compared to a gyro-radius (a by-now familiar condition).

Consider a differential cylindrical volume centered around a magnetic field line somewhere in the system where the field lines are converging (such as, but not necessarily, along the z axis of Figure 3.3). Now choose a new local cylindrical coordinate system (r, θ, z) centered on this cylinder, with $\hat{\mathbf{z}} = \hat{\mathbf{b}}$ as shown in Figure 3.4.

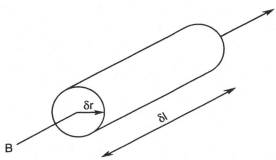

Figure 3.4. Geometry for calculating parallel acceleration.

We must have $\nabla \cdot \mathbf{B} = 0$ everywhere in the volume, so Gauss's law implies that there can be no net flux out of our differential volume. The net flux out through the end faces is $\pi(\delta r)^2\delta\ell(\mathrm{d}B/\mathrm{d}z)$. This plus the net flux out through the sides of the cylinder must be zero. Thus, the *average* radially directed **B** field in the local coordinate system, $\langle B_r \rangle \equiv \langle \mathbf{B} \cdot \hat{\mathbf{r}} \rangle$, is determined by

$$\pi(\delta r)^2\delta\ell(\mathrm{d}B/\mathrm{d}z) + 2\pi\,\delta r\delta\ell\langle B_r \rangle = 0 \qquad (3.17)$$

or

$$\langle B_r \rangle = -\frac{\delta r}{2}\frac{\mathrm{d}B}{\mathrm{d}z}. \qquad (3.18)$$

Let us now suppose that the radius of the cylinder, δr, is chosen to be the gyro-radius of a particle whose guiding center lies on the axis of the cylinder. In this case, it is just this $\langle B_r \rangle\hat{\mathbf{r}}$ which must be crossed with the azimuthal velocity

$v_\perp$ to give an average Lorentz force directed back along the magnetic field line. Averaging around a gyro-orbit we obtain (again assuming r_L is small)

$$\langle F_\parallel \rangle = -\frac{|q|v_\perp^2}{2\omega_c}\frac{dB}{dz} = -\frac{W_\perp}{B}\frac{dB}{dz}. \tag{3.19}$$

Equation (3.19) gives a force in the direction opposite to the field gradient for both electrons *and* ions.

It is convenient at this point to note that the quantity $mv_\perp^2/2B = W_\perp/B$ is the magnetic moment, μ, of the gyrating particle, because it is indeed equal to IA, the current represented by the moving charged particle times the area of the loop it circumnavigates. The current is I (amperes = coulombs per second) $= |q|\omega_c/2\pi$, while the area is $A = \pi r_L^2 = \pi v_\perp^2/\omega_c^2$, so $\mu = |q|v_\perp^2/(2\omega_c) = mv_\perp^2/2B = W_\perp/B$.

Thus, we obtain

$$m\frac{dv_\parallel}{dt} = -\mu\frac{dB}{ds} \tag{3.20}$$

where we have transformed back into a general coordinate system, in which we parameterize distance along the field line by the variable s.

We can next use equation (3.20) to determine if μ changes in time to zeroth order in kr_L. Multiplying both sides of this equation by $v_\parallel (= ds/dt)$ in order to obtain an energy-conservation-type of equation, we have

$$\frac{d}{dt}\left(\frac{mv_\parallel^2}{2}\right) = -\mu\frac{dB}{ds}\frac{ds}{dt} = -\mu\frac{dB}{dt} \tag{3.21}$$

where dB/dt is the total derivative, meaning the time-derivative as felt by the particle, due to its motion in the static magnetic field: specifically, $dB/dt = \partial B/\partial t + v_\parallel \nabla_\parallel B$, if we only include the zeroth-order guiding-center motion. (The partial time derivative, $\partial B/\partial t$ at any fixed position, is zero because of our assumption of a static **B** field.) We also know, however, that in the presence of only a static magnetic field, the total kinetic energy of the particle must be separately conserved at each order in kr_L, since higher-order terms cannot correct a mismatch in energy in a lower-order equation. Thus, we can ignore any energy in the ∇B and curvature drifts as being second order (v_{gc}^2), and any dB/dt due to them as being first order ($v_{gc}\cdot\nabla$), giving at zeroth order

$$\frac{d}{dt}\left(\frac{mv_\parallel^2}{2} + \frac{mv_\perp^2}{2}\right) = \frac{d}{dt}\left(\frac{mv_\parallel^2}{2} + \mu B\right) = 0. \tag{3.22}$$

Substituting equation (3.21) into the last part of this equation, we have

$$-\mu\frac{dB}{dt} + \frac{d}{dt}(\mu B) = 0 \tag{3.23}$$

which reduces to

$$d\mu/dt = 0. \qquad (3.24)$$

This invariance of μ implies that the velocity component $v_\perp$ increases as the particle moves along magnetic field lines into a region of higher magnetic field strength, in just such a way as to maintain $W_\perp/B$ constant. Since the particle's energy is also constant, it follows that the parallel component $v_\parallel$ decreases as $v_\perp$ increases. As a particle moves into a region of higher field-strength, e.g. towards an end of the solenoid shown in Figure 3.3, its velocity along the field decreases.

3.4 MAGNETIC MIRRORS

With the understanding of particle motion in static non-uniform **B** fields that we have now developed, we are able to understand the basic principle of one of the primary 'magnetic traps' for confining plasma, both in the laboratory and in nature: the 'magnetic mirror'. Since a particle's kinetic energy, W, and magnetic moment, μ, are both constant (in the absence of **E** fields) the particle's parallel velocity will vary as it moves into regions of different field strength according to

$$\frac{mv_\parallel^2}{2} = W - \mu B. \qquad (3.25)$$

As the particle moves from a weak-field region to a strong-field region in the course of its motion along a field line, it sees an increasing B, and therefore its parallel velocity $v_\parallel$ decreases. If B is high enough in the 'throat' of the mirror (see Figure 3.3), $v_\parallel$ becomes zero, and the particle is 'reflected back' toward the weak-field region, a process that serves to hold both electrons and ions within the solenoidal field structure. Mirror trapping does not work, however, for all values of the ratio $v_\parallel/v$. For instance, a particle with $v_\parallel = v$, $v_\perp = 0$, has a zero magnetic moment μ, and will not experience any decelerating force at all as it approaches a high-field region.

If the minimum field along a field line at the midplane is defined to be B_{min} and that at the mirror throat B_{max}, it is clear from the constancy of particle energy that all particles with $\mu > W/B_{max}$ are trapped, because, if such a particle were to reach the mirror throat with μ conserved, this would imply $\mu B_{max} = W_\perp > W$, which is not possible. Translating to $v_\parallel/v$ at the midplane we have, for the *marginally trapped* particles,

$$W_\perp(\text{midplane}) = \mu B_{min} = W B_{min}/B_{max}$$
$$W_\parallel(\text{midplane})/W = (1 - B_{min}/B_{max})$$

that is,

$$v_\perp(\text{midplane})/v = (B_{min}/B_{max})^{1/2}$$
$$v_\parallel(\text{midplane})/v = (1 - B_{min}/B_{max})^{1/2}. \tag{3.26}$$

Particles with lower $v_\parallel/v$ at the midplane are trapped by the mirror field, while those with greater $v_\parallel/v$ are in a 'loss cone' in velocity space, defined by

$$v_\parallel/v > (1 - B_{min}/B_{max})^{1/2} \tag{3.27}$$

or equivalently

$$v_\parallel/v_\perp > (B_{max}/B_{min} - 1)^{1/2} \tag{3.28}$$

as shown in Figure 3.5.

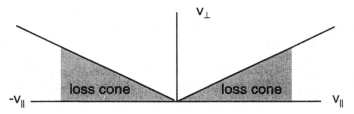

Figure 3.5. Velocity-space 'loss cones' in a magnetic mirror. The angle of the loss cone is given by equation (3.26).

The concept of a loss 'cone' derives from recognizing that the $v_\perp$ axis really represents two dimensions, and Figure 3.5 can be rotated around the $v_\parallel$ axis to represent a fully three-dimensional velocity space. Since all particles below the diagonal lines are rapidly lost from the system, a mirror-trapped plasma is never isotropic in velocity space. The magnetic loss cone is independent of the charge and mass of the particles. However, when particles collide with each other, they change the direction of their velocity vectors and can 'scatter' into the loss cone. Thus the species that collides more frequently (the electrons for $T_e \approx T_i$, as we will see in Chapter 11) will be lost preferentially. An electric field will then build up, corresponding to a positive electric potential in the central region, holding back the electrons along the magnetic field lines, and so electrostatically 'plugging up' the low-energy portion of the electron loss cone. This electric potential builds up to the point where it keeps the net outflux of electrons balanced with the slower outflux of ions. However, more energetic electrons will still escape over the 'top' of the electrostatic 'plug' in velocity space, so electron thermal losses still tend to dominate the energy balance of mirror-trapped plasmas, while ion scattering sets the pace for the particle loss.

Problem 3.5: Assume $\mathbf{B} = \hat{z}B_0(1 + \gamma z^2)$. To lowest order in kr_L (i.e. only the $v_\parallel \hat{b}$ motion), calculate the bounce period for a particle moving back and forth in this magnetic well. Note that $ds = v_\parallel dt$.

Problem 3.6: Consider a 10 keV energetic ion in the Van Allen belts $\sim 10^6$ m above the Earth's surface, in a dipole magnetic field of $\sim 10^{-6}$ T. Estimate the curvature and ∇B drift speeds of this particle. Compare them with the gravitational drift speed.

3.5 ENERGY AND MAGNETIC-MOMENT CONSERVATION TO FIRST ORDER FOR STATIC FIELDS*

Although we have made use of the constancy of a particle's kinetic energy in our proof of the constancy of μ, it is important to note that μ-invariance is valid in much more general circumstances, and to higher order, including cases where the kinetic energy is *not* constant (see Chapter 4, where we treat time-dependent fields). It is also important to understand how total energy conservation (kinetic plus potential) works to first order in kr_L. For this purpose, let us consider the case where there is a static electric field, as well as the static magnetic field, with arbitrary gradients but with kr_L small as usual.

We will start by re-examining energy conservation in zeroth order, including $\mathbf{E} \neq 0$. The parallel equation of motion is simply

$$m\frac{dv_\parallel}{dt} = qE_\parallel - \mu\frac{dB}{ds} = -q\frac{d\phi}{ds} - \mu\frac{dB}{ds} \qquad (3.29)$$

obtained by adding a parallel electric field to equation (3.20), and using $E_\parallel = -d\phi/ds$, which is possible because we have assumed $\partial B/\partial t = 0$. Again we multiply by $v_\parallel = ds/dt$ to see energy conservation from a 'kinematic' (following the trajectory) point of view, to zeroth order in kr_L:

$$\frac{d}{dt}\left(\frac{mv_\parallel^2}{2}\right) = -q\frac{d\phi}{dt} - \mu\frac{dB}{dt} \qquad (3.30)$$

where we are still ignoring any contribution of the ∇B and curvature drifts to dB/dt and to $d\phi/dt$, as being higher order in kr_L. The sum of the particle's kinetic plus potential energy must be conserved at each order in kr_L (and so specifically at zeroth order); thus we have a zeroth-order energy conservation equation from first-principles considerations:

$$\frac{d}{dt}\left(\frac{mv_\parallel^2}{2} + \mu B + q\phi\right) = 0. \qquad (3.31)$$

Using equation (3.30) to substitute for the first term in equation (3.31), we again obtain equations (3.23) and (3.24), and so μ is conserved at zeroth order in kr_L, even in the presence of a static electric field.

Next we move to first order. We will see that adding first-order effects to the analysis generates a flurry of additional terms, which reveal interesting transfers of energy between potential, perpendicular kinetic, and parallel kinetic energy, but all of which cancel perfectly in their effect on our proof of μ conservation.

We must begin by including a missing first-order term in the parallel acceleration equation, namely equation (3.29). This term arises from the Coriolis pseudo-force, for the case where $\omega \times v_{gc} \neq 0$. (Here we introduce a vector $\omega \equiv -v_\parallel \hat{b} \times R_c / R_c^2$ whose magnitude is $v_\parallel / R_c$ and whose direction points along the z axis of the local coordinate system we established in deriving the curvature drift.) The curvature drift, v_{curv}, as we noted when we derived it, satisfies $\omega \times v_{gc} = 0$, but the $E \times B$ drift, v_E, may not, and the result will be a parallel acceleration. (The case where $\omega \times v_{grad} \neq 0$ is handled in Problem 3.7.) We calculated the curvature drift by locally approximating B as lying on a circle with radius-of-curvature R_c, and transforming to the rotating frame where $v_\parallel = 0$. The E field in this frame is equal to the stationary-frame E field, since the local frame velocity is parallel to B. However, due to the v_E drift, there is also a Coriolis pseudo-force in this frame, $F_c = -2m (\omega \times v_E)$. This shows up as a new first-order term for $m(dv_\parallel / dt)$ *in the rotating frame*:

$$m \left. \frac{dv_\parallel}{dt} \right|_{\omega \times v_E} = -2m(\omega \times v_E) \cdot \hat{b} = -\frac{2mv_\parallel}{R_c^2} v_E \cdot R_c. \qquad (3.32)$$

The $\omega \times v_E$ subscript indicates the component of the total derivative due to the Coriolis force which arises from the $E \times B$ drift. The second equality follows from writing $(\omega \times v_E) \cdot \hat{b} = (\hat{b} \times \omega) \cdot v_E$ and noting that $\hat{b} \times \omega = v_\parallel R_c / R_c^2$. To translate this result to the laboratory frame, we note $v_\parallel (\text{lab}) = v_\parallel (\text{rotating}) + \omega R_c$, and so

$$\begin{aligned}
\left. \frac{dv_\parallel}{dt} \right|_{lab} &= \left. \frac{dv_\parallel}{dt} \right|_{rot} + \omega \frac{dR_c}{dt} \\
&= \left. \frac{dv_\parallel}{dt} \right|_{rot} + \omega v_E \cdot R_c / R_c \\
&= \left. \frac{dv_\parallel}{dt} \right|_{rot} + \frac{v_\parallel}{R_c^2} v_E \cdot R_c. \qquad (3.33)
\end{aligned}$$

Thus, in the laboratory frame, we have

$$m \left. \frac{dv_\parallel}{dt} \right|_{\omega \times v_E} = -\frac{mv_\parallel}{R_c^2} v_E \cdot R_c. \qquad (3.34)$$

This is a new first-order term that must be added to the right-hand side of equation (3.29).

Note that by conservation of angular momentum, $mv_\parallel R$ must be constant if **E** is locally perpendicular to **B** and $dB/ds = 0$. Equation (3.34) nicely shows that the first-order drifts do indeed conserve angular momentum in this situation, when the parallel Coriolis force is included. Multiplying through by $v_\parallel$ as we did to derive equation (3.30), and substituting the full expression for $\mathbf{v}_E$ we have

$$\left.\frac{dW_\parallel}{dt}\right|_{\omega \times \mathbf{v}_E} = mv_\parallel \frac{dv_\parallel}{dt} = \frac{(\nabla\phi \times \mathbf{B}) \cdot \mathbf{R}_c}{B^2 R_c} \frac{mv_\parallel^2}{R_c} \tag{3.35}$$

(where the $\omega \times \mathbf{v}_E$ subscript again indicates the component of the total derivative *due to* the Coriolis force—now in the laboratory frame). This is a new first-order term that must be added to the right-hand side of equation (3.30), the zeroth-order 'kinematic' energy balance equation, in order to make it correct to first order when $\omega \times \mathbf{v}_E \neq 0$. Note that the total derivatives in that equation were explicitly evaluated only in terms of the zeroth-order motion along $\hat{\mathbf{b}}$.

Now we will find the elegant result that a *first-order interpretation* of the total derivatives on the right-hand side of equation (3.30) will provide just this new first-order term! In going from equation (3.29) (parallel acceleration) to equation (3.30) (kinematic energy balance), at lowest order in kr_L, we assumed $d/dt = v_\parallel d/ds$, with $\partial/\partial t = 0$ (static fields). Including terms at *first order* in kr_L—interpreting equation (3.30) *as it stands* as including first-order terms as well as zeroth-order terms—we take the total derivative d/dt as

$$d/dt = \partial/\partial t + \mathbf{v}_{gc} \cdot \nabla \tag{3.36}$$

where the first-order terms of $\mathbf{v}_{gc}$ (the ∇B drift, the $\mathbf{v}_E$ drift and the curvature drift) are now included. Thus the terms $qd\phi/dt$ and $\mu dB/dt$ have *first-order* components that need to be evaluated in order to obtain all the first-order components in the right-hand side of equation (3.30). Surprisingly, this first-order interpretation of equation (3.30) will provide just the term we need to represent the new first-order term due to the Coriolis force.

The extra first-order term on the right-hand-side of equation (3.30) due to the curvature drift dotted with $\nabla\phi$ is

$$-q\mathbf{v}_{curv} \cdot \nabla\phi = \frac{-mv_\parallel^2(\mathbf{R}_c \times \mathbf{B})}{B^2 R_c^2} \cdot \nabla\phi. \tag{3.37}$$

This term is exactly equal to the Coriolis force term (equation (3.35)) in its effect on the kinematic energy balance. Thus, equation (3.30), when interpreted as including $\mathbf{v}_{curv}$ in the total derivative, gives the correct answer for $dW_\parallel/dt$, including the first-order effect we just calculated from the Coriolis acceleration.

Insofar as kinetic and potential energy are exchanged due to curvature drifts in the direction of $\nabla\phi$, the particle absorbs this in the $W_\parallel$ component, without changing μ or $W_\perp$.

Problem 3.7: Calculate the analog to equation (3.35) where the ∇B drift causes the Coriolis force (this requires $\omega \times v_{\text{grad}} \neq 0$). In this case $v_{\text{curv}} \cdot \nabla B \neq 0$ also, so calculate the effect of the curvature drift on $\mu dB/dt$. Then show that the first-order interpretation of d/dt on the right-hand side of equation (3.30) including $\mu v_{\text{grad}} \cdot \nabla B$ provides just the required effect of the Coriolis force on $W_\parallel$. In this case $W_\parallel$ is exchanged with $W_\perp$, again without changing μ. (This particular situation can only arise when $\nabla \times \mathbf{B} \neq 0$, i.e. when there are volume currents.)

The ∇B drift also gives rise to a first-order $q d\phi/dt$ term

$$-q\mathbf{v}_{\text{grad}} \cdot \nabla\phi = \frac{-W_\perp(\mathbf{B} \times \nabla B)}{B^3} \cdot \nabla\phi = \frac{-\mu(\mathbf{B} \times \nabla B)}{B^2} \cdot \nabla\phi \qquad (3.38)$$

but no additional $\mu dB/dt$ term, since $\mathbf{v}_{\text{grad}} \cdot \nabla B = 0$. The $\mathbf{v}_E$ drift gives rise to an additional first-order $\mu dB/dt$ term, but no $q d\phi/dt$ term, since $\mathbf{v}_E \cdot \nabla\phi = 0$. We have

$$-\mu\mathbf{v}_E \cdot \nabla B = \mu\frac{(\nabla\phi \times \mathbf{B})}{B^2} \cdot \nabla B. \qquad (3.39)$$

We see from equations (3.38) and (3.39) that the contributions to the first-order interpretation of equation (3.30) from $\mathbf{v}_{\text{grad}}$ to $q d\phi/dt$ and from $\mathbf{v}_E$ to $\mu dB/dt$ sum to zero, so there is no net change in equation (3.30) when we take into account the total derivative to first-order in kr_L due to the sum of the ∇B and $\mathbf{v}_E$ drifts. The change in kinetic energy, W, due to the ∇B drift along $\nabla\phi$ is absorbed into μ times a change in B (i.e. a change in $W_\perp$), with no required change in μ and no change in $W_\parallel$. Thus equation (3.30) is *fully correct to first order (including the Coriolis effects) when* d/dt *is interpreted as including all of the first-order guiding-center drifts.* Energy conservation (equation (3.31)) is correct as it stands to first order, since the first-order guiding center drifts would contribute to energy terms only at second order. The proof of μ-conservation, now to first order, follows directly from these two equations as shown before at zeroth order in equations (3.23) and (3.24).

In summary, when a particle's curvature drift carries it across an electric potential, its $\mathbf{v}_E$ drift moves it in the direction of the local radius-of-curvature. The particle balances the change of potential energy with a change in $W_\parallel$, thereby conserving angular momentum. When a particle's ∇B drift carries it across an electric potential, it balances the change in potential energy with a change in

$W_\perp$, because at the same time its $\mathbf{v}_E$ drift is carrying it to a region of changed B—and μ is conserved. In the case where $\nabla \times \mathbf{B} \neq 0$, curvature drifts can also move particles to regions of changed B, and then $W_\perp$ and $W_\parallel$ are exchanged, conserving energy, angular momentum and magnetic moment, μ.

We have shown now that the nominally zeroth-order kinematic energy balance, equation (3.30), is also perfectly correct *when the convective derivative is interpreted at first order*. We were able to show that the net effect of the extra terms which arose from considering the first-order interpretation of d/dt (i.e. including $\mathbf{v}_{gc} \cdot \nabla$ in d/dt) is equal to the physical effect of the Coriolis force on $W_\parallel$. Thus, equation (3.30) is a reliable basis for calculating $v_\parallel(t)$ for the purpose of evaluating particle drifts in time-independent fields. Note, however, that equation (3.29) for parallel acceleration is *not* accurate to first order. The correct result, starting from kinematic energy balance, equation (3.30), is

$$m\frac{dv_\parallel}{dt} = -\frac{q}{v_\parallel}\frac{d\phi}{dt} - \frac{\mu}{v_\parallel}\frac{dB}{dt} \tag{3.40}$$

where the total derivatives include the contribution of all the first-order drifts. As we have seen, the effects of the ∇B and $\mathbf{v}_E$ drifts cancel perfectly on the right-hand side of this equation, but the curvature drift can have a net effect on $-q d\phi/dt - \mu dB/dt$. Equation (3.40) can be difficult to evaluate, so in time-independent situations it is usually easier to obtain $v_\parallel$ from energy conservation: $mv_\parallel^2/2 + \mu B + e\phi = $ constant. The simplest way to calculate $W_\perp$ is then from μ conservation.

Problem 3.8: Consider a particle orbiting at radius r in the magnetic field from an infinite wire carrying current I in the z direction. Imagine there is also a constant electric field of magnitude E pointing in the z direction. At $t = 0$, evaluate dr/dt, dz/dt, $dW_\perp/dt$ and $dW_\parallel/dt$ for this particle's guiding center, in terms of E, I, r, the particle's mass m, charge q, and its initial parallel and perpendicular velocities $v_{\parallel 0}$, $v_{\perp 0}$.

3.6 DERIVATION OF DRIFTS: GENERAL CASE*

In this Chapter, in order to provide the clearest possible derivations of the ∇B and curvature drifts, we have used special geometries, in which the magnetic field had either a gradient but no curvature, or curvature but no gradient. In fact these two drifts describe completely the lowest-order cross-field guiding-center motion in *any* static non-uniform magnetic field. The addition of a perpendicular electric field perpendicular to the magnetic field simply adds the $\mathbf{v}_E$ drift to the other first-order drifts, provided that the electric field is small enough for $\mathbf{v}_E$

to be a first-order velocity. These results can be derived formally by extending the methods used earlier in this Chapter to general field geometry, although the analysis requires more sophisticated vector manipulations.

We begin with the particle's equation of motion, which to zeroth order is simply

$$m\frac{d\mathbf{v}_0}{dt} = q\mathbf{v}_0 \times \mathbf{B}_{gc} \qquad (3.41)$$

and describes gyration about a fixed guiding center, plus parallel motion. Note that, in equation (3.41), we have taken care to employ the magnetic field at the particle's 'average' position, i.e. its guiding center. In this order, the vector relationship between the particle's position $\mathbf{x}_0$ and its guiding center $\mathbf{x}_{gc,0}$ can be written (for a positively charged particle)

$$\mathbf{x}_0 = \mathbf{x}_{gc,0} - \mathbf{v}_0 \times \hat{\mathbf{b}}/\omega_c. \qquad (3.42)$$

Moving now to first order in our expansion in kr_L, the equation of motion must take account of the difference between $\mathbf{B}$ at the particle's position $\mathbf{x}$ and $\mathbf{B}_{gc}$, i.e.

$$\mathbf{B} = \mathbf{B}_{gc} - \frac{1}{\omega_c}[(\mathbf{v}_0 \times \hat{\mathbf{b}}) \cdot \mathbf{\nabla}]\mathbf{B}_{gc} \qquad (3.43)$$

and it must also allow for the first-order time dependences of the zeroth-order guiding-center velocity (i.e. time dependences much slower than gyro-motion), which we denote $(d/dt)_1$. Keeping all these terms and also the electric field, the first-order equation of motion after time-averaging over many gyro-periods is

$$m\left(\frac{d}{dt}\right)_1 (v_{\parallel}\hat{\mathbf{b}}) = q(\mathbf{E} + \langle \mathbf{v}_1 \rangle \times \mathbf{B}) - \frac{m}{B}\langle \mathbf{v}_0 \times [(\mathbf{v}_0 \times \hat{\mathbf{b}}) \cdot \mathbf{\nabla}]\mathbf{B}\rangle \qquad (3.44)$$

where $\langle \; \rangle$ denotes the time average. The time-average guiding-center velocity $\langle \mathbf{v}_1 \rangle$ is simply the drift across the magnetic field, which we denote $\mathbf{v}_d$. Since we are considering here only time-independent fields, the left-hand side of equation (3.44) arises only from the spatial derivative following the guiding-center motion, i.e.

$$\left(\frac{d}{dt}\right)_1 (v_{\parallel}\hat{\mathbf{b}}) = v_{\parallel}\hat{\mathbf{b}} \cdot \mathbf{\nabla}(v_{\parallel}\hat{\mathbf{b}}) \qquad (3.45)$$

(where we have implicitly included any time variation of $v_{\parallel}$ as a spatial derivative).

Taking the cross-product of equation (3.44) with $\mathbf{B}$, we obtain

$$\mathbf{v}_d = \langle \mathbf{v}_{1\perp} \rangle = \mathbf{v}_E \pm \frac{1}{\omega_c B^2}\mathbf{B} \times \langle \mathbf{v}_0 \times [(\mathbf{v}_0 \times \hat{\mathbf{b}}) \cdot \mathbf{\nabla}]\mathbf{B}\rangle$$

$$\pm \frac{1}{\omega_c B}\mathbf{B} \times [v_{\parallel}\hat{\mathbf{b}} \cdot \mathbf{\nabla}(v_{\parallel}\hat{\mathbf{b}})] \qquad (3.46)$$

where $\mathbf{v}_E = \mathbf{E} \times \mathbf{B}/B^2$ and $\pm$ denotes the sign of the particle's charge. The second term on the right-hand side will give rise to the ∇B drift. The last term, which arose from the slow change in $m\mathbf{v}_0$ (equivalent to the centrifugal force in the rotating frame discussed in Section 3.2) will give rise to the curvature drift.

First consider the second term on the right-hand side of equation (3.46). Since $\mathbf{v}_0$ describes gyro-motion, the time-average of an element of the tensor formed by the product of two $\mathbf{v}_0$-vectors will be given, in index notation (see Appendix D), by

$$\langle v_{0i} v_{0j} \rangle = \frac{v_\perp^2}{2} \delta_{ij} + \left(v_\parallel^2 - \frac{v_\perp^2}{2} \right) \hat{b}_i \hat{b}_j. \tag{3.47}$$

This result can be derived by considering a local coordinate system $\hat{\mathbf{b}}, \hat{\mathbf{e}}_\perp, \hat{\mathbf{b}} \times \hat{\mathbf{e}}_\perp$, where $\hat{\mathbf{e}}_\perp$ is a unit vector perpendicular to $\hat{\mathbf{b}}$ at an arbitrary angle. Then we have

$$\mathbf{v} = v_\parallel \hat{\mathbf{b}} + v_\perp^2 \cos\omega t \ \hat{\mathbf{e}}_\perp + v_\perp^2 \sin\omega t \ \hat{\mathbf{b}} \times \hat{\mathbf{e}}_\perp.$$

Thus

$$\begin{aligned}
\langle \mathbf{vv} \rangle &= v_\parallel^2 \hat{\mathbf{b}}\hat{\mathbf{b}} + (v_\perp^2/2)\hat{\mathbf{e}}_\perp\hat{\mathbf{e}}_\perp + (v_\perp^2/2)(\hat{\mathbf{b}} \times \hat{\mathbf{e}}_\perp)(\hat{\mathbf{b}} \times \hat{\mathbf{e}}_\perp) \\
&= (v_\parallel^2 - v_\perp^2/2)\hat{\mathbf{b}}\hat{\mathbf{b}} + (v_\perp^2/2)[\hat{\mathbf{b}}\hat{\mathbf{b}} + \hat{\mathbf{e}}_\perp\hat{\mathbf{e}}_\perp + (\hat{\mathbf{b}} \times \hat{\mathbf{e}}_\perp)(\hat{\mathbf{b}} \times \hat{\mathbf{e}}_\perp)].
\end{aligned}$$

Using index notation, the second term on the right-hand side in equation (3.46) can now be simplified as follows:

$$\begin{aligned}
\langle \mathbf{v}_0 \times [(\mathbf{v}_0 \times \hat{\mathbf{b}}) \cdot \nabla]\mathbf{B} \rangle_i &= \langle \epsilon_{ijk} v_{0j} \epsilon_{lmn} v_{0m} \hat{b}_n \frac{\partial B_k}{\partial x_l} \rangle \\
&= \frac{v_\perp^2}{2} \epsilon_{ijk}\epsilon_{ljn} \hat{b}_n \frac{\partial B_k}{\partial x_l} \\
&= \frac{v_\perp^2}{2} (\delta_{il}\delta_{kn} - \delta_{in}\delta_{kl}) \hat{b}_n \frac{\partial B_k}{\partial x_l} \\
&= \frac{v_\perp^2}{2} \left(\hat{b}_k \frac{\partial B_k}{\partial x_i} - \hat{b}_i \frac{\partial B_k}{\partial x_k} \right) \\
&= \frac{v_\perp^2}{2} \left(\frac{\partial B}{\partial x_i} - \hat{b}_i (\nabla \cdot \mathbf{B}) \right) \\
&= \frac{v_\perp^2}{2} (\nabla B)_i \tag{3.48}
\end{aligned}$$

where we have used the expression for the product of two Levi-Civita symbols in terms of Kronecker delta functions given in Appendix D. We have also used $\partial B_k / \partial x_k = \nabla \cdot \mathbf{B} = 0$. This term will clearly give rise to the ∇B drift.

In the third term on the right-hand side of equation (3.46), only the term with the gradient operator applied to $\hat{\mathbf{b}}$ and not to $v_\|$ will survive, since the other term will contain $\hat{\mathbf{b}} \times \hat{\mathbf{b}} = 0$. Thus

$$\mathbf{B} \times \left[v_\| \hat{\mathbf{b}} \cdot \nabla(v_\| \hat{\mathbf{b}}) \right] = v_\|^2 \mathbf{B} \times (\hat{\mathbf{b}} \cdot \nabla)\hat{\mathbf{b}}. \tag{3.49}$$

This term evidently will give rise to the curvature drift.

Substituting equations (3.48) and (3.49) into equation (3.46), we obtain our final expression for the guiding-center drift

$$\mathbf{v}_d = \mathbf{v}_E + \frac{W_\perp}{q} \frac{\mathbf{B} \times \nabla B}{B^3} + \frac{2W_\|}{q} \frac{\mathbf{B} \times (\hat{\mathbf{b}} \cdot \nabla)\hat{\mathbf{b}}}{B^2} \tag{3.50}$$

which is the v_E drift, together with the sum of the ∇B drift given in equation (3.9) and the curvature drift given in equation (3.13). This completes our formal proof that the previously derived drifts apply generally to non-uniform fields with gradient scale lengths $\gg r_L$.

Problem 3.9: For a field with $\nabla \times \mathbf{B} = 0$, prove that the relationship

$$\mathbf{B} \times (\hat{\mathbf{b}} \cdot \nabla)\hat{\mathbf{b}} = \hat{\mathbf{b}} \times \nabla B$$

holds generally, thereby formally demonstrating that the curvature drift takes the form given in equation (3.15) for vacuum fields. (Hint: start with $0 = \hat{\mathbf{b}} \times (\nabla \times \mathbf{B})$ written in index notation using the Levi-Civita symbols. Reduce this to a different vector equation and take its cross-product with $\hat{\mathbf{b}}$ to obtain the desired result.)

Chapter 4

Particle drifts in time-dependent fields

So far we have considered the guiding-center drifts that arise from perpendicular electric fields and from various types of non-uniformity of the magnetic field, all with gradient scale-lengths long compared to a gyro-radius. In these cases, the electric and magnetic fields were assumed to be constant in time. Now we complete the analysis by considering the effect of time-dependences of these fields, where we will consider only changes slow compared to a gyro-period.

4.1 TIME-VARYING B FIELD

First, let us consider the case of a time-varying **B** field, with a characteristic time variation $\partial/\partial t \sim \omega \ll \omega_c$. From the point of view of a moving particle, this slowness requirement is similar to the requirement on the spatial variation $\partial/\partial x \sim k \ll 1/r_L$. Since d/dt gives the time derivative at the particle's changing position, the requirement $k \ll 1/r_L$ already implies that the convective part of d/dt satisfies the slowness requirement, because $v \cdot (\partial/\partial x) \ll v/r_L \sim \omega_c$. Thus we are requiring that the **B** field does not change much during a single gyro-orbit, either due to its intrinsic time variation *or* due to the particle's motion.

For simplicity, consider the case of a spatially homogeneous magnetic field, changing in time. The equation for the parallel particle velocity $v_\parallel \hat{b}$ goes through as before. There is an interesting consequence for the perpendicular velocity, however. With any time-changing magnetic field, Maxwell's equations tell us that there must be a curl to the electric field:

$$\nabla \times \mathbf{E} = -\partial \mathbf{B}/\partial t \tag{4.1}$$

or using Stokes's theorem:

$$\oint \mathbf{E} \cdot d\mathbf{l} = \int \nabla \times \mathbf{E} \cdot d\mathbf{S} = -\frac{\partial}{\partial t} \left(\int \mathbf{B} \cdot d\mathbf{S} \right) \tag{4.2}$$

where dl is an element of arc length along the perimeter of an area and dS is an element of that area. The vector signs of dl and dS are determined by the right-hand rule (fingers following dl and thumb pointing in direction of dS). If we imagine following a negatively charged particle ($q < 0$, right-hand sense of gyration) around its gyro-orbit, when $\partial B/\partial t > 0$, we see that $q\mathbf{v} \cdot \mathbf{E}$ is everywhere positive, so that the particle will be accelerated steadily in $v_\perp$ as it gyrates. Similarly, if we imagine following a positively charged particle around its orbit (in the left-hand direction now) both q and $\mathbf{v}$ have changed sign, so again the particle is steadily accelerated in $v_\perp$ as it gyrates. Taking the time average (indicated by $\langle \rangle$) around many gyro-orbits in order to determine the average $q\mathbf{v} \cdot \mathbf{E}$ and consequent time-averaged change in $W_\perp$, we have

$$\frac{\mathrm{d}}{\mathrm{d}t}\langle W_\perp \rangle = q\langle \mathbf{v} \cdot \mathbf{E} \rangle = |q|v_\perp \frac{\pi r_L^2}{2\pi r_L}\frac{\partial B}{\partial t} = \frac{|q|v_\perp^2}{2\omega_c}\frac{\partial B}{\partial t} = \frac{W_\perp}{B}\frac{\partial B}{\partial t} = \mu\frac{\partial B}{\partial t}. \quad (4.3)$$

So, as noted before, $W_\perp$ grows steadily as B increases. Interestingly, $W_\perp$ grows in just such a way that $\mu(\equiv W_\perp/B)$ is still conserved (again so long as the characteristic timescale of changes in B is very slow compared to Larmor gyration):

$$\frac{\mathrm{d}\mu}{\mathrm{d}t} = \frac{1}{B}\frac{\mathrm{d}W_\perp}{\mathrm{d}t} - \frac{W_\perp}{B^2}\frac{\partial B}{\partial t} = 0. \quad (4.4)$$

Thus the magnetic moment is conserved in slowly time-varying magnetic fields. Taking this together with the results of Chapter 3, we conclude that so long as $\omega \ll \omega_c$ and $k \ll 1/r_L$, the magnetic moment, μ, is a good constant of motion in essentially all cases. It turns out that this implies that the magnetic flux enclosed in a gyro-orbit, $\pi r_L^2 B$, is also conserved, since

$$\pi r_L^2 B = \frac{\pi v_\perp^2 B}{\omega_c^2} = \frac{\pi m^2 v_\perp^2}{q^2 B} = \frac{2\pi m}{q^2}\mu. \quad (4.5)$$

By including the new energy source associated with $\partial B/\partial t$ into equation (3.30) we can now construct the full energy-conservation equation appropriate for the guiding-center drift equations to first order in kr_L, where we assume that all the drifts (including $\mathbf{v}_E$) are of order kr_L compared to the particle velocities:

$$\frac{\mathrm{d}}{\mathrm{d}t}\left(\tfrac{1}{2}mv_\parallel^2 + \mu B\right) = q\mathbf{v}_{gc} \cdot \mathbf{E} + \mu\frac{\partial B}{\partial t}. \quad (4.6)$$

Here $\mathbf{v}_{gc}$ is the sum of the ∇B drift, the $\mathbf{v}_E$ drift, the curvature drift and the $v_\parallel \hat{\mathbf{b}}$ guiding-center motion. The $\mathbf{v}_E$ drift of course does not contribute to $\mathbf{v}_{gc} \cdot \mathbf{E}$. (We are ignoring gravitational drifts, and gravitational potential energy.) Casting equation (4.6) as an equation for $v_\parallel$ we have

$$mv_\parallel\frac{\mathrm{d}v_\parallel}{\mathrm{d}t} = \mathbf{v}_{gc} \cdot (q\mathbf{E} - \mu\nabla B). \quad (4.7)$$

Thus the **B** field acts as a potential field for the parallel energy, even when $\partial \mathbf{B}/\partial t \neq 0$. For practical applications, equation (4.7) can be further simplified. From the considerations in the previous Chapter (equations (3.38) and (3.39)), the $\mathbf{v}_E$ drift contribution to equation (4.7) cancels the ∇B drift contribution, so that all that is needed of $\mathbf{v}_{\text{gc}}$ in this equation is the zeroth-order parallel motion ($\propto v_\parallel$) and the first-order curvature drift ($\propto v_\parallel^2$). This result that the relevant part of $\mathbf{v}_{\text{gc}}$ scales at least linearly with $v_\parallel$ is necessary in order that $dv_\parallel/dt$ remain finite as $v_\parallel \to 0$ at mirror reflection points, in the presence of arbitrary field gradients. When equation (4.7) is evaluated numerically, typically both sides are first divided by $v_\parallel$, so that instead of evaluating $\mathbf{v}_{\text{gc}} \cdot (q\mathbf{E} - \mu\nabla B)$ and later dividing by $v_\parallel$ (giving 0/0 at mirror reflection points), one evaluates $(\mathbf{v}_{\text{gc}}/v_\parallel) \cdot (q\mathbf{E} - \mu\nabla B)$ to obtain $m dv_\parallel/dt$. The relevant form of $\mathbf{v}_{\text{gc}}/v_\parallel$ needed for this equation is just $\hat{\mathbf{b}} + m v_\parallel \mathbf{B} \times (\hat{\mathbf{b}} \cdot \nabla)\hat{\mathbf{b}}/(q B^2)$.

Problem 4.1: Show explicitly that the $\mathbf{v}_E$ drift contribution to equation (4.7) precisely cancels the ∇B drift contribution.

The full set of equations required to solve for guiding-center particle motion in slowly time- and space-varying magnetic and electric fields, up to first order in $k r_{\text{L}}$ and in ω/ω_{c}, are thus

$$\mathbf{v}_{\text{gc}} = v_\parallel \hat{\mathbf{b}} + \frac{\mathbf{E} \times \mathbf{B}}{B^2} + \frac{W_\perp \mathbf{B} \times \nabla B}{q B^3} + \frac{2 W_\parallel \mathbf{B} \times (\hat{\mathbf{b}} \cdot \nabla)\hat{\mathbf{b}}}{q B^2} \tag{4.8}$$

together with $W_\perp = \mu B$ (μ constant) and the evolution of $v_\parallel$ given by equation (4.7).

4.2 ADIABATIC COMPRESSION

The conservation of magnetic moment, μ, means that a changing magnetic field will heat (or cool) a plasma. Consider a cylindrical plasma in a solenoidal magnetic field. If the field is ramped up in time, the perpendicular energies $W_\perp$ of all the particles will rise as well. It is interesting to note that the plasma will be driven in towards the center of the solenoid, compressed away from the coils. By equation (4.2), we have

$$2\pi r E_\theta = -\pi r^2 \partial B_z/\partial t \tag{4.9}$$

and the radial drift velocity is

$$\frac{dr}{dt} = \mathbf{v}_E \cdot \hat{\mathbf{r}} = \frac{E_\theta}{B_z} = -\frac{r}{2 B_z}\frac{\partial B_z}{\partial t}. \tag{4.10}$$

If we track any annulus of plasma inwards in time, we can evaluate the time derivative of the amount of magnetic flux enclosed by this annulus:

$$\frac{d}{dt}\left(\pi r^2 B_z\right) = 2\pi r B_z \frac{dr}{dt} + \pi r^2 \frac{\partial B_z}{\partial t} = 0 \tag{4.11}$$

where the last step is made by substituting dr/dt from equation (4.10). Thus the entire plasma conserves magnetic flux as it moves in radially, just as the gyro-orbits conserve flux. This property of plasma drifting at velocity $\mathbf{v}_E$ is sometimes called being 'frozen' to the magnetic flux lines and is considered again in more detail in Chapter 8. (It is entertaining to observe that a plasma needs to be very *hot*, so that its collision frequency is very low, in order for it to be *frozen* to field lines.) As we will learn later, Coulomb collisions, whose rate drops rapidly with increasing temperature, allow plasmas to become 'unfrozen' and to diffuse slowly across magnetic field lines.

Problem 4.2: Imagine that you have an isotropic magnetized plasma with $T_{\|0} = T_{\perp0} = T_0$. Double the magnetic field slowly compared to a gyro-period, but fast compared to the energy transfer time between $T_\|$ and $T_\perp$. What are the new values of $T_\|$ and $T_\perp$ (call them $T_{\|1}$ and $T_{\perp1}$)? Now let the plasma sit long enough for $T_{\|1}$ and $T_{\perp1}$ to mix by collisions and come to an isotropic temperature T_1, but not long enough for the plasma to exchange energy with the outside world. What is T_1? Reduce the magnetic field back down to its original value slowly compared to a gyro-period, but fast compared to the energy transfer time between $T_\|$ and $T_\perp$. What are $T_{\|2}$ and $T_{\perp2}$? And after the plasma becomes isotropic, what is T_2? This process is called 'magnetic pumping'.

4.3   TIME-VARYING E FIELD

In order to understand plasma dynamics reasonably well from a particle-drift point of view, it is necessary to know about one further drift motion, the polarization drift, which is *second order* in ω/ω_c. Consider a situation with a uniform **B** field pointing in the z direction, and a time-varying spatially uniform **E** field, pointing in the x direction. Starting from the Lorentz force equation

$$\dot{v}_x = (qB/m)v_y + (q/m)E_x(t) = \pm\omega_c v_y \pm (\omega_c/B)E_x(t)$$
$$\dot{v}_y = -v_x(qB/m) = \mp\omega_c v_x$$

where the $\pm$ indicates the sign of q, and differentiating once with respect to time, we obtain

$$\frac{d^2 v_x}{dt^2} = -\omega_c^2 v_x \pm \frac{\omega_c}{B}\frac{\partial E_x(t)}{\partial t} \tag{4.12}$$

$$\frac{d^2 v_y}{dt^2} = -\omega_c^2 v_y - \omega_c^2 \frac{E_x(t)}{B}. \tag{4.13}$$

If we assume that the characteristic time of variation of the electric field is long compared to a gyro-period, and take v_E to be of order kr_L compared to v, the terms on the right-hand side of equation (4.12) are, respectively, zeroth order and second order ($\partial/\partial t$ is higher order than ω_c, and E/B is higher order than v). The terms on the right-hand side of equation (4.13) are, respectively, zeroth order and first order. Ignoring the second-order term, we have just the equations we solved originally for the v_E drift, using the Lorentz transformation. Thus the solution for v_0 is just the usual gyration and parallel motion, and v_1 is just the v_E drift, in this case in the y direction. If we substitute for v our formal expansion in kr_L, the second-order parts of equation (4.12) are

$$\frac{d^2 v_{x2}}{dt^2} = -\omega_c^2 v_{x2} \pm \frac{\omega_c}{B} \frac{\partial E_x}{\partial t}. \tag{4.14}$$

Since we are only interested in the gyro-averaged drift, we can 'annihilate' the first term by averaging over many gyro-periods. If we do this average, then $d^2 \langle v_{x2} \rangle / dt^2$ becomes much smaller than $\omega_c^2 v_{x2}$—because we have smoothed out any fast time variation—so the term on the left-hand side becomes higher order than the other terms. It follows that $v_{x2} = \pm (\omega_c B)^{-1} \partial E_x / \partial t$. Equation (4.13), at second order and averaged over many gyro-periods, just gives $v_{y2} = 0$, by the same argument. Since we could have chosen the x direction arbitrarily (but perpendicular to B), we can express our result more generally, i.e.

$$v_{\perp 2} = \pm \frac{dE_\perp/dt}{\omega_c B} = \frac{m}{q B^2} \frac{dE_\perp}{dt}. \tag{4.15}$$

The direction of $v_{\perp 2}$ depends on the sign of q, and its magnitude depends on m, so ions and electrons do not have equal velocities, and a net current is driven in the plasma. This current is analogous to the polarization current in dielectric materials (which is also proportional to dE/dt), so this drift is referred to as the 'polarization drift', v_{pol}. Note that it is by far dominated by ions compared to electrons.

Problem 4.3: Our energy conservation equation, equation (4.6), is consistent to order kr_L, but no higher. In other words, energy in the drift motion such as $m(v_{grad})^2/2$ is not included. However, as was mentioned before, one can also derive guiding-center drift equations in the case where v_E is not assumed to be small compared to v_0. In this case one has to include $m v_E^2 / 2$ in the energy equation, as well as the polarization drift in the first-order $\mathbf{v}_{\text{gc}} \cdot \mathbf{E}$. For the simplest geometry—a uniform time-independent $\mathbf{B}$ field in the z direction and a uniform, perpendicular, time-dependent $\mathbf{E}$ field in the x direction—show that $(\mathrm{d}/\mathrm{d}t)mv_E^2/2 = q\mathbf{v}_{\text{pol}} \cdot \mathbf{E}$. Draw what the drift orbits look like for ions and electrons in the case of $\dot{E}_x$ constant and positive and E_x always greater than zero. Note that we have not calculated all the other drifts for the case where v_E can be comparable to v—and there are indeed other terms which come into the complete calculation—so this is only an exercise. Equation (4.6) is as far as we will go self-consistently for energy conservation in the drift equations.

It is interesting to use equation (4.15) to derive a low-frequency perpendicular dielectric constant, $\epsilon_\perp$, for a plasma, where the polarization current is considered an 'internal' current, in contrast to an 'external' current density $\mathbf{j}_{\text{ext}}$. We can write

$$\nabla \times \mathbf{B} = \mu_0(\mathbf{j}_{\text{ext}} + \mathbf{j}_{\text{pol}} + \epsilon_0\dot{\mathbf{E}}) = \mu_0(\mathbf{j}_{\text{ext}} + \epsilon\dot{\mathbf{E}}) \qquad (4.16)$$

or

$$\epsilon\dot{\mathbf{E}} \equiv (\mathbf{j}_{\text{pol}} + \epsilon_0\dot{\mathbf{E}}). \qquad (4.17)$$

The polarization current density $\mathbf{j}_{\text{pol}}$ carried by each species (ions or electrons) is just

$$\mathbf{j}_{\text{pol}} = nq\mathbf{v}_{\text{pol}} = nm\dot{\mathbf{E}}/B^2 \qquad (4.18)$$

where n is the density of the species. So we obtain

$$\epsilon_\perp = \epsilon_0 + \rho/B^2 \qquad (4.19)$$

where $\rho = n_i m_i + n_e m_e$ is the total mass density of the plasma. For typical plasma parameters $\epsilon \gg \epsilon_0$ by a factor of $\sim 10^3$. Note that the plasma is a highly anisotropic medium: this dielectric constant only characterizes the plasma response perpendicular to the magnetic field. We will encounter $\epsilon_\perp$ again when we discuss propagation of low-frequency electromagnetic waves in a plasma.

The result for $\epsilon_\perp$ can also be used to solve the problem of a plasma in a gravitational field, thereby gaining some insight into how the perpendicular dielectric property affects plasma motion. Imagine that we have a slab of magnetized proton–electron plasma at density $n_e = n_i = n$ in a gravitational field, as shown in Figure 4.1. We choose this slab-like geometry to simplify the calculation of the electric field which is created by charge separation. The gravitational drift gives rise to a net current perpendicular both to $\mathbf{B}$ and to the force of gravity, $m\mathbf{g}$. The gravitational drift velocity is

$$\mathbf{v}_g = m(\mathbf{g} \times \mathbf{B})/qB^2 \qquad (4.20)$$

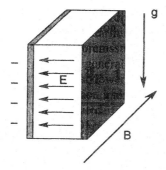

Figure 4.1. Geometry for calculating plasma motion in crossed magnetic and gravitational fields.

which is just equation (2.23). The total current density $\mathbf{j} = \sum nq\mathbf{v}$ (where the summation is over species) due to the gravitational force is $j_{\text{ext}} = \rho g/B$ directed rightward in Figure 4.1, where ρ is the mass density of the plasma. We are taking the gravitational-drift current to be an 'external' current in the sense of not being part of the polarization current of the medium. Since this current density stops at the faces of our slab of plasma, we will assume that a 'free' charge density steadily builds up there (as a consequence of the 'external' current). If we take σ_s to be the free surface charge density (in units of coulombs per square meter), we can easily convince ourselves that $d\sigma_s/dt = j$. This is most easily seen by recognizing that j is in units of coulombs per square meter per second, representing the coulombs per second that would pass through a surface of 1 square meter presented at right angles to the direction of j. The relation $d\sigma_s/dt = j$ just expresses the fact that charge is conserved, and we are not letting it pass through the surface, but rather it must accumulate there. Given $d\sigma_s/dt$ at the faces, we can calculate $dE_\perp/dt$ at each face, assuming that $E_\perp$ outside the plasma is negligible, as in an infinite parallel-plate capacitor. Integrating Poisson's equation across the plasma surface we obtain

$$\epsilon_\perp E_\perp = \sigma_s$$

and so

$$\frac{dE_\perp}{dt} = \frac{d\sigma_s/dt}{\epsilon_\perp} = \frac{j_{\text{ext}}}{\epsilon_\perp} = \frac{\rho g}{\epsilon_\perp B}. \tag{4.21}$$

The resulting *downward directed* accelerating $\mathbf{v}_E$ drift is:

$$\frac{dv_E}{dt} = \frac{\rho g}{\epsilon_\perp B^2} = \frac{g\rho/B^2}{(\epsilon_0 + \rho/B^2)} = \frac{g}{(1 + \epsilon_0 B^2/\rho)}. \tag{4.22}$$

Galileo would be satisfied with this solution (since he noted in Pisa that all bodies fall with the same acceleration), except for the small second term in the denominator, which reduces the downward acceleration slightly. The gravitational potential energy is being turned into kinetic energy as the plasma falls downward, as usual, with the exception of a small part of it. If we multiply both sides by $(1 + \epsilon_0 B^2/\rho)\rho v_E$, we obtain

$$\left(1 + \frac{\epsilon_0 B^2}{\rho}\right) \frac{d}{dt} \left(\frac{\rho v_E^2}{2}\right) = g\rho v_E. \tag{4.23}$$

In this form, the equation is particularly troublesome, since it looks as if it violates energy conservation. However, let us evaluate the ratio of the energy density stored in the electric field, $\epsilon_0 E^2/2$, to the plasma kinetic energy density of drift motion, i.e.

$$\frac{\epsilon_0 E^2/2}{\rho E^2/2B^2} = \epsilon_0 B^2/\rho. \tag{4.24}$$

This is just the size of the apparent 'error' in energy conservation. We have tricked Galileo by transferring a small fraction of the gravitational potential energy to the plasma's internal electric field, charging the plasma capacitor. For typical laboratory or geophysical plasma parameters, this is a very small fraction of the energy indeed (i.e. $\epsilon_\perp \gg \epsilon_0$), and in many plasma physics calculations the contribution from ϵ_0 can be ignored.

This is a good example of how the energetics of a plasma's perpendicular dielectric constant works. This forms the basis for a 'plasma capacitor'. In a laboratory experiment, if a surface charge density is built up externally, which in vacuum would have stored energy in the perpendicular electric field, the presence of a plasma causes this field to be shielded out via the polarization current, greatly reducing the stored electric field energy and putting the bulk of the energy into the kinetic energy associated with the $\mathbf{v}_E$ drift. As with any capacitor, a high dielectric constant allows a larger free charge and a larger stored energy to be built up at a given electric field strength.

Problem 4.4: For this calculation, we chose to consider $\epsilon_\perp$ a property of the medium, and thus to take the polarization drift into account implicitly through $\epsilon_\perp$ (and taking v_g as causing a j_{ext}). We could instead have calculated the polarization drift's contribution to j as part of j_{ext}, and also its self-consistent contribution to $d\sigma_s/dt$ (where σ_s would now be the total surface charge density), and then used the vacuum ϵ_0 to characterize the remaining vacuum 'medium'. Show that both approaches give the same answer for $dE_\perp/dt$.

4.4   ADIABATIC INVARIANTS

It is valuable at this point to consider our results for the drift equations in the wider context of Hamiltonian classical mechanics. Hamiltonian systems are those in which the equations of motion can be expressed in the form

$$\dot{q}_i = \frac{\partial H}{\partial p_i} \qquad \dot{p}_i = -\frac{\partial H}{\partial q_i} \qquad (4.25)$$

where $H(q_1, \ldots, q_i, p_1, \ldots, p_i)$ is called the 'Hamiltonian' of the system. The p_i are generalized momenta, and the q_i are generalized positions—so-called 'canonical' variables. The classical equations of motion of a charged particle in the presence of electric and magnetic fields are Hamiltonian. The guiding-center drift equations we have derived, which are only correct to first order in kr_L, can also be cast in a strictly Hamiltonian form, with $H = mv_{\parallel}^2/2 + \mu B + q\phi$. The Hamiltonian is evidently very simple (even obvious)—the key is determining what to use as the canonical variables. To be precise, the Hamiltonian form does not give exact equations of the 'true' guiding-center particle motion to all orders in kr_L, nor does it give precisely the equations we have derived, but it is strictly Hamiltonian to all orders and it also agrees with our equations to order kr_L, which is as far as our equations are valid. This Hamiltonian nature of the drift equations justifies taking over results from classical mechanics (or even quantum mechanics!), and applying them to guiding-center drifts.

One important result of the classical mechanics of Hamiltonian systems is that the *action*, defined as $\int p\,dq$ around a loop which represents nearly periodic motion, is adiabatically invariant. The magnetic moment, μ, is an adiabatic invariant of the basic Lorentz force equations. In this case, the nearly periodic motion is the Larmor gyration with frequency ω_c. The appropriate momentum for this case, p, is the particle's angular momentum, $mr_L v_\perp$, and q is its angular position, θ. Adiabatic invariance means that if the trajectory changes slowly, either because the fields are changing slowly, or because the loop is slowly drifting into a region of different field geometry, then the action changes much less, proportionally, than the field geometry. 'Slow' here means that the characteristic time of variation of the field is long compared to the oscillation period of the basic periodic motion, and the space scale of variation is large compared to the distance the loop drifts in one period. In the case of μ conservation, for example, the relative change in the adiabatic invariant compared to the change in the magnetic field due to some time-dependent perturbation with frequency ω is $\exp(-\omega_c/\omega)$. This means that for ω_c/ω of order unity, there are order unity changes in μ. However, as ω_c/ω becomes much larger than unity, the changes in the adiabatic invariant become *exponentially* small. Since $\exp(-\omega_c/\omega)$ cannot be expressed in a Taylor series of ω/ω_c, we say that the adiabatic invariant is conserved 'to all orders'. (Speaking precisely, for this to be true it has been shown that the actual adiabatic invariant is μ to lowest order but has corrections of higher order in the ratio of the Larmor radius to the scale-length of field variation.)

Note that when we say that the scale-length or time of variation must be long, this is not only that

$$\frac{1}{X}\frac{dX}{dt} \ll \frac{1}{\tau} \tag{4.26}$$

where X is any field quantity, and τ is the period of oscillation. If we impose a small-amplitude high-frequency oscillation on **B** or **E**, with a frequency greater than $1/\tau$, equation (4.26) may be satisfied, but the appropriate adiabatic invariant will not be well-conserved compared with the amplitude of that high-frequency component. Thus, if we require exponentially good conservation of μ, for example, the high-frequency components in the range $\omega \approx \omega_c$ or greater must be exponentially small.

4.5   SECOND ADIABATIC INVARIANT: *J* CONSERVATION

Let us now go on to consider an adiabatic invariant of the guiding-center motion, rather than of the particle motion, which arises when the guiding-center parallel motion is of a periodic nature: for example, a trapped particle bouncing in a magnetic mirror. This is usually called the 'second adiabatic invariant' and is given by

$$J \equiv \oint_a^b v_\parallel \cdot ds \tag{4.27}$$

i.e. the loop integral of the parallel velocity along a particle trajectory. The endpoints of the integral are taken at the two turning points, a and b, where $v_\parallel = 0$, but the total value of J is typically defined as the integral from a to b, and then back again to a. A simple proof of J invariance can be constructed by invoking the correspondence principle, the Hamiltonian nature of the drift equations, and a basic understanding of quantum mechanics. In the action integral, equation (4.27), the momentum p *corresponds* to k, the quantum-mechanical wave number, so J is proportional to the phase integral, $\int k dl$, along the trajectory. A *quantized* solution to the orbit would require the phase integral to equal some integer, n. (In any macroscopic case, n is very large.) The conservation of J then *corresponds* to the quantum mechanical requirement that a perturbation with frequency of order the bounce frequency is required to cause a state transition to a different n. (This is because the beat frequency between the nth and $(n-1)$th states is just the bounce frequency.) Thus J conservation *corresponds* to the quantum mechanical result that if a potential well is transformed adiabatically (slowly compared to a bounce time), the quantum number of a particle trapped in the well is not altered.

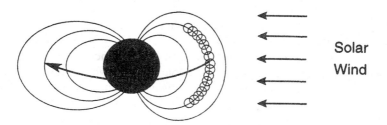

Figure 4.2. Particles trapped in the Earth's dipole magnetic field, precessing around the Earth. The Earth's field is shown schematically, distorted by pressure of the solar wind.

For a specific example of J invariance, we might consider the motion of a high-energy particle trapped in the Earth's magnetic field, which is dominantly a dipole field modified by the pressure of the solar wind, as illustrated in Figure 4.2. Protons with MeV energies arise, for example, from the decay of neutrons created by cosmic ray collisions. If we assume that the Earth's magnetic field is essentially static, and the electric fields are modest, we may use in equation (4.27) simply

$$v_\parallel = [2(W - \mu B)/m]^{1/2}. \tag{4.28}$$

High-energy particles 'bounce' between higher-field points near the North and South poles, and slowly precess around the Earth due to the ∇B and curvature drifts. It is the 'bounce' motion from North to South and back again that defines the trajectory for determining the invariant J. The Earth's field is distorted away from axisymmetry, however, by the action of the solar wind. Because of this asymmetry, there is no *a priori* reason to believe that a particle should return to its earlier trajectory as it makes a full turn around the globe. It might—as far as we know so far—return to its initial longitude (East–West location) but be at a new altitude. However, at a given longitude, each field line as a function of altitude above the equator has a different effective length ($\int ds$) between turning points for a particle with given magnetic moment μ and energy W, and different values of the field strength, B, along the field line. Each line thus represents a different J for that particle, and so if J is conserved, as well as μ and W, then the particle must return to the same altitude after precessing around the Earth. Thus it cannot spiral in or out of the Van Allen belts (for example) without the presence of some fast-time-scale or short-space-scale perturbation. This explains the persistence of these radiation belts.

The proper first-principles proof of J conservation, without simply taking over a result from Hamiltonian mechanics (or quantum mechanics), is very lengthy. It was first published by T G Northrop and E Teller (1960 *Phys. Rev.* **117** 215). Unfortunately, elementary textbooks abound with poor (but quick) pseudo-proofs. While it is beyond our scope here to give the complete proof of J invariance for all cases, it *is* worth our while here to outline the classic proof for the case of *time-independent* fields.

Northrop and Teller start by noting that J is a function of field line (as discussed above), and as such is a function of only two spatial dimensions. They introduce spatial coordinates, α and β, which do not vary along field lines, but rather *distinguish between* field lines. They show that α and β can be chosen such that

$$\mathbf{B} = \nabla\alpha \times \nabla\beta. \tag{4.29}$$

Note that by construction $\nabla \cdot \mathbf{B} = \nabla\beta \cdot (\nabla \times \nabla\alpha) - \nabla\alpha \cdot (\nabla \times \nabla\beta) = 0$. The fact that α and β are constant along field lines can be seen from noting that $\mathbf{B} \cdot \nabla\alpha = \mathbf{B} \cdot \nabla\beta = 0$. Now we are in a position to assert that $J = J(\alpha, \beta)$, because J characterizes a field line, for fixed μ and W. ($J(\alpha, \beta)$ might also depend explicitly on time, in the presence of time-varying fields, but we will not discuss that case here.)

If $J = J(\alpha, \beta)$, then to evaluate dJ/dt, averaged around a zeroth-order orbit (i.e. one with no motion other than $v_\parallel$), is a well-defined operation. We average first-order guiding-center drifts around the zeroth-order orbit, symbolized by the $\langle\ \rangle$ operator, to obtain the first-order change in J:

$$\langle dJ/dt \rangle = \langle d\alpha/dt \rangle (\partial J/\partial\alpha) + \langle d\beta/dt \rangle (\partial J/d\beta). \tag{4.30}$$

The proof then proceeds by noting that

$$\left\langle \frac{d\alpha}{dt} \right\rangle = \frac{\int \mathbf{v}_{gc} \cdot \nabla\alpha\, dt}{\int dt} = \frac{\int [(\mathbf{v}_{gc} \cdot \nabla\alpha)/v_\parallel] ds}{\int (1/v_\parallel) ds} \tag{4.31}$$

together with an equivalent expression for $\langle d\beta/dt \rangle$, and

$$\frac{\partial J}{\partial\alpha} = \frac{\partial \int v_\parallel ds}{\partial\alpha} = \int \frac{\partial v_\parallel}{\partial\alpha} ds = -\int \frac{\mu}{m} \frac{\partial B}{\partial\alpha} \frac{ds}{v_\parallel} \tag{4.32}$$

together with an equivalent expression for $\partial J/\partial\beta$. The second step in equation (4.32) follows because $v_\parallel = 0$ at the endpoints, so $\partial/\partial\alpha$ of the endpoint positions gives no contribution. The third step follows because

$$\frac{\partial v_\parallel}{\partial\alpha} = \frac{\partial [2(W - \mu B)/m]^{1/2}}{\partial\alpha} = \frac{-\mu\partial B/\partial\alpha}{m v_\parallel}. \tag{4.33}$$

Thus we see that $\partial J/\partial\alpha$ depends on the bounce-integral of $(1/v_\parallel)\partial B/\partial\alpha$. Equation (4.31) shows that $\langle \partial\alpha/\partial t \rangle$ depends on a similar integral of $\mathbf{v}_{gc} \cdot \nabla\alpha$. The guiding-center drift in the $\nabla\alpha$ direction is, however, closely related to $\partial B/\partial\beta$. Northrop and Teller indeed find that the bounce average of $\mathbf{v}_{gc} \cdot \nabla\alpha$

can be expressed in terms of a bounce-integral over $\partial B/\partial \beta$ only. The result is that the first term on the right-hand side of equation (4.30) is a product of identical bounce-integrals over $\partial B/\partial \alpha$ and $\partial B/\partial \beta$. The second term, of course, is structurally identical, and its sign turns out to be reversed, with the result that it cancels the first term, proving conservation of J to first order.

Problem 4.5: Now we examine a case with time-dependent **B** fields, but ignoring spatial drifts. Assume $\mathbf{B} = \hat{z}B_0(1 + \gamma z^2)$, where $\gamma = \gamma(t)$. For a particle bouncing back and forth in the magnetic well defined by this field (cf Problem 3.5), show that the second adiabatic invariant, J, is conserved if γ varies slowly compared to a bounce time. To do this, first evaluate J as a function of γ, W, and μ. Next evaluate $\langle \dot{W} \rangle$ as a function of γ, $\dot{\gamma}$, W and μ where $\langle\ \rangle$ indicates the time average around a zeroth-order orbit, i.e. an orbit over which γ is taken to be constant for calculating the trajectory, but $\dot{\gamma}$ is finite for evaluating $\dot{W} = \mu \partial B/\partial t$. For any quantity x, $\langle x \rangle \equiv \int x\,dt / \int dt = \int (x/v_\parallel)ds / \int(1/v_\parallel)ds$. Finally show that $\dot{J} = (\partial J/\partial \gamma)\dot{\gamma} + (\partial J/\partial W)\dot{W} = 0$.

The conservation of J is a very powerful tool for calculating particle trajectories in complex geometries. For numerical evaluation of the properties of different magnetic geometries, it is often superior to integrating the guiding-center equations of motion. For developing a qualitative understanding of particle orbits in moderately complex geometries, it is unsurpassed. One must be cautious, however, about assuming its validity in all cases. A particle can drift 'over' its trapping barrier, in cases where B has a local maximum which decreases in the direction that the 'bounce' orbit is drifting. The particle then undergoes a 'jump' in J. Up to the jump, J is nicely conserved; after the jump quite a different value of J is also nicely conserved. *However these jumps can often be the most important ingredient in determining particle transport in the given situation.* The key requirement for J conservation is that the variation of the **B** field geometry (e.g. ∇B, $(\hat{\mathbf{b}} \cdot \nabla)\hat{\mathbf{b}}$) *along* a bounce trajectory must in some sense accurately 'predict' the **B** field which will be experienced along the next bounce, for J to be conserved.

4.6 PROOF OF J CONSERVATION IN TIME-INDEPENDENT FIELDS*

In the previous Section, we only *outlined* Northrop and Teller's proof of J conservation in time-independent fields. In this Section, we will construct the proof itself.

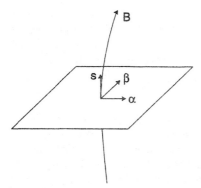

Figure 4.3. Geometry for establishing an α, β, s coordinate system.

First we must prove that we can generally express **B** in the form $\mathbf{B} = \nabla\alpha \times \nabla\beta$. It is only necessary to prove this locally in the close vicinity of a given field line of interest, where we are planning to perform the bounce-integrals required for the proof of J conservation. We have already shown that α and β are constant along field lines if $\mathbf{B} = \nabla\alpha \times \nabla\beta$. Thus let us choose to develop a coordinate system in the vicinity of a specific field line, to which we assign the values ($\alpha = 0$, $\beta = 0$). Let us arbitrarily choose a starting plane, perpendicular to this field line, where we will define $s = 0$. Now we have an origin for our coordinate system, $\alpha = \beta = s = 0$ (see Figure 4.3). Next we assign to a differentially nearby field line the values ($\alpha = \delta\alpha$, $\beta = 0$) in order to define a direction for the α axis in the $s = 0$ plane. Let us then take a third field line piercing the $s = 0$ plane to define the β axis on that plane and assign it the values ($\alpha = 0$, $\beta = \delta\beta$). For simplicity, we choose the location of this third field line such that the β axis it defines is perpendicular to the α axis in the $s = 0$ plane. Furthermore, if we choose the location of the field line (i.e. the distance in the $s = 0$ plane away from the origin in the β direction) such that $|\nabla\beta| = B/|\nabla\alpha|$ at the origin, then $\mathbf{B} = \nabla\alpha \times \nabla\beta$ at the origin, by construction. Obtaining the right answer at a single point may seem a meager result, but we will see that this sets the cornerstone for our coordinate system.

To complete the coordinate system, we denote the length along any field line from the $s = 0$ plane by s. Now we have a fully defined three-dimensional coordinate system, (α, β, s), in the vicinity of the field line ($\alpha = 0$, $\beta = 0$). This coordinate system is curvilinear and non-orthogonal, except at the origin where we have made the coordinate system orthogonal by construction. To prove that $\mathbf{B} = \nabla\alpha \times \nabla\beta$ will first require some basic results about such coordinate systems.

The vector $\partial\mathbf{x}/\partial\alpha$ is defined as the variation of the usual cartesian position vector, $\mathbf{x}$, with respect to α, at fixed β and s. The quantity $\partial\mathbf{x}$ in this case is just the differential vector that connects the field line ($\alpha = 0$, $\beta = 0$) to the field line ($\alpha = \partial\alpha$, $\beta = 0$) at a given fixed value of s. The vector $\partial\mathbf{x}/\partial\beta$ is defined similarly. The vector $\partial\mathbf{x}/\partial s$ is the variation of $\mathbf{x}$ with s at fixed α and β, i.e. along a given field line. Thus $\partial\mathbf{x}/\partial s$ is just the unit vector along $\mathbf{B}$, the familiar $\hat{\mathbf{b}}$. By analogy we define $\hat{\alpha} \equiv (\partial\mathbf{x}/\partial\alpha)/|\partial\mathbf{x}/\partial\alpha|$ and $\hat{\beta} \equiv (\partial\mathbf{x}/\partial\beta)/|\partial\mathbf{x}/\partial\beta|$. For completeness we also define $\hat{\mathbf{s}} \equiv (\partial\mathbf{x}/\partial s)/|\partial\mathbf{x}/\partial s| = (\partial\mathbf{x}/\partial s) = \hat{\mathbf{b}}$. In a non-orthogonal coordinate system, the dot products of the unit vectors along the axes (e.g. $\hat{\alpha} \cdot \hat{\mathbf{b}}$) are *not* necessarily zero, as they are in an orthogonal system. Another important set of vectors $\nabla\alpha$, $\nabla\beta$, and ∇s can be constructed from the equations

$$\nabla\alpha = \hat{\mathbf{x}}(\partial\alpha/\partial x) + \hat{\mathbf{y}}(\partial\alpha/\partial y) + \hat{\mathbf{z}}(\partial\alpha/\partial z)$$
$$\nabla\beta = \hat{\mathbf{x}}(\partial\beta/\partial x) + \hat{\mathbf{y}}(\partial\beta/\partial y) + \hat{\mathbf{z}}(\partial\beta/\partial z) \qquad (4.34)$$
$$\nabla s = \hat{\mathbf{x}}(\partial s/\partial x) + \hat{\mathbf{y}}(\partial s/\partial y) + \hat{\mathbf{z}}(\partial s/\partial z)$$

where the partial derivative $\partial/\partial x$, for example, means the derivative with respect to x at fixed y and z. It will be useful to note that, while $\nabla\alpha$, $\nabla\beta$ and ∇s are not orthogonal, it is nonetheless the case that

$$\frac{\partial\mathbf{x}}{\partial\alpha} \cdot \nabla\beta = 0 \qquad \frac{\partial\mathbf{x}}{\partial\alpha} \cdot \nabla s = 0$$
$$\frac{\partial\mathbf{x}}{\partial\beta} \cdot \nabla\alpha = 0 \qquad \frac{\partial\mathbf{x}}{\partial\beta} \cdot \nabla s = 0 \qquad (4.35)$$
$$\frac{\partial\mathbf{x}}{\partial s} \cdot \nabla\alpha = 0 \qquad \frac{\partial\mathbf{x}}{\partial s} \cdot \nabla\beta = 0$$

because in each case the partial derivative explicitly points in a direction along which the other coordinates do not vary. Furthermore, we will find it helpful to use

$$\nabla\alpha \cdot \left(\frac{\partial\mathbf{x}}{\partial\alpha}\right) = \nabla\beta \cdot \left(\frac{\partial\mathbf{x}}{\partial\beta}\right) = \nabla s \cdot \left(\frac{\partial\mathbf{x}}{\partial s}\right) = \nabla s \cdot \hat{\mathbf{b}} = 1. \qquad (4.36)$$

To prove this, consider, for example, the vector $\nabla\alpha$. It points in the direction normal to a surface of $\alpha = $ constant, and its magnitude is $\partial\alpha/\partial l$, where ∂l is the length element along the normal to that surface. The component of $\partial\mathbf{x}/\partial\alpha$ in that same direction normal to the surface is just the reciprocal value, $(\partial\alpha/\partial l)^{-1}$ (see Figure 4.4).

A final set of relations we will find helpful is

$$\nabla\alpha = \left(\frac{\partial\mathbf{x}}{\partial\beta}\right) \times (\nabla\alpha \times \nabla\beta)$$

$$\nabla\beta = -\left(\frac{\partial\mathbf{x}}{\partial\alpha}\right) \times (\nabla\alpha \times \nabla\beta). \qquad (4.37)$$

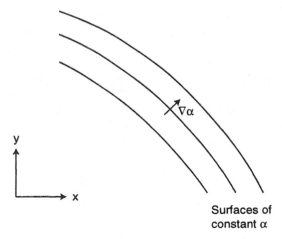

Figure 4.4. Geometry for seeing that $\nabla\alpha \cdot (\partial\mathbf{x}/\partial\alpha) = 1$.

Expanding the double cross-product in the expression for $\nabla\alpha$ we have

$$\frac{\partial\mathbf{x}}{\partial\beta} \times (\nabla\alpha \times \nabla\beta) = -\left(\nabla\alpha \cdot \frac{\partial\mathbf{x}}{\partial\beta}\right)\nabla\beta + \left(\nabla\beta \cdot \frac{\partial\mathbf{x}}{\partial\beta}\right)\nabla\alpha. \qquad (4.38)$$

Using equations (4.35) and (4.36) we see that this reduces to $\nabla\alpha$, as required. A similar analysis leads to the expression for $\nabla\beta$.

We have now described some of the important properties of the curvilinear non-orthogonal coordinate system that we will use. (The use of such coordinates is often convenient in plasma physics, due to the special importance of $\hat{\mathbf{b}}$, which is generally curvilinear.) This puts us in a position to prove in general that $\mathbf{B} = \nabla\alpha \times \nabla\beta$ in the vicinity of $\alpha = \beta = 0$ for the $\alpha(\mathbf{x})$, $\beta(\mathbf{x})$ and $s(\mathbf{x})$ we have constructed. First we show that the *direction* of $\nabla\alpha \times \nabla\beta$ is correct. Our coordinate system satisfies $\hat{\mathbf{s}} \cdot \nabla\alpha = \hat{\mathbf{s}} \cdot \nabla\beta = 0$ (see equation (4.35)) and $\hat{\mathbf{s}} = \hat{\mathbf{b}}$ by construction. The direction $\hat{\mathbf{b}}$ is uniquely defined by being perpendicular to both $\nabla\alpha$ and $\nabla\beta$. Thus $\nabla\alpha \times \nabla\beta$ certainly points in the same direction as $\hat{\mathbf{b}}$ everywhere in the vicinity of $\alpha = \beta = 0$. This leaves open only the question of the *magnitude* of $\nabla\alpha \times \nabla\beta$. From $\nabla \cdot \mathbf{B} = 0$, we can write

$$B(\nabla \cdot \hat{\mathbf{b}}) + (\hat{\mathbf{b}} \cdot \nabla)B = 0 \qquad (\hat{\mathbf{b}} \cdot \nabla)\ln(B) = -(\nabla \cdot \hat{\mathbf{b}}). \qquad (4.39)$$

Since we showed just after equation (4.29) that $\nabla \cdot (\nabla\alpha \times \nabla\beta) = 0$, the same as equation (4.39) asserts about B and $\hat{\mathbf{b}}$ can be asserted about the magnitude of $\nabla\alpha \times \nabla\beta$, and the unit vector pointing in its direction. Furthermore, since we have just seen that the unit vector pointing in the direction of $\nabla\alpha \times \nabla\beta$ is identical to $\hat{\mathbf{b}}$, we may conclude that

$$(\hat{\mathbf{b}} \cdot \nabla)\ln(|\nabla\alpha \times \nabla\beta|) = (\hat{\mathbf{b}} \cdot \nabla)\ln(B). \tag{4.40}$$

In the vicinity of the origin, by construction, $|\nabla\alpha \times \nabla\beta| = B$, so $|\nabla\alpha \times \nabla\beta|$ gives the proper magnitude of B *all along* s. Thus we have proven that $\mathbf{B} = \nabla\alpha \times \nabla\beta$ in the vicinity of $\alpha = \beta = 0$ in our coordinate system. It is not the case, however, that the functions α and β are unique, as can be seen, for example, from how they were constructed.

Now we proceed to give the heart of the Northrop and Teller proof of J invariance, specialized to the case of time-independent magnetic fields, and no electric fields. Let us start by focusing on $\mathbf{v}_{gc} \cdot \nabla\alpha$, the key ingredient in equation (4.31):

$$\mathbf{v}_{gc} \cdot \nabla\alpha = \left(\frac{W_\perp}{q} \frac{\mathbf{B} \times \nabla B}{B^3} + \frac{2W_\parallel}{q} \frac{\mathbf{B} \times (\hat{\mathbf{b}} \cdot \nabla)\hat{\mathbf{b}}}{B^2} \right) \cdot \nabla\alpha. \tag{4.41}$$

The vector ∇B can be expressed, in our coordinate system, by

$$\nabla B = \left(\frac{\partial B}{\partial\alpha} \right) \nabla\alpha + \left(\frac{\partial B}{\partial\beta} \right) \nabla\beta + \left(\frac{\partial B}{\partial s} \right) \nabla s. \tag{4.42}$$

This can be seen to be true, even though $\nabla\alpha$, $\nabla\beta$ and ∇s are not orthogonal, by dotting both sides of this equation with $\hat{\mathbf{x}}$, $\hat{\mathbf{y}}$ or $\hat{\mathbf{z}}$. The $\hat{\mathbf{x}}$ components of ∇B, for example, is clearly given by $\partial B/\partial x = (\partial B/\partial\alpha)(\partial\alpha/\partial x) + (\partial B/\partial\beta)(\partial\beta/\partial x) + (\partial B/\partial s)(\partial s/\partial x)$. The first term in equation (4.42) does not contribute to $(\mathbf{B} \times \nabla B) \cdot \nabla\alpha$ in equation (4.41). The contribution from the second term can be simplified by noting that

$$(\mathbf{B} \times \nabla\beta) \cdot \nabla\alpha = \mathbf{B} \cdot (\nabla\beta \times \nabla\alpha) = -B^2. \tag{4.43}$$

Finally we note that $\hat{\mathbf{b}} \cdot \nabla$ is the same as $\partial/\partial s$. Equation (4.41) then becomes

$$\mathbf{v}_{gc} \cdot \nabla\alpha = -\frac{W_\perp}{qB} \frac{\partial B}{\partial\beta} + \frac{W_\perp}{q} \frac{(\mathbf{B} \times \nabla s) \cdot \nabla\alpha}{B^3} \frac{\partial B}{\partial s} + \frac{2W_\parallel}{qB^2} \left(\mathbf{B} \times \frac{\partial\hat{\mathbf{b}}}{\partial s} \right) \cdot \nabla\alpha. \tag{4.44}$$

We will now work on the second and third terms on the right-hand side, and show that they cancel when they are integrated around a bounce orbit, as required for the evaluation of $\langle d\alpha/dt \rangle$ in equation (4.31).

Using the expression for $\nabla\alpha$ in equation (4.37), we can simplify the second term on the right-hand side by writing

$$(\mathbf{B} \times \nabla s) \cdot \nabla\alpha = (\nabla\alpha \times \mathbf{B}) \cdot \nabla s$$

$$= \left[\left(\frac{\partial \mathbf{x}}{\partial \beta} \times \mathbf{B} \right) \times \mathbf{B} \right] \cdot \nabla s$$

$$= \left(\mathbf{B} \cdot \frac{\partial \mathbf{x}}{\partial \beta} \right) (\mathbf{B} \cdot \nabla s) - B^2 \nabla s \cdot \frac{\partial \mathbf{x}}{\partial \beta}$$

$$= B^2 \hat{\mathbf{b}} \cdot \frac{\partial \mathbf{x}}{\partial \beta}. \tag{4.45}$$

In the last step, we used $\hat{\mathbf{b}} \cdot \nabla s = 1$ from equation (4.36) and $\nabla s \cdot (\partial x/\partial \beta) = 0$ from equation (4.35).

Now we turn our attention to the third term on the right-hand side of equation (4.44). This can be put into a more useful form by writing

$$\left(\mathbf{B} \times \frac{\partial \hat{\mathbf{b}}}{\partial s} \right) \cdot \nabla \alpha = (\nabla \alpha \times \mathbf{B}) \cdot \frac{\partial \hat{\mathbf{b}}}{\partial s}$$

$$= \left[\left(\frac{\partial \mathbf{x}}{\partial \beta} \times \mathbf{B} \right) \times \mathbf{B} \right] \cdot \frac{\partial \hat{\mathbf{b}}}{\partial s}$$

$$= \left(\mathbf{B} \cdot \frac{\partial \mathbf{x}}{\partial \beta} \right) \left(B \hat{\mathbf{b}} \cdot \frac{\partial \hat{\mathbf{b}}}{\partial s} \right) - B^2 \frac{\partial \mathbf{x}}{\partial \beta} \cdot \frac{\partial \hat{\mathbf{b}}}{\partial s}$$

$$= -B^2 \frac{\partial \mathbf{x}}{\partial \beta} \cdot \frac{\partial \hat{\mathbf{b}}}{\partial s}. \tag{4.46}$$

In the last step, we have used of $\hat{\mathbf{b}} \cdot (\partial \hat{\mathbf{b}}/\partial s) = (1/2)(\partial |\hat{\mathbf{b}}|^2/\partial s) = 0$. We can simplify this result further by using

$$\frac{\partial \mathbf{x}}{\partial \beta} \cdot \frac{\partial \hat{\mathbf{b}}}{\partial s} = \frac{\partial}{\partial s} \left(\hat{\mathbf{b}} \cdot \frac{\partial \mathbf{x}}{\partial \beta} \right) - \left(\hat{\mathbf{b}} \cdot \frac{\partial^2 \mathbf{x}}{\partial \beta \partial s} \right)$$

$$= \frac{\partial}{\partial s} \left(\hat{\mathbf{b}} \cdot \frac{\partial \mathbf{x}}{\partial \beta} \right) - \hat{\mathbf{b}} \cdot \frac{\partial \hat{\mathbf{b}}}{\partial \beta}$$

$$= \frac{\partial}{\partial s} \left(\hat{\mathbf{b}} \cdot \frac{\partial \mathbf{x}}{\partial \beta} \right) \tag{4.47}$$

where, in the second step, we have used $\hat{\mathbf{b}} = \partial \mathbf{x}/\partial s$, and in the third step we have again used the fact that $|\hat{\mathbf{b}}|^2$ is constant everywhere.

It is now time to collect our results, equations (4.45), (4.46) and (4.47) in order to rewrite equation (4.44) in a new form

$$\mathbf{v}_{gc} \cdot \nabla \alpha = -\frac{\mu}{q} \frac{\partial B}{\partial \beta} + \frac{\mu}{q} \left(\hat{\mathbf{b}} \cdot \frac{\partial \mathbf{x}}{\partial \beta} \right) \frac{\partial B}{\partial s} - \frac{2W_{\parallel}}{q} \frac{\partial}{\partial s} \left(\hat{\mathbf{b}} \cdot \frac{\partial \mathbf{x}}{\partial \beta} \right). \tag{4.48}$$

The trick now is to note that the last two terms on the right can be combined into a single term, which will vanish when we average equation (4.48) over a complete 'bounce' orbit. To do this, we observe that $W_\parallel = W - \mu B$, where both W and μ are constants of the particle's motion, and we note that

$$\frac{\partial W_\parallel^{1/2}}{\partial s} = -\frac{\mu}{2W_\parallel^{1/2}} \frac{\partial B}{\partial s}. \tag{4.49}$$

Equation (4.48) can then be written

$$\mathbf{v}_{\text{gc}} \cdot \nabla \alpha = -\frac{\mu}{q} \frac{\partial B}{\partial \beta} - \frac{2W_\parallel^{1/2}}{q} \frac{\partial}{\partial s}\left(W_\parallel^{1/2} \hat{\mathbf{b}} \cdot \frac{\partial \mathbf{x}}{\partial \beta}\right). \tag{4.50}$$

Recalling that averaging over the bounce orbit requires that we divide by $v_\parallel$ (which is proportional to $W_\parallel^{1/2}$) and then integrate over all s, *around a closed loop*, we see that the second term in equation (4.50) will integrate to zero, and we obtain our final result, namely

$$\int (\mathbf{v}_{\text{gc}} \cdot \nabla \alpha) \frac{\mathrm{d}s}{v_\parallel} = -\frac{\mu}{q} \int \frac{\partial B}{\partial \beta} \frac{\mathrm{d}s}{v_\parallel} = \frac{m}{q} \frac{\partial J}{\partial \beta} \tag{4.51}$$

where, in the last step, we have used equation (4.32). A similar result, but with a reversed sign, can be derived for the other cross-field component of $\mathbf{v}_{\text{gc}}$, i.e.

$$\int (\mathbf{v}_{\text{gc}} \cdot \nabla \beta) \frac{\mathrm{d}s}{v_\parallel} = -\frac{m}{q} \frac{\partial J}{\partial \alpha}. \tag{4.52}$$

Substituting equation (4.51) into equation (4.31) (and equation (4.52) into the equivalent expression for $\langle \mathrm{d}\beta/\mathrm{d}t \rangle$), and then substituting these into equation (4.30), we obtain our final result

$$\left\langle \frac{\mathrm{d}J}{\mathrm{d}t} \right\rangle = \frac{m}{q} \left(\frac{\partial J}{\partial \beta} \frac{\partial J}{\partial \alpha} - \frac{\partial J}{\partial \alpha} \frac{\partial J}{\partial \beta}\right) \left(\int \frac{\mathrm{d}s}{v_\parallel}\right)^{-1} = 0. \tag{4.53}$$

Thus we have shown that, in the case of time-independent magnetic fields, the quantity J is an invariant of the particle's drift motion to first order in the guiding-center drifts. Perhaps not surprisingly, the quantities α and β used here are closely related to the canonical variables used in the explicitly Hamiltonian formulation of drift equations.

If a static electric field is added, the generalization is straightforward. The constant of motion is no longer the particle's kinetic energy, but rather the sum of its kinetic and potential energy, $W = mv^2/2 + q\phi$. In calculating for this case, one must use

$$v_\parallel = [2(W - \mu B - q\phi)/m]^{1/2} \tag{4.54}$$

in the integral. The adiabatic invariance of J holds even in the case of slowly time-dependent fields, as Northrop and Teller showed in their more general proof, but it should be noted that the electric field is then not derivable from a scalar potential, and so there is no simple energy-quantity W that is a constant of the particle's bounce motion.

Those interested in pursuing further the topic of single-particle motion in slowly varying electric and magnetic fields are referred to a monograph by T G Northrop (1963 *The Adiabatic Motion of Charged Particles* New York: Interscience).

Chapter 5

Mappings

This Chapter uses non-J-conserving particle orbits to introduce the theory of Hamiltonian maps and chaos in dynamical systems. These powerful and elegant concepts are fundamental to many areas of modern plasma physics research, and indeed are now also used in such widely disparate fields as nonlinear mechanics and population ecology. The Chapter includes homework problems and two longer computational physics exercises based on the program ERGO provided with the text. The computational exercises can be used as independent work projects.

5.1 NON-CONSERVATION OF J: A SIMPLE MAPPING

It is fun now to consider situations where J is not conserved. Imagine that the dipole magnetic field of the Earth (or other suitable source) has superimposed upon it some $\cos(n\theta)$ electric or magnetic perturbation—rather powerful aliens trying to meddle with the Van Allen belts, for example. Now we can imagine that if there is a component of the perturbation such that $2\pi/n$ is comparable to or less than the circumferential angular motion (or 'precession') per bounce of fast particles, we will obtain substantial perturbations to the particle trajectories, which will not conserve J.

A powerful modern technique to study problems of this sort is to consider the particle trajectories as 'mappings'. Any possible trajectory (equator → over Northern hemisphere → over equator → over Southern hemisphere → over equator) is viewed as 'mapping' a particle from one location to another around in the circumferential direction, θ, and possibly up or down in altitude, r, but at fixed latitude—the equator. Iteration of this map can be used to create a 'puncture plot' (or 'Poincaré plot') in the (r, θ) plane, where we mark a dot every time a particle crosses the equator passing, for example, from South to North. In a pure dipole case without any 'alien' (or natural) perturbations, we have a fairly simple map

$$\theta_{j+1} = \theta_j + \theta_p\big|_{r_0} + \theta_p'(r_j - r_0)$$
$$r_{j+1} = r_j$$

(5.1)

where $\theta_p\big|_{r_0}$ is the angular precession per bounce at the initial radius, r_0, and the successive j's represent successive full particle cycles, or 'bounces'. (A full particle cycle is referred to as a 'bounce', although strictly speaking it involves two mirror reflections or bounces, one near the North pole and one near the South pole.) In general, θ_p will depend on r, for a given W and μ, since the gradient and curvature angular drift speeds depend on altitude. The quantity θ_p' is $d\theta_p/dr$, to represent a linear approximation to the variation of precession speed with altitude. This map, as presented, is rather uninteresting. Particles do not move in r (we ignore the effect of the solar wind on $\mathbf{B}$ here) and they just precess around azimuthally at different rates, depending on r.

The map becomes more interesting if we assume that the 'alien' perturbation modifies r each bounce with a $\cos(n\theta)$-dependent 'kick' and that these kicks are *not* J-conserving, so the orbits do not necessarily return to the same altitude, r_0, when they return to longitude θ_0. This would occur, for example, if $n\,\theta_p\big|_{r_0} \sim 1$. Then we have

$$\theta_{j+1} = \theta_j + \theta_p\big|_{r_0} + \theta_p'(r_j - r_0)$$
$$r_{j+1} = r_j + \epsilon\cos(n\theta_{j+1})$$

(5.2)

where ϵ represents the 'alien' perturbation. By starting particles at various r and θ, and following the mapping for many iterations, we can get a good sense for the nonlinear dynamics of this system, and for the types of effects ϵ will have. The map is *very* much more complex than when $\epsilon = 0$.

5.2 EXPERIMENTING WITH MAPPINGS

This text includes a graphic program, ERGO, for the purpose of letting you experiment with mappings for yourself, so that you can develop a feel for their properties. It is provided in both Macintosh[1] and IBM PC-compatible versions. Instructions on how to use this program are included in text files labeled README-ERGO on the Macintosh disk and ERGO.WRI on the IBM PC disk. (Computer source code is included as well.) The program allows you to vary the precession per bounce at $r = 0$, the radial gradient of the precession per bounce, the amplitude of the perturbation, and the mode number of the perturbation. The display can be modified by varying the minimum and maximum radii shown on

---

[1] Macintosh is a registered trademark of Apple Computer, Inc.

the screen. The computer beeps if the particle goes beyond these values. To see the process more clearly, you can vary the maximum rate at which the computer plots points. You can also toggle between the map we have been discussing here, the Chirikov–Taylor map, and a more complex two-step map treated in Problems 5.2 and 5.4. The Chirikov–Taylor map is named after its discoverers, B V Chirikov (1969 *Research Concerning the Theory of Nonlinear Resonances and Stochasticity*, translated by A T Sanders, CERN Translation 71-40, Geneva; *USSR Academy of Sciences Report 267*, Novosibirsk) and J B Taylor (1969 *Investigation of Charged Particle Invariants* in UKAEA Culham Laboratory Progress Report CLM-PR 12). In the wider world of nonlinear mechanics, this has come to be known as the 'standard' map.

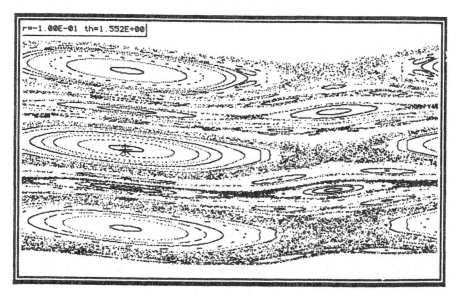

Figure 5.1. Sample ERGO output. *r* and *th* indicate location of pointer: $r = -0.100$; $th = 1.552$.

Figure 5.1 shows a sample ERGO output. Twenty-five mappings (each with lots of iterations) were used to make Figure 5.1.

Problem 5.1: Experiment with ERGO and find out what you can, qualitatively, about what goes on. Determine the effect of each parameter on the resulting map. For example, measure the variation of island width versus ϵ, at $n = \theta'_p = 1$. (From looking at the picture, you can guess that the 'islands' are the large elliptical regions containing fairly orderly trajectories.) How does the topology change as ϵ increases? Study the variation of island width and topology with n and θ'_p. Are there parameters you can vary together which expand or contract the map, without changing the overall topology? (Note: This problem should be attempted before proceeding to the next Section.)

5.3 SCALING IN MAPS

In the last Section, we discussed an iterative map of the form

$$\theta_{j+1} = \theta_j + \theta_p|_{r_0} + \theta'_p(r_j - r_0)$$

$$r_{j+1} = r_j + \epsilon\cos(n\theta_{j+1})$$

(5.3)

and Problem 5.1 asked you, among other things, to find a combination of parameters which just sets the scale of the plots without changing the topology. This kind of activity, where we pare down a set of physical equations to their essentials, is sometimes called a 'scaling analysis'. When we successfully complete a scaling analysis, even if we do not know the complete solution of the equations for any particular case, we do know what combination of parameters is important, so that we can reduce—in effect—the dimensionality of the problem. At first glance, it looks in our present case as though we have to understand a map that can be described in terms of five control parameters: $\theta_p|_{r_0}, \theta'_p, r_0, \epsilon$ and n. If we can reduce this to a simpler problem, with fewer control parameters, then a solution in terms of these fewer control parameters can be simply transformed into a solution for any values of $\theta_p|_{r_0}, \theta'_p, r_0, \epsilon$ and n.

Generally we make progress in this direction by recasting the equations in terms of those dimensionless variables that give the simplest possible equations. The coordinate θ is already dimensionless, but since we suspect that we will see periodicity in θ of $2\pi/n$, let us see if the equation simplifies when we define a new angular coordinate $\varphi \equiv n\theta$, which will have periodicity 2π, independent of n. If we do this, and then examine the first line of the transformed equation (5.3), we see that it will become very much simpler with a linear transformation from r_j to a dimensionless x_j:

$$x_j \equiv n\theta_p|_{r_0} + n\theta'_p(r_j - r_0).$$

(5.4)

Now we have a much simpler set of equations:

$$\varphi_{j+1} = \varphi_j + x_j$$

$$x_{j+1} = x_j + \Delta\cos\varphi_{j+1}$$

(5.5)

where $\Delta \equiv \epsilon n\theta'_p$. By transforming the variables to dimensionless forms that make the equations as simple as possible, we have managed to lump *all the control parameters* into the single parameter Δ. As a result, any answer we find in terms of the variables x and φ can be transformed algebraically into the answer for any member of the class of problems that have the same value of $\epsilon n \theta_p'$. Alternatively, for any values of $\theta_p|_{r_0}$, θ_p', r_0, ϵ and n, we simply evaluate Δ, look up a solution of equation (5.5) for that value of Δ, and scale the coordinates to the problem that interested us. Thus the quantity Δ must set all the topology of the map—for example, what fraction of space is filled by islands, and how chaotic the mapping looks. Note, also, that if we add 2π to x, and iterate the map, the 2π will simply function as a one-time additive term in x and a sequentially additive 2π term in φ, but it will have no effect on the map, since φ is a periodic variable with period 2π. This is to say that the map repeats in the φ direction with period 2π (implying that it does indeed repeat with period $2\pi/n$ in the θ coordinate), but it also repeats in the x direction, with period 2π as well! This implies that it repeats in the r direction, with period $2\pi/n\theta_p'$.

5.4   HAMILTONIAN MAPS AND AREA PRESERVATION

The reason that φ_{j+1} comes into the cosine term of equation (5.5), rather than φ_j, is interesting: it is because the map is constructed to maintain one of the basic features of a Hamiltonian system—it obeys Liouville's theorem. Liouville's theorem states that a Hamiltonian system (which, for present purposes, could be considered simply to be a system with no energy input and no dissipation) preserves phase-space density. If we follow a group of particle's trajectories in the phase space defined by their positions and momenta we find that, when they group together in position, they spread out in momentum, and *vice versa*, so as to preserve their total phase-space density. This is a result from classical mechanics, which we will actually re-derive in the course of developing the Vlasov equation in Chapter 22. The equivalent statement for a position mapping (where we are not anticipating changes in momentum) is that it should be area-preserving. Consider the differential square defined by (x_j, φ_j), $(x_j + \delta x, \varphi_j)$, $(x_j, \varphi_j + \delta\varphi)$, $(x_j + \delta x, \varphi_j + \delta\varphi)$ in Figure 5.2. We would like the map to carry this square over to a new quadrilateral with the same area, to assure that our 'particles' will not bunch together. The area of our original square was (in appropriate units) $\delta\varphi\delta x$. The first three corners of our new quadrilateral will be located at

$$[x_{j+1}, \varphi_{j+1}]$$
$$[x_{j+1} + (\partial x_{j+1}/\partial x_j)\delta x, \varphi_{j+1} + (\partial\varphi_{j+1}/\partial x_j)\delta x]$$
$$[x_{j+1} + (\partial x_{j+1}/\partial\varphi_j)\delta\varphi, \varphi_{j+1} + (\partial\varphi_{j+1}/\partial\varphi_j)\delta\varphi].$$

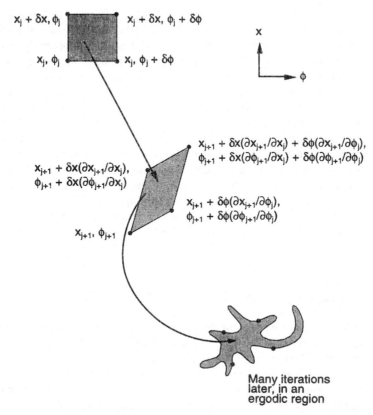

Figure 5.2. Area preservation in a Hamiltonian map.

These corners are enough to define two vectors, the magnitude of whose cross-product gives us the area of our new quadrilateral (a parallelogram). The result is

$$\left|\left[(\partial x_{j+1}/\partial x_j)\,\delta x\hat{\mathbf{x}} + (\partial \varphi_{j+1}/\partial x_j)\delta x\hat{\varphi}\right]\right.$$
$$\times \left[(\partial x_{j+1}/\partial \varphi_j)\delta\varphi\hat{\mathbf{x}} + (\partial \varphi_{j+1}/\partial \varphi_j)\delta\varphi\hat{\varphi}\right]\left|\right.$$
$$= \left|\left[(\partial x_{j+1}/\partial x_j)(\partial \varphi_{j+1}/\partial \varphi_j) - (\partial \varphi_{j+1}/\partial x_j)(\partial x_{j+1}/\partial \varphi_j)\right]\delta\varphi\delta x\right|. \quad (5.6)$$

Those familiar with coordinate transformations will recognize the term in square brackets as the Jacobian of the transformation, and it is a general result that its determinant must equal unity to assure that a transformation is area-preserving. However, it may not be completely obvious how to take the indicated partial derivatives through one full mapping step. Referring to equation (5.5), we see that

$$\frac{\partial \varphi_{j+1}}{\partial \varphi_j} = 1 \qquad \frac{\partial \varphi_{j+1}}{\partial x_j} = 1. \tag{5.7}$$

To find the partial derivatives of x_{j+1} with respect to x_j and φ_j (at fixed φ_j and x_j, respectively), however, requires substituting for φ_{j+1} in terms of φ_j and x_j (just *because* we have used φ_{j+1} here, as mentioned before). We obtain

$$x_{j+1} = x_j + \Delta\cos(\varphi_j + x_j) \tag{5.8}$$

so that

$$\frac{\partial x_{j+1}}{\partial x_j} = 1 - \Delta\sin\varphi_{j+1} \tag{5.9}$$

where the second term would not have been present if we had used φ_j, rather than φ_{j+1}, in the second step of the mapping. We also obtain

$$\frac{\partial x_{j+1}}{\partial \varphi_j} = -\Delta\sin(\varphi_{j+1}). \tag{5.10}$$

The argument of the sine function would be φ_j in this equation if we had used φ_j in the second step of the mapping, but this term itself would still have been present. Now evaluating the determinant of the Jacobian, we obtain

$$1 - \Delta\sin(\varphi_{j+1}) + \Delta\sin(\varphi_{j+1}) = 1 \tag{5.11}$$

as desired. The choice of φ_{j+1} in the second term was crucial for this result: otherwise the determinant would not have reduced to unity and the map would not have been area-preserving. Without this term, for example, in the Chirikov–Taylor map particle orbits flow rapidly out of the island structures and accumulate in the region between islands, violating the known underlying physics of the full system of equations of the drift orbits. Finding an area-preserving map is sometimes one of the main challenges in defining a mapping to correctly represent a Hamiltonian process.

As we saw by experimenting with ERGO, this apparently very deterministic map gives what appears to be random or chaotic behavior, for large enough values of $\Delta \equiv \epsilon n \theta_p'$. What does area-preservation mean in this case? It means that an original compact area spreads out like a drop of ink in a glass of water into a more and more spidery shape, but with the same original area.

It is important to recognize that there are also maps that are not area-preserving—and should *not* be, in order to represent non-Hamiltonian systems. In 'dissipative maps', which represent systems with energy input and dissipation, one finds that particle trajectories tend to collapse to 'attractors'—patterns in phase space that 'attract' trajectories from a 'basin' of initial positions. In some ways, these maps seem to represent the creation of order out of chaos, which is made possible by the flow of energy through the systems they represent. These dissipative maps are proving to be useful representations of fluid turbulence, as well as useful models for other turbulent nonlinear systems, just as Hamiltonian maps have proven useful for understanding nonlinear energy-conserving systems.

Problem 5.2: Consider a case where our dipole Earth has a line current driven through its center! Then the particle bounce orbits are not trajectories at fixed longitude, but they curve across the Earth at an angle (which itself depends on altitude), due to the θ directed magnetic field. Let us assume we still have the 'alien' period-n perturbation, but now we have to consider the fact that the bounce angle of the orbits may arrange things so that the 'kicks' on each half-orbit add or cancel, or something in between. The new map becomes

$$\theta_{j+1} = \theta_j + \theta_p|_{r_0} + \theta_p' r_j + \theta_b|_{r_0} + \theta_b' r_j$$
$$r_{j+1} = r_j + \epsilon\cos(n\theta_{j+1})$$
$$\theta_{j+2} = \theta_{j+1} + \theta_p|_{r_0} + \theta_p' r_{j+1} - \theta_b|_{r_0} - \theta_b' r_{j+1}$$
$$r_{j+2} = r_{j+1} + \epsilon\cos(n\theta_{j+2})$$

where the two-step feature comes from separately considering the kick on the top half and on the bottom half of the bounce. The θ_b term represents the particle's bounce motion, which takes it forward $(+)$ on one half of the bounce, and backwards $(-)$ on the other. Show that this map is area preserving.

5.5   PARTICLE TRAJECTORIES

Next, of course, it is interesting to try to determine how far the particles in our map depart from their original unperturbed trajectories. (Remember that we began this Chapter by being concerned about aliens bombarding us by perturbing the Van Allen belts. We might also be considering the perturbed orbits of energetic particles we want to contain in a fusion plasma.) As a first approximation, we assume that the particles do not move far enough in x that their precession speed changes significantly, so we can estimate

$$\varphi_j \approx \varphi_0 + jx_0$$
$$x_{j+1} \approx x_j + \Delta\cos(\varphi_{j+1})$$

(5.12)

where x_0 is the initial x coordinate. In the transformed equation (5.5), it also represents the initial precession rate in φ. Recognizing that we are only considering the real part of x we can write

$$x_{j+1} \approx x_j + \Delta \exp(i\varphi_{j+1}) \tag{5.13}$$

and substituting for φ_{j+1} we obtain

$$x_{j+1} \approx x_j + \Delta \exp(i\varphi_0)\exp[i(j+1)x_0]$$

or

$$x_j \approx x_{j-1} + \Delta \exp(i\varphi_0)\exp(ijx_0). \tag{5.14}$$

This is a simple recursion relation, which can be solved explicitly:

$$x_m \approx x_0 + \Delta \exp(i\varphi_0) \sum_{j=1}^{m} \exp(ijx_0) \tag{5.15}$$

and the summation is a well-known result, i.e.

$$\sum_{j=1}^{m} \exp(ijx_0) = \left(\exp(ix_0) \sum_{j=1}^{m} \exp(ijx_0) \right) + \exp(ix_0) - \exp[i(m+1)x_0]$$

so

$$\sum_{j=1}^{m} \exp(ijx_0) = \frac{\exp(ix_0) - \exp[i(m+1)x_0]}{1 - \exp(ix_0)} \tag{5.16}$$

so that

$$x_m - x_0 \approx \Delta \exp(i\varphi_0) \frac{\exp(ix_0) - \exp[i(m+1)x_0]}{1 - \exp(ix_0)}. \tag{5.17}$$

Thus, we would seem to have obtained our required answer, apparently in general. The right-hand side of equation (5.17) is of order Δ multiplied by a set of terms each of order unity. If Δ is small enough, we would expect $x_m - x_0$ to be small, as assumed. The terms multiplying Δ represent the fact that sequential radial kicks tend to cancel each other, rather than steadily accumulate. However, because the denominator can go to zero, we should carefully check our initial assumption that the particles do not drift far in the x direction from their original locations, *no matter where they begin*. A little analysis will show that there is trouble in small regions where the precession 'resonates' with the perturbation, i.e. x_0 is close to $2\pi k$ where k is any integer. In these regions, on each bounce the particle gets the same radial kick that it got the last time, so there is no sequential cancellation. Assume that $x_0 = 2\pi k + \delta$, where δ is a small enough number that even $m\delta \ll 1$. Then equation (5.17) becomes, approximately

$$x_m - x_0 \approx \Delta \exp(i\varphi_0) \frac{i\delta - i(m+1)\delta}{-i\delta} = \Delta m \exp(i\varphi_0). \tag{5.18}$$

This result implies that, if we start on a resonance, the particle position will go to infinity as $m \to \infty$. However, this violates the assumption made in deriving equation (5.17), that the precession speed will stay approximately constant $\approx x_0$. Thus equation (5.17) is suspect for starting points in a narrow region $\pm\delta(< 1/m)$ away from resonance, for large enough m. It is, of course, always suspect for large Δ.

5.6 RESONANCES AND ISLANDS

As a result of the breakdown of equation (5.17), we must try another approximation close to resonances. Specifically, we will assume that we are close enough to the resonance that φ changes very little (modulo 2π) on each bounce, but we will allow the precession speed to be a function of x, as x changes due to the accumulation of radial kicks. The physical effect we will find is that the particle will be 'kicked' radially away from the resonance by the perturbation, but then the subsequent kicks will no longer sequentially accumulate, so the particle will not drift off to infinity. Indeed it will eventually *move in* φ to the location where the radial kicks have the reverse sign, so it will drift back to the resonant radius, and continue to oscillate around the resonance in this manner.

For a small enough change in $\delta\varphi$ (modulo 2π) per iteration, our mapping equations can be 'reconstructed' into *differential* equations, where the 'unit of time' is an iteration of the map:

$$d\varphi/dt = x - x_s \qquad dx/dt = \Delta\cos\varphi \qquad (5.19)$$

where x_s is the resonant surface location, i.e. the location where the φ precession angle per bounce is exactly $2\pi k$. Differentiating the first equation with respect to 'time', we obtain

$$\frac{d^2\varphi}{dt^2} = \Delta\cos\varphi \qquad (5.20)$$

which is the equation for a ball rolling in a sinusoidal well. We can find some of its important properties from the conservation equations

$$\frac{d\varphi}{dt}\frac{d^2\varphi}{dt^2} = \frac{d\varphi}{dt}\Delta\cos\varphi$$
$$\frac{d}{dt}\left[\frac{1}{2}\left(\frac{d\varphi}{dt}\right)^2\right] = \frac{d}{dt}(\Delta\sin\varphi) \qquad (5.21)$$

or

$$\frac{1}{2}\left(\frac{d\varphi}{dt}\right)^2 - \Delta\sin\varphi = \text{constant} \qquad (5.22)$$

where d/dt indicates the total derivative along the particle's trajectory.

If the constant in equation (5.22) is chosen to be greater than Δ, then any value of φ corresponds to a positive value of $(d\varphi/dt)^2$, and so φ will increase or decrease indefinitely. If $d\varphi/dt$ starts positive, there is no place where it goes to zero, so it will stay positive for its whole trajectory; by the same argument, if $d\varphi/dt$ starts negative, it will stay negative. For values of the constant less than Δ, the value of φ is trapped and oscillates between the zeros of $(d\varphi/dt)^2$. The combination of the oscillation in φ and the associated oscillation in x creates a closed trajectory referred to as an 'island'—by now you have seen plenty of these in ERGO. It is interesting to calculate the width of this island. From equation (5.19), we see that a particle reaches its maximum $(x - x_s)$ at the same time that it reaches its maximum $d\varphi/dt$. In the case where the constant in equation (5.22) is just equal to Δ (the barely trapped orbit, corresponding to the 'separatrix' between trapped and passing orbits in the ERGO plots), the maximum value obtained by $d\varphi/dt$ is

$$(d\varphi/dt)_{max} = (x - x_s)_{max} = 2\sqrt{\Delta} \qquad (5.23)$$

so, by the definition of $\Delta \equiv \epsilon n\theta_p'$ and of $x_j \equiv n\theta_p|_{r_0} + n(r_j - r_0)\theta_p'$, we have

$$(r - r_s)_{max} = 2\sqrt{\epsilon/n\theta_p'}. \qquad (5.24)$$

The island width is proportional to the square root of the perturbation strength, and inversely proportional to the square root of the gradient in the precession speed as a function of altitude, sometimes called the 'shear', in the precession.

5.7 ONSET OF STOCHASTICITY

So far, we have been talking as if all of the trajectories were nicely bounded. In fact, as Δ grows we have seen from ERGO that some of the trajectories eventually become 'stochastic' or apparently random in their behavior. Trajectories that randomly fill a region of space are called 'ergodic'— hence the name ERGO. The transition to stochastic behavior is fascinating. The separatrix trajectories become stochastic first, long before the regions inside the islands. The stochasticity is sometimes characterized as being due to *overlap* of the islands which occur at radii with different resonances. The thought is that as Δ grows, trajectories cannot tell which island they are on.

But what has really gone wrong with the derivation above, from which we obtained a simple non-stochastic island structure? Basically, the island we have been calculating has reached out so far from the resonant surface that the precession is significantly different from 2π, so the jumps in the φ direction (modulo 2π), namely $\delta\varphi$, are no longer small, and we cannot legitimately model the mapping as a differential equation—thus, just like equation (5.17), our solution near the resonances is not valid in a situation with large Δ. A simple estimate of when the stochasticity enters would follow from noting that when $\delta\varphi$ approaches unity the steps are no longer small relative to the scale length of variation of the potential well described by equation (5.22). Thus when the island width is equal to unity, in the scaled map, we should expect to begin to see breakdown of our differential-equation model near the separatrices. Using equation (5.23) for the island width, this condition can be written

$$2\sqrt{\Delta} \approx 1$$

or

$$\Delta \approx 1/4. \tag{5.25}$$

There are further nonlinear features that enter as Δ grows to ~ 1, and these are not modeled by our differential equation, also because the jump in x (as well as in φ) is now no longer small. An order unity jump in x means an order unity change in the precession speed per jump, in which case the differential-equation model fails totally—on both variables! As Δ grows, secondary islands with periodicity p can be seen at $x = 2\pi k + 2\pi n/p$, because at these radii trajectories repeat themselves after p iterations rather than average smoothly over all φ, so finite 'kicks' due to the nonlinearity of the map accumulate, and the resulting islands contribute to the island-overlap process (see Problem 5.5). Furthermore, the trajectories circulating *around in the volume within the islands* have 'shear'—the bounce time in a sinusoidal well depends on the depth into the well. Thus there are trajectories inside the primary islands which are resonant (for example, it takes six or seven circuits to go completely around the island at some distance from its center). Again orbits repeat over themselves and kicks accumulate.

The primary islands therefore have island 'daisy-chains' within them! If you blow up the scale in ERGO you can see these chains. These island chains even have their own island daisy-chains inside, *ad infinitum*. Furthermore, if more than one 'alien' perturbation is present (e.g. there are various perturbations with different values of n), then the islands from the different perturbations can overlap. In the pure Chirikov–Taylor map, trajectories become free to jump across periods of 2π in the x direction (a crucial moment) when $\Delta = \epsilon n\theta'_p = 0.989\ldots$, where this numerical value is taken from Chirikov's classic monograph (1979 *Phys. Rep.* **52** 265). In this critical vicinity, however, it takes of order 10^7 mappings for a typical orbit to 'jump' periods, so for your level of patience (and your computer's speed), you may find the practical limit is around 1.05. In the range of $\Delta \gg 1$, the orbits end up with individual steps in radius which are effectively uncorrelated from one to the next. Using a 'random walk' model, we can estimate a radial diffusion coefficient of $\approx \Delta^2/4$ in this regime.

Problem 5.3: From the definition of the random-walk diffusion coefficient, $D = \langle \Delta x^2 \rangle / 2\tau$, derive the radial diffusion coefficient for the standard map in the limit of large Δ.

Problem 5.4: Toggle ERGO to the two-step map introduced in Problem 5.2 and study the onset of stochasticity in this map—investigate if it is roughly consistent with the following criterion for stochastic onset:

$$\epsilon > \frac{1}{|n\theta_b'|} + \frac{1}{|n\theta_p'|} \tag{5.26}$$

which was initially derived by R J Goldston, R B White and A H Boozer (1981 *Phys. Rev. Lett.* **47** 647), both when $\theta_b' > \theta_p'$ and when $\theta_b' < \theta_p'$. Explain the basis for this criterion. Hint: it is not exactly island overlap. (Note: this problem requires extensive independent work and has typically been set as a semester project.)

Problem 5.5: There is an island chain that grows at $r = \pi$, when the parameters in the Chirikov–Taylor map are set at their defaults. You can see it in Figure 5.1 as a chain of two islands located vertically about halfway between the largest islands. Empirically determine the scaling of its width at constant θ, as a function of ϵ. Then explain why this scaling is observed. Hint: use the differential-equation approach developed in this Chapter, but consider a unit of 'time' to be *two* iterations of the map. Consider that Δ and $(x - x_s)$ are similarly small quantities, so terms of order $\Delta(x - x_s)$ and $(x - x_s)^2$ can be neglected compared to terms of order Δ or $(x - x_s)$. Finally, note that you want to know the island's full width at fixed θ, not its total extent in the x direction, which is different, since the line it is 'riding' oscillates in x. (Note: like Problem 5.4, this problem has also typically been set as a semester project, as an alternative to Problem 5.4.)

UNIT 2

PLASMAS AS FLUIDS

As an alternative to analysing the motion of individual charged particles in electric and magnetic fields, we could have begun our study of plasmas by introducing the concept of an 'electrically conducting fluid'. Even a primitive knowledge of hydrodynamics would allow us to write down equations of continuity and momentum transfer, which we could adapt to the plasma case by introducing effects specific to an electrically charged current-carrying fluid, such as electric and magnetic forces. This would produce a set of 'magnetohydrodynamic' equations.

In this Unit we will derive these equations from the basic conservation laws, i.e. conservation of the total number of particles of each species and of the total momentum, for a plasma containing two particle species, namely electrons and ions. In the momentum transfer equations, we will encounter a force arising from the gradient of the plasma pressure, but we will note that, in the plasma case, the 'pressure' is not necessarily isotropic (i.e. the same in all three directions), as it generally is in the case of a hydrodynamic fluid. Since the single-particle (i.e. guiding center) and fluid models seem at first to offer alternative, even contradictory, approaches to solving problems in plasma physics, we will take care to describe how these two approaches are related and to demonstrate how they must give the same answer to any given physical problem.

The limit in which the electrical resistivity of the plasma is vanishingly small is of particular interest and has important and quite general consequences, which we will prove, in regard to the 'tying' of plasma to magnetic field-lines. The 'ideal' magnetohydrodynamic equations that follow in this limit will be shown to be suitable tools for analysing a number of different plasma equilibria.

Chapter 6

Fluid equations for a plasma

We have been looking at plasmas as collections of individual charged particles. It is time now to consider the behavior of an ensemble of charged particles which we will find acts as a special kind of fluid. In this Chapter, we will derive fluid equations for each species of particles (i.e. ions and electrons) separately, and later we will see how to treat the whole plasma as a single fluid.

6.1 CONTINUITY EQUATION

Consider a differential element of volume in the shape of a cube whose sides are parallel to the coordinate surfaces, as shown in Figure 6.1. In order to display the similarity of each of the three coordinate directions, and especially because we will later want to sum various expressions over all three coordinate directions, we have labelled the coordinates x_1, x_2 and x_3, rather than the more familiar x, y and z.

The number of particles flowing out of this volume element across the surface shown in the figure per second will be $n\langle v_1 \rangle dx_2 dx_3$ evaluated at $x_1 + dx_1$, where $\langle v_1 \rangle$ is the average velocity of our species of particles in the x_1-direction. (In the more conventional notation, the component v_1 would be written v_x, and similarly v_2 and v_3 would be written v_y and v_z.) The number of particles flowing *into* the volume element per second will be the same expression evaluated at x_1. Assuming that no particles are gained or lost from the volume element other than by flow across its boundaries, we can express the rate of change of the number of particles in the cube shown in Figure 6.1 in terms of the flow of particles across each of the six sides, i.e.

$$\frac{\partial n}{\partial t} d^3 x = -[n\langle v_1 \rangle dx_2 dx_3]_{x_1+dx_1} + [n\langle v_1 \rangle dx_2 dx_3]_{x_1} \cdots \qquad (6.1)$$

85

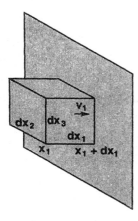

Figure 6.1. Differential element of volume with sides parallel to the coordinate surfaces.

where $d^3x = dx_1 dx_2 dx_3$. Dividing by d^3x and taking the limit as the size of the cube shrinks, we obtain the 'continuity equation'

$$\frac{\partial n}{\partial t} = -\sum_i \frac{\partial}{\partial x_i}(n \langle v_i \rangle). \tag{6.2}$$

For clarity, we display the summation sign explicitly in equation (6.2) and elsewhere in this Chapter although, even without it, index notation under the Einstein convention would imply summation over the repeated suffix i. Changing to vector notation and writing $\mathbf{u}$ for the average velocity, $\langle \mathbf{v} \rangle$, the continuity equation may be written

$$\frac{\partial n}{\partial t} + \nabla \cdot (n\mathbf{u}) = S \tag{6.3}$$

where we have added a volume source rate of particles S. For the charged particles of a plasma, a volume source term would arise from the ionization of neutral atoms; recombination would give rise to a corresponding volume sink term. For the present, however, we will generally neglect ionization and recombination, but we should be aware that sources and sinks of particles *do* arise in plasmas and give additional terms in all of the fluid equations.

6.2 MOMENTUM BALANCE EQUATION

We consider next the rate of change of momentum density in a differential element of volume. First, we will ignore the fact that particles move in and out of this volume, and consider just the macroscopic forces on the element. The familiar Lorentz force, extended to all the particles of a given species per unit volume, is just

$$\mathbf{F} = nq(\mathbf{E} + \mathbf{u} \times \mathbf{B}) \tag{6.4}$$

where again $\mathbf{u}$ is the average velocity, $\langle \mathbf{v} \rangle$, of our species of particles (for example the electrons) and q is their charge per particle. The force density $\mathbf{F}$ represents a local source rate of momentum density; the change in momentum density due to this force density is just

$$\frac{\partial(nm\mathbf{u})}{\partial t} = \mathbf{F} = nq(\mathbf{E} + \mathbf{u} \times \mathbf{B}). \tag{6.5}$$

Next we have to consider the momentum density changes arising from particle motion carrying momentum with it. We have so far included momentum changes due to external forces acting on the particles that are 'members' of the element, but as particles move into or out of the element carrying momentum with them, this gives another contribution to the time derivative of the momentum in the element. Let us begin by considering the flux in the x_1 direction of x_2-directed momentum. This flux is just the number of particles per unit area per second passing through a surface of constant x_1, times the momentum in the x_2 direction, mv_2, carried by each particle.

In Chapter 1 we introduced the distribution function, $f(\mathbf{x}, \mathbf{v})$. We recall that $f(\mathbf{x}, \mathbf{v})$ is the relative probability of finding a particle with a given velocity vector $\mathbf{v}$, normalized such that the integral of $f(\mathbf{x}, \mathbf{v})$ over all $\mathbf{v}$ gives the density $n(\mathbf{x})$. Thus the differential number of particles in a phase space element $d^3v d^3x$ located at $(\mathbf{x}, \mathbf{v})$ is simply $f(\mathbf{x}, \mathbf{v})d^3v d^3x$. This element of phase space 'empties out' in the x_1 direction in a time interval $dt = dx_1/v_1$. The differential number of particles carried per second across the surface of constant x_1 by this element of phase space is $f d^3v d^3x/dt = v_1 f d^3v dx_2 dx_3$. These specific particles each carry x_2-directed momentum mv_2, so the differential amount of momentum in this direction carried per second across a surface of constant x_1 by this element of phase space is $mv_2 v_1 f d^3v dx_2 dx_3$. To obtain the total momentum flux, i.e. the total quantity of x_2-directed momentum crossing a surface of constant x_1 per unit area per second, we divide out the differential area $dx_2 dx_3$ and integrate over velocity space. The total flux of x_2-directed momentum in the x_1 direction becomes $\int mv_2 v_1 f d^3v$, which, by the definition of f, can be written simply $mn\langle v_2 v_1 \rangle$. The rate of change of x_2-directed momentum, averaged over all of the particles, can now be expressed in terms of the divergence of fluxes of momentum across the various surfaces:

$$\frac{\partial(nmu_2)}{\partial t} = -\frac{\partial}{\partial x_1}(mn\langle v_2 v_1 \rangle) - \frac{\partial}{\partial x_2}(mn\langle v_2 v_2 \rangle) - \frac{\partial}{\partial x_3}(mn\langle v_2 v_3 \rangle). \tag{6.6}$$

Since we have not used any special properties of x_2, we can replace it with an arbitrary suffix i. Furthermore, it now becomes clear that this momentum flux is closely related to a generalized definition of pressure, in the case where pressure is viewed as a tensor quantity, the 'pressure tensor'. This pressure tensor P (we use boldface italics for tensors) is defined in index form by

$$P_{ij} \equiv mn\langle (v_i - u_i)(v_j - u_j) \rangle$$
$$= mn(\langle v_i v_j \rangle - u_i u_j) \tag{6.7}$$

where we have used the definition of mean velocity $\mathbf{u}$, namely $u_i = \langle v_i \rangle$. Thus, the flux in the i direction of j-directed momentum is just $P_{ij} + mnu_i u_j$. Incidentally, this derivation makes clear that the familiar concept of pressure is, in a more fundamental sense, a momentum flux. Note that it is only the *divergence* of this flux that results in acceleration.

For the special case of a Maxwellian distribution, drifting at a given velocity $\mathbf{u}$ (i.e. $f(\mathbf{v} - \mathbf{u}) =$ Maxwellian), then $P_{ij} = 0$ for $i \neq j$, and $P_{ij} = nT$ for $i = j$, where T is the temperature (measured in energy units, i.e. joules, so that the Boltzmann constant may be omitted). For a case where the plasma is characterized by different perpendicular and parallel temperatures, relative to the direction of the local magnetic field, then P_{ij} still is zero for $i \neq j$, but now, if $i = j$ and is perpendicular to $\mathbf{B}$, then $P_{ij} = nT_\perp$, while if $i = j$ and is parallel to $\mathbf{B}$, then $P_{ij} = nT_\parallel$. If we take the direction of $\mathbf{B}$ to be the x_3 direction, we obtain

$$P_{ij} = \begin{bmatrix} nT_\perp & 0 & 0 \\ 0 & nT_\perp & 0 \\ 0 & 0 & nT_\parallel \end{bmatrix}. \tag{6.8}$$

It is interesting to note that, in all cases, $P_{ij} = P_{ji}$. Furthermore, the off-diagonal elements $(i \neq j)$ constitute the flow in one direction of momentum in another direction. This is the mechanism of viscosity, whereby if a fluid has flow, for example in the x_1 direction, but that flow has a gradient in the x_2 direction, then x_1 momentum is transferred in the x_2 direction to the slower moving fluid, acting so as to speed it up. However, in a plasma, if velocity-gradient scale-lengths, L, are much greater than a gyro-radius, r_L, these off-diagonal elements of the pressure tensor are smaller than the diagonal elements, being higher-order by at least one power of r_L/L.

Problem 6.1: Show that

$$P = nT_\perp I + (nT_\parallel - nT_\perp)\hat{\mathbf{b}}\hat{\mathbf{b}}$$

is an equivalent way of writing equation (6.8). Evaluate all of the nine elements P_{ij} for the case where the magnetic field is oriented at 45° to the x_1 direction, in the (x_1, x_2) plane.

Returning to our derivation of the momentum balance equation for a plasma, we now know that the flux in the x_1 direction of x_2 momentum is given by $P_{12} + mnu_1u_2$. If we want to know the time derivative of x_2 momentum density due to this flux, we need only use equation (6.6), which involves the divergence of this flux of momentum. Thus, the contribution to the rate of change of x_2 momentum density from the motion of particles is

$$\frac{\partial(mnu_2)}{\partial t} = -\frac{\partial P_{12}}{\partial x_1} - \frac{\partial P_{22}}{\partial x_2} - \frac{\partial P_{32}}{\partial x_3} - m\left(\frac{\partial(nu_1u_2)}{\partial x_1} + \frac{\partial(nu_2u_2)}{\partial x_2} + \frac{\partial(nu_3u_2)}{\partial x_3}\right)$$
(6.9)

with similar equations for the change of momentum in the x_1 and x_3 directions. In index notation, this becomes

$$\frac{\partial(mnu_j)}{\partial t} = -\sum_i \frac{\partial P_{ij}}{\partial x_i} - m\sum_i \frac{\partial}{\partial x_i}(nu_iu_j).$$
(6.10)

Reverting to vector notation (where we do not need to specify a coordinate system) and denoting the pressure tensor by P, we can combine the two contributions to the rate of change of momentum density that we have evaluated, to obtain the 'momentum balance equation':

$$\frac{\partial(mn\mathbf{u})}{\partial t} = nq(\mathbf{E} + \mathbf{u} \times \mathbf{B}) - \nabla \cdot P - \nabla \cdot (mn\mathbf{uu}).$$
(6.11)

There are various alternative forms in which we can put the momentum balance equation. We can substitute for $\partial n/\partial t$ from the continuity equation, equation (6.3), and we can expand the last term on the right-hand side in equation (6.11), by using

$$m\nabla \cdot (n\mathbf{uu}) = m\mathbf{u}(\nabla \cdot n\mathbf{u}) + mn(\mathbf{u} \cdot \nabla)\mathbf{u}.$$
(6.12)

In index notation and cartesian coordinates, this is equivalent to

$$m\sum_i \frac{\partial}{\partial x_i}(nu_iu_j) = mu_j\sum_i \frac{\partial}{\partial x_i}(nu_i) + mn\sum_i u_i\frac{\partial u_j}{\partial x_i}.$$
(6.13)

In this way, the complete momentum balance equation can be reduced to its most standard form:

$$mn\left(\frac{\partial \mathbf{u}}{\partial t} + (\mathbf{u} \cdot \nabla)\mathbf{u}\right) = nq(\mathbf{E} + \mathbf{u} \times \mathbf{B}) - \nabla \cdot P - m S\mathbf{u}.$$
(6.14)

(Often, the term in S associated with particle sources and sinks is neglected. Furthermore, another term must be added if the ionization or recombination processes involve a net loss or gain of momentum.)

Problem 6.2: Provide a simple physical picture of the appearance of the last term on the right-hand side of equation (6.14). For example, consider a boy standing on a bridge dropping bricks onto trucks passing underneath. What happens to the velocities of the bricks, and also of the trucks?

Equation (6.14) is the 'momentum balance equation', but it is also sometimes referred to as the 'fluid equation of motion', since it equates an acceleration to the sum of a number of forces. Very often, we express it in terms of the total derivative, which gives the time rate of change in an element of fluid moving with the local flow

$$\frac{\mathrm{d}}{\mathrm{d}t} = \frac{\partial}{\partial t} + \mathbf{u} \cdot \nabla \tag{6.15}$$

in which case equation (6.14), neglecting the term due to a source S, can be written in the transparent form

$$mn\frac{\mathrm{d}\mathbf{u}}{\mathrm{d}t} = nq(\mathbf{E} + \mathbf{u} \times \mathbf{B}) - \nabla \cdot P. \tag{6.16}$$

In the case where a plasma is nearly Maxwellian (or at least nearly isotropic), $\nabla \cdot P$ can be replaced by the gradient of a scalar pressure, ∇p. If this is not the case, the more complete form for the pressure tensor must be retained. The fluid form of the plasma equations of motion can handle fairly complex situations, but it is too much to expect that any simple closed form for the pressure tensor can represent all the effects associated with the full distribution function of particles. In fact, we could now proceed to generate a higher-rank tensor (the 'heat-flux tensor'), involving the divergence of averages of quantities like $v_i v_j v_k$, and we could generate fluid-like equations for the evolution of the pressure tensor in terms of this heat-flux tensor. Frequently people have the fortitude to maintain some of the off-diagonal elements of the pressure tensor and even some elements of the heat-flux tensor in their calculations, but much of the heat-flux tensor generally falls by the wayside. This amounts to assuming that the pressure tensor, and the velocity dependences in its underlying distribution function, $f(\mathbf{x}, \mathbf{v})$, have fairly simple symmetrical forms. However the fluid equations cannot handle very complex features of $f(\mathbf{x}, \mathbf{v})$, such as subgroups of suprathermal particles, or complex anisotropies. Under these circumstances, the more complete kinetic theory (which will be introduced in Chapter 22) is required.

It is somewhat surprising, however, that even in a fairly collisionless plasma the distribution function can be close to Maxwellian, and fluid concepts can apply. This is because the magnetic field prevents particles from free-streaming and accelerating across **B**, and so they are forced to remain close to their original neighbors in the same fluid element. Along the **B** field, it is easier for particles to stream and mix (and run in and out of the fluid elements rapidly), but for just this reason gradients *along* **B** tend to be very gentle, and situations usually do not arise, except transiently, where a very hot region is feeding extremely energetic particles into a cold region on the same field line and creating an anisotropic suprathermal tail on the distribution of particles. This does happen sometimes, however, and then kinetic theory is required to calculate even simple things like momentum balance. Waves that have a finite wavelength along the direction of **B** are especially likely to be subject to such an effect. Later we will learn about Landau damping, which is essentially a phenomenon of energy and momentum transfer between particles and waves, due to kinetic effects.

6.3 EQUATIONS OF STATE

Even in the very simplest cases where the pressure tensor is isotropic, some additional relationship must be introduced to describe how the scalar pressure p varies in time. To avoid dealing with the heat-flux tensor explicitly, we will approximate the heat flow by introducing a thermodynamic equation of state for a plasma. This is an equation of the form $p = Cn^\gamma$, which relates the scalar pressure p to the density n. The quantity γ expresses how much the temperature of a plasma increases as it is compressed, since $pV^\gamma = $ constant, where V is the plasma volume. As such, the equation of state constitutes a simple (and therefore only approximate) statement about the heat flow.

For the case of compression that is slow compared to thermal conduction, we have $\gamma = 1$, i.e. 'isothermal' compression. The pressure goes up only because the density goes up. In many cases, because particles can freely stream along a magnetic field **B**, conduction parallel to **B** provides an avenue for the plasma to remain isothermal, if the compression is, for example, periodic or wave-like along **B**.

On the other hand, if the compression is fast enough to be 'adiabatic' (faster than heat conduction), but slow enough that energy is collisionally exchanged between the three degrees of freedom, then $\gamma = 5/3$, the usual result for a three-dimensional ideal gas. This is a special case of the more general result for an ideal gas, $\gamma = (2 + N)/N$, where N denotes the number of degrees of freedom. Later, we will see that a plasma can support a number of different types of waves, some of which compress the plasma isothermally, while others compress it adiabatically, and this has a significant effect on the wave dynamics.

An important third case can arise in a plasma if adiabatic compression is both fast compared to collisions and also anisotropic. In this case, the parallel and perpendicular degrees of freedom are separated, so that $T_\parallel$ can be heated very effectively ($N = 1$, $\gamma = 3$), as can, to a lesser degree, $T_\perp$ ($N = 2$, $\gamma = 2$).

The adiabatic invariants of particle motion in a strong magnetic field can be used to derive generalizations of these relationships for the case of compression that involves components both parallel and perpendicular to the magnetic field, as we will now see.

The perpendicular pressure can be expressed in terms of an average of the particles' invariant magnetic moments μ by means of the relation

$$p_\perp = mn \left\langle \frac{v_\perp^2}{2} \right\rangle = n \langle \mu \rangle B. \tag{6.17}$$

If the compression is fast compared to collisions but slow compared to the Larmor gyration of the particles, then the μ value of each particle will be conserved, which leads to the adiabatic relation

$$\frac{d}{dt} \left(\frac{p_\perp}{nB} \right) = 0. \tag{6.18}$$

For the case of pure perpendicular compression, which would typically be accomplished by increasing the strength of the magnetic field, conservation of particles and of magnetic flux as the area A of the plasma cross section is changed implies that $nA = $ constant and $BA = $ constant, so that n is proportional to B. We then see that equation (6.18) reduces to the simplest adiabatic relationship $p/n^\gamma = $ constant, with $\gamma = 2$, as is appropriate for two-dimensional adiabatic compression. (The conservation of magnetic flux BA was demonstrated in Chapter 4 for the simple case of a straight cylinder of Larmor gyrating particles: an increase in B results in a decrease in the area A of each of the Larmor orbits, such that $BA = $ constant. The more general result is demonstrated in Chapter 8.)

Similarly, the parallel pressure can be expressed in terms of the particles' J invariants by means of the relations

$$p_\parallel = mn \langle v_\parallel^2 \rangle \qquad J \approx v_\parallel L \tag{6.19}$$

where L is some measure of the length of the plasma along the field lines. If the compression is slow compared to the motion of particles to-and-fro along field lines, then the J value of each particle will be conserved. Again using the conservation of particles and of magnetic flux, in this case as the length L, the cross sectional area A and the volume V all change, we obtain the relationships $V = AL$, $nV = $ constant and $BA = $ constant. Using these relationships to express L in terms of the physical quantities n and B, we obtain $L = V/A \propto B/n$; substituting this into equation (6.19), we obtain the adiabatic relationship

$$\frac{d}{dt} \left(\frac{p_\parallel B^2}{n^3} \right) = 0. \tag{6.20}$$

For the case of pure parallel compression, in which B is unchanged, this reduces to the simpler adiabatic relationship $p/n^\gamma = $ constant, with $\gamma = 3$, again as appropriate for one-dimensional compression.

The two adiabatic relationships that we have derived here, equations (6.18) and (6.20), are called the 'double adiabatic' equations of state.

6.4 TWO-FLUID EQUATIONS

So far, we have derived fluid equations by considering only one species of particle at a time. In a plasma, there can be many different species of particle, and there will always be at least *two* species (ions and electrons) in any neutral plasma. The continuity equation (6.3) will, of course, apply separately to each of the different species. However, in applying the momentum balance equation (6.14) to the separate species, allowance must be made for the fact that particles of one species can collide with particles of another species, thereby transferring momentum between the different species.

In the fluid approximation, the effect of collisions between particles of different species is often simply modeled by means of a set of 'collision frequencies', $\nu_{\alpha\beta}$, that express the rate at which the momentum of species α is transferred by collisions to species β. Since it is reasonable to estimate that the transfer of momentum will be proportional to the difference in the mean velocities of the two species, the rate at which momentum per unit volume is gained by species α due to collisions with species β is given by

$$\mathbf{R}_{\alpha\beta} = -m_\alpha n_\alpha \nu_{\alpha\beta}(\mathbf{u}_\alpha - \mathbf{u}_\beta). \tag{6.21}$$

This gain (or loss) in momentum must be included in the momentum balance equation for species α, which now becomes

$$m_\alpha n_\alpha \left(\frac{\partial \mathbf{u}_\alpha}{\partial t} + (\mathbf{u}_\alpha \cdot \nabla)\mathbf{u}_\alpha \right) = n_\alpha q_\alpha(\mathbf{E} + \mathbf{u}_\alpha \times \mathbf{B}) - \nabla \cdot P_\alpha + \sum_\beta \mathbf{R}_{\alpha\beta} \tag{6.22}$$

where the summation is over all species β, not equal to α, with which particles of species α can collide. The quantity $\nu_{\alpha\beta}$ is called the 'collision frequency' of species α on species β. For the case where $\mathbf{u}_\beta = 0$, the quantity $\nu_{\alpha\beta}$ is simply the rate at which the momentum of species α is lost due to the presence of another species β of particles that are at rest.

Since the momentum density transferred to species α from species β, namely $\mathbf{R}_{\alpha\beta}$, and the momentum density transferred to species β from species α, namely $\mathbf{R}_{\beta\alpha}$, must obey momentum conservation, we can deduce that

$$\mathbf{R}_{\beta\alpha} = -\mathbf{R}_{\alpha\beta}. \tag{6.23}$$

From this it follows that $\nu_{\alpha\beta}$ and $\nu_{\beta\alpha}$ must obey a symmetry relation:

$$m_\alpha n_\alpha \nu_{\alpha\beta} = m_\beta n_\beta \nu_{\beta\alpha}.$$

6.5 PLASMA RESISTIVITY

Collisions between electrons and ions in a plasma will impede the acceleration of electrons in response to an electric field applied along (or in the absence of) a magnetic field. Without such collisions, electrons would be accelerated indefinitely by an applied electric field, so that an infinitesimal voltage would be sufficient to drive a large current through a plasma, at least in the direction along a magnetic field. In practice, the acceleration of electrons is impeded by collisions with non-accelerated particles, in particular the ions, which, because of their much larger mass, are relatively unresponsive to the applied electric field. Collisions between electrons and ions, acting in this way to limit the current that can be driven by an electric field, give rise to an important plasma quantity, namely its *electrical resistivity*, usually denoted η.

We will conclude this Chapter by deriving a simple expression for the resistivity in the case of a hydrogen plasma, in which the electrons have charge $-e$ and the ions are protons, with charge e. The resistivity may be expressed in terms of the electron–ion collision frequency, ν_{ei}, by applying equations (6.21) and (6.22) to the case of electrons that have reached a steady-state equilibrium in the presence of an electric field $E_\parallel$, applied either parallel to a magnetic field **B** or in the absence of any magnetic field. Because of the very small mass, and therefore negligible inertia, of the electrons, such an equilibrium will be reached relatively rapidly. Assuming that the electrons are homogeneous and therefore neglecting also in equation (6.22) the electron pressure and velocity gradients along **B**, we obtain

$$\mathbf{R}_{ei} = -m_e n_e \nu_{ei}(\mathbf{u}_e - \mathbf{u}_i) \qquad 0 = -n_e e E_\parallel + R_{ei\parallel}. \tag{6.24}$$

Introducing the current density

$$j_\parallel = -n_e e(u_{e\parallel} - u_{i\parallel}) \tag{6.25}$$

we obtain

$$E_\parallel = -\frac{m_e \nu_{ei}}{e}(u_{e\parallel} - u_{i\parallel}) = \frac{m_e \nu_{ei}}{n_e e^2} j_\parallel = \eta j_\parallel. \tag{6.26}$$

By analogy with electrical properties of normal matter, we call the constant of proportionality between the applied electric field **E** and the current density **j** the resistivity η, which we see to be given by $m_e \nu_{ei}/n_e e^2$.

Since the actual frequency with which electrons collide with ions will depend on the electron velocities, the collision frequency ν_{ei} appearing in the resistivity and in all of the above expressions should be an average over an appropriate distribution of electron velocities, in which case the collision frequency should more correctly be written $\langle \nu_{ei} \rangle$. Accordingly, the resistivity η becomes

$$\eta = \frac{m_e \langle \nu_{ei} \rangle}{n_e e^2}. \tag{6.27}$$

The momentum gained by electrons due to collisions with ions, $\mathbf{R}_{ei}$, may now be expressed in terms of the resistivity η and the current density $\mathbf{j}$. We obtain

$$\begin{aligned} \mathbf{R}_{ei} &= -m_e n_e \langle \nu_{ei} \rangle (\mathbf{u}_e - \mathbf{u}_i) \\ &= -\eta n_e^2 e^2 (\mathbf{u}_e - \mathbf{u}_i) \\ &= \eta \, n_e e \mathbf{j}. \end{aligned} \tag{6.28}$$

For the case of an electron–proton plasma, we may substitute equation (6.28) into equation (6.22), to obtain two momentum balance equations, one for the electron mean velocity $\mathbf{u}_e$, and the other for the proton mean velocity $\mathbf{u}_i$. Alternatively, the two equations can be added together, making use of the fact that $\mathbf{R}_{ei} + \mathbf{R}_{ie} = 0$, with the result that the collisional terms vanish from this summed momentum balance equation, which determines the total momentum density of the plasma. However, since it is generally necessary to distinguish between the mean velocities $\mathbf{u}_i$ and $\mathbf{u}_e$ in order to evaluate the plasma current density, we will still need to work with two separate momentum balance equations, in some form. Equation (6.26) is an example of a momentum balance equation taking the form of a simplified 'Ohm's Law' parallel to the magnetic field, involving the *difference* of the two mean velocities, which determines the current density $\mathbf{j}$. We will return to this topic in more detail in Chapter 8.

Although the resistivity was derived for the case of an electric field applied parallel to a magnetic field (or in the case where there is no magnetic field), the collisional transfer of momentum between electrons and ions due to the presence of an electrical current depends only moderately on the direction of the current. Specifically, we see from equation (6.27) that the resistivity is a quantity that is proportional to the mean electron–ion collision frequency $\langle \nu_{ei} \rangle$. The expression for $\mathbf{R}_{ei}$ given in equation (6.28) can be used in the momentum balance equation, for example equation (6.22), not only for the case where the electric field $\mathbf{E}$ and current density $\mathbf{j}$ are parallel to $\mathbf{B}$, but also for arbitrary orientation of $\mathbf{E}$, $\mathbf{j}$ and $\mathbf{B}$. This requires, however, that the resistivity be expressed as a tensor quantity, specifically a diagonal tensor with diagonal elements $(\eta_\perp, \eta_\perp, \eta_\parallel)$, where we have taken $\mathbf{B}$ to be in the z direction. As we will see in Chapters 10 and 11, this is because the magnitude of the resistivity is not the same parallel and perpendicular to a magnetic field. Indeed, these two resistivities differ by about a factor two. This is because the electron distribution becomes significantly distorted from Maxwellian in the case of an electric field parallel to the magnetic field, which can accelerate the electrons relatively freely, thereby reducing the resistivity significantly. Although an accurate knowledge of the resistivity is often important, for example when verifying the consistency of current and voltage measurements in experimental plasmas, there are other resistive plasma phenomena, such as large-scale instabilities, for which factors of two are unimportant, in which case it is sufficient to treat the resistivity as a scalar quantity, as is implied by equation (6.28).

To obtain the *magnitude* of the resistivity of a plasma, it is clearly necessary to know something about the magnitude of the electron–ion collision frequency ν_{ei}, and the distribution function f over which the collision frequency must be averaged. These topics are taken up in Chapters 10 and 11. For present purposes, it is sufficient to note that the resistivity of a plasma can be very small. Indeed, plasmas in fusion experiments can have resistivities lower than that of pure copper, implying that very large currents will be produced by quite small voltage differences. Resistivities of naturally occurring plasmas are generally somewhat higher, but this is more than compensated by the large size of these plasmas so that, again, small electric fields produce large total currents.

Problem 6.3: An applied electric field does work against the ions and electrons of a plasma at the rate $nq\mathbf{u} \cdot \mathbf{E}$ per unit volume per second. By adding over both species, show that an electric field, $E_\parallel$ (parallel to the magnetic field), driving a current, $j_\parallel$, produces resistive heating of a plasma. Do you expect this resistive heating to heat mainly the electrons or the ions?

Chapter 7

Relation between fluid equations and guiding-center drifts

For the purpose of establishing the relationship between the guiding-center drifts discussed in Chapters 2–4 and the fluid equations derived in Chapter 6, it is sufficient to consider a single species of particles and to ignore both collisions with other species and sources and sinks of particles. In this case, we may begin with the standard form of the momentum balance equation:

$$mn \left(\frac{\partial \mathbf{u}}{\partial t} + (\mathbf{u} \cdot \nabla)\mathbf{u} \right) = nq(\mathbf{E} + \mathbf{u} \times \mathbf{B}) - \nabla \cdot P. \tag{7.1}$$

7.1 DIAMAGNETIC DRIFT

If we imagine that the bulk velocities which would be obtained from solving equation (7.1) are of the same order as the guiding-center drift velocities that we obtained in Chapters 2–4 (implying that the flows are far subsonic), then we can say that the order of magnitude of u is roughly $v_t(kr_L)$. Here, v_t is a thermal velocity, r_L is the Larmor radius and k is an inverse distance (or wavenumber) that characterizes the spatial scale-length of variation of the plasma. Time derivatives of any quantity can be estimated as having a size determined from $\partial/\partial t \sim ku \sim v_t k^2 r_L$. As a result, the two terms on the left-hand side in equation (7.1) can be estimated to be of size $mnv_t^2 k^3 r_L^2$. If we assume E to be of order uB, the first two terms on the right-hand side of equation (7.1) are of order $mn\omega_c v_t k r_L = mnkv_t^2$, which is the same order as the last term on the right-hand side, and larger by $(kr_L)^{-2}$ than the terms on the left-hand side. Thus, in the spirit of expanding in kr_L, we can simplify the equation even further:

$$nq(\mathbf{E} + \mathbf{u} \times \mathbf{B}) \approx \nabla \cdot P. \tag{7.2}$$

This approximate equation is valid provided that u is of order $v_t(kr_L)$ and $kr_L \ll 1$. Thus, this equation is *not* valid for near-sonic flows, and we should make the cautionary remark that the flows arising from some of the more violent instabilities in a plasma can, at least in principle, approach sonic speeds; in addition, some laboratory-made and naturally occurring plasmas can have near-sonic equilibrium flows.

For the more usual situation where our approximations (equivalent to those made in our treatment of single-particle motion in Chapters 2–4) are valid, however, we can now solve for the lowest-order perpendicular fluid velocity by taking the cross product of this equation with $\mathbf{B}$:

$$nq[\mathbf{E} \times \mathbf{B} + (\mathbf{u} \times \mathbf{B}) \times \mathbf{B}] = (\nabla \cdot P) \times \mathbf{B}. \tag{7.3}$$

Using the vector identity for the triple vector product (see Appendix D), we obtain

$$nq[\mathbf{E} \times \mathbf{B} - \mathbf{u}B^2 + \mathbf{B}(\mathbf{u} \cdot \mathbf{B})] = (\nabla \cdot P) \times \mathbf{B}. \tag{7.4}$$

If we only consider the components of equation (7.4) perpendicular to $\mathbf{B}$, we obtain

$$\mathbf{u}_\perp = \frac{\mathbf{E} \times \mathbf{B}}{B^2} + \frac{\mathbf{B} \times (\nabla \cdot P)}{nqB^2}. \tag{7.5}$$

The first term on the right-hand side of this equation is clearly the familiar $\mathbf{E} \times \mathbf{B}$ drift. The second term is something new, which is generally referred to as the 'diamagnetic drift'. If we imagine a cylindrical plasma in an essentially uniform magnetic field, with the high pressure in the center of the plasma, it is easy to see that both the electron ($q = -e$) and the ion ($q = e$) diamagnetic drifts as derived here give rise to currents in the plasma that serve to reduce the magnetic field inside the plasma. Hence the name 'diamagnetic' drift. (Note, however, that in a *non-uniform* magnetic field the diamagnetic drift includes a component of guiding-center motion. We will examine this case later in this Chapter.)

Curiously, we did not find a 'diamagnetic' *guiding-center* drift in Chapters 2–4. We did note, however, that the ion and electron Larmor orbits themselves were intrinsically diamagnetic. The diamagnetic drift in a uniform magnetic field is the result of adding together these Larmor orbits in the presence of a density or temperature gradient. The effect of the density gradient is especially easy to see by examining Figure 7.1, which shows the Larmor orbits of positively charged particles (ions) about a magnetic field directed into the paper.

It is clear that, in the shaded area of Figure 7.1, there is a greater current going to the left than to the right, despite the fact that the guiding centers are stationary. This can be made quantitative, starting with the particle picture, by introducing the 'distribution function for guiding centers', in addition to the distribution function for particles, with which we are now familiar.

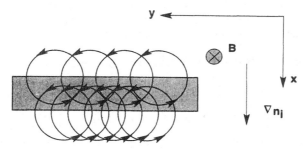

Figure 7.1. Larmor orbits of ions in the presence of a density gradient. In the shaded region there is a net current to the left, even though the guiding centers have no net motion.

To do this, we consider the various contributions to the mean drift in the y direction of particles located in a small element dx of the shaded area in Figure 7.1. If $f(x, \mathbf{v})$ is the distribution function of *particles* with velocity vectors $\mathbf{v}$ located at x, then the mean y-directed drift, u_y, at location x due to particles in a differential length dx is obtained from

$$nu_y dx = \int v_y f(x, \mathbf{v}) d^3 v dx \qquad (7.6)$$

where the integral is over the velocity variables only. Now the particles that are to be found at location x with velocity v_y are those that have guiding centers at a location x_{gc} that is related to x by the equation

$$x = x_{gc} - \frac{v_y}{\omega_c} \qquad (7.7)$$

where ω_c is the cyclotron (Larmor) frequency of the particles. The quantity v_y/ω_c is, in magnitude, closely related to the Larmor radius (see equation (2.9)), but it takes into account also the gyration phase of the particle. It is easy to verify by examination of Figure 7.1 that the signs in equation (7.7) work out correctly (but note carefully the direction of the coordinate axes in Figure 7.1). When a positively charged particle (as illustrated in Figure 7.1) has a positive value of v_y, it is displaced in the negative-x direction (i.e. upward in Figure 7.1) relative to its guiding center. The cyclotron frequency ω_c appearing in equation (7.7) is to be evaluated at the *guiding-center* location but, for the present discussion, we assume that the magnetic field is uniform, so that it does not matter where ω_c is evaluated (see later in this Chapter for the case of a non-uniform field).

We can also define a distribution function of guiding centers, $f_{gc}(x_{gc}, \mathbf{v})$, where $\mathbf{v}$ remains the particle velocity, *not* the guiding-center velocity. Then, the distribution function of particles located within an element dx at x can be expressed in terms of the distribution function of guiding centers located within a differential element dx_{gc} at x_{gc}. Specifically

$$f(x, \mathbf{v})dx = f_{gc}(x_{gc}, \mathbf{v})dx_{gc}$$

$$= f_{gc}\left(x + \frac{v_y}{\omega_c}, \mathbf{v}\right)\frac{dx_{gc}}{dx}dx. \tag{7.8}$$

For our case of a uniform magnetic field, there will be no tendency for particles to be 'crowded' in the x direction any more or less than are their guiding centers. Expressed formally, we can differentiate equation (7.7) with respect to x_{gc}, taking ω_c to be a constant, and we find simply

$$dx = dx_{gc}. \tag{7.9}$$

(Again, see later in this Chapter for the case of a non-uniform field, where this is no longer true.)

Substituting equations (7.8) and (7.9) into equation (7.6), we now have an expression for the mean drift u_y, namely

$$u_y = \frac{1}{n}\int v_y f_{gc}\left(x + \frac{v_y}{\omega_c}, \mathbf{v}\right)d^3v$$

$$\approx \frac{1}{n}\int v_y\left(f_{gc}(x, \mathbf{v}) + \frac{v_y}{\omega_c}\frac{\partial f_{gc}(x, \mathbf{v})}{\partial x}\right)d^3v. \tag{7.10}$$

Taking a Maxwellian distribution of velocities for $f_{gc}(x, \mathbf{v})$, the first term in the integral will vanish, because contributions from positive and negative v_y values will cancel each other. The second term in the integral will give a non-zero contribution. Allowing for both density and temperature gradients, this may easily be evaluated, and it leads to the diamagnetic drift

$$u_y = \frac{1}{nqB}\frac{dp}{dx} \tag{7.11}$$

which is, of course, the same as equation (7.5) for an isotropic pressure and a geometry as shown in Figure 7.1. The pressure could be anisotropic, in which case p would be replaced by $p_\perp$ in equation (7.11).

The diamagnetic drifts of ions and electrons given in equation (7.5) can be summed to form a 'diamagnetic current':

$$\mathbf{j}_d = \sum nq\mathbf{u}_\perp = \mathbf{B} \times [\boldsymbol{\nabla} \cdot (P_i + P_e)]/B^2. \tag{7.12}$$

In Figure 7.1, the diamagnetic current will be to the left, and it will tend to reduce the magnetic field in the higher-density part of the plasma, i.e. below the shaded region. It should not really be surprising that a diamagnetic current exists in a plasma even in the absence of guiding-center motion. Atoms in a piece of iron are essentially tiny 'paramagnets' (as opposed to Larmor orbit 'diamagnets' in a plasma) and, when they are aligned by an external field, they give rise to a net magnetization current. This current exists, of course, without any steady net motion of the iron atoms. (Note again, however, that the diamagnetic current derived in equation (7.12) is identical to that coming from this 'fixed diamagnets' picture only in a uniform magnetic field. As we will see later, for magnetic fields with gradients and curvature, the diamagnetic current includes other contributions—and is not always even divergence-free.)

7.2 FLUID DRIFTS AND GUIDING-CENTER DRIFTS

What might seem strange at this point is the fact that we have not yet found in the fluid picture the familiar ∇B and curvature drifts. At first sight, it appears that the fluid theory is ignoring certain essential physical effects, but this is not so. It is legitimate to think of a plasma as a collection of particles moving according to their guiding-center velocities, and it is also correct to think in the fluid terms that we are now developing. The crucial point is that, while the two approaches are very different, if they are each carried through consistently to the same order, they give the same answer for any observable quantity at that order. In particular, if it is desired to calculate a *fluid* quantity such as a local current density using the guiding-center picture, it is not sufficient just to use the spatial density of guiding centers. It is also necessary to take the correct average of contributions from particles whose guiding centers are separated by a distance of order the Larmor radius, as we did in the previous Section. In all cases, it is necessary to be very careful not to *mix* the two approaches—guiding-center picture and fluid picture—in one calculation. For example, if the guiding-center ∇B drift is added to the fluid diamagnetic drift, the answer will not be physical.

In order to focus on this concept of two independent but correct ways to view a plasma (and in order to find the equivalent of the ∇B and curvature drifts in the fluid picture), it is interesting to consider a specific situation that can be fully analyzed with the tools that we have developed. Consider a plasma confined by a purely θ-directed magnetic field, and which is also finite in the r and z directions, as shown in Figure 7.2. This field B_θ is produced by an external current-carrying conductor, as shown in Figure 7.2, and its strength must decrease with radius like r^{-1}. For simplicity, we suppose that the plasma density and pressure are uniform throughout the main body of the toroidal ('doughnut'-shaped) plasma, falling to zero in a narrow edge layer at the boundary of the plasma.

In this configuration, there are both ∇B and curvature guiding-center drifts. Since these guiding-center drifts are both in the z direction, charge builds up on the 'top' and 'bottom' surfaces of the plasma, giving rise to a z-directed

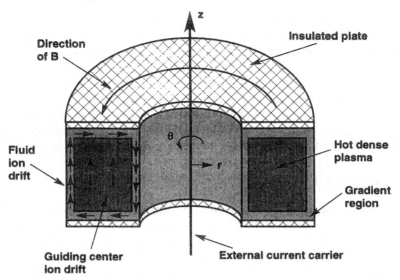

Figure 7.2. Illustration of guiding-center and fluid drifts for the case of a toroidal ('doughnut-shaped') plasma in a purely toroidal θ-directed magnetic field **B**. The plasma pressure is assumed to be uniform in the 'hot dense region', falling to zero in a 'gradient region' at the boundary.

(downward in Figure 7.2) time-dependent **E** field. This **E** field is largely, but not completely, shielded out by the plasma dielectric constant, or—equivalently—the polarization drift. The residual vertical electric field causes the plasma to **E** × **B** drift radially outward.

Let us start by performing the guiding-center drift calculation in detail, considering first the case where $p_\perp = p_\parallel = p$, for simplicity. Then the two guiding-center drifts can be added together and, averaging over a Maxwellian distribution of velocities, the sum is given by

$$\mathbf{v}_{gc} = \frac{m\langle v_\parallel^2 + v_\perp^2/2\rangle}{qB}\frac{\mathbf{B}\times\boldsymbol{\nabla}B}{B^2} = \frac{T_\parallel + T_\perp}{q}\frac{\mathbf{B}\times\boldsymbol{\nabla}B}{B^3} = \frac{2T}{qBr}\hat{\mathbf{z}} \qquad (7.13)$$

where, as usual, $\hat{\mathbf{z}}$ is the unit vector in the z direction. The rate of accumulation of surface charge density σ_s (i.e. charge per unit area) at the top and bottom surfaces of the plasma is just equal to the vertical current density and so, further assuming $n_e \approx n_i \approx n$, we have

$$\frac{d\sigma_s}{dt} = \pm\frac{2n(T_e + T_i)}{rB} = \pm\frac{2p}{r_0 B_0} \qquad (7.14)$$

where the $\pm$ apply to the top and bottom, respectively, of the plasma in the geometry shown in Figure 7.2. Note that rB is a constant (take $rB = r_0 B_0$), since the magnetic field in this geometry must fall off as $1/r$. The accumulation of surface charge due to the *guiding-center* drifts is thus straightforward.

The question to be answered now is: what, in the fluid picture, gives rise to this charge accumulation? Figure 7.2 shows that the *fluid* ion drift (and by analogy the electron drift) is predominantly circulatory in character, and is confined to the region where the ion and electron pressures drop from their constant values in the main body of the plasma to zero at the boundary. However, charge accumulation will arise in the fluid picture if the diamagnetic current is not divergence-free. For simplicity, let us take the fall-off rate in pressure to be a constant, $|\nabla p|$, throughout an edge layer of uniform width all around the plasma boundary. (Since $T_\parallel = T_\perp$, we are dealing with a scalar pressure, and we do not need to consider the full pressure tensor.) The diamagnetic current flowing along the vertical sides of the plasma is constant, and thus divergence-free. However, in the gradient region at the top and bottom of the plasma, while we have taken $|\nabla p|$ to be constant, B is falling off like $1/r$, and a non-zero divergence of the horizontally flowing diamagnetic current, $j_r = \mp|\nabla p|/B = \mp r|\nabla p|/(r_0 B_0)$ arises. We evaluate this divergence of the diamagnetic current to find the rate of accumulation of volume charge density σ:

$$\frac{d\sigma}{dt} = -\nabla \cdot \mathbf{j} = \frac{1}{r}\frac{d}{dr}\left(r^2 \frac{\pm|\nabla p|}{r_0 B_0}\right) = \frac{\pm 2|\nabla p|}{r_0 B_0} \tag{7.15}$$

where the $\pm$ signs again indicate the top and bottom of the plasma. In order to obtain an equivalent surface charge density, in the limit that the fall-off scale-length of the plasma pressure becomes very short, we integrate the volume charge density σ in the vertical direction to obtain the surface charge density σ_s:

$$\frac{d\sigma_s}{dt} = \pm\frac{2p}{r_0 B_0} \tag{7.16}$$

which is exactly the same result as was obtained from the guiding-center drifts (cf equation (7.14)). Thus, the guiding-center and the fluid picture give the same answers for a physically measurable quantity—in this case the surface charge density.

7.3 ANISOTROPIC-PRESSURE CASE

Now suppose that $T_\parallel \neq T_\perp$, and let us see if the correspondence between the guiding-center and the fluid pictures still holds up. In this case, the *guiding-center* drifts indicate that the rate of charge accumulation is given by

$$\frac{d\sigma_s}{dt} = \pm\frac{n(T_{e\parallel} + T_{i\parallel} + T_{e\perp} + T_{i\perp})}{rB} = \pm\frac{(p_\parallel + p_\perp)}{r_0 B_0} \tag{7.17}$$

where the $\pm$ indicates top and bottom of the plasma again.

In the *fluid* picture, if $p_\parallel \neq p_\perp$, we must consider pressure as a tensor. To leading order, the tensor is diagonal (off-diagonal terms, i.e. viscosity, enter at higher order in kr_L). For this geometry, in tensor notation, we have $P = p_\perp \hat{r}\hat{r} + p_\parallel \hat{\theta}\hat{\theta} + p_\perp \hat{z}\hat{z}$, where $\hat{r}, \hat{\theta}$ and $\hat{z}$ are the three unit vectors in the coordinate directions. From our knowledge of the form of the divergence operator it is clear that, in cylindrical coordinates, the meaning of $\nabla \cdot P$ is

$$\nabla \cdot P = \left(\frac{1}{r}\frac{\partial}{\partial r}r\hat{r} + \frac{1}{r}\frac{\partial}{\partial \theta}\hat{\theta} + \frac{\partial}{\partial z}\hat{z} \right) \cdot (p_\perp \hat{r}\hat{r} + p_\parallel \hat{\theta}\hat{\theta} + p_\perp \hat{z}\hat{z})$$

$$= \frac{1}{r}\frac{\partial}{\partial r}(rp_\perp \hat{r}) + \frac{1}{r}\frac{\partial}{\partial \theta}(p_\parallel \hat{\theta}) + \frac{\partial}{\partial z}(p_\perp \hat{z}). \tag{7.18}$$

Notice that this formalism takes into account explicitly the fact that $\hat{\theta}$ is a function of θ. Now, using

$$\hat{r} = \hat{x}\cos\theta + \hat{y}\sin\theta \qquad \hat{\theta} = \hat{y}\cos\theta - \hat{x}\sin\theta \tag{7.19}$$

we easily see that $\partial\hat{\theta}/\partial\theta = -\hat{r}$, giving finally

$$\nabla \cdot P = \frac{\partial p_\perp}{\partial r}\hat{r} + \frac{\partial p_\parallel}{r\partial\theta}\hat{\theta} + \frac{\partial p_\perp}{\partial z}\hat{z} + \frac{(p_\perp - p_\parallel)}{r}\hat{r}. \tag{7.20}$$

(The complete cylindrical-coordinates form for the divergence of a general tensor, including off-diagonal elements, is given in Appendix E.)

The first three terms on the right-hand side are as expected. The last term arises, physically, from proper application of geometry to the basic idea of momentum flux, which was behind the derivation of the pressure tensor. Since in our problem $\partial p_\parallel/\partial\theta$ is zero, this means that particles are streaming into a differential volume carrying the same parallel momentum with which other particles are streaming out. However, since $\partial\hat{\theta}/\partial\theta$ is not zero, this is equivalent to a group of particles 'turning' within the differential volume, thereby *leaving behind some radial momentum*. It is easy to see that the $p_\parallel$ contribution to the last term on the right in equation (7.20) reflects this. The $p_\perp$ contribution to this term is even more basic. If we only consider the perpendicular contribution to the momentum flux into the differential volume, we must take account of the fact that the box has a larger side at large r than at small r, so there is a net flux of radial momentum out of the differential volume across the large-r surface, even for uniform $p_\perp$. For $p_\perp = p_\parallel$ these two effects cancel, as they must, since there can be no net divergence of momentum flux for an everywhere-Maxwellian uniform-pressure plasma.

We are still not exactly where we want to be. Our previous calculation of the divergence of the diamagnetic current depended only on the z component of

∇p, with p taken as a scalar. From equation (7.20), it is clear that this part of the derivation will go through unchanged when $\nabla \cdot P$ replaces ∇p, and it will produce a surface charge-density build-up as given by equation (7.16), except that p will be replaced by $p_\perp$. Now, however, we have a new force density (a divergence of momentum flux), $(p_\parallel - p_\perp)/r$ in the r direction, which we must include. Substituting this part of equation (7.20) into equation (7.5), we obtain a new part of the perpendicular *fluid* drift

$$\mathbf{u}_\perp = \frac{(p_\parallel - p_\perp)\hat{\mathbf{r}} \times \mathbf{B}}{nqr B^2} \qquad (7.21)$$

which gives rise to a vertical current, and a rate of charge accumulation

$$\frac{d\sigma_s}{dt} = \pm \frac{p_\parallel - p_\perp}{r_0 B_0} \qquad (7.22)$$

with the $\pm$ sign for top and bottom, respectively. As stated above, our derivation of the divergence of the diamagnetic current, which led to equation (7.16), is simply modified in the case of a tensor pressure by using $p_\perp$ rather than p. Adding this contribution to the charge accumulation to that given in equation (7.22), we obtain altogether

$$\frac{d\sigma_s}{dt} = \pm \frac{2p_\perp}{r_0 B_0} \pm \frac{p_\parallel - p_\perp}{r_0 B_0} = \pm \frac{p_\perp + p_\parallel}{r_0 B_0} \qquad (7.23)$$

which is consistent with equation (7.17), again demonstrating the equivalence of the guiding-center and fluid pictures.

Problem 7.1: Applied to the isotropic and uniform-pressure core of the plasma shown in Figure 7.2, our results from the fluid picture mean that, despite the ∇B and curvature drifts, there is no net current in this region. What happens in the case of an anisotropic-pressure plasma, with uniform but unequal $p_\perp$ and $p_\parallel$? Is there a net current in this case? *Using the fluid picture*, calculate its magnitude.

7.4 DIAMAGNETIC DRIFT IN NON-UNIFORM B FIELDS*

Our discussion of guiding-center versus fluid drifts has not fully explained why there are no net currents in the isotropic and uniform-pressure core of the plasma shown in Figure 7.2 due to the ∇B and curvature drifts. The fluid picture clearly predicts that the current vanishes, because the diamagnetic drift given in equation (7.5) is zero for an isotropic, uniform-pressure plasma. On the other hand, the particle picture has non-vanishing ∇B and curvature drifts. For an isotropic pressure, we will see that the net current properly derived from the guiding-center picture is also zero, because the volume currents due to the ∇B and curvature drifts are exactly cancelled by additional order-kr_L terms introduced by going from the guiding-center density to the actual particle density in the averaging procedure illustrated in Figure 7.1. The net volume current is a physically measurable quantity, and must be the same in the two pictures.

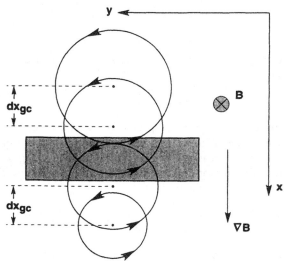

Figure 7.3. Larmor orbits of ions in the presence of a field gradient. For guiding centers with equal spacing dx_{gc}, there are more particles with $v_y < 0$ falling in the shaded region, and fewer particles with $v_y > 0$, leading to a net current to the right.

To see this, we generalize Figure 7.1 to the case where B has a gradient, which we take to be in the x direction. (For the time being, however, we will assume that the field has no curvature.) The new case is illustrated in Figure 7.3, which shows that the Larmor radii of particles with guiding centers above the shaded region are larger than those of particles with guiding centers below the shaded region. A particle's instantaneous position x is still related to its guiding center x_{gc} by equation (7.7), but it now becomes important that ω_c is to be evaluated at the particle's average position, i.e. at its guiding center. Differentiating equation (7.7), we obtain

$$dx = \left(1 + \frac{v_y}{\omega_c}\frac{1}{B}\frac{dB}{dx}\right) dx_{gc}. \tag{7.24}$$

If B increases with x, as in Figure 7.3, the particles with $v_y > 0$ in the shaded region (guiding centers below the shaded region) are *less densely packed* than are the guiding centers themselves, i.e. $dx > dx_{gc}$. Correspondingly, particles with $v_y < 0$ are *more* densely packed than are their guiding centers. The result will be a preponderance of negative-v_y particles, leading to an average drift in the negative-y direction, even in the case of a completely uniform plasma in which the guiding centers themselves are uniformly spaced. Figure 7.3 shows a case where, for guiding centers that are uniformly spaced in the x direction, there are two particles with $v_y < 0$ falling in the shaded region (guiding centers above the shaded region) for every one particle with $v_y > 0$ (guiding center below the shaded region). This average drift is in addition to any drift of the guiding centers themselves.

Equation (7.24) replaces equation (7.9) and may be substituted into equation (7.8) to give

$$f(x, \mathbf{v}) = f_{gc}\left(x + \frac{v_y}{\omega_c}, \mathbf{v}\right)\left(1 + \frac{v_y}{\omega_c}\frac{1}{B}\frac{dB}{dx}\right)^{-1}$$

$$\approx \left(f_{gc} + \frac{v_y}{\omega_c}\frac{df_{gc}}{dx}\right)\left(1 - \frac{v_y}{\omega_c}\frac{1}{B}\frac{dB}{dx}\right)$$

$$\approx f_{gc} + \frac{v_y}{\omega_c}\frac{df_{gc}}{dx} - \frac{v_y}{\omega_c}\frac{1}{B}\frac{dB}{dx}f_{gc}. \tag{7.25}$$

The average velocity u_y is given by

$$u_y = \frac{1}{n}\int v_y f(x, \mathbf{v})\mathrm{d}^3 v. \tag{7.26}$$

Substituting for $f(x, \mathbf{v})$ from equation (7.25) and noting that only the terms quadratic in v_y will survive in the integration, we obtain

$$u_y = \frac{1}{nqB}\frac{dp}{dx} - \frac{T}{qB^2}\frac{dB}{dx} \tag{7.27}$$

where we have assumed a Maxwellian distribution, so that we can write $\langle v_y^2 \rangle = T/m$. Equation (7.27), by construction, gives the average velocity ignoring any guiding-center velocity, i.e. *in the frame in which the guiding centers have no motion.*

For this geometry, the ∇B drift is also in the y direction and is given by

$$v_{Dy} = \frac{\langle v_\perp^2/2 \rangle}{\omega_c B}\frac{dB}{dx} = \frac{T}{qB^2}\frac{dB}{dx} \tag{7.28}$$

where we have again averaged over Maxwellian particles, writing $\langle v_\perp^2/2 \rangle = T/m$.

Thus, if we add the average guiding-center velocity, equation (7.28), to the average velocity in the frame in which the guiding centers are at rest, equation (7.27), we recover, for the total fluid velocity, the diamagnetic drift; in this case

$$u_y = \frac{1}{nqB}\frac{dp}{dx}. \tag{7.29}$$

Equation (7.29) applies to both ions and electrons. Thus, the current perpendicular to a magnetic field in a plasma arises only from the diamagnetic drift, i.e. it requires a gradient in the plasma pressure. There is no current in the uniform-pressure core of the plasma shown in Figure 7.2, despite the ∇B guiding-center drifts that carry ions upward and electrons downward. The ∇B drifts are cancelled by additional terms that arise from the field gradient in the averaging process used to obtain the fluid velocity.

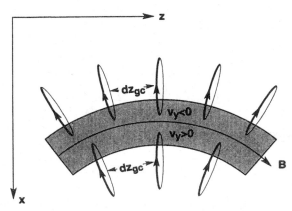

Figure 7.4. Larmor orbits of ions in the presence of field curvature. For guiding centers with equal spacing dz_{gc}, the particles with $v_y < 0$ are more crowded than those with $v_y > 0$, leading to a net current in the negative-y direction.

The calculation that we have just given treats the case of a field *gradient*, but not the case of field *curvature*. If the **B** field is curved, Figures 7.1 and 7.3 are not very useful, since they do not show the essential effect, which is that, if the guiding centers are equally spaced in the z direction, the particles themselves are unequally spaced in z in the shaded region. Figure 7.4 is an end view of Figure 7.1, looking from the right, for the case of a field that is concave downward. For this case, we see that the particles with $v_y < 0$ in the shaded region are more densely spaced than those with $v_y > 0$. Thus, the average drift will be in the negative-y direction.

To analyze this case quantitatively, we start by observing that the spacing in the z direction of particles in the shaded region, relative to the spacing of the particles' guiding centers, is given by

$$dz = \frac{R_c}{R_c - v_y/\omega_c} dz_{gc} \tag{7.30}$$

where R_c is the radius of curvature. Generalizing equation (7.8), the distribution function for particles will be related to the distribution function of guiding centers by

$$f(x, \mathbf{v})dx dz = f_{gc}(x_{gc}, \mathbf{v})dx_{gc}dz_{gc}$$

$$= f_{gc}\left(x + \frac{v_y}{\omega_c}, \mathbf{v}\right) \frac{dx_{gc}}{dx} \frac{dz_{gc}}{dz} dx dz. \tag{7.31}$$

The new effect will arise from

$$\frac{dz_{gc}}{dz} \approx 1 - \frac{v_y}{R_c \omega_c}. \tag{7.32}$$

The calculation now proceeds as before, and the average velocity in the guiding-center rest-frame is again obtained by substituting for $f(x, \mathbf{v})$ from equation (7.31) into equation (7.26). The additional term in the fluid velocity is now found to *precisely* cancel the average curvature drift in an isotropic plasma, which for this geometry is given by

$$v_{Dy} = \frac{\langle v_\parallel^2 \rangle}{\omega_c R_c}. \tag{7.33}$$

The details are left as an exercise (see Problem 7.2). In an anisotropic plasma ($p_\parallel \neq p_\perp$), this cancellation does not occur, but the net drift calculated in this way from the particle picture exactly equals the fluid drift for the anisotropic case, i.e. equation (7.21) (see again Problem 7.2).

It should be noted that these currents associated with finite gyro-radius, but stationary guiding centers, are of necessity divergence-free, so they do not affect the previous guiding-center calculation of charge accumulation. For the plasma shown in Figure 7.2, these currents are uniform and vertically directed in the core plasma.

Problem 7.2: Complete the calculation of the diamagnetic drift in an isotropic plasma for a curved magnetic field, showing that the average velocity in the guiding-center frame, obtained from equations (7.26), (7.31) and (7.32), contains an additional term from field curvature that exactly cancels the average guiding-center curvature drift. For an anisotropic plasma, in which both $p_\parallel$ and $p_\perp$ are uniform, show from this guiding-center picture that there is a net drift similar to that given in equation (7.21).

7.5 POLARIZATION CURRENT IN THE FLUID MODEL

The next question to be addressed is how to obtain the plasma dielectric effect in the fluid model. We derived this from guiding-center motion, and it is crucial for evaluating the rate of change of the electric field due to the charge build-up at the top and bottom of the torus in the problem illustrated in Figure 7.2 and discussed in the previous three Sections. Can we obtain the dielectric effect—or equivalently the polarization drift—from the fluid equations?

Once again, we must apply our ideas about ordering procedures. In the presence of an electric field, in lowest order, we obtain a simple $\mathbf{E} \times \mathbf{B}$ drift by balancing the dominant terms in equation (7.1):

$$\mathbf{u}_E = \frac{\mathbf{E} \times \mathbf{B}}{B^2}. \tag{7.34}$$

(For simplicity, we will ignore $\nabla \cdot P$ and any other spatial non-uniformity in this derivation of the polarization drifts.) Using the method of successive approximation, we simply substitute this drift velocity, assumed first-order in kr_L, into the small term on the left-hand side of equation (7.1). We then obtain an equation for the second-order correction to the velocity, which we will for now call $\mathbf{u}_p$:

$$mn\dot{\mathbf{u}}_E = nq\mathbf{u}_p \times \mathbf{B} \tag{7.35}$$

or

$$\frac{mn\dot{\mathbf{E}} \times \mathbf{B}}{B^2} = nq\mathbf{u}_p \times \mathbf{B}. \tag{7.36}$$

This equation says nothing about the parallel component of $\mathbf{u}_p$, but the perpendicular component has a unique solution

$$\mathbf{u}_p = \frac{m\dot{\mathbf{E}}_\perp}{qB^2} = \pm\frac{\dot{\mathbf{E}}_\perp}{\omega_c B} \tag{7.37}$$

where $\pm$ now stands for the sign of q. Thus we can identify $\mathbf{u}_p$ as the low-frequency 'polarization drift' that was calculated in Section 4.3, and so the previous arguments constructing a plasma dielectric constant go through exactly as before. Once again, the fluid calculation, carried out to the appropriate order, gives the same result as the guiding-center calculation for any physically measurable quantity.

Problem 7.3: Working in the guiding-center drift formulation, calculate the outward acceleration of the plasma shown in Figure 7.2 due to the $\mathbf{E} \times \mathbf{B}$ drift that is created by the combination of ∇B, curvature and polarization drifts. Consider both isotropic and anisotropic pressure cases. In the anisotropic-pressure case, can you come up with a physical reason why parallel energy density is twice as effective as perpendicular energy density in driving outward acceleration? (Hint: think of conservation of angular momentum and of magnetic moment μ as constraints, which determine the energy available to the system as it moves outwards.)

7.6 PARALLEL PRESSURE BALANCE

We have now demonstrated a correspondence between each of the terms in the *perpendicular* components of equation (7.1) and one or more of the guiding-center drifts. Before leaving the fluid equation of motion for an individual species of particles, it is useful to consider the physical content of the *parallel* component of equation (7.1).

In the scalar pressure situation, if we assume zero flow velocity and slow time derivatives, we have in dominant order

$$nqE_\parallel = \nabla_\parallel p \qquad (7.38)$$

or

$$nq\nabla_\parallel\phi + \nabla_\parallel p = 0 \qquad (7.39)$$

where ϕ is the electrostatic potential. If we assume that the parallel thermal conductivity is very rapid, so that the temperature T is constant along a field line, we can write

$$nq\nabla_\parallel\phi + T\nabla_\parallel n = 0 \qquad (7.40)$$

or equivalently

$$\ln n + q\phi/T = \text{constant.} \qquad (7.41)$$

Taking the exponential of both sides of equation (7.41), we obtain a relation for the variation of the density along a field line in equilibrium:

$$n \propto \exp(-q\phi/T). \qquad (7.42)$$

This is just the Boltzmann relation for a system in contact with a heat bath, which we derived from fundamental principles of statistical mechanics in Chapter 1. We see here how the same result follows from our fluid equations.

Obviously, both electrons and ions cannot simultaneously be in Boltzmann equilibrium in the presence of an electric potential that varies along the field lines, or else charge neutrality would be violated in the absence of some externally introduced charge. The only charge-neutral equilibrium under normal circumstances is one in which the electric potential and ion and electron densities are constant along the field lines. However, if a density variation, say a density 'hump', is created dynamically along a field line in a charge-neutral plasma, electron and ion flow velocities parallel to the magnetic field will arise. The larger mass of the ions results in them responding relatively slowly to the presence of the 'hump' in n (and therefore a non-zero $\nabla_\parallel p_i$) along a field line, with the characteristic time-scale being set by some $\tau \approx L/(T/m_i)^{1/2}$. Meanwhile the lighter electrons respond much more quickly, $\tau \approx L/(T/m_e)^{1/2}$, and rapidly set up a Boltzmann distribution in the presence of the density hump. This means that the electron force balance dictates that

$$\nabla_\parallel p_e = e n_e \nabla_\parallel \phi \qquad (7.43)$$

since the electrons come to equilibrium on a time-scale much faster than the ions. In a plasma with strong parallel thermal conductivity (and therefore uniform T_e), this implies that

$$n_e \propto \exp(e\phi/T_e) \qquad (7.44)$$

which determines the variation in electric potential ϕ that will arise:

$$\phi \propto (T_e/e)\ln n_e. \qquad (7.45)$$

When this electric field has been created, the electrons are in force balance. Provided the scale-length of the density 'hump' is much larger than the Debye length, charge-neutrality will be maintained by n_e remaining almost equal to n_i.

Problem 7.4: Suppose a small varying electric potential $\phi(x) = \phi_1 \sin kx$ is created in an initially uniform, neutral plasma ($e\phi_1 \ll T_e$). Show that the electrons will come to equilibrium with $n_e(x) = n_0 + n_{e1}\sin kx$ where $n_{e1}/n_0 = e\phi_1/T_e$. Using Poisson's equation, show that the ion density will be given by $n_i(x) = n_0 + n_{i1}\sin kx$, where $(n_{i1} - n_{e1})/n_{e1} = k^2 \lambda_D^2$.

Now, examining the ion equation of motion parallel to the field, we see that the electric field pulls the ions in the same direction as their own pressure gradient pushes them, tending to cause the ions to smooth out the original 'hump' more rapidly. In effect, the electrons contribute their pressure gradient to the force on the ions, via the Boltzmann electric field. The parallel force-balance equation for the ions, on the time-scale required for them to respond, becomes

$$m_i n_i \dot{u}_{i\parallel} = -n_i e \nabla_\parallel \phi - \nabla_\parallel p_i = -T_e \nabla_\parallel n_e - \nabla_\parallel p_i \approx -(T_e + T_i)\nabla_\parallel n \qquad (7.46)$$

the last step being valid if the ion temperature is also smoothed out along the field.

An important feature of this result is that the electric field is derived from the electron Boltzmann relation, given the density perturbation. We assumed, for time-scales where the electrons had plenty of time to come into equilibrium, that they would immediately set up force balance along the field line, thereby satisfying the Boltzmann relation without delay. An alternative way to work this problem would have been to solve for the electron dynamics as well as for the ion dynamics, and then to calculate the very small charge separation, $e(n_i - n_e)$, and use Poisson's equation to find $E_\parallel$. This would have been much more cumbersome, and insignificantly more accurate if we are only interested in the time-scale at which the overall density hump relaxes via ion motion and so long as the spatial scale-length greatly exceeds the Debye length. In this situation, we say that 'the electron inertia is negligible'. We will find that this trick of circumventing Poisson's equation via the Boltzmann relation for the electrons is of great use in many problems of 'low-frequency' plasma dynamics.

Chapter 8

Single-fluid magnetohydrodynamics

We are now in a position to formulate a 'single-fluid' model of a fully ionized plasma, in which the plasma is treated as a single hydrodynamic fluid acted upon by electric and magnetic forces. This is called the 'magnetohydrodynamic' (MHD) model. The attraction of this model, relative to the more complex 'two-fluid' models, is that it provides a somewhat more tractable set of equations while retaining much of the important physics. Historically, this was one of the earliest plasma models to be developed and used, because it allowed application of many of the techniques of ordinary hydrodynamics to the plasma case, even though a plasma is much more complicated because of the variety of electric and magnetic forces that are possible.

8.1 THE MAGNETOHYDRODYNAMIC EQUATIONS

We will limit our derivation of the equations of magnetohydrodynamics to the case of a hydrogen plasma, in which the ions and electrons have charges $\pm e$, respectively. We will also assume that charge neutrality is at least approximately satisfied, so that $n_i \approx n_e \approx n$, but we will allow the possibility of a *small* non-vanishing charge density. The final equations, however, apply just as well to the case of a plasma in which the ions are multiply charged, i.e. have charges Ze, in which case the charge-neutrality condition will be $n_e \approx Zn_i$. The assumption of approximate charge neutrality will be valid whenever the spatial scale-lengths of the phenomena of interest greatly exceed the Debye length. We will here denote the electron and ion masses by m and M respectively.

The magnetohydrodynamic model treats the plasma as a single fluid with mass density

$$\rho = n_i M + n_e m \approx n(M + m) \approx nM \tag{8.1}$$

charge density

$$\sigma = (n_i - n_e)e \tag{8.2}$$

mass velocity

$$\mathbf{u} = (n_i M \mathbf{u}_i + n_e m \mathbf{u}_e)/\rho \approx (M \mathbf{u}_i + m \mathbf{u}_e)/(M + m) \approx \mathbf{u}_i + (m/M) \mathbf{u}_e \quad (8.3)$$

and current density

$$\mathbf{j} = e(n_i \mathbf{u}_i - n_e \mathbf{u}_e) \approx ne(\mathbf{u}_i - \mathbf{u}_e). \quad (8.4)$$

These may be solved to obtain expressions for $\mathbf{u}_i$ and $\mathbf{u}_e$ in terms of $\mathbf{u}$ and $\mathbf{j}$:

$$\mathbf{u}_i \approx \mathbf{u} + \frac{m}{M} \frac{\mathbf{j}}{ne} \qquad \mathbf{u}_e \approx \mathbf{u} - \frac{\mathbf{j}}{ne} \quad (8.5)$$

where we have dropped terms that are unambiguously small in m/M.

The single-fluid magnetohydrodynamic equations can be obtained by taking various linear combinations of the individual ion and electron equations. In particular, the two individual continuity equations

$$\frac{\partial n_{i,e}}{\partial t} + \boldsymbol{\nabla} \cdot (n_{i,e} \mathbf{u}_{i,e}) = 0 \quad (8.6)$$

may be multiplied by the ion and electron masses M and m, respectively, and added together to produce a 'mass continuity equation':

$$\frac{\partial \rho}{\partial t} + \boldsymbol{\nabla} \cdot (\rho \mathbf{u}) = 0. \quad (8.7)$$

The individual continuity equations may be subtracted from one another, to produce the 'charge continuity equation':

$$\frac{\partial \sigma}{\partial t} + \boldsymbol{\nabla} \cdot \mathbf{j} = 0. \quad (8.8)$$

In a similar way, the two individual momentum balance equations, which we will tend to refer to here as the individual fluid 'equations of motion'

$$\begin{aligned}
M n_i \frac{d\mathbf{u}_i}{dt} &= e n_i (\mathbf{E} + \mathbf{u}_i \times \mathbf{B}) - \boldsymbol{\nabla} p_i + \mathbf{R}_{ie} \\
m n_e \frac{d\mathbf{u}_e}{dt} &= -e n_e (\mathbf{E} + \mathbf{u}_e \times \mathbf{B}) - \boldsymbol{\nabla} p_e + \mathbf{R}_{ei}
\end{aligned} \quad (8.9)$$

(where $\mathbf{R}_{ie}$ and $\mathbf{R}_{ei}$ describe collisional transfer of momentum between the two species) may be added together, to produce the combined 'single-fluid equation of motion':

$$\rho \frac{d\mathbf{u}}{dt} = \rho \left(\frac{\partial \mathbf{u}}{\partial t} + \mathbf{u} \cdot \boldsymbol{\nabla} \mathbf{u} \right) = \sigma \mathbf{E} + \mathbf{j} \times \mathbf{B} - \boldsymbol{\nabla} p \quad (8.10)$$

where $p = p_e + p_i$ is the total pressure. Additional non-electromagnetic forces, e.g. gravitational forces ρg, may be included on the right-hand side of the single-fluid equation of motion, if necessary. In adding the two individual fluid equations of motion, the collision terms cancel each other, since $\mathbf{R}_{ei} = -\mathbf{R}_{ie}$. Although equations (8.9) and (8.10) have assumed that the electron and ion pressures are isotropic, this is not essential to the magnetohydrodynamic model; indeed, there are important cases where the electron pressure is isotropic, whereas the ion pressure, because of the larger ion Larmor orbits, must be taken as a tensor.

Strictly, the convected derivative terms $\mathbf{u} \cdot \nabla \mathbf{u}$ in the individual fluid equations of motion, being nonlinear in $\mathbf{u}$, do not add together as simply as we have suggested. Moreover, the individual-species pressures were defined in terms of the random motion of particles of each species relative to the species' own mean velocity. Thus, again, the two individual pressure gradients do not add together so simply, because there is an ambiguity about which mean velocity the random motions are measured against. This difficulty can be removed by redefining the pressure of each species in terms of random motion about the *mass* velocity. Equation (8.10), where $\mathbf{u}$ is this mass velocity and ∇p involves the pressure defined in this way, then becomes *exactly* correct, including the convective derivative. In practice, however, the mass velocity of a plasma is generally dominated by the ions, being much heavier than the electrons, so there is no distinction between $\mathbf{u}$ and the ion mean velocity $\mathbf{u}_i$. Moreover, the electron random motions are so rapid compared with any mean velocity that it does not matter which mean velocity is used in the definition of electron pressure. It is evident that equation (8.10) is valid in this approximate sense also, where $\mathbf{u}$ in both terms on the left-hand side is interpreted as the ion velocity.

To obtain a second single-fluid equation from the two individual fluid equations of motion, we must invoke two approximations. First, we must express the momentum transfer from ions to electrons in terms of the velocity difference and average collision frequency (or, equivalently, the resistivity η), as we already did in Chapter 6, namely

$$\mathbf{R}_{ei} = mn\langle \nu_{ei}\rangle(\mathbf{u}_i - \mathbf{u}_e) = \eta n^2 e^2(\mathbf{u}_i - \mathbf{u}_e) = \eta n e \mathbf{j}. \tag{8.11}$$

Second, we must *neglect electron inertia entirely*. This will be valid for phenomena that are sufficiently slow that electrons have time to reach dynamical equilibrium in regard to their motion along the magnetic field. With these two approximations, the single-fluid *electron* equation of motion may be rewritten

$$\mathbf{E} + \mathbf{u}_e \times \mathbf{B} = \eta \mathbf{j} - \frac{\nabla p_e}{ne} \tag{8.12}$$

or

$$\mathbf{E} + \mathbf{u} \times \mathbf{B} = \eta \mathbf{j} + \frac{\mathbf{j} \times \mathbf{B} - \nabla p_e}{ne} \tag{8.13}$$

where, in obtaining the second form, we have substituted from equation (8.5) for $\mathbf{u}_e$ in terms of $\mathbf{j}$ and $\mathbf{u}$. Equation (8.13) is usually called the 'generalized Ohm's law' for a plasma. If it is important to retain the distinction between resistivity perpendicular and parallel to a magnetic field, then the scalar η can be replaced by a diagonal tensor, with diagonal elements $\eta_\perp, \eta_\perp$ and $\eta_\parallel$, for the case where the magnetic field is in the z direction.

To provide a complete set of equations, some kind of 'equation of state' must be added to describe how the plasma pressure p changes in time. As we have seen in Chapter 6, the *adiabatic law* is often assumed, i.e.

$$\frac{\mathrm{d}}{\mathrm{d}t}\left(\frac{p}{\rho^\gamma}\right) = 0 \tag{8.14}$$

but the isothermal law, $p = n(T_e + T_i)$ with $T_e, T_i = $ constant, provides an alternative model that is sometimes more appropriate. For magnetohydrodynamic phenomena that are rapid compared to collisions, the plasma pressure may become significantly anisotropic, in which case the *double adiabatic* laws introduced in Chapter 6 are employed.

The system of equations is closed by including the four Maxwell equations:

$$\nabla \times \mathbf{B} = \mu_0 \mathbf{j} + \frac{1}{c^2}\frac{\partial \mathbf{E}}{\partial t} \tag{8.15}$$

$$\nabla \times \mathbf{E} = -\frac{\partial \mathbf{B}}{\partial t} \tag{8.16}$$

$$\nabla \cdot \mathbf{B} = 0 \tag{8.17}$$

$$\nabla \cdot (\epsilon_0 \mathbf{E}) = \sigma. \tag{8.18}$$

In these equations, we consider the plasma polarization current as external, so we use the vacuum form for the permittivity ϵ. Equations (8.7), (8.8), (8.10), and (8.13)–(8.18) constitute a full set of single-fluid equations for a plasma treated as an electrically conducting fluid. We will now examine various limiting forms of these equations.

8.2 THE QUASI-NEUTRALITY APPROXIMATION

So far, we have retained in our equations certain terms describing the effects of a non-zero charge density σ. In particular, we have retained the electric force $\sigma\mathbf{E}$ in the equation of motion and the charge separation $\partial\sigma/\partial t$ in the charge continuity equation. Often, neither of these terms is very important.

To see this, we adopt an estimation procedure that compares the size of the term in question with the size of another term (anticipated to be more important) in the same equation. For example, we compare the size of the electric force

$\sigma\mathbf{E}$ to the size of the inertial term $\rho\mathbf{u}\cdot\nabla\mathbf{u}$ in the equation of motion using the Maxwell equation $\nabla\cdot(\epsilon_0\mathbf{E}) = \sigma$ to estimate σ: we find

$$\frac{\sigma E}{\rho\mathbf{u}\cdot\nabla\mathbf{u}} \sim \frac{\epsilon_0 E^2/L}{\rho u^2/L} \sim \frac{\epsilon_0 E^2}{\rho u^2} \sim \frac{\epsilon_0 B^2}{\rho} \qquad (8.19)$$

where we have introduced a characteristic length scale L and have assumed that the plasma velocity $\mathbf{u}$ is of order $\mathbf{E}\times\mathbf{B}/B^2 \sim E/B$. The dimensionless parameter $\epsilon_0 B^2/\rho$ is a small quantity in almost all plasmas of interest. (Equivalently, the plasma 'dielectric constant' $1 + \rho/\epsilon_0 B^2$ is large—usually 10^2–10^3.) Thus, the electric force is negligible.

The charge separation $\partial\sigma/\partial t$ in the charge continuity equation may be estimated in a similar way, by comparing its size to the size of the term $\nabla\cdot\mathbf{j}$. We find

$$\frac{\partial\sigma/\partial t}{\nabla\cdot\mathbf{j}} \sim \frac{\epsilon_0 E/L\tau}{j/L} \sim \frac{\epsilon_0 u B/\tau}{\rho u/B\tau} \sim \frac{\epsilon_0 B^2}{\rho} \qquad (8.20)$$

where we have introduced a characteristic time-scale τ and have estimated $\mathbf{j}$ by equating the magnetic force $\mathbf{j}\times\mathbf{B}$ to the inertial term $\rho\partial\mathbf{u}/\partial t$. Since this is again a small quantity and yet the terms in the numerator and denominator must, by equation (8.8), be equal and opposite to each other, it follows that $\mathbf{j}$ must be almost divergence-free, in which case the denominator is greatly overestimated in equation (8.20).

In the limit $\rho/\epsilon_0 B^2 \gg 1$, the terms $\sigma\mathbf{E}$ and $\partial\sigma/\partial t$ may both be omitted from their respective equations. This is known as the 'quasi-neutrality approximation'. It is important to note that the quasi-neutrality approximation does *not* mean that the charge density σ can be omitted from the Maxwell equation $\nabla\cdot(\epsilon_0\mathbf{E}) = \sigma$. What it means is that this Maxwell equation serves to *define* the magnitude of σ, which turns out to be *too small to be of importance elsewhere*. This particular Maxwell equation can then be dropped from the system, since σ does not appear anywhere else. We only use it when we want to evaluate σ. However, it is *not* correct to write $\nabla\cdot(\epsilon_0\mathbf{E}) = 0$. In particular, there will often be electric fields in the plasma, with finite divergence, such as those perpendicular to a magnetic field that define the perpendicular plasma motion.

Problem 8.1: A similar estimation procedure leads to the conclusion that the displacement current can also be omitted from the Maxwell equations as part of the same quasi-neutrality approximation. Show this.

Usually the term 'magnetohydrodynamics' *refers to the case where the quasi-neutrality approximation is invoked.* In this limit, the term $\partial\sigma/\partial t$ may be omitted from equation (8.8), the electric force $\sigma\mathbf{E}$ may be omitted from equation (8.10), and the displacement current $\partial(\epsilon_0 \mathbf{E})/\partial t$ may be omitted from equation (8.15).

8.3   THE 'SMALL LARMOR RADIUS' APPROXIMATION

We will now show that the second and third terms on the right-hand side in the generalized Ohm's law, equation (8.13), are negligible in a special example of the 'small Larmor radius approximation'. We adopt the same estimation procedure used above, i.e. we compare the size of the term anticipated to be unimportant with the size of another term in the same equation thought to be more important. In a typical plasma-dynamical situation, the fluid motion **u** will be driven by pressure gradients and by magnetic forces and, in the limit where the motion is fully developed, for example as a result of a strong magnetohydrodynamic instability, the fluid velocity will reach an order-of-magnitude given by

$$\rho \mathbf{u} \cdot \nabla \mathbf{u} \sim \nabla p \sim \mathbf{j} \times \mathbf{B}. \tag{8.21}$$

The $\mathbf{E} \times \mathbf{B}$ drift, which will provide the dominant contribution to the fluid velocity **u**, is now taken to be much larger than the other drifts (such as the diamagnetic drift) unlike many of the cases considered in previous Chapters; these large fluid velocities are the result of large unbalanced forces or strong instabilities.

Writing $p = nT$ and $\rho = nM$, we obtain for this case $u \sim v_{\mathrm{t,i}} \sim (T/M)^{1/2}$, the ion thermal velocity. Of course, not all plasma-dynamical phenomena produce fluid velocities as large as the ion thermal speed. However, in fully developed magnetohydrodynamic flows, where unbalanced ∇p and $\mathbf{j} \times \mathbf{B}$ forces are reacted only by the inertia of the plasma, the plasma fluid velocity can, and does, attain such large values. It is in such situations that the 'small Larmor radius' approximation can be appropriate; it is usually not valid for weaker magnetohydrodynamic phenomena, for example phenomena where the fluid velocity, i.e. the $\mathbf{E} \times \mathbf{B}$ drift, is not larger than the diamagnetic drift or the guiding-center ∇B or curvature drifts. We will use the relation $u \sim v_{\mathrm{t,i}}$ to estimate the magnitude of the second and third terms on the right-hand side in the generalized Ohm's law, equation (8.13), compared to one of the terms on the left-hand side, in particular the term $\mathbf{u} \times \mathbf{B}$. Starting with the last term on the right-hand side in equation (8.13), the generalized Ohm's law, we find

$$\frac{\nabla p_e/ne}{\mathbf{u} \times \mathbf{B}} \sim \frac{T}{euBL} \sim \frac{Mv_{\mathrm{t,i}}^2}{euBL} \sim \frac{Mv_{\mathrm{t,i}}}{eBL} \sim \frac{v_{\mathrm{t,i}}}{\omega_{\mathrm{ci}}L} \sim \frac{r_{\mathrm{Li}}}{L} \tag{8.22}$$

where r_{Li} is the ion Larmor radius. Since $\mathbf{j} \times \mathbf{B} \sim \nabla p$ (for the most general case where both pressure gradients and magnetic forces are about equally important in driving the fluid motion), a similar estimation of the second term on the right-hand side of equation (8.13) will give exactly the same result. We conclude that the second and third terms on the right in equation (8.13) can be neglected *if the ion Larmor radius is very small compared to the scale-length of the fluid motion*, i.e. $r_{Li}/L \ll 1$ and we are considering fluid velocities of order $v_{t,i}$. Treatments that take these additional terms into account are often called 'finite Larmor radius' treatments.

In the 'small Larmor radius' approximation, the Ohm's law is simply

$$\mathbf{E} + \mathbf{u} \times \mathbf{B} = \eta \mathbf{j}. \tag{8.23}$$

Magnetohydrodynamics (often abbreviated to 'MHD') usually refers to the set of equations with this 'simple' Ohm's law replacing the 'generalized' Ohm's law. Summarizing the other 'MHD equations', they are:

$$\begin{aligned}
&\frac{\partial \rho}{\partial t} + \boldsymbol{\nabla} \cdot (\rho \mathbf{u}) = 0 \\
&\boldsymbol{\nabla} \cdot \mathbf{j} = 0 \\
&\rho \frac{d\mathbf{u}}{dt} = -\boldsymbol{\nabla} p + \mathbf{j} \times \mathbf{B}
\end{aligned} \tag{8.24}$$

together with the required versions of the Maxwell equations:

$$\begin{aligned}
&\boldsymbol{\nabla} \times \mathbf{B} = \mu_0 \mathbf{j} \\
&\boldsymbol{\nabla} \times \mathbf{E} = -\frac{\partial \mathbf{B}}{\partial t} \\
&\boldsymbol{\nabla} \cdot \mathbf{B} = 0.
\end{aligned} \tag{8.25}$$

Henceforth in this book, the term 'magnetohydrodynamics' (MHD) will mean the plasma model described by equations (8.23)–(8.25).

The physical significance of each of these equations is evident by analogy with the usual equations of fluid mechanics and electrodynamics. The most novel feature is the replacement of $\mathbf{E}$ by $\mathbf{E} + \mathbf{u} \times \mathbf{B}$ in Ohm's law. The quantity $\mathbf{E} + \mathbf{u} \times \mathbf{B}$ is the effective electric field seen by a fluid element moving with velocity $\mathbf{u}$ across a magnetic field $\mathbf{B}$, taking into account the Lorentz transformation for $u \ll c$.

The MHD equations are standard tools for treating problems of large-scale plasma motion. Before giving some examples of their use, we must first discuss one final approximation, namely the one in which the resistivity is unimportant in regard to its effect on large-scale plasma motion.

8.4 THE APPROXIMATION OF 'INFINITE CONDUCTIVITY'

Since the magnitude of resistivity in a high-temperature plasma is very small, dynamical phenomena in a plasma can often be described in the approximation of 'infinite conductivity', in which the plasma Ohm's law is simply

$$\mathbf{E} + \mathbf{u} \times \mathbf{B} = 0. \qquad (8.26)$$

The approximation in which this infinite-conductivity form of the Ohm's law is employed is usually called 'ideal magnetohydrodynamics'. Later in this Chapter we will derive a dimensionless quantity, the 'magnetic Reynolds number', the magnitude of which will tell us whether or not this is a good approximation.

The consequence of 'infinite conductivity' is that *the plasma is tied to the magnetic field lines*, in a sense that we will now describe.

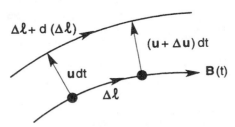

Figure 8.1. Motion of two elements of plasma that lie initially on the same field line $\mathbf{B}(t)$ separated by a distance $\Delta\boldsymbol{\ell}$. We prove that the final position of the elements lie also on the same field line $\mathbf{B}(t + \mathrm{d}t)$.

We will show that all fluid elements initially located on any given field line will still be located on the same field line after an arbitrary motion of an infinitely conducting plasma. Consider two neighboring fluid elements on some particular field line at time t; the two elements are separated by a vector differential length $\Delta\boldsymbol{\ell}$, which must, by assumption, lie in the direction parallel to $\mathbf{B}(t)$. In a time interval $\mathrm{d}t$, the two elements move $\mathbf{u}\mathrm{d}t$ and $(\mathbf{u} + \Delta\mathbf{u})\mathrm{d}t$, respectively, as shown in Figure 8.1. To prove our assertion, we must show that $\Delta\boldsymbol{\ell} + \mathrm{d}(\Delta\boldsymbol{\ell})$ is parallel to $\mathbf{B}(t + \mathrm{d}t)$. (In this analysis, note that both d and Δ are used to denote differentials. The differential d is associated with the differential time-step $\mathrm{d}t$, which carries the plasma from its initial location on a given field line to a subsequent displaced location. The differential Δ is associated with the differential length $\Delta\boldsymbol{\ell}$ between two plasma elements initially on the same field line.)

First, let us calculate the quantity $\mathrm{d}(\Delta\boldsymbol{\ell})$, meaning the differential change of $\Delta\boldsymbol{\ell}$ following the motion of the plasma for the differential time interval $\mathrm{d}t$. The Taylor expansion for $\mathbf{u}$ gives

$$\Delta\mathbf{u} = (\Delta\boldsymbol{\ell} \cdot \boldsymbol{\nabla})\mathbf{u} \qquad (8.27)$$

and, by reference to Figure 8.1, we see that

$$\Delta\boldsymbol{\ell} + \mathrm{d}(\Delta\boldsymbol{\ell}) = \Delta\boldsymbol{\ell} + (\mathbf{u} + \Delta\mathbf{u})\mathrm{d}t - \mathbf{u}\mathrm{d}t \qquad (8.28)$$

since the combination of the three vectors on the right-hand side traces out the same path as the vector on the left-hand side, so that

$$\frac{d(\Delta \boldsymbol{\ell})}{dt} = \Delta \mathbf{u} = (\Delta \boldsymbol{\ell} \cdot \nabla)\mathbf{u}. \tag{8.29}$$

Next, let us consider how $\mathbf{B}$ itself changes. By combining Faraday's law with Ohm's law, we have

$$\frac{\partial \mathbf{B}}{\partial t} = -\nabla \times \mathbf{E}$$

$$= \nabla \times (\mathbf{u} \times \mathbf{B})$$

$$= (\mathbf{B} \cdot \nabla)\mathbf{u} - (\mathbf{u} \cdot \nabla)\mathbf{B} - \mathbf{B}(\nabla \cdot \mathbf{u}) \tag{8.30}$$

where, in the final form, we have used a familiar expansion for the curl of a vector product (see Appendix D) and have also made use of $\nabla \cdot \mathbf{B} = 0$ to eliminate one of the four terms. The total derivative of $\mathbf{B}$, following the motion of the plasma, may now be written

$$\frac{d\mathbf{B}}{dt} = \frac{\partial \mathbf{B}}{\partial t} + (\mathbf{u} \cdot \nabla)\mathbf{B}$$

$$= (\mathbf{B} \cdot \nabla)\mathbf{u} - \mathbf{B}(\nabla \cdot \mathbf{u}). \tag{8.31}$$

Let us now evaluate

$$\frac{d}{dt}(\Delta \boldsymbol{\ell} \times \mathbf{B}) = \frac{d(\Delta \boldsymbol{\ell})}{dt} \times \mathbf{B} + \Delta \boldsymbol{\ell} \times \frac{d\mathbf{B}}{dt}$$

$$= [(\Delta \boldsymbol{\ell} \cdot \nabla)\mathbf{u}] \times \mathbf{B} + \Delta \boldsymbol{\ell} \times [(\mathbf{B} \cdot \nabla)\mathbf{u} - \mathbf{B}(\nabla \cdot \mathbf{u})] \tag{8.32}$$

where we have substituted from equations (8.29) and (8.31). The third term on the right-hand side in equation (8.32) must vanish for, initially, $\Delta \boldsymbol{\ell}$ is parallel to $\mathbf{B}$, so that

$$\Delta \boldsymbol{\ell} \times \mathbf{B} = 0. \tag{8.33}$$

Moreover, when we examine the first two terms on the right-hand side in the second form of equation (8.32), we see that they cancel each other if $\Delta \boldsymbol{\ell}$ and $\mathbf{B}$ are parallel. For, if $\Delta \boldsymbol{\ell}$ and $\mathbf{B}$ are parallel, they may be interchanged in the first term on the right-hand side in equation (8.32), which then becomes $[(\mathbf{B} \cdot \nabla)\mathbf{u}] \times \Delta \boldsymbol{\ell}$, which is equal and opposite to the second term on the right-hand side. Thus, altogether,

$$\frac{d}{dt}(\Delta \boldsymbol{\ell} \times \mathbf{B}) = 0 \tag{8.34}$$

which shows that $\Delta \boldsymbol{\ell}$ moves so as to remain parallel to $\mathbf{B}$.

Thus, our assertion is proved: *any two elements of an ideal plasma that lie initially on a given field line will still lie on the same field line after an arbitrary motion of the plasma.* The field configuration itself will, of course, have changed, and the plasma elements and their field line may have moved to a completely different location in physical space. But, in any such motion, however complicated, each field line retains its identity, since the plasma elements themselves must retain their identity. If one could 'paint' with a color those plasma elements that lie initially on some field line, this line of 'colored' plasma would move around in some complicated way in physical space, but it would always remain on a field line: i.e. the 'colored' elements of plasma would always find themselves aligned along the same field line.

Of course, this otherwise very general result applies only to the extent that the simple Ohm's law is valid. This means not only that the plasma resistivity must be negligible, but also that the fluid velocities arise from $\mathbf{E} \times \mathbf{B}$ drifts that are much larger than diamagnetic or guiding-center drifts.

8.5 CONSERVATION OF MAGNETIC FLUX

The tying of plasma to magnetic field lines has another important consequence: *the magnetic flux through any closed contour that moves with the plasma is constant.* By magnetic flux, we mean the integral of $\mathbf{B}$ over the area enclosed by some closed contour, i.e.

$$\Phi = \int \mathbf{B} \cdot d\mathbf{S} \tag{8.35}$$

where $d\mathbf{S}$ is a (vector) element of area. For the closed contour, we choose any closed line 'painted' on the plasma, not lying along a field line, and we imagine this line to move as the plasma moves. The change in Φ is made up of two parts: the change due to time variation of $\mathbf{B}$ integrated over the area within the original closed contour, and the change due to the movement of the contour that bounds the area of integration. Using, from equation (8.30), $\partial \mathbf{B} / \partial t = \nabla \times (\mathbf{u} \times \mathbf{B})$, we obtain

$$\frac{d\Phi}{dt} = \int \nabla \times (\mathbf{u} \times \mathbf{B}) \cdot d\mathbf{S} + \int \mathbf{B} \cdot \frac{d(\Delta \mathbf{S})}{dt}. \tag{8.36}$$

We may use Stokes' theorem to transform the first term on the right-hand side into a line integral around the contour bounding the area of integration; we denote the element of length along this contour by $\Delta \boldsymbol{\ell}$. (As in the previous Sections, both d and Δ are used to denote differentials, the former with respect to the time step dt and the latter with respect to the length element along the contour. Since $\Delta \boldsymbol{\ell}$ is a legitimate vector differential length element, we can have a line integral over $\Delta \boldsymbol{\ell}$.) Referring to Figure 8.2, we also see that the increment

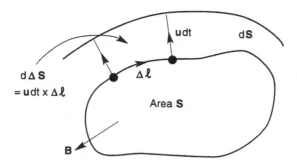

Figure 8.2. Area S bounded by some closed contour in the plasma. After a time increment d*t*, the plasma elements on the contour have moved to encompass an additional area dS, made up of pieces such as dΔS, as shown. We prove that the new contour encloses the same amount of magnetic flux $\int \mathbf{B} \cdot \mathrm{dS}$.

in the (vector) element of area produced by the plasma motion is given by

$$d(\Delta S)/dt = \mathbf{u} \times \Delta \boldsymbol{\ell}. \tag{8.37}$$

Thus, altogether, we obtain

$$\frac{d\Phi}{dt} = \int (\mathbf{u} \times \mathbf{B}) \cdot \Delta \boldsymbol{\ell} + \int \mathbf{B} \cdot (\mathbf{u} \times \Delta \boldsymbol{\ell}) = 0 \tag{8.38}$$

the two terms cancelling by the well-known property of the triple scalar product. Thus, our assertion is proved: *the magnetic flux through an area bounded by any closed contour 'painted' on the plasma is unchanged in any motion of the plasma.* The conditions for the validity of this result are the same as those for the result of the preceding Section—negligible resistivity and dominantly $\mathbf{E} \times \mathbf{B}$ drifts.

8.6 CONSERVATION OF ENERGY

Using the equation of motion (8.10), and the equation for the convection of magnetic field **B** in a perfectly conducting plasma, equation (8.30), together with Maxwell equations as needed, we can derive the equation of energy conservation for a perfectly conducting plasma that obeys an adiabatic equation of state. The equation to be derived is

$$\frac{dW}{dt} = 0 \tag{8.39}$$

where the total energy W is given by

$$W = \int \left(\frac{\rho |\mathbf{u}|^2}{2} + \frac{p}{\gamma - 1} + \frac{\epsilon_0 |\mathbf{E}|^2}{2} + \frac{|\mathbf{B}|^2}{2\mu_0} \right) d^3 x. \tag{8.40}$$

We start by taking the dot product of equation (8.10) with **u** and then integrating over all space. The contribution from the terms on the left-hand side in equation (8.10) is relatively straightforward to obtain. We have

$$
\int \rho \mathbf{u} \cdot \left(\frac{\partial \mathbf{u}}{\partial t} + \mathbf{u} \cdot \nabla \mathbf{u} \right) \mathrm{d}^3 x = \frac{1}{2} \int \left(\rho \frac{\partial |\mathbf{u}|^2}{\partial t} + \rho \mathbf{u} \cdot \nabla |\mathbf{u}|^2 \right) \mathrm{d}^3 x
$$

$$
= \frac{1}{2} \int \left(\rho \frac{\partial |\mathbf{u}|^2}{\partial t} - |\mathbf{u}|^2 \nabla \cdot (\rho \mathbf{u}) \right) \mathrm{d}^3 x
$$

$$
= \frac{1}{2} \int \left(\rho \frac{\partial |\mathbf{u}|^2}{\partial t} + |\mathbf{u}|^2 \frac{\partial \rho}{\partial t} \right) \mathrm{d}^3 x
$$

$$
= \frac{\partial}{\partial t} \int \frac{\rho |\mathbf{u}|^2}{2} \mathrm{d}^3 x \tag{8.41}
$$

where we have used Gauss's theorem and assumed that the surface integral at infinity vanishes in going from the first line to the second line. This gives the first term in the expression for W, which is the energy of directed motion.

The treatment of the third term on the right-hand side of equation (8.10) from the pressure gradient is a little tricky and goes as follows. First we rewrite the adiabatic equation of state as

$$
0 = \frac{\mathrm{d}}{\mathrm{d}t} \left(\frac{p}{\rho^\gamma} \right) = \frac{1}{\rho^{\gamma-1}} \frac{\mathrm{d}}{\mathrm{d}t} \left(\frac{p}{\rho} \right) - \frac{(\gamma-1)p}{\rho^{\gamma+1}} \frac{\mathrm{d}\rho}{\mathrm{d}t}
$$

$$
= \frac{1}{\rho^{\gamma-1}} \frac{\mathrm{d}}{\mathrm{d}t} \left(\frac{p}{\rho} \right) + \frac{(\gamma-1)p}{\rho^\gamma} \nabla \cdot \mathbf{u}. \tag{8.42}
$$

Using Gauss's theorem and assuming that the surface integral at infinity vanishes, because either p or **u** vanish there, and then using the above relation to substitute for $\nabla \cdot \mathbf{u}$, the term from the pressure gradient can be transformed as follows:

$$
\int \mathbf{u} \cdot \nabla p \, \mathrm{d}^3 x = - \int p \nabla \cdot \mathbf{u} \, \mathrm{d}^3 x
$$

$$
= \frac{1}{\gamma-1} \int \rho \frac{\mathrm{d}}{\mathrm{d}t} \left(\frac{p}{\rho} \right) \mathrm{d}^3 x
$$

$$
= \frac{1}{\gamma-1} \int \left[\rho \frac{\partial}{\partial t} \left(\frac{p}{\rho} \right) + \rho \mathbf{u} \cdot \nabla \left(\frac{p}{\rho} \right) \right] \mathrm{d}^3 x
$$

$$
= \frac{1}{\gamma-1} \int \left[\rho \frac{\partial}{\partial t} \left(\frac{p}{\rho} \right) - \frac{p}{\rho} \nabla \cdot (\rho \mathbf{u}) \right] \mathrm{d}^3 x
$$

$$
= \frac{1}{\gamma-1} \int \left[\rho \frac{\partial}{\partial t} \left(\frac{p}{\rho} \right) + \frac{p}{\rho} \frac{\partial \rho}{\partial t} \right] \mathrm{d}^3 x
$$

$$
= \frac{\partial}{\partial t} \int \frac{p}{\gamma-1} \mathrm{d}^3 x. \tag{8.43}
$$

This provides the essentials of the derivation of the second term in the expression for W, which is the energy of random motion.

Problem 8.2: Complete the derivation of the other two terms in W. The plasma should be assumed to be an isolated system, so that fluxes of electromagnetic energy into or out of the system at its boundary may be taken to be zero. (Hint: in considering the magnetic force term in the equation of motion, you might find it helpful to use the identity $(\nabla \times \mathbf{B}) \times \mathbf{B} = (\mathbf{B} \cdot \nabla)\mathbf{B} - \nabla(B^2/2)$.) Although the derivation of this part of the energy conservation equation requires some lengthy analysis, which will take several pages of calculations, the final answer is revealing: in particular, it shows that the total energy is made up of the kinetic energy of directed motion, plus the energy associated with random motion, plus the usual electric and magnetic field 'energy densities' of $\epsilon_0|\mathbf{E}|^2/2$ and $|\mathbf{B}|^2/2\mu_0$, respectively.

8.7 MAGNETIC REYNOLDS NUMBER

The tying of plasma to magnetic field lines is a property peculiar to the limit of infinite conductivity. The question naturally arises: how good must the conductivity be for this approximation to be valid?

Including resistivity, the time variation of the magnetic field is given by

$$
\begin{aligned}
\frac{\partial \mathbf{B}}{\partial t} &= -\nabla \times \mathbf{E} \\
&= \nabla \times (\mathbf{u} \times \mathbf{B}) - \nabla \times (\eta \mathbf{j}) \\
&= \nabla \times (\mathbf{u} \times \mathbf{B}) + (\eta/\mu_0)\nabla^2 \mathbf{B}
\end{aligned}
\tag{8.44}
$$

where we have used Ampere's law for $\mathbf{j}$, assumed η to be constant, and made use of the familiar identity (see Appendix D)

$$
\nabla \times \nabla \times \mathbf{B} = \nabla(\nabla \cdot \mathbf{B}) - \nabla^2 \mathbf{B}
\tag{8.45}
$$

remembering also that $\nabla \cdot \mathbf{B} = 0$. Note that the meaning of $\nabla^2 \mathbf{B}$ in index notation is $\partial^2 B_i / \partial x_j \partial x_j$. The first term on the right-hand side in equation (8.44) describes convection of the field with the plasma (and its amplification or reduction due to compressive motion perpendicular to the magnetic field), while the second (resistive) term describes diffusion of the field across the plasma.

For some general magnetohydrodynamic motion with a characteristic scale-length L and a characteristic plasma velocity u, the ratio of the convection term to the (resistive) diffusion term is a dimensionless quantity

$$R_M = \mu_0 u L / \eta \tag{8.46}$$

which is sometimes called the 'magnetic Reynolds number'. If the magnetic Reynolds number is sufficiently large, the infinite-conductivity assumption is valid. It is evident that the magnetic Reynolds number depends on the velocity of the plasma motion, and therefore its magnitude depends on the nature of the dynamical phenomenon under investigation. For fully developed magnetohydrodynamic motion, the characteristic velocities are very large, approaching the Alfvén speed (see Problem 8.3). When the Alfvén speed is used for the velocity u in equation (8.46), the magnetic Reynolds number is called the 'Lundquist number', after its discoverer. Lundquist numbers in low-resistivity plasmas can range up to 10^8, or higher.

Problem 8.3: Estimate the magnetic Reynolds numbers for two of the typical plasmas with magnetic fields discussed in Chapter 1, namely the solar corona, in which the magnetic field may be taken to be about 10^{-8} T, and an experimental fusion plasma, in which the magnetic field is about 5 T. The physical dimensions can be taken to be about 10^8 m and 1 m, respectively. In both cases, you may assume that the plasma resistivity is about the same as that of copper, namely 2×10^{-8} Ω m. In each case, an estimate will be needed of the velocity u of magnetohydrodynamic motion. This can be provided by balancing the inertia in the equation of motion against either the pressure gradient, ∇p, or the magnetic force, $\mathbf{j} \times \mathbf{B}$. In the former case, the velocity of magnetohydrodynamic motion is approximately the sound speed, $v_{t,i}$. In the latter case, which is the one that you should assume here, the magnetic force should first be expressed in terms of the gradient of the magnetic pressure, $\nabla(B^2/2\mu_0)$, which results in the velocity of magnetohydrodynamic motion being of order $B/(\mu_0\rho)^{1/2}$, which is called the 'Alfvén speed'. (When the plasma pressure equals the magnetic pressure, the Alfvén speed and the sound speed are essentially equal.) You should use this Alfvén speed in your estimation of the magnetic Reynolds numbers.

Chapter 9

Magnetohydrodynamic equilibrium

In Chapters 2–4, we discussed the orbits of individual charged particles in various types of electric and magnetic fields. In particular, in a strong static magnetic field, we found that charged particles gyrate in tight spirals about the magnetic field lines. If the magnetic field is non-uniform or curved—as it necessarily is in any real situation—charged particles also drift *across* the magnetic field lines. In our discussion of particle orbits, however, we analyzed this situation as if the magnetic field was externally generated, i.e. unaffected by the presence of the plasma particles.

However, as more and more charged particles are added to a plasma, the currents that flow along the magnetic field, as well as the diamagnetic current perpendicular to the magnetic field arising from pressure non-uniformity, can become large enough to modify the externally created magnetic field. The *plasma equilibrium* must then be determined *self-consistently*: the presence of the plasma itself modifies the magnetic field configuration.

The fluid equations that we have derived in the previous three Chapters are well-suited for addressing this problem. Even in the simplest 'ideal magnetohydrodynamic' approximation, these equations contain the essential ingredient, which is the requirement that the plasma currents needed for force balance be consistent with those required to form the magnetic field configuration.

9.1 MAGNETOHYDRODYNAMIC EQUILIBRIUM EQUATIONS

For a steady-state solution of the magnetohydrodynamic (MHD) equations for the special case with $\mathbf{u} = 0$ and isotropic pressure, the plasma and magnetic field must satisfy the three equations

$$\nabla p = \mathbf{j} \times \mathbf{B} \qquad \nabla \cdot \mathbf{B} = 0 \qquad \nabla \times \mathbf{B} = \mu_0 \mathbf{j}. \qquad (9.1)$$

The charge continuity equation, in the quasi-neutral approximation, $\nabla \cdot \mathbf{j} = 0$, is redundant with the third of these; Ohm's law provides no useful information, since both $\mathbf{u}$ and $\mathbf{E}$ are zero in this *static* equilibrium and resistivity is neglected.

The first of these relations states that the plasma pressure gradient and the Lorentz force must be in balance. Consider, for example, a cylindrical plasma with the maximum pressure on the axis of the cylinder, so that the vector ∇p is directed inwards. In the case where the magnetic field is directed along the axis of the cylinder, i.e. in the z direction, in the fluid picture the outward force of expansion must be counteracted by the Lorentz force arising from an azimuthal current flowing in the negative-θ direction in the plasma, as shown in Figure 9.1.

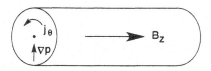

Figure 9.1. Cylindrical plasma equilibrium in which the inward $\mathbf{j} \times \mathbf{B}$ force from the azimuthal current balances the outward force from the pressure gradient. Note that the azimuthal current flows in the negative-θ direction.

From the point of view of a differential volume element, the net momentum flow associated with ∇p arises from the fact that more radial momentum flows into the box from the small-r side than flows out of the box on the large-r side; $\langle v_r v_r \rangle|_r > \langle v_r v_r \rangle|_{r+dr}$. On the other hand, the pressure gradient gives rise to a net diamagnetic current, and the $\mathbf{j} \times \mathbf{B}$ force that arises provides the needed force balance. In the particle picture, fluid force-balance does not play a role. The orbits spiral along $\mathbf{B}$, and for the case of scalar pressure there is no net current due to gradient or curvature of the magnetic field. However the diamagnetic current arises as an observable current via the orbit overlaps and, in a self-consistent particle-picture of the plasma, would affect the magnetic field.

In the general case, the fluid plasma current required for equilibrium can be found by taking the cross-product of the force-balance equation with $\mathbf{B}$ and using the well-known identity for the triple cross-product; this determines the component of the current density perpendicular to the magnetic field, i.e.

$$\mathbf{j}_\perp = \frac{\mathbf{B} \times \nabla p}{B^2}. \tag{9.2}$$

We of course encountered exactly this current in Chapter 7 and called it the 'diamagnetic current'.

The component of current density along the magnetic field, $j_\parallel$, is not determined by the force-balance equation. However, in the general case where

$\mathbf{j}_\perp$ is not divergence-free, a non-zero $j_\parallel$ is needed to satisfy the quasi-neutrality requirement

$$\nabla \cdot \mathbf{j} = 0. \tag{9.3}$$

Specifically, writing

$$\mathbf{j} = \mathbf{j}_\perp + j_\parallel \hat{\mathbf{b}} \tag{9.4}$$

and using $\nabla \cdot \mathbf{B} = 0$, we obtain

$$\mathbf{B} \cdot \nabla(j_\parallel/B) + \nabla \cdot \mathbf{j}_\perp = 0 \tag{9.5}$$

from which $j_\parallel$ can, in principle, be obtained, except for an arbitrary $j_\parallel$ proportional everywhere to B. Often, however, the $\mathbf{j}_\perp$ vector turns out to be divergence-free by itself, in which case $j_\parallel$ can be zero. This is certainly the case, for example, in the equilibrium shown in Figure 9.1.

9.2 MAGNETIC PRESSURE: THE CONCEPT OF BETA

If we substitute for $\mathbf{j}$ from Ampere's law into the force-balance equation, we obtain

$$\begin{aligned}
\nabla p &= (\nabla \times \mathbf{B}) \times \mathbf{B}/\mu_0 \\
&= [(\mathbf{B} \cdot \nabla)\mathbf{B} - \nabla(B^2/2)]/\mu_0
\end{aligned} \tag{9.6}$$

using the vector identity for $(\nabla \times \mathbf{A}) \times \mathbf{B}$ (see Appendix D). This may be rewritten

$$\nabla(p + B^2/2\mu_0) = (\mathbf{B} \cdot \nabla)\mathbf{B}/\mu_0 \tag{9.7}$$

which is known as the 'pressure-balance condition'. The terms on the left in equation (9.7) indicate that the magnetic field may be considered to have a 'magnetic pressure' given by $B^2/2\mu_0$. The term on the right-hand side of equation (9.7) comes from *bending* and *parallel compression* of the field, producing perpendicular and parallel forces, respectively. This may be seen by writing

$$\begin{aligned}
(\mathbf{B} \cdot \nabla)\mathbf{B} &= B(\hat{\mathbf{b}} \cdot \nabla)(B\hat{\mathbf{b}}) \\
&= B^2(\hat{\mathbf{b}} \cdot \nabla)\hat{\mathbf{b}} + \hat{\mathbf{b}}(\hat{\mathbf{b}} \cdot \nabla)B^2/2.
\end{aligned}$$

Here, the first term on the second line is perpendicular to the magnetic field (taking the dot product with $\hat{\mathbf{b}}$ gives $(\hat{\mathbf{b}} \cdot \nabla)|\hat{\mathbf{b}}|^2/2 = 0$), and represents the 'bending' force. The second term on the second line is parallel to the magnetic field and represents a force due to parallel 'compression' of the field lines.

In some interesting cases, the field lines may be taken as approximately *straight and parallel*, in which case the term on the right-hand side of equation (9.7) vanishes identically. In these cases, the pressure balance condition becomes simply

$$p + B^2/2\mu_0 = \text{constant} \tag{9.8}$$

i.e. the sum of the plasma pressure and the magnetic-field pressure is constant. One example of an equilibrium of this type is the cylindrical plasma with a magnetic field directed along the axis of the cylinder as shown in Figure 9.1. From the pressure-balance condition, we see that the magnetic field is lowered at the center of the plasma, where the plasma pressure is highest—a particular example of plasma diamagnetism.

The ratio of the plasma pressure to the magnetic-field pressure (normally taken *outside* the plasma) is usually denoted by β, i.e.

$$\beta = 2\mu_0 p/B^2. \tag{9.9}$$

The quantity β is a measure of the degree to which the magnetic field is holding a non-uniform plasma in equilibrium. In a low-β plasma, the force balance is mainly a matter of different magnetic forces in balance with each other. At $\beta \sim 1$, the magnetic and pressure forces are largely balancing, whereas for $\beta \gg 1$ the magnetic field plays a minor role in the dynamics of the plasma. Astrophysical plasmas can have β approaching unity, sometimes even $\beta \gg 1$. Laboratory plasmas tend to have β values in the range of a few per cent at most, although it is possible in special configurations to create laboratory plasmas with β values near unity.

9.3 THE CYLINDRICAL PINCH

Another interesting configuration is the 'cylindrical pinch' in which the magnetic field is *azimuthal* (i.e. B_θ only), while the plasma current is axial (i.e. j_z only), as shown in Figure 9.2. In this case, the magnetic field is curved, and the pressure-balance condition must include the term on the right-hand side of equation (9.7), which in this case describes tension of the field lines holding the plasma in. Noting that the θ-derivative of the unit azimuthal vector $\hat{\theta}$ is the inwardly directed unit radial vector, $-\hat{r}$, we obtain

$$\frac{\partial}{\partial r}\left(p + \frac{B_\theta^2}{2\mu_0}\right) = -\frac{B_\theta^2}{\mu_0 r} \tag{9.10}$$

which may be integrated from 0 to r to give

$$p(r) = p_0 - \frac{B_\theta^2(r)}{2\mu_0} - \frac{1}{\mu_0}\int_0^r \frac{B_\theta^2}{r}\,\mathrm{d}r \tag{9.11}$$

where p_0 is the (peak) pressure, assumed in this case to occur at $r = 0$.

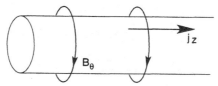

Figure 9.2. Cylindrical pinch equilibrium in which the azimuthal magnetic field is created by the axial plasma current j_z.

There are infinitely many possible equilibria of this sort. For an illustration, we might consider the case of a plasma carrying a uniformly distributed current surrounded by a vacuum:

$$j_z(r) = j_{z0} \qquad r < a$$
$$j_z(r) = 0 \qquad r > a \tag{9.12}$$

giving a total plasma current $I = \pi a^2 j_{z0}$. Since the current density within the plasma is uniform, then by Ampere's law the magnetic field strength B_θ must be proportional to the radius r. Thus, we may write $B_\theta = B_{\theta a} r / a$. Carrying out the integral over r in equation (9.11), we find that the third term on the right-hand side makes a contribution equal to that of the second term on the right. Thus, within $r < a$, the pressure is given by

$$p(r) = p_0 - \frac{B_{\theta a}^2 r^2}{\mu_0 a^2} \tag{9.13}$$

where $B_{\theta a}$ is the azimuthal field at the edge of the plasma, related to the total current by $B_{\theta a} = \mu_0 I / 2\pi a$. We see that the pressure profile has a parabolic dependence on radius r. Since the pressure must vanish at the edge of the plasma, i.e. $p(a) = 0$, we have

$$p_0 = \frac{B_{\theta a}^2}{\mu_0} = \frac{\mu_0 I^2}{4\pi^2 a^2}. \tag{9.14}$$

This is known as the 'pinch condition', describing a magnetically self-constricted current-carrying plasma.

For this equilibrium, the plasma current comes entirely from the plasma diamagnetic current, and this current provides the entire magnetic field. The pressure gradient is proportional to r, as is the only field-component B_θ, consistent with a constant diamagnetic current j_z. The pinch can be established dynamically, by applying a very large voltage difference across a pair of electrodes to drive a large plasma current j_z. However, we will see in Chapter 19 that this plasma is strongly unstable.

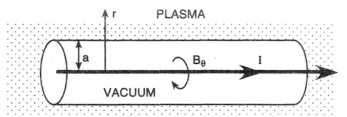

Figure 9.3. A plasma equilibrium in which a uniform plasma fills all space except for an evacuated cylindrical hole produced by a current-carrying conductor (see Problem 9.1).

Problem 9.1: A plasma of uniform pressure p fills all of space except for an evacuated infinitely long cylindrical 'hole' of radius a, which is produced by a conductor carrying a current I placed on the axis of the cylindrical hole. The conductor produces an azimuthal field B_θ, as shown in Figure 9.3. Show that the maximum radius of the vacuum hole that can be created in equilibrium in this way is given by $a^2 = \mu_0 I^2 / 8\pi^2 p$. Is there a z-directed current in the plasma? What is its magnitude and where is it located?

9.4   FORCE-FREE EQUILIBRIA: THE 'CYLINDRICAL' TOKAMAK

Equilibria with small or negligible plasma pressure are of interest, since they describe magnetic configurations containing 'low-β' plasmas. If the pressure gradient is negligible, the Lorentz force must vanish, i.e.

$$0 = \mathbf{j} \times \mathbf{B}. \tag{9.15}$$

Such equilibria are called 'force-free'.

Non-trivial force-free cylindrical equilibria are possible if both axial (B_z) and azimuthal (B_θ) field-components are present. In this case, the pressure-balance condition becomes

$$\frac{\partial}{\partial r}\left(\frac{B_\theta^2}{2} + \frac{B_z^2}{2}\right) = -\frac{B_\theta^2}{r}. \tag{9.16}$$

There are again infinitely many solutions of this equation, one of which is illustrated in Figure 9.4, which applies to a current-carrying plasma cylinder of radius a.

For an example of a configuration of this type, we might assume again that the current density j_z is uniformly distributed within the plasma, so that

$B_\theta(r) = B_{\theta a} r/a$. This allows us to integrate equation (9.16) to obtain

$$B_z(r)^2 = B_{z0}^2 - B_\theta(r)^2 - 2 \int_0^r \frac{B_\theta^2}{r} \mathrm{d}r = B_{z0}^2 - 2B_\theta(r)^2. \tag{9.17}$$

Here, B_{z0} is the longitudinal field at the center of the plasma cylinder. Equation (9.17) applies in the region $r < a$. Outside the plasma cylinder, i.e. for $r > a$, there can be no current j_z. The azimuthal field B_θ must then decrease with r like r^{-1}. Equation (9.16) then shows that B_z must be constant in this region. The radial profiles of the B_z and B_θ fields are shown in Figure 9.4.

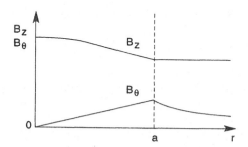

Figure 9.4. Cylindrical force-free equilibrium. The radius $r = a$ is the edge of the plasma, outside of which no currents flow, so that $B_z = $ constant and $B_\theta \propto r^{-1}$.

This is a cylindrical approximation to the 'tokamak' configuration at very low values of β (imagining the torus of the tokamak to be straightened out into an infinitely long cylinder). The strong field B_z is produced by external coils; the weaker field B_θ is produced by currents in the plasma. At low β, the tokamak is seen to be *paramagnetic*, in the sense that the B_z field is actually stronger at the center than at the edge. When a moderate amount of plasma pressure is added (specifically, sufficient that $p > B_\theta^2/2\mu_0$), the usual diamagnetic effect appears, and the B_z field is lowered at the center of the plasma.

Problem 9.2: Add a small amount of plasma pressure $p(r)$ to the cylindrical tokamak equilibrium, assuming that this pressure falls to zero at the plasma edge. Show that the generalization of equation (9.17), still allowing an arbitrary current distribution within the plasma, is

$$p(r) + \frac{B_z(r)^2}{2\mu_0} = p_0 + \frac{B_{z0}^2}{2\mu_0} - \frac{B_\theta(r)^2}{2\mu_0} - \frac{1}{\mu_0}\int_0^r \frac{B_\theta^2}{r}\mathrm{d}r$$

where p_0 is the pressure at the center of the plasma cylinder. For the case of uniform current distribution, i.e. $B_\theta(r) = B_{\theta a} r/a$, show that the plasma is

diamagnetic, i.e. $B_z(a) > B_{z0}$, if $p_0 > B_\theta(a)^2/\mu_0$. Is this result dependent, or not, on the current and pressure distributions in the plasma?

9.5   ANISOTROPIC PRESSURE: MIRROR EQUILIBRIA*

Consider the component of the force-balance equation along the direction of the magnetic-field vector $\hat{\mathbf{b}}$

$$\hat{\mathbf{b}} \cdot \nabla p = \hat{\mathbf{b}} \cdot (\mathbf{j} \times \mathbf{B}) = 0 \qquad (9.18)$$

or, if l denotes a length coordinate along field lines,

$$\frac{\partial p}{\partial l} = 0. \qquad (9.19)$$

This relation shows that plasma pressure must be constant along field lines in any plasma equilibrium. This implies that a magnetic configuration that has 'open' field lines (i.e. lines that go out of the plasma), such as the 'magnetic mirror' configuration introduced in Chapter 3 and shown again in Figure 9.5, is incapable of confining a plasma with isotropic pressure. Nonetheless, we know that this type of magnetic field *can* contain individual ions and electrons.

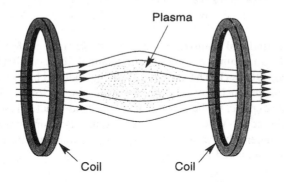

Figure 9.5. Plasma equilibrium in a 'magnetic mirror' configuration.

This paradox is resolved, however, by recognizing that the plasma pressure in this case is by necessity *anisotropic*. Although our magnetohydrodynamic model has, thus far, assumed the pressure p to be a scalar, we have seen in Chapter 6 that the plasma pressure *along* the magnetic field, $p_\parallel$, can be different from the plasma pressure *across* the field, $p_\perp$. For a field in the z direction, the pressure would then be a *diagonal tensor*, with diagonal elements $p_\perp, p_\perp$ and $p_\parallel$. We can write such a tensor as

$$\mathbf{P} = p_\perp \mathbf{I} + (p_\parallel - p_\perp)\hat{\mathbf{b}}\hat{\mathbf{b}} \qquad (9.20)$$

where, as usual, $\hat{\mathbf{b}}$ is the unit vector along $\mathbf{B}$ and I is the unit tensor (the tensor which, expressed in matrix form, would have diagonal elements all unity, and off-diagonal elements all zero). The force-balance relation $\hat{\mathbf{b}} \cdot (\nabla \cdot P) = 0$ is most easily manipulated using index notation, as follows:

$$0 = \hat{b}_i \frac{\partial}{\partial x_j} \left[p_\perp \delta_{ij} + (p_\parallel - p_\perp)\hat{b}_i \hat{b}_j \right]$$

$$= \hat{b}_j \frac{\partial p_\perp}{\partial x_j} + \hat{b}_i B_j \frac{\partial}{\partial x_j} \left(\frac{(p_\parallel - p_\perp)\hat{b}_i}{B} \right)$$

$$= \hat{b}_j \frac{\partial p_\perp}{\partial x_j} + B_j \frac{\partial}{\partial x_j} \left(\frac{p_\parallel - p_\perp}{B} \right)$$

$$= \hat{b}_j \frac{\partial p_\perp}{\partial x_j} + \hat{b}_j \frac{\partial(p_\parallel - p_\perp)}{\partial x_j} - \frac{(p_\parallel - p_\perp)}{B}\hat{b}_j \frac{\partial B}{\partial x_j}$$

$$= \hat{b}_j \frac{\partial p_\parallel}{\partial x_j} + \frac{(p_\perp - p_\parallel)}{B}\hat{b}_j \frac{\partial B}{\partial x_j}$$

$$= \hat{\mathbf{b}} \cdot \nabla p_\parallel + \frac{p_\perp - p_\parallel}{B}\hat{\mathbf{b}} \cdot \nabla B \qquad (9.21)$$

where in going from the first line to the second line we have used $\nabla \cdot \mathbf{B} \equiv \partial B_j/\partial x_j = 0$, and in going from the second line to the third line we have used $\hat{b}_i(\partial \hat{b}_i/\partial x_j) = \partial|\hat{\mathbf{b}}|^2/\partial x_j = 0$. Equation (9.21) can be written

$$\frac{\partial p_\parallel}{\partial l} + \frac{p_\perp - p_\parallel}{B}\frac{\partial B}{\partial l} = 0. \qquad (9.22)$$

Here, l is again a coordinate measuring the distance along a field line.

Many solutions of equation (9.22) are possible, corresponding to mirror-confined equilibria. These solutions have $p_\perp > p_\parallel$, with both $p_\perp$ and $p_\parallel$ decreasing with l as one moves from the center of the mirror-confined plasma to the ends. (The field strength B obviously increases with l.) Thus, both $p_\perp$ and $p_\parallel$ can have their peak values at the center of the mirror-confinement region. A particularly simple class of solutions of the above relation is obtained by assuming that both $p_\perp$ and $p_\parallel$ depend on the *magnitude* of the field strength only, i.e. $p_\perp \equiv p_\perp(B)$ and $p_\parallel \equiv p_\parallel(B)$. Using $\partial p_\parallel/\partial l = (\mathrm{d}p_\parallel/\mathrm{d}B)(\partial B/\partial l)$ in this case, we see that these functions must be related by

$$\frac{\mathrm{d}p_\parallel}{\mathrm{d}B} + \frac{p_\perp - p_\parallel}{B} = 0. \qquad (9.23)$$

This special class of solutions is of particular interest because it also provides for a simple solution of the *perpendicular* force balance equation

$$\nabla \cdot P = \mathbf{j} \times \mathbf{B} \qquad (9.24)$$

in the case of an anisotropic plasma in a mirror magnetic field.   If we limit ourselves to the case of a low-β plasma, in which the plasma currents make a negligible perturbation of an essentially vacuum magnetic field, the only requirement for MHD equilibrium is that the plasma currents derived from equation (9.24) be divergence-free.   Although equation (9.24) provides information only on the perpendicular component of the plasma current, allowing parallel currents to arise to help satisfy the quasi-neutrality condition, $\nabla \cdot \mathbf{j} = 0$, an important special case is where the perpendicular currents are themselves everywhere divergence-free, i.e. $\nabla \cdot \mathbf{j}_\perp = 0$, so that parallel currents are not required for equilibrium.   Such equilibria are of particular relevance to mirror configurations, where plasma currents cannot normally flow out of the ends of the mirror.   For such cases, the condition for equilibrium is

$$
\begin{aligned}
0 &= \nabla \cdot \mathbf{j}_\perp \\
&= \nabla \cdot \left(\frac{1}{B} \hat{\mathbf{b}} \times \nabla \cdot P \right) \\
&= \nabla \cdot \left(\frac{1}{B} \hat{\mathbf{b}} \times \{ \nabla p_\perp + \nabla \cdot [(p_\parallel - p_\perp)\hat{\mathbf{b}}\hat{\mathbf{b}}] \} \right) \\
&= \nabla \cdot \left(\frac{1}{B} [\hat{\mathbf{b}} \times \nabla p_\perp + (p_\parallel - p_\perp)\hat{\mathbf{b}} \times (\hat{\mathbf{b}} \cdot \nabla)\hat{\mathbf{b}}] \right) \\
&= \nabla \cdot \left[\frac{1}{B} \left(\hat{\mathbf{b}} \times \nabla p_\perp + \frac{p_\parallel - p_\perp}{B} \hat{\mathbf{b}} \times \nabla B \right) \right] \\
&= 0.
\end{aligned}
\tag{9.25}
$$

In the next-to-last step here, we have made use of a property of vacuum magnetic fields that was derived during our discussion of ∇B and curvature drifts in Chapter 3, namely

$$
\hat{\mathbf{b}} \times (\hat{\mathbf{b}} \cdot \nabla)\hat{\mathbf{b}} = (\hat{\mathbf{b}} \times \nabla B)/B.
\tag{9.26}
$$

In the last step of equation (9.25), we have taken the particular case where $p_\perp = p_\perp(B)$ and $p_\parallel = p_\parallel(B)$.   For this special case, examination of the final expression on the right-hand side of equation (9.25) shows that all terms are of the form $\nabla \cdot [f(B)\mathbf{B} \times \nabla B]$ for various functions $f(B)$. All such terms must be zero, because the contribution from the gradient of $f(B)$ will vanish when dotted with $\mathbf{B} \times \nabla B$, and the contribution from the divergence of $\mathbf{B} \times \nabla B$ will expand into terms involving the curls of $\mathbf{B}$ and ∇B, both of which vanish.   Thus, all solutions of form $p_\perp \equiv p_\perp(B)$ and $p_\parallel \equiv p_\parallel(B)$ that satisfy equation (9.23) describe low-β plasma equilibria with $\nabla \cdot \mathbf{j}_\perp = 0$ in a mirror magnetic field.

Such solutions would not, however, describe useful equilibria in the *simple axisymmetric* mirror configuration, because the field strength B *decreases* as one moves *radially* away from the central confinement region. To make use of these solutions, assuming that they can be realized in a practical situation, special non-axisymmetric mirror configurations must be designed that have the field strength B increasing outward in *every* direction.

9.6 RESISTIVE DISSIPATION IN PLASMA EQUILIBRIA

In this Chapter, we have discussed equilibria in which magnetic fields are embedded in the plasma. However, since a perfectly conducting plasma will act to exclude magnetic flux that it does not already contain, the question arises as to how these magnetic fields succeed in penetrating into a plasma if they are not present at the time of formation of the plasma. To address this question, we must restore plasma resistivity into the magnetohydrodynamic equations. A more general question is how dissipation allows our plasma to return to full thermodynamic equilibrium with uniform pressure; this also requires plasma resistivity.

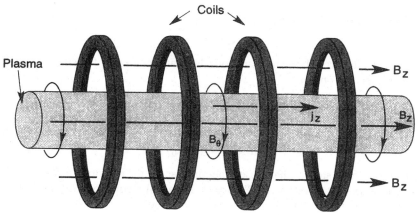

Figure 9.6. The tokamak equilibrium in the cylindrical approximation. The approximately uniform field B_z is dominantly produced by external coils; the much weaker field B_θ is produced by currents flowing in the plasma.

The tokamak configuration in the cylindrical approximation, which has already been introduced and is illustrated again in Figure 9.6, provides a good example to consider. The actual tokamak geometry is toroidal, and the main magnetic field is toroidally directed, with a smaller field directed the short way around the torus. In the 'cylindrical tokamak', the main field B_z (corresponding to the toroidal field in the actual tokamak geometry) is produced like that in the actual geometry largely by external coils surrounding the plasma. It is present before the initial formation of the plasma, which is normally accomplished by ionizing a rarefied neutral gas that is fed into the containment vessel, using an induced toroidal voltage to produce a cascade of breakdown and ionization. When the plasma becomes well-conducting, the magnetic flux associated with the B_z field is already embedded in it.

However, the B_θ field within the plasma requires currents that flow in the z direction in the plasma itself. Since such currents can be increased to significant values only after the plasma becomes reasonably well-conducting, there will be a tendency for the plasma to exclude the magnetic flux of the B_θ field produced by the current j_z, as this current rises. By considering a closed loop drawn in some constant-θ surface within the plasma, it is clear that no B_θ flux could ever penetrate into a fixed plasma in the perfectly conducting case. For present purposes we can consider the plasma to be fixed in place by the strong *toroidal* magnetic field while we examine the penetration of the *poloidal* field into this fixed plasma.

When plasma resistivity is included in the analysis, the magnetic flux of the B_θ field is allowed to diffuse slowly into the plasma. To see this, we restore resistivity to the plasma Ohm's law, i.e.

$$\mathbf{E} + \mathbf{u} \times \mathbf{B} = \eta\mathbf{j}. \tag{9.27}$$

Combining Ampere's law and Faraday's law, and assuming for simplicity that $\eta = $ constant, we obtain

$$
\begin{aligned}
\frac{\partial \mathbf{B}}{\partial t} &= \nabla \times (\mathbf{u} \times \mathbf{B}) - \nabla \times (\eta\mathbf{j}) \\
&= \nabla \times (\mathbf{u} \times \mathbf{B}) - (\eta/\mu_0)\nabla \times \nabla \times \mathbf{B} \\
&= \nabla \times (\mathbf{u} \times \mathbf{B}) + (\eta/\mu_0)\nabla^2 \mathbf{B}
\end{aligned} \tag{9.28}
$$

where in the final step we have used the familiar vector identity for $\nabla \times \nabla \times \mathbf{B}$ (see Appendix D) together with $\nabla \cdot \mathbf{B} = 0$. Physically, equation (9.28) describes a magnetic field that changes partly by convection (the first term on the right-hand side) and partly by diffusion (the second term on the right-hand side).

In our present cylindrical tokamak example, we have a situation where the plasma is held in approximate equilibrium by the pressure of a very strong constant and near-uniform field B_z, as shown in Figure 9.6. The z component of equation (9.28), coupled with the pressure-balance condition, will tell us exactly how long-lasting this equilibrium will be, i.e. at what rate the plasma can still leak across the confining B_z field. Specifically, since the externally produced B_z field is held constant, the z component of equation (9.28) gives

$$0 = -\frac{1}{r}\frac{\partial}{\partial r}(r u_r B_z) + \frac{\eta}{\mu_0}\frac{1}{r}\frac{\partial}{\partial r}\left(r\frac{\partial B_z}{\partial r}\right) \tag{9.29}$$

or

$$u_r = \frac{\eta}{\mu_0 B_z}\frac{\partial B_z}{\partial r} \approx -\frac{\eta}{B_z^2}\frac{\partial p}{\partial r} \tag{9.30}$$

the latter arising from pressure balance, $p + B_z^2/2\mu_0 \approx$ constant, where the approximation reflects the neglect of small contributions from the B_θ field. The expression for the plasma velocity u_r given in equation (9.30) describes a process of slow 'leakage' of plasma across a magnetic field due to resistivity—a topic to be taken up at length in Chapter 12. For now, we note simply that the plasma will leak only very slowly across the B_z field, with a radial velocity that is of order $u_r \sim \eta p/B_z^2 L$ where L is a radial scale length. For present purposes, it is sufficient to note that this leakage rate can be made arbitrarily small by increasing the magnitude of B_z.

The θ component of equation (9.28) tells us how the *azimuthal* field can leak *into* the plasma. Using the appropriate expressions for curl and ∇^2 of a vector in cylindrical coordinates (see Appendix E), we obtain

$$\frac{\partial B_\theta}{\partial t} = -\frac{\partial}{\partial r}(u_r B_\theta) + \frac{\eta}{\mu_0}\frac{\partial}{\partial r}\left(\frac{1}{r}\frac{\partial(r B_\theta)}{\partial r}\right) \tag{9.31}$$

but, for u_r of a magnitude given by equation (9.30), the first term on the right-hand side is too small to be of significance in a low-β tokamak (it is of order $\beta \sim 2\mu_0 p/B_z^2$ relative to the second term on the right-hand side), and the equation reduces to

$$\frac{\partial B_\theta}{\partial t} \approx \frac{\eta}{\mu_0}\frac{\partial}{\partial r}\left(\frac{1}{r}\frac{\partial(r B_\theta)}{\partial r}\right). \tag{9.32}$$

(The expression used in equation (9.31) for ∇^2 of a vector in cylindrical coordinates, which is given in general in Appendix E, can be obtained for our present case, where it acts only on a θ-directed field $B_\theta(r)$, as follows. The operator ∇^2 means the successive application of, first, the gradient operator

$$\nabla \equiv \hat{\mathbf{r}}\frac{\partial}{\partial r} + \hat{\boldsymbol{\theta}}\frac{\partial}{r\,\partial\theta} + \hat{\mathbf{z}}\frac{\partial}{\partial z}$$

and, second, the divergence operator

$$\nabla\cdot \equiv \left(\frac{1}{r}\frac{\partial}{\partial r}r\,\hat{\mathbf{r}} + \frac{1}{r}\frac{\partial}{\partial\theta}\hat{\boldsymbol{\theta}} + \frac{\partial}{\partial z}\hat{\mathbf{z}}\right).$$

When these are applied to the vector $B_\theta\hat{\boldsymbol{\theta}}$, remembering that $\partial\hat{\mathbf{r}}/\partial\theta = \hat{\boldsymbol{\theta}}$ and $\partial\hat{\boldsymbol{\theta}}/\partial\theta = -\hat{\mathbf{r}}$, we obtain

$$\left[\frac{1}{r}\frac{\partial}{\partial r}\left(r\frac{\partial B_\theta}{\partial r}\right) - \frac{B_\theta}{r^2}\right]\hat{\boldsymbol{\theta}} = \left(\frac{\partial^2 B_\theta}{\partial r^2} + \frac{1}{r}\frac{\partial B_\theta}{\partial r} - \frac{B_\theta}{r^2}\right)\hat{\boldsymbol{\theta}}$$

$$= \frac{\partial}{\partial r}\left(\frac{1}{r}\frac{\partial(r B_\theta)}{\partial r}\right)\hat{\boldsymbol{\theta}}$$

which is the expression used in equation (9.31).)

The general nature of the solutions of equation (9.32) may be inferred from its mathematical form, which is very similar to that of a diffusion equation. The azimuthal magnetic field penetrates into the plasma by a diffusive process in a characteristic time $\tau \sim \mu_0 L^2/\eta$, where L is the scale size of the plasma—in this case, its radius. Thus the plasma behaves *almost* as if it were a solid conductor of resistivity η. Although the characteristic time for penetration of the azimuthal field, i.e. $\mu_0 L^2/\eta$, may be quite long, it is short compared to the time for which the equilibrium holds together against 'leakage' of plasma across the B_z field, which is a time of order $L^2 B_z^2/p\eta$, i.e. longer by a factor β^{-1}. (In practice, especially when the leakage of plasma is 'anomalously' rapid, it is usually necessary to maintain the plasma density and pressure by means of particle and heat sources.)

This principle is employed in the tokamak to induce a current to flow around a toroidal plasma, which may be approximated by the cylindrical configuration shown in Figure 9.6, where the toroidal direction is represented by the z axis. The plasma is held in place by a strong externally generated axial field B_z. The axial current j_z (and its associated azimuthal field B_θ) is then established by induction (i.e. using the plasma as the secondary of a transformer). After the initial breakdown cascade in the induced electric field, as the current is inductively increased, at first the increased plasma current j_z flows entirely on the outermost surface of the plasma in the form of a 'skin current'. Subsequently, the axial current j_z and the azimuthal field B_θ distribute themselves within the plasma in a characteristic time $\tau \sim \mu_0 a^2/\eta$, where a is the radius of the plasma. Sometimes, in order to defeat this 'skin effect' and encourage more rapid penetration of current to the plasma interior, the radius of the plasma is 'grown' along with the rise of current so that a distributed current may be created layer by layer. For *quasi-steady* operation, the transformer can continue to apply magnetic flux to the surface of the plasma. This flux diffuses inward, continually replenishing the flux that is 'disappearing' at $r = 0$, due to the non-zero value of $E_z = \eta j_z$ at $r = 0$. For *completely steady* operation, the current in a tokamak must be sustained by other means, since the transformer primary will not be able to maintain a voltage indefinitely.

Problem 9.3: A cylindrical plasma with radius a in a strong longitudinal field B_z (such that $p \ll B_z^2/2\mu_0$) has a finite and uniform resistivity η. A current in the z direction is induced in the plasma. The total induced current I_z is then held constant in time, but initially it flows entirely in a thin skin at the surface of the plasma $r = a$. Sketch the radial profiles for $j_z(r)$ and $B_\theta(r)$ in the plasma at three different times: (i) just after $t = 0$; (ii) some intermediate time (i.e. $t \sim \mu_0 a^2/\eta$); and (iii) after a very long time (i.e. $t \gg \mu_0 a^2/\eta$). At very long times, the difference between the asymptotic steady-state field B_θ and the actual B_θ field will be a term that decays like $\exp(-t/\tau)$, where τ is a time constant. If you are familiar with Bessel functions, try to solve equation (9.32) for the asymptotic time dependence and show that the decay time constant τ is given by $\tau = \mu_0 a^2 / \eta \lambda_1^2$ where λ_1 is the first zero of the Bessel function $J_1(\lambda)$.

UNIT 3

COLLISIONAL PROCESSES IN PLASMAS

In Unit 1, we considered the motion of single charged particles in electric and magnetic fields. Throughout this analysis of particle orbits, it was assumed that particles never 'collide' with other particles. If such a collision does occur, certainly a particle will be deflected from its initial orbit and, to the extent that the initial orbit exhibits a regular or 'cyclic' pattern (e.g. Larmor gyration or closed constant-J drift surfaces), this pattern will be disrupted, at least to some degree.

When fluid equations are formulated for a 'collisional' plasma, new force-terms will appear, arising from the exchange of momentum between colliding particles. Indeed, we have already encountered one such effect, namely the electrical resistance of a plasma, which we described in terms of a resistivity η. Although we succeeded in expressing η in terms of an electron–ion collision frequency ν_{ei}, we have not yet obtained any expression for ν_{ei} itself.

In this Unit, we will consider various effects arising from collisions of the charged particles in a plasma, beginning with an analysis of whether collisions with neutral atoms or with other charged particles are more important. We will then consider in more detail the particular case of collisions with other charged particles, where the interaction is described by the Coulomb force law. We will see that collisions give rise to various processes of spatial 'diffusion' in plasmas, including the important case of diffusion across a magnetic field, which we have already encountered, indirectly, as an example of the effect of plasma resistivity. A more formal treatment of Coulomb collisions via the Fokker–Planck equation is then introduced, and collisional effects on suprathermal ions moving through a plasma are discussed.

Chapter 10

Fully and partially ionized plasmas

Collisions of charged particles in a plasma are of two types: collisions with other charged particles and collisions with neutral atoms and molecules. To most plasma physicists, the collisions with other charged particles are by far the more interesting, because they are dominant in high-temperature plasmas where the degree of ionization is high. Indeed, we will see in this Chapter that collisions with other charged particles tend to dominate over collisions with neutral particles even if the degree of ionization is only a few per cent. The opposite case—where the degree of ionization is so low that collisions with neutral particles are dominant—is usually called a 'partially ionized plasma' (or, better, a 'weakly ionized gas'). Of course, weakly ionized gases are also of practical interest: high-pressure arcs, ionospheric plasmas, process plasmas and most low-current gas discharges fall into this category.

Before we can estimate the relative importance of collisions of charged particles with other charged particles versus collisions with neutral particles, we must first estimate the density of neutral particles in a plasma—i.e. the degree of ionization.

10.1 DEGREE OF IONIZATION OF A PLASMA

Atomic processes determine the degree of ionization of a partially ionized gas. Depending on the average energy of the free electrons, the range of possibilities extends from cases where only a very small fraction of the particles are ionized to cases where the ionization is essentially complete (often with the remaining neutrals constituting only one part in about 10^6).

Recalling elementary quantum mechanics, we obtain a measure of both the 'size' of the atom, and the energy needed to ionize it. The Bohr radius of the hydrogen atom is given by

$$a_0 = \hbar^2 \frac{4\pi\epsilon_0}{me^2} \approx 5 \times 10^{-11} \text{ m} \qquad (10.1)$$

where $\hbar = h/2\pi$ is Planck's constant. The energy needed to ionize the hydrogen atom (the Rydberg) is the work needed to remove the electron from its negative-potential-energy bound state minus the kinetic energy of the bound electron, namely

$$E_i = \frac{e^2}{4\pi\epsilon_0 a_0} - \frac{mv^2}{2} = \frac{e^2}{8\pi\epsilon_0 a_0} = 13.6 \text{ eV} \qquad (10.2)$$

where we have determined the velocity of an orbiting bound electron by balancing the outward centrifugal force, mv^2/a_0, against the inward electrostatic force, $e^2/4\pi\epsilon_0 a_0^2$, which gives the well-known result that the kinetic energy of the orbiting electron is exactly half its negative potential energy.

There are two basic processes of ionization, satisfying the conditions for conservation of momentum and energy: (a) impact ionization, where an electron strikes an atom, so that an ion and two electrons come off; and (b) radiative ionization, where a photon with sufficient energy (often in the ultraviolet range) is absorbed by an atom, dissociating it into an ion and an electron. Ions can recombine into atoms by the reverse of these processes: (a) three-body recombination, where two electrons and an ion join to make a neutral atom plus a free electron; and (b) radiative recombination, where an electron and an ion combine into an atom, and a photon is emitted. These processes are illustrated in Figure 10.1.

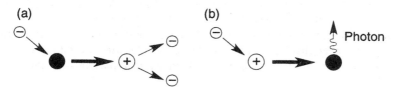

Figure 10.1. Ionization and recombination processes in a plasma: (a) electron-impact ionization, and (b) radiative recombination. The processes of three-body recombination and radiative ionization are the inverse of these processes and are obtained by reversing the direction of the arrows in (a) and (b), respectively. Neutral atoms are represented by black dots, while electrons and protons are represented by open circles labeled ($-$) and ($+$), respectively.

In strict thermodynamic equilibrium, the competing processes of ionization and recombination produce a certain ratio n_i/n_n of ions relative to neutrals, which can be calculated by statistical mechanics using the ratio of free to bound electron states. The ratio n_i/n_n is found to depend on both electron temperature and density. However this equilibrium applies only to large dense plasmas, such as stellar plasmas, where both the particles and the radiation are sufficiently well trapped that thermodynamic equilibrium is achieved between the particles and the radiation field. Most plasmas, certainly including all laboratory plasmas, are much too small to trap ultraviolet radiation. In this case, one can still have 'local thermodynamic equilibrium' in plasmas with very high density, where impact ionization and three-body recombination are more important then either of the radiative ionization or recombination processes, provided of course that the plasma particles are themselves in thermodynamic equilibrium. However, for three-body recombination to exceed radiative recombination, the plasma density must exceed a critical density that is about 10^{22} m^{-3} in the few-eV temperature range, and is even larger at higher temperatures.

At lower densities, radiative recombination is larger than three-body recombination, and a different steady-state arises, known as 'coronal equilibrium' because of its occurrence in the solar corona, in which impact ionization and radiative recombination are in balance. Here, the degree of ionization is a function only of the electron temperature, not of the density. We will give a semi-quantitative treatment of this case later in this Chapter: we find that the degree of ionization becomes very high at electron temperatures above a few eV.

In still other cases, the charged particles and neutral atoms do not reach a state of local coronal equilibrium, often because of a continuous influx of new neutrals into the plasma from the outside. In these cases, the neutral density is set by balancing ionization against the external source of neutrals, rather than against recombination. We will also give a semi-quantitative treatment of this case later in this Chapter. The neutral density is, of course, much higher in this case than if recombination were the only source of neutrals. Nonetheless, it is still generally true that at temperatures of more than a few eV, the degree of ionization is very high.

10.2 COLLISION CROSS SECTIONS, MEAN-FREE PATHS AND COLLISION FREQUENCIES

Before proceeding with quantitative treatments of these effects, we must introduce the idea of a collision 'cross section'. A cross section can be defined for any kind of collision, but for present purposes it is sufficient to consider the case of an electron colliding with a neutral atom. Even in this restricted case there can be two types of collisions: (i) 'elastic collisions' in which the electron essentially 'bounces' off the atom, with the two particles retaining their identities as electron and atom, and the atom remaining in the same energy state; and (ii) 'inelastic collisions', such as ionization or excitation, in which one or more of the particles changes its identity or internal energy state. In the first case, the electron may lose any fraction of its initial momentum, depending on the angle at which it rebounds. The probability of momentum loss can be expressed in terms of the equivalent cross section σ that the atoms would have if they were perfect absorbers of momentum. In the second case, the probability of ionization, for example, can be expressed in terms of the equivalent cross section σ that an atom would have if it were ionized by all electrons striking within this cross sectional area.

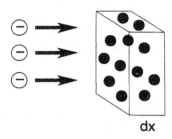

Figure 10.2. Electrons incident on a thin slab of thickness dx containing neutral atoms of density n_n.

In Figure 10.2, electrons are incident upon a thin slab of thickness dx containing n_n neutral atoms per unit volume. The atoms are imagined to be opaque spheres of cross sectional area σ: i.e. every time an electron strikes the area blocked by the atom, either it loses all of its momentum (elastic collision) or it ionizes the atom (inelastic collision). The number of atoms per unit area of the slab is $n_n dx$, and the fraction of the slab blocked by atoms is $n_n \sigma dx$. If a flux Γ of electrons is incident on the slab, the flux emerging on the other side is $\Gamma + d\Gamma = \Gamma(1 - n_n \sigma dx)$, so that the change of flux Γ with distance x is given by

$$\frac{d\Gamma}{dx} = -n_n \sigma \Gamma \qquad (10.3)$$

which has the solution

$$\Gamma = \Gamma_0 \exp(-n_n \sigma x) = \Gamma_0 \exp(-x/\lambda_{mfp}) \qquad (10.4)$$

where

$$\lambda_{mfp} = (n_n \sigma)^{-1}. \qquad (10.5)$$

The quantity λ_{mfp} is called the mean-free path for collisions. In a distance λ_{mfp}, the flux would be decreased to $1/e$ of its initial value. In other words, an electron travels a distance λ_{mfp} before it has a reasonable probability of colliding with an atom. For electrons of velocity v, the mean time between collisions is given by

$$\tau = \lambda_{mfp}/v. \qquad (10.6)$$

The 'collision frequency', namely the inverse of τ, is usually defined in terms of an average over all velocities in the Maxwellian distribution (which may have different individual collision frequencies), namely

$$v = \langle \tau^{-1} \rangle = n_n \langle \sigma v \rangle = (n_n/n_e) \int f_e(v) \sigma(v) v \mathrm{d}^2 v. \qquad (10.7)$$

As is implied by this formula, for more complex collisional processes than that illustrated in Figure 10.2, the cross section σ is often itself a function of the velocity v of the incident particle.

10.3 DEGREE OF IONIZATION: CORONAL EQUILIBRIUM

In the case where the collision of the electron with the atom results in ionization of the atom, we may calculate the rate of production of new electrons per unit volume simply by multiplying the ionization collision frequency of the electrons, equation (10.7), by the electron density, n_e. This 'source rate' S_e of electrons is given by

$$S_e = n_e n_n \langle \sigma_{ion} v_e \rangle \qquad (10.8)$$

where σ_{ion} is the cross section for electron-impact ionization and where we assume that the electron velocities v_e greatly exceed the neutral velocities v_n, so that the velocity of impact comes mainly from the electron's motion. This cross section is definitely a strong function of electron velocity, at least below energies of about 30 eV, so the averaging over the Maxwellian distribution of electrons is necessary. There is, of course, an equal and opposite 'sink rate' for neutral atoms, i.e. neutral atoms are lost by ionization at the same rate per unit volume, S_e.

The dependence of the ionization cross section σ_{ion} for hydrogen atoms on the energy of the bombarding electron is shown in Figure 10.3, and the ionization rate $\langle \sigma_{ion} v_e \rangle$ averaged over a Maxwellian distribution of electrons is shown in Figure 10.4. The maximum cross section σ_{ion} is reached for electrons with energies somewhat above E_i (the Rydberg ionization energy, which is about 13.6 eV for hydrogen) and is in the neighborhood of 10^{-20} m^2, the 'size' of the hydrogen atom. However, the ionization rate is significant even for electron temperatures well below E_i, because a Maxwellian distribution still contains a few energetic electrons that are efficient ionizers. A good approximation to the data is given by the simple formula

$$\langle \sigma_{ion} v_e \rangle = \frac{2.0 \times 10^{-13}}{6.0 + T_e(\mathrm{eV})/13.6} \left(\frac{T_e(\mathrm{eV})}{13.6} \right)^{1/2} \exp\left(-\frac{13.6}{T_e(\mathrm{eV})} \right) \mathrm{m}^3 \mathrm{s}^{-1}. \qquad (10.9)$$

The source rate for neutrals (corresponding to a sink term for electrons) in a plasma in coronal equilibrium is given by

$$S_n = n_e n_i \langle \sigma_{rec} v_e \rangle \qquad (10.10)$$

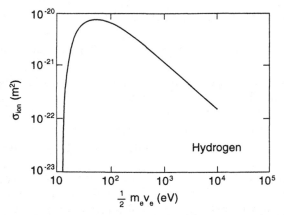

Figure 10.3. Ionization cross section σ for hydrogen atoms as a function of the energy of the bombarding electron.

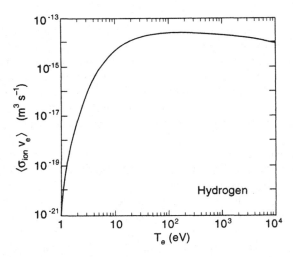

Figure 10.4. Ionization rate $\langle \sigma_{ion} v_e \rangle$ for electron-impact ionization of hydrogen atoms averaged over a Maxwellian distribution of electrons, temperature T_e.

where σ_{rec} is the cross section for radiative recombination. For a neutral hydrogen plasma, $n_i = n_e$. A good approximation to the data on radiative recombination in the relevant temperature regime is given by the simple formula

$$\langle \sigma_{rec} v_e \rangle = 0.7 \times 10^{-19} \left(\frac{13.6}{T_e(eV)} \right)^{1/2} \text{m}^3 \text{s}^{-1}. \tag{10.11}$$

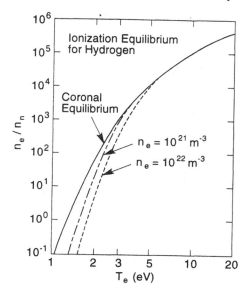

Figure 10.5. Ionization equilibrium for hydrogen in the coronal equilibrium model, and at higher electron densities with three-body recombination included.

(The formulae given in equations (10.9) and (10.11) are taken from R W P McWhirter (1965 *Spectral Intensities* in *Plasma Diagnostic Techniques* edited by R H Huddlestone and S L Leonard, New York: Academic).)

The degree of ionization of a homogeneous hydrogen plasma in coronal equilibrium is given by balancing the source of electrons by collisional ionization against the sink of electrons by radiative recombination. We find that, at an electron temperature of approximately the ionization potential, i.e. 13.6 eV, the plasma is almost fully ionized so that the neutrals constitute only about one part in 10^5. Only at electron temperatures below about 1.5 eV is the plasma less than 50% ionized. Figure 10.5 shows the degree of ionization, i.e. n_e/n_n, against electron temperature for the coronal equilibrium model, and also for higher-density plasmas where three-body recombination has been included.

The concept of coronal equilibrium can be generalized to the case of a plasma composed of, or containing an admixture of, high-Z ions. In such cases, depending mainly on the electron temperature, the ions will be stripped of their outer-shell electrons but will retain some bound inner-shell electrons. An equilibrium distribution among the various ionization states arises which, in the solar corona for example, is determined by balancing the processes of impact ionization and radiative recombination for each ionization state. A particular case is illustrated in Figure 10.6, which shows the fractional abundances in the

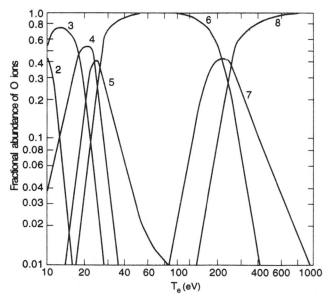

Figure 10.6. Fractional abundances in the various ionization levels for oxygen ions as a function of electron temperature in coronal equilibrium. Fully ionized oxygen has $Z = 8$. (Dielectronic recombination has been neglected in calculating these fractional abundances.)

various ionization levels of oxygen as a function of electron temperature in coronal equilibrium. We see that oxygen ions are stripped of all six outer-shell electrons (giving an ion with charge-number $Z = 6$) at electron temperatures of about 30 eV, but to remove the final two inner-shell electrons to produce fully stripped oxygen with $Z = 8$ requires temperatures in excess of about 200 eV. The validity of any coronal equilibrium model depends on the time-scale for reaching ionization/recombination balance (for the *slowest* such process, generally at the highest relevant ionization state, in the case of high-Z ions) being much shorter than the timescale on which particles are introduced into, or lost from, the plasma. If this 'confinement' time begins to be comparable to the slowest atomic processes, the ionization balance shifts towards lower charge states. If hydrogen neutrals are present, there is also the possibility of a 'charge-exchange' event, in which the electron of a neutral hydrogen atom is captured by a high-Z ion; this process also lowers the charge-state balance of the high-Z ions.

For high-Z ions, another process known as 'dielectric recombination' can play a significant role in the charge-state balance. In this recombination process, a free electron is captured into an excited state and the excess energy that is available is invested in excitation of a different bound electron to a higher state. Both electrons then decay to the ground state, emitting photons. Dielectronic recombination has not been included in calculating the charge-state distribution shown in Figure 10.6.

10.4 PENETRATION OF NEUTRALS INTO PLASMAS

To complete our discussion of neutrals in plasmas, we should consider what happens at the edge of a hot dense plasma that is enveloped by neutral gas. This situation arises in many laboratory plasmas—magnetically confined fusion plasmas or low-pressure arc discharges, for example. In such cases, the plasma is often hot and dense enough to be fully ionized, but the electrons and ions that diffuse out of the plasma recombine into neutral atoms when they strike the containing vessel. The neutral atoms thus formed are often reflected back into the plasma (or other neutrals are desorbed from 'saturated' vessel walls), where they are ionized again. Depending on the surface material of the containing vessel (and whether its surface is already saturated with a layer of hydrogen molecules), this process of recycling can be almost 'perfect', i.e. the plasma density is maintained almost indefinitely despite diffusive losses of charged particles, because the lost particles reappear one-for-one as neutrals which are readily ionized again by the plasma. For hot dense laboratory plasmas, this recycling process occurs entirely at the plasma edge, because the main body of the plasma is 'opaque' to neutrals, i.e. a neutral atom has almost no chance of reaching the center of the plasma before being ionized.

Recombination in the plasma (as distinct from at the vessel surface itself) is usually unimportant in this situation: the neutral density in the edge region of the plasma is set by a balance between the influx from outside and the ionization within the plasma.

Neutral atoms entering the plasma with velocity v_n will penetrate a distance given by the neutral 'mean-free path' for ionization, i.e.

$$\lambda_n = \frac{v_n}{n_e \langle \sigma_{ion} v_e \rangle}. \tag{10.12}$$

This can be derived by noting that the volumetric ionization rate is given by $n_e n_n \langle \sigma_{ion} v_e \rangle$ for $v_e \gg v_n$, implying that the effective 'collision frequency' of the neutrals must be $n_e \langle \sigma_{ion} v_e \rangle$. The thermal velocity of neutral hydrogen atoms at 'room temperature' is about $2 \times 10^3 \, m \, s^{-1}$. If the electron temperature in the edge-region of the plasma reaches 10–20 eV, the ionization rate $\langle \sigma_{ion} v_e \rangle$ is about $10^{-14} \, m^3 \, s^{-1}$. Thus, we have $\lambda_n(m) = 2 \times 10^{17}/n_e(m^{-3})$. For example, if the density of the edge-region plasma is about $10^{19} \, m^{-3}$, typical of many magnetically confined fusion plasmas, the neutrals will only penetrate about

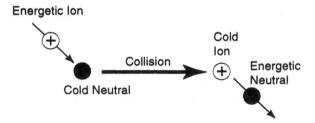

Figure 10.7. Charge-exchange process in which an energetic ion takes an electron from a cold neutral, thereby becoming an energetic neutral. A time just before the charge-exchange collision is shown to the left of the thick black arrow; a time just after the collision is shown to the right.

2 cm into the plasma. In many practical cases, such as neutrals re-emerging from a saturated surface, the hydrogen appears initially in molecular, rather than atomic, form. In such cases, the first effect of electron impact is molecular dissociation, which produces two atoms with equal and opposite momenta and each with energy of about 3 eV; the atom with momentum directed toward the plasma can penetrate somewhat further into the plasma.

A second atomic process—known as 'charge exchange'—allows much deeper penetration of hot dense plasmas by neutrals. In hydrogen charge-exchange, an energetic plasma proton captures the electron from a lower-energy

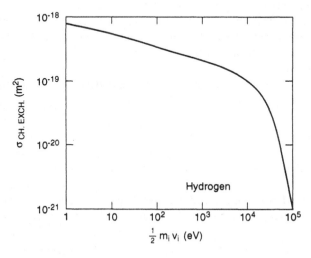

Figure 10.8. Cross section for charge exchange in hydrogen against the energy of the bombarding ion.

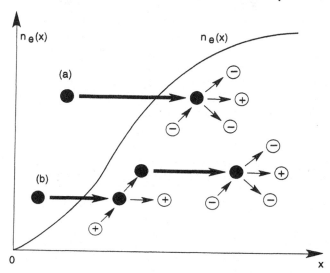

Figure 10.9. Trajectories of two neutrals incident (thick arrows on the left) upon a plasma of increasing density. Neutral (a) is ionized. Neutral (b) undergoes charge exchange, producing a more energetic neutral that penetrates (thick arrows on the left) further into the plasma before being ionized.

neutral. As a result, it can escape from the plasma, or move further into the plasma, as an energetic neutral, as illustrated in Figure 10.7. Not much energy is exchanged by the charge-exchange collision itself: the emerging neutral has about the same energy as the incident plasma ion.

The cross section for charge-exchange of hydrogen atoms bombarded by protons of various energies is shown in Figure 10.8. In the energy range of most interest for laboratory plasmas and the edge-region of fusion plasmas (10–100 eV), the cross section is seen to be quite large ($\sim 4 \times 10^{-19}\,\mathrm{m}^2$), almost a hundred times larger than the ionization cross section. (The cross section is large because charge exchange is a *resonant* process, where the initial and final quantum mechanical states have no difference in energy.) For a plasma with $T_i \approx T_e$, the charge-exchange rate $\langle \sigma_{cx} v_i \rangle$ is usually two-to-three times larger than the ionization rate $\langle \sigma_{ion} v_e \rangle$.

The process of charge exchange essentially prevents a *hot-ion* plasma from ever being formed with an appreciable neutral-gas density in the hot region. The cross section for charge exchange is so large that, if this were to occur, each energetic ion would readily turn into an energetic neutral, which would escape, so that the hot plasma would quickly be converted into cold plasma.

A low-energy neutral atom injected into the edge-region of a plasma has a somewhat higher probability of undergoing a charge-exchange event than of being ionized. Thus, it might appear that charge exchange would *reduce* the penetration of neutrals into hot dense plasmas. In fact the opposite is true, because charge exchange produces a second generation of more energetic neutrals, with energies comparable to the *ion* energies in the region of the plasma where charge exchange occurs. While some of these more energetic second-generation charge-exchange neutrals will escape from the plasma, others will penetrate much more deeply into the plasma interior than did the first-generation neutrals, until these neutrals themselves are ionized or produce a third generation of charge-exchange neutrals with even higher energies. Two neutral trajectories—one ionized and one charge-exchanged—are illustrated in Figure 10.9.

The production of multiple generations of increasingly penetrating neutral atoms by charge exchange is primarily responsible for the presence of any neutrals at all in the center of a hot dense plasma. In the core of the plasma, these neutrals have 'thermalized' with the plasma ions—i.e. they have about the same average energy. However, charge-exchange transport still provides an avenue for ion energy loss from the plasma.

10.5 PENETRATION OF NEUTRALS INTO PLASMAS: QUANTITATIVE TREATMENT*

An approximate analytic treatment of the penetration of hydrogen neutrals into hot dense plasmas, including the combined effects of ionization and charge exchange, can be given. We consider the case where the neutral mean-free path for charge-exchange collisions, namely $\lambda_{cx} \sim v_n/n_i\langle\sigma_{cx}v_i\rangle$, is quite short compared to the plasma size. Viewing charge exchange as simply a 'direction-randomizing' collision, in the sense that one incident neutral produces one charge-exchanged neutral with little correlation between the two velocities, the migration of neutrals due to successive charge-exchanges can be treated as a diffusive process, i.e. a random walk with step size λ_{cx} and frequency of steps $v_{cx} \sim n_i\langle\sigma_{cx}v_i\rangle$. (The reader who is unfamiliar with the concept of a random walk and its description by means of a diffusion coefficient is referred to the discussion at the beginning of Chapter 12.) Thus, the diffusion coefficient for neutrals is essentially

$$D_n \approx v_{cx}\lambda_{cx}^2 \approx v_{t,i}^2/n_i\langle\sigma_{cx}v_i\rangle \tag{10.13}$$

where we have replaced v_n by $v_{t,i}$, the thermal velocity of ions since, after successive charge exchanges, the neutral energies will have reached approximate thermal equilibrium with ion energies. To consider the penetration distance in a particular case, we will take account of the fact that $D_n \propto n_i^{-1}$, but we will otherwise neglect the spatial dependence of the various quantities. In particular, we will treat the ion temperature and the quantity $\langle \sigma_{cx} v_i \rangle$ as essentially constant throughout the edge-region of the plasma under consideration.

We consider a simple one-dimensional case, in which a plasma occupies the entire region $x > 0$. The plasma is in contact with a material surface at $x = 0$, at which charged particles recombine into neutrals, which are then re-injected back into the plasma. When an equilibrium has been reached between the flow of plasma to the material surface at $x = 0$ and the flux of neutrals into the plasma, the density of charged particles, say the ion density $n_i(x)$, will be an increasing function of x, with $n_i(0) = 0$, while the density of neutrals $n_n(x)$ will be a decreasing function of x, with a finite value at $x = 0$, as shown in Figure 10.10. In order to obtain a specific form for the profiles $n_i(x)$ and $n_n(x)$, it is necessary to make some assumption about what governs the rate of leftward plasma flow. For our calculation here, we assume that the plasma flow is *diffusive*, i.e. the particle flux is proportional (and opposite) to the density gradient, with a constant of proportionality which is called the 'plasma diffusion coefficient'. Moreover, for present purposes we will simply take the diffusion coefficient D in the edge-region of the plasma to be a constant independent of the plasma parameters such as density and temperature. This contradicts the predictions of the theory of collisional plasma diffusion to be presented in Chapter 12. However, for practical cases where diffusion is dominated by turbulent processes, the choice of a constant D may be a reasonable approximation to the actual physical situation.

The diffusion equations for ion density (equal to the electron density, by assumption) and neutral density, with appropriate source and sink terms from ionization, are

$$\frac{\partial n_i}{\partial t} = D \frac{\partial^2 n_i}{\partial x^2} + n_i n_n \langle \sigma_{ion} v_e \rangle \tag{10.14}$$

$$\frac{\partial n_n}{\partial t} = \frac{\partial}{\partial x} \left(D_n \frac{\partial n_n}{\partial x} \right) - n_i n_n \langle \sigma_{ion} v_e \rangle \tag{10.15}$$

where D is the assumed plasma diffusion coefficient. Assuming steady state ($\partial/\partial t = 0$), adding the two equations and integrating once, we obtain

$$D \frac{\partial n_i}{\partial x} + D_n \frac{\partial n_n}{\partial x} = 0. \tag{10.16}$$

(The constant of integration is chosen to be zero to express the fact that there is no net flux of particles to the wall, i.e. 'perfect' recycling.) Integrating once more, taking account of the inverse dependence of D_n on n_i, we obtain

$$\tfrac{1}{2} D(n_{i\infty}^2 - n_i^2) = D_{n\infty} n_{i\infty} n_n \tag{10.17}$$

where the suffix '∞' refers to values in the plasma, in the region sufficiently deep into the plasma that it is not penetrated by neutrals and that corresponds to

$x \rightarrow \infty$ in our present analysis, where the following boundary conditions apply:

$$n_i \rightarrow n_{i\infty} \qquad n_n \rightarrow 0 \qquad D_n \rightarrow D_{n\infty}. \qquad (10.18)$$

Substituting for n_n into the equation for n_i, we obtain

$$\frac{\partial^2 n_i}{\partial x^2} + \frac{\langle \sigma_{ion} v_e \rangle}{2 D_{n\infty} n_{i\infty}} n_i (n_{i\infty}^2 - n_i^2) = 0. \qquad (10.19)$$

Writing

$$\frac{\partial^2 n_i}{\partial x^2} = \left(\frac{\partial}{\partial n_i} \right) \frac{1}{2} \left(\frac{\partial n_i}{\partial x} \right)^2$$

and neglecting any spatial variation of $\langle \sigma_{ion} v_e \rangle$, we can integrate this equation once more to produce

$$\frac{\partial n_i}{\partial x} = \left(\frac{\langle \sigma_{ion} v_e \rangle}{4 D_{n\infty} n_{i\infty}} \right)^{1/2} (n_{i\infty}^2 - n_i^2) \qquad (10.20)$$

which has the following solutions:

$$
\begin{aligned}
n_i(x) &= n_{i\infty} \tanh(x/x_0) \\
n_n(x) &= n_{no} \operatorname{sech}^2(x/x_0) \\
x_0 &= \left[4 D_{n\infty} / (\langle \sigma_{ion} v_e \rangle) n_{i\infty} \right]^{1/2} \\
&= 2 v_{t,i} / \left[n_{i\infty} (\langle \sigma_{ion} v_e \rangle \langle \sigma_{cx} v_i \rangle)^{1/2} \right] \\
n_{n0} &= D n_{i\infty} / 2 D_{n\infty}.
\end{aligned}
\qquad (10.21)
$$

These solutions for the ion density profile $n_i(x)$ and the neutral density profile $n_n(x)$ have the shapes shown in Figure 10.10. The effective penetration distance of neutrals x_0, is the *geometric mean* of the mean-free paths for ionization and charge-exchange for a neutral with a velocity of the order the *thermal velocity of ions in the edge region of the plasma*—much larger, of course, than the velocity of neutrals in the 'room-temperature' gas that is assumed to surround the plasma. Nonetheless, for many high-temperature dense plasmas, the penetration distance is still small, with the result that the core of the plasma is almost fully ionized.

Problem 10.1: Estimate the penetration distance of neutral atoms (including charge-exchange processes) into a thermonuclear plasma with a central density of 10^{20} m^{-3}, assuming that the ion temperature in the edge region is about 100 eV. In our quantitative treatment of neutral penetration, we assumed that $\langle \sigma_{ion} v_e \rangle$ and $\langle \sigma_{cx} v_i \rangle$ are roughly constant. In the temperature range of interest, examine the data in Figures 10.4 and 10.8 to assess how good an approximation this is.

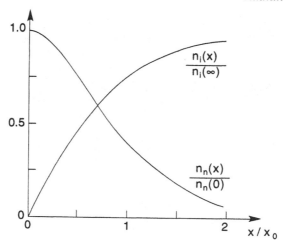

Figure 10.10. Profiles of ion (or electron) density and neutral density in the edge region of a plasma of sufficient density to be opaque to neutrals. Both charge exchange and ionization are included.

Problem 10.2: Consider neutral penetration into plasmas as discussed in the preceding section, but ignore diffusion of the ions, simply assuming a uniform charged-particle density, i.e. n_i = constant. Show that the neutral density profile shape in this case is proportional to $\exp(-2x/x_0)$, with x_0 as given in equation (10.21). Why is the typical neutral penetration distance, i.e. $x_0/2$ versus x_0, shorter in this case?

10.6 RADIATION

Certain inelastic scattering processes involving radiation can be important in plasmas, especially those composed of ions other than hydrogen, as well as hydrogen plasmas containing small admixtures of higher-Z ions. As we have already seen, in these cases there are collisional and radiative processes of ionization and recombination among the various partially stripped states (i.e. high-Z ions still with some bound electrons), and there are also collisional processes of excitation of ions to higher energy levels and the associated process of 'line radiation'. For such processes, the total power radiated per unit volume is proportional to the product of the electron density and the high-Z ion density and is also a strong function of electron temperature. Typically, line radiation is somewhat larger than recombination radiation, although both processes contribute importantly. Line radiation is relatively intense when there is a high fractional abundance in those ionization levels corresponding to partially filled shells. Thus, the dependence of the total radiated power on electron temperature is non-monotonic, reflecting the temperature dependence of the fractional abundances in these ionization levels (see, for example, Figure 10.6). The total radiated power for oxygen ions in coronal equilibrium (including both recombination and line radiation) as a function of electron temperature is shown in Figure 10.11. Figure 10.11 should be compared with Figure 10.6, which shows the fractional abundances among the various ionization levels for the same case.

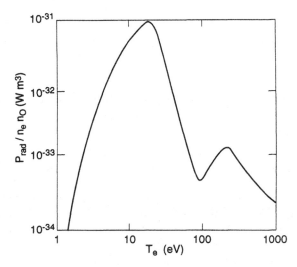

Figure 10.11. Power radiated per unit volume in line and recombination radiation, P_{rad}(W m^{-3}), in coronal equilibrium for oxygen. All densities are in m^{-3}; n_O refers to the total density of oxygen in all ionization states.

Even Coulomb collisions in a fully ionized plasma give rise to radiation, called 'bremsstrahlung' or sometimes 'free–free bremsstrahlung', which comes from electromagnetic waves emitted by the accelerating/decelerating electrons as they are deflected by the Coulomb attraction of the ions. A derivation of bremsstrahlung is given in the next Chapter after the specific properties of Coulomb collisions have been analyzed. Bremsstrahlung from Coulomb collisions of electrons with high-Z ions is generally important relative to recombination and line radiation only when the electron temperature is high enough for the ion to be fully stripped. For example, the power radiated by bremsstrahlung from electron collisions with oxygen ions is insignificant across the range of electron temperatures depicted in Figure 10.11.

Another type of radiation arises in *magnetized* plasmas even without collisional effects, because the electrons are continuously accelerating (i.e. they have a $\dot{v}$ because of the changing direction of the velocity vector) as they execute their Larmor orbits. This is called 'cyclotron radiation' and occurs primarily at the electron cyclotron frequency and its low harmonics. At sufficiently high electron temperatures that relativistic effects become important, this radiation shifts to higher harmonics and, eventually, these harmonics overlap so much that the spectrum forms a continuum. In this case, the radiation is called 'synchrotron radiation'.

Except for this brief introduction to the topic, atomic processes involving high-Z ions and plasma radiation processes are outside the scope of this book. The reader interested in these topics is referred to the excellent texts by G Bekefi (1966 *Radiation Processes in Plasmas* New York: Wiley) and by H R Griem (1964 *Plasma Spectroscopy* New York: McGraw-Hill). Computations of the power radiated in coronal equilibrium by recombination and line radiation for a variety of high-Z ions have been given by D E Post and R V Jensen (1977 *At. Data Nucl. Data Tables* **20** 5).

10.7 COLLISIONS WITH NEUTRALS AND WITH CHARGED PARTICLES: RELATIVE IMPORTANCE

Finally, we return to the question raised of the beginning of this Chapter: what is the relative importance of collisions of charged particles in a plasma with other charged particles versus collisions with neutral particles?

The cross section for elastic scattering of an electron by a neutral atom may be estimated very roughly as

$$\sigma_n \sim \pi a_0^2 \sim 10^{-20}\,\text{m}^2. \tag{10.22}$$

At the distance a_0, an incoming electron has a substantial chance of undergoing a large-angle collision. On the other hand, when an electron comes within a distance r of a singly charged (e.g. hydrogen) ion, it experiences an attractive Coulomb force:

$$F_r = -e^2/4\pi\epsilon_0 r^2 \tag{10.23}$$

which tends to deflect the electron orbit toward the ion. When the angle of deflection is as much as 90°, the electron's initial momentum is mostly lost. Thus, from the viewpoint of momentum exchange, a 'close encounter' with the Coulomb force of another charged particle is essentially the same as a 'collision'. The angle of deflection will be large when the potential energy of the Coulomb interaction equals the kinetic energy of the colliding electron, i.e.

$$e^2/4\pi\epsilon_0 b \sim mv^2/2 \sim T_e \tag{10.24}$$

where m and v are the mass and velocity of the electron, and where b is the distance of closest approach of the electron to the ion. This serves to define an effective 'Coulomb cross section' of the ion, namely

$$\sigma_i \sim \pi b^2 \sim \frac{\pi e^4}{(4\pi \epsilon_0)^2 T_e^2} \sim 10^{-17}/T_e(\text{eV})^2 \, \text{m}^2 \qquad (10.25)$$

where $T_e(\text{eV})$ denotes the electron temperature measured in eV. (In fact, we will see in the next Chapter that the effective Coulomb cross section is actually almost two orders of magnitude larger than this because of the cumulative effect of multiple small-angle deflections.)

Comparing σ_n with σ_i, simply using equation (10.25) for the latter, and consulting Figure 10.5 to relate the degree of ionization to T_e, we see that *Coulomb collisions will dominate over collisions with neutrals in any plasma that is even just a few per cent ionized.* Only if the ionization level is very low ($< 10^{-3}$) can neutral collisions dominate. Moreover, a plasma becomes almost fully ionized at electron temperatures above about 1 eV. Thus, the case of collisions with neutrals is not of much concern to the physicist interested in high-temperature plasmas. Not only are high-temperature plasmas almost fully ionized, but the dynamical behavior of charged particles even in partially ionized plasmas with more than very small ionization levels tends to be dominated by Coulomb collisions with other charged particles, rather than by collisions with neutrals. Of course, the various inelastic scattering processes involving high-Z ions discussed in the previous Section, i.e. ionization, recombination and excitation, are still more important than Coulomb collisions in determining the radiation from high-temperature plasmas, provided only that there remains a sufficient fraction of these ions in partially stripped ionization levels.

Chapter 11

Collisions in fully ionized plasmas

When an electron collides with an ion, the electron is gradually deflected by the long-range Coulomb field of the ion. It is still possible to think in terms of a cross section for this kind of collision. At the end of Chapter 10, we derived an estimate for the effective cross section of a hydrogen ion, namely

$$\sigma_i \sim \frac{\pi e^4}{(4\pi\varepsilon_0)^2 m^2 v^4} \tag{11.1}$$

which was obtained by calculating how close the electron must come to the ion for the potential energy of the Coulomb interaction to be comparable to the electron's kinetic energy. In fact, the effective cross section for Coulomb scattering is *considerably larger than this*, as we shall see from the following more detailed analysis that takes into account the effects of multiple small-angle deflections of the electron.

11.1 COULOMB COLLISIONS

We again consider an electron of mass m, charge $-e$ and velocity v approaching a fixed ion of charge Ze. To obtain the most general result possible, we will allow $Z \neq 1$, thereby including multiply charged ions as well as hydrogen. In the absence of Coulomb forces, the electron would have a distance of closest approach b, called the *impact parameter*, as illustrated in Figure 11.1. In the presence of the Coulomb attraction, the electron will be deflected through an angle θ, which will of course be related to the impact parameter b.

It is well known that a particle acted upon by an inverse-square-law force will execute a hyperbolic orbit. It is shown in standard textbooks on classical mechanics (see also Problem 11.1) that the angle of deflection for a light particle

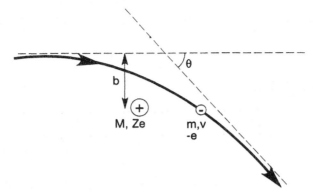

Figure 11.1. The orbit of an electron undergoing a Coulomb collision with a fixed ion of charge Ze.

colliding with a much heavier (infinitely massive) stationary particle is given by

$$\tan\frac{\theta}{2} = \frac{Ze^2}{4\pi\epsilon_0 mv^2 b}. \tag{11.2}$$

Problem 11.1: Prove relationship (11.2) for the deflection of an electron of mass m in a Coulomb collision with a much heavier ion, charge Ze. (Hint: use polar coordinates centered on the scattering ion and remember that energy and angular momentum are conserved.)

For scattering through $90°(\theta/2 = 45°, \tan(\theta/2) = 1)$, the impact parameter b must have the value

$$b_0 = \frac{Ze^2}{4\pi\epsilon_0 mv^2} \tag{11.3}$$

and equation (11.2) for the angle of deflection for a general impact parameter becomes $\tan(\theta/2) = b_0/b$. Thus the cross section of the ion for $90°$-scattering is

$$\sigma_i = \pi b_0^2 = \frac{\pi Z^2 e^4}{(4\pi\epsilon_0)^2 m^2 v^4} \tag{11.4}$$

which agrees with the rough estimate given previously. However, as we stated before, the effective cross section for Coulomb scattering is considerably larger than this. The reason is that the cross section given above is based on large-angle collisions alone. In practice, because of the long-range nature of the Coulomb force, small-angle collisions are much more frequent than large-angle collisions, and *the cumulative effect of many small-angle deflections turns out to be larger than the effect of the relatively fewer large-angle deflections.*

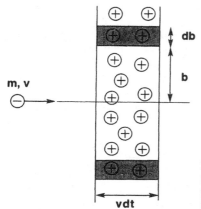

Figure 11.2. Electron Coulomb scattering by ions in an annular element of volume with impact parameters between b and $b + db$ as the electron moves a distance vdt.

To see this, we must consider the cumulative effect of many scatterings by many different ions with different values of the impact parameter b. Consider an electron with initial velocity v in the z direction, and suppose that it undergoes a large number of small-angle scattering events. In each event, the electron will be given small incremental velocity components Δv_x and Δv_y, but since there is no preferred direction for scattering (i.e. the electron is just as likely to be given a negative Δv_x as a positive one), the averages must vanish, i.e.

$$\langle \Delta v_x \rangle = \langle \Delta v_y \rangle = 0. \tag{11.5}$$

However, the mean square deflections do not vanish, so that

$$\langle (\Delta v_x)^2 \rangle = \langle (\Delta v_y)^2 \rangle = \tfrac{1}{2} \langle (\Delta v_\perp)^2 \rangle \neq 0 \tag{11.6}$$

where $\perp$ (and later $\parallel$) are relative to the particle's initial direction of motion, here taken to be in the z direction. For Coulomb collisions, we have seen that

$$\tan\frac{\theta}{2} = \frac{b_0}{b} \tag{11.7}$$

so that, using the trigonometric identities

$$\sin\theta = 2\sin(\theta/2)\cos(\theta/2) = 2\tan(\theta/2)\cos^2(\theta/2) = 2\tan(\theta/2)/[1 + \tan^2(\theta/2)]$$

we see that

$$\sin\theta = \frac{2(b/b_0)}{1 + (b/b_0)^2}. \tag{11.8}$$

For a single scattering event, i.e. a single electron passing a single ion, we have

$$(\Delta v_\perp)^2 = v^2 \sin^2\theta = \frac{4v^2(b/b_0)^2}{[1 + (b/b_0)^2]^2}. \tag{11.9}$$

Consider the average behavior of an electron passing many ions, as shown in Figure 11.2. In a time dt, the electron will move a distance $v\,dt$, and the number of ions in a scattering element defined by having impact parameter between b and $b + db$ is obtained by multiplying the ion density n_i by the volume of the scattering element, $2\pi b\,db\,v\,dt$, giving a number of ions $2\pi n_i b v\,db\,dt$. Integrating over impact parameters and differentiating with respect to time, we find that, on the average, the electron is deflected so that its perpendicular velocity changes at the rate

$$\frac{d\langle(\Delta v_\perp)^2\rangle}{dt} = 2\pi n_i v \int (\Delta v_\perp)^2 b\,db = 8\pi n_i v^3 \int \frac{(b/b_0)^2 b\,db}{[1 + (b/b_0)^2]^2}. \tag{11.10}$$

In principle, the integral should be taken over all values of b, from $b = 0$ to $b = \infty$. However, although the integral is well defined at the lower limit of integration, it diverges logarithmically at large values of b. For the moment, we will avoid this problem by simply introducing an 'ad hoc' cut-off at $b = b_{max}$. Evaluating the integral explicitly by substituting $y = 1 + (b/b_0)^2$, we obtain

$$\frac{d\langle(\Delta v_\perp)^2\rangle}{dt} = 4\pi n_i v^3 b_0^2 \left\{ \ln\left[1 + \left(\frac{b_{max}}{b_0}\right)^2\right] + \frac{1}{1 + (b_{max}/b_0)^2} - 1 \right\}$$

$$= 8\pi n_i v^3 b_0^2 \ln\Lambda = \frac{n_i Z^2 e^4 \ln\Lambda}{2\pi \epsilon_0^2 m^2 v} \tag{11.11}$$

where the final two forms are for the case where

$$\Lambda \equiv b_{max}/b_0 \gg 1. \tag{11.12}$$

Since the electron energy is nearly conserved in the collision (a light particle scattering off a heavy particle loses its momentum but not much of its energy), there is a reduction $\Delta v_\parallel$ in the velocity parallel to the original direction of motion. Noting that the initial velocity v is, by definition, entirely in the parallel direction and that perpendicular velocities arise only from the collisions, the energy conservation equation $(v + \Delta v_\parallel)^2 + (\Delta v_\perp)^2 = v_\parallel^2$ tells us that

$$v(\Delta v_\parallel) + \tfrac{1}{2}(\Delta v_\perp)^2 = 0 \tag{11.13}$$

showing that $\Delta v_\parallel$ is second order in $\Delta v_\perp$ and thus justifying the neglect of the fourth-order term from $(\Delta v_\parallel)^2$. We then obtain

$$\frac{d\langle\Delta v_\parallel\rangle}{dt} = -4\pi n_i v^2 b_0^2 \ln\Lambda = -\frac{n_i Z^2 e^4 \ln\Lambda}{4\pi \epsilon_0^2 m^2 v^2}. \tag{11.14}$$

This relation allows us to define a collision rate ν_{ei} (dimensions of a frequency, i.e. inverse time) for loss of electron momentum, i.e.

$$\frac{d\langle \Delta v_{\parallel} \rangle}{dt} = -\nu_{ei} v \tag{11.15}$$

$$\begin{aligned} \nu_{ei} &= 4\pi n v b_0^2 \ln \Lambda \\ &= \frac{n_i Z^2 e^4 \ln \Lambda}{4\pi \epsilon_0^2 m^2 v^3}. \end{aligned} \tag{11.16}$$

Note that the 'collision frequency' ν_{ei} varies inversely with the cube of the electron velocity v.

An estimate for the quantity Λ may be obtained by noting that a charged particle will interact weakly with particles further removed from it than the Debye length, λ_D. As was discussed in Chapter 1, the charged particle produces an electrostatic potential $\phi = e/4\pi\epsilon_0 r$, which perturbs the density of neighboring particles. The effect of this local charge separation is to shield out the electric potential at distances r greater than a Debye length, $\lambda_D \approx (\epsilon_0 T/ne^2)^{1/2}$. Indeed, in Problem 1.3, it was shown that Debye shielding results in an *exponential* decrease in the electric potential for $r > \lambda_D$. Thus, the maximum impact parameter should be taken to be λ_D, because Debye shielding suppresses the Coulomb field at larger distances. Accordingly

$$\begin{aligned} \Lambda &\sim b_{max}/b_0 \sim \lambda_D/b_0 \\ b_0 &\sim Ze^2/12\pi\epsilon_0 T \sim (Z/12\pi)(n\lambda_D^2)^{-1} \end{aligned} \tag{11.17}$$

where, in evaluating b_0 as an average over a Maxwellian distribution of electrons, we have taken $mv^2 \sim 3T$. We see that $\Lambda \sim (12\pi/Z)n\lambda_D^3$, which shows that our definition of a plasma, i.e. $n\lambda_D^3 \gg 1$ (see Chapter 1) *implies* that Λ must also be a large number.

Problem 11.2: Defining an electron mean-free path for Coulomb collisions with ions by $\lambda_{mfp} = v/\nu_{ei}$, show that the ratio of this length to the Debye length λ_D is given by $\lambda_{mfp}/\lambda_D \sim \Lambda/\ln\Lambda \gg 1$.

Although Λ depends on n and T, its logarithm is fairly insensitive to the exact values of these parameters. Typical values of $\ln\Lambda$ are given in Table 11.1. It is evident that $\ln\Lambda$ varies by not more than a factor of two as the plasma parameters range over many orders of magnitude. For rough estimates of collision rates, it is usually sufficient to consult a table such as this, rather than evaluate $\ln\Lambda$ directly.

Table 11.1. Values of $\ln\Lambda$ for naturally occurring and laboratory plasmas.

	$n(\mathrm{m}^{-3})$	$T\,(\mathrm{eV})$	$\ln\Lambda$
Solar wind	10^7	10	26
Van Allen belts	10^9	10^2	26
Earth's ionosphere	10^{11}	10^{-1}	14
Solar corona	10^{13}	10^2	21
Gas discharge	10^{16}	10^0	12
Process plasma	10^{18}	10^2	15
Fusion experiment	10^{19}	10^3	17
Fusion reactor	10^{20}	10^4	18

Problem 11.3: At high electron temperatures, the minimum impact parameter b_0 appearing in the Coulomb logarithm becomes so small that quantum mechanical effects must be included. Show that for this case, b_0 should be taken to be the de Broglie wavelength $\hbar/mv$. For Maxwellian electrons, we take $v \sim (3T/m)^{1/2}$. For electron collisions, at what temperature will these quantum mechanical effects become important in determining b_0? Which of the values of $\ln\Lambda$ given for typical laboratory and naturally occurring plasmas in Table 11.1 have involved this quantum mechanical correction?

We are now in a position to compare the total multiple-small-angle-collision Coulomb cross section with the 90°-scattering Coulomb cross section. The total cross section for scattering of electrons by massive stationary ions can be obtained from the usual relation between collision frequency ν_{ei} and cross section σ_{ei}, namely

$$\nu_{ei} = n_i \sigma_{ei} v \tag{11.18}$$

giving

$$\sigma_{ei} = \frac{Z^2 e^4 \ln\Lambda}{4\pi \epsilon_0^2 m^2 v^4}. \tag{11.19}$$

We see that the actual cross section exceeds the 90°-scattering cross section by a factor $4\ln\Lambda \sim 70$. The large size of the Coulomb cross section arises from the cumulative effect of very many small-angle scatterings. It is peculiar to the r^{-2} force law: it does not happen for force laws with a sharper drop-off with increasing r. As we have stated previously, this effect increases even further the ratio of the Coulomb cross section of an ion to the cross section of a neutral atom.

11.2 ELECTRON AND ION COLLISION FREQUENCIES

We have obtained an expression for the collision frequency for (light) electrons striking (heavy) ions. The collision frequency varies with electron velocity as v^{-3}, i.e. the more fast-moving the electron the less frequently it collides with ions (in contrast, of course, to a 'hard-sphere' collisional model). In order to define an *average* electron collision frequency, it is useful to evaluate the frictional force on a distribution of electrons drifting through essentially stationary ions, namely

$$\mathbf{F} = -n_e m \langle \nu_{ei} \mathbf{v} \rangle \tag{11.20}$$

where the average is over the distribution of electron velocities. For present purposes, we suppose that the drifting electrons have a 'shifted Maxwellian' distribution, i.e. a Maxwellian distribution relative to a non-zero mean velocity $\mathbf{u}$, which we take to be in the z direction, i.e. $\mathbf{u} = u_z \hat{\mathbf{z}}$. We assume also that $u_z \ll v_{t,e}$, where $v_{t,e}$ is the electron thermal velocity, $(T_e/m)^{1/2}$, and we expand the distribution function retaining terms up to first order in $u_z/v_{t,e}$. We then obtain

$$f_e(v) = \frac{n_e}{(2\pi)^{3/2} v_{t,e}^3} \exp\left(-\frac{|\mathbf{v} - \mathbf{u}|^2}{2 v_{t,e}^2}\right)$$

$$\approx \frac{n_e}{(2\pi)^{3/2} v_{t,e}^3} \left(1 + \frac{\mathbf{u} \cdot \mathbf{v}}{v_{t,e}^2}\right) \exp\left(-\frac{v^2}{2 v_{t,e}^2}\right)$$

$$\approx \left(1 + \frac{u_z v_z}{v_{t,e}^2}\right) f_{e0}(v)$$

where f_{e0} is the 'unshifted' Maxwellian distribution. Using this distribution in equation (11.20), we obtain

$$F_z = -m \int \nu_{ei} v_z f_e \mathrm{d}^3 v = -m u_z \int \frac{v_z^2}{v_{t,e}^2} \nu_{ei} f_{e0} \mathrm{d}^3 v$$

$$= -\frac{m u_z}{3} \int \frac{v^2}{v_{t,e}^2} \nu_{ei} f_{e0} \mathrm{d}^3 v$$

where in the final step we have noted that f_{e0} is spherically symmetric in velocity space, so the integral over v_z^2 must be one third of the integral over v^2. Substituting equation (11.16) for ν_{ei} as a function of v, we see that there arises an integral which may be evaluated as follows:

$$\int \frac{f_{e0}(v)}{v} \mathrm{d}^3 v = 4\pi \int_0^\infty f_{e0}(v) v \mathrm{d}v = 2\pi \int_0^\infty f_{e0}(v) \mathrm{d}(v^2)$$

$$= \left(\frac{2}{\pi}\right)^{1/2} \frac{n_e}{v_{t,e}}. \tag{11.21}$$

Our final expression for the frictional force becomes

$$F_z = -n_e m \langle \nu_{ei} \rangle u_z$$

where

$$\langle \nu_{ei} \rangle = \frac{2^{1/2} n_i Z^2 e^4 \ln \Lambda}{12 \pi^{3/2} \epsilon_0^2 m^{1/2} T_e^{3/2}}. \tag{11.22}$$

As we will see later in this Chapter, and again in Chapters 13 and 14, different collisional processes introduce different averagings over the Maxwellian distribution of colliding particles, each of which introduces a different numerical factor in the applicable collision frequency. However, it is useful to have a standardized definition of average electron collision frequency, and this is what is given in equation (11.22). Note that the ion mass does not appear in the expression for $\langle \nu_{ei} \rangle$; for practical purposes, it can be taken to be infinite.

In addition to their collisions with ions, electrons also collide with other electrons. In this case, the Coulomb force is repulsive, and the impinging electron is deflected away from the scattering electron. Electron–electron collisions are more complicated to analyze, since the scattering particle may no longer be taken to be fixed. However, since the Coulomb force has the same magnitude, an electron is deflected about the same amount in a collision with another electron as in a collision with a hydrogenic ion (at the same impact parameter). Thus, to within factors of order unity, we have

$$\langle \nu_{ee} \rangle \approx \frac{n_e e^4 \ln \Lambda}{\epsilon_0^2 m^{1/2} T_e^{3/2}} \approx \frac{\langle \nu_{ei} \rangle}{n_i Z^2 / n_e}. \tag{11.23}$$

In a hydrogenic plasma ($Z = 1$), electrons collide with other electrons as frequently as they collide with ions. However, in a plasma containing many different ions with differing Z values, the effective electron–ion collision frequency is higher than the electron–electron collision frequency by a factor of about $Z_{eff} = \Sigma_i n_i Z_i^2 / n_e$, where the sum is over the ion species present.

Ions make Coulomb collisions with other ions and with electrons. From the viewpoint of the relatively massive ion, momentum exchange through collisions with electrons is generally not very important, since the momentum gained or lost by the ion in such a collision is relatively small. Indeed, ion collisional processes are generally dominated by collisions of ions with other ions. Although the calculation presented above (for electrons) is not strictly applicable to this case (since we can no longer treat the scattering particle as infinitely heavy relative to the scattered particle), nonetheless it gives the correct result to within factors of order unity.

In order to define an *average* ion–ion collision frequency, as we have just done for the electron–ion collision frequency, we consider the frictional force on a population of ions drifting through another population of ions of the same species. Relative to the case of electrons colliding with massive ions, we might expect a somewhat smaller frictional force in this case (at the same relative drifting speed and the same collision frequency), because the scattering ion can take up some finite fraction of the momentum of the scattered ion. For both populations of ions added together, the total momentum must of course be conserved in ion–ion collisions. The formal method for treating the dynamics of a collision between two ions would be to go to the center-of-mass frame, in which an ion pair is 'replaced' by a particle with the combined mass moving at the mass velocity, together with a particle with the reduced mass, $M_1 M_2/(M_1 + M_2)$, moving at the relative velocity, $\mathbf{v}_{rel} = \mathbf{v}_1 - \mathbf{v}_2$. For the case of two populations of ions of the same species, the reduced mass in $M/2$. The calculation of the frictional force between the two populations of ions due to ion–ion collisions will go through just as for the case of electron–ion friction, except that the momentum transfer will be proportional to $\mathbf{v}_1 - \mathbf{v}_2$ and the relevant collision frequency will vary as $|\mathbf{v}_1 - \mathbf{v}_2|^{-3}$. However, since the frictional force was found to be proportional to the square-root of the mass and the relevant mass here is the reduced mass, $M/2$, an additional numerical factor of $2^{-1/2}$ arises in the ion–ion case, as well as the change $m \rightarrow M$, relative to the electron–ion case given in equation (11.22). Thus, we may define an average frequency for ions colliding with other similar ions, namely

$$\langle \nu_{ii} \rangle = \frac{n_i Z^4 e^4 \ln \Lambda}{12\pi^{3/2} \epsilon_0^2 M^{1/2} T_i^{3/2}}. \tag{11.24}$$

Equation (11.24) gives the standard expression for the average ion collision frequency. Although individual ions scatter with frequency ν_{ii}, it is important to remember that the total momentum and total energy of the ion population cannot be changed by ion–ion collisions alone, since momentum and energy are conserved in Coulomb collisions when both scattering and scattered particles are summed together.

Comparing electron and ion collision frequencies in a plasma with $T_e \sim T_i$, we see that

$$\nu_{ei}/\nu_{ii} \sim (M/m)^{1/2}. \tag{11.25}$$

Thus, electrons scatter about 40 times faster than ions in a hydrogen plasma.

For a hydrogen plasma with an electron and ion density n (in particles per m^3) and electron and ion temperatures $T_{e,i}$ (here the temperatures are in eV), the collision frequencies given by equations (11.22), (11.23) and (11.24) are

$$\langle \nu_{ei} \rangle \sim \langle \nu_{ee} \rangle \sim 5 \times 10^{-11} n/T_e^{3/2} \quad (\text{s}^{-1})$$
$$\langle \nu_{ii} \rangle \sim 10^{-12} n/T_i^{3/2} \quad (\text{s}^{-1}).$$

Numerical values for collision frequencies vary enormously, depending on the plasma density and temperature.

Consistent with the collision frequency varying as $nT^{-3/2}$, the cross section σ for Coulomb scattering varies as T^{-2} and is independent of density. This allows us, for example, to compare Coulomb cross sections with the deuterium–tritium fusion cross section, as is done in Figure 11.3. We see that the Coulomb cross section is always much larger than the fusion cross section. Thus, ions in a fusion reactor must be confined for many collision times for them to have a good chance of fusing. To maximize fusion reactivity at a fixed value of the plasma pressure p (since the beta value, $\beta = 2\mu_0 p/B^2$, is limited by plasma physical constraints, and the field strength B is limited by technological constraints), the optimum plasma temperature in a fusion reactor is in the range 10–30 keV. In this range, Figure 11.3 shows that there will be of order ten thousand or more scattering events per fusion reaction. Although only a fraction of the 'fuel' (i.e. the deuterium–tritium ions) in a fusion reactor need actually fuse before being lost, since the energy produced per reaction is a thousand times larger than the average energy of the ions, it is clear nonetheless that concepts for fusion reactors must involve confinement of ions for many collision times, so that the ion velocity distribution must necessarily be essentially Maxwellian.

11.3 PLASMA RESISTIVITY

When an electric field is applied to a fully ionized plasma, the electrons are accelerated in one direction (opposite to **E** since their charge is negative) and the ions are accelerated in the other direction (along **E**). The increasing relative motion between electrons and ions produces an increasing electrical current in the direction of **E**. However, Coulomb collisions between electrons and ions impede this relative motion, and a steady state is reached after a few electron–ion collision times. In equilibrium, the electric field **E** and the plasma current density **j** are proportional to one another, i.e.

$$\mathbf{E} = \eta \mathbf{j}. \tag{11.26}$$

The constant of proportionality η is the resistivity. So far, we have paid some attention to the effects arising from resistivity, but we have investigated neither the magnitude nor dependences of the resistivity itself.

The resistivity was obtained in Chapter 6 by considering the equation of motion for electrons in a uniform plasma (no pressure gradients) either along a magnetic field **B** or with no magnetic field:

$$mn_e d\mathbf{u}_e/dt = -en_e\mathbf{E} + \mathbf{R}_{ei}. \tag{11.27}$$

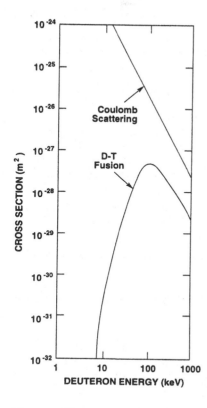

Figure 11.3. Coulomb and fusion cross sections compared for a deuterium ion (deuteron) in a deuterium–tritium plasma.

The term $\mathbf{R}_{ei}$, representing the momentum gain or loss of the electrons caused by collisions with ions, was written

$$\mathbf{R}_{ei} = -mn_e\langle\nu_{ei}\rangle(\mathbf{u}_e - \mathbf{u}_i) \tag{11.28}$$

where we assumed that the momentum exchange between the two species was proportional to the relative velocity $\mathbf{u}_e - \mathbf{u}_i$. Neglecting the inertia of the relatively light electrons, and expressing the current density $\mathbf{j}$ as $-n_e e(\mathbf{u}_e - \mathbf{u}_i)$, the resistivity was found to be

$$\eta = \frac{m\langle\nu_{ei}\rangle}{n_e e^2}. \tag{11.29}$$

Substituting our previous expression, equation (11.22), for the average electron–ion collision frequency $\langle\nu_{ei}\rangle$ and using $n_e = Zn_i$, we obtain an approximate value for the plasma resistivity, namely

$$\eta = \frac{2^{1/2}m^{1/2}Ze^2\ln\Lambda}{12\pi^{3/2}\epsilon_0^2 T_e^{3/2}}. \tag{11.30}$$

This simple calculation overestimates the true resistivity of a hydrogen plasma by a factor of about two. The weakness in the present calculation lies in using a 'standardized' average electron–ion collision frequency $\langle \nu_{ei} \rangle$ that was obtained by using a 'shifted Maxwellian' electron distribution, without taking into account the specific distortion of the electron velocity distribution that arises due to the presence of the electric field.

In the real situation, electrons with different velocities respond differently to the combined effects of a driving electric field and collisions with ions. Specifically, electrons with higher velocities are accelerated more readily by the electric field, since their collision frequencies are smaller than those of lower-velocity electrons. This tends to distort the electron distribution function, allowing more current to be carried by the faster electrons. Electron–electron collisions tend to pull these high-velocity electrons back into the Maxwellian distribution, however, thereby reducing this distortion. When all these effects are included, the net result for a hydrogen plasma is a resistivity that is about two times smaller than that given above. We will return to this topic in more detail in Chapter 13.

Our expression for η shows that the resistivity of a fully ionized plasma is independent of its density. This is a rather surprising result since, with a given **E** field, we might have expected the current density to increase if the number of charge carriers per unit volume, n_e, increases. The reason this does not happen is that the collisional friction force on the electrons also increases with the number of scatterers n_i. At fixed **E**, the current **j** is proportional to n_e but inversely proportional to n_i. Since $n_e = Z n_i$, the two dependences cancel each other. Note that a fully ionized plasma behaves quite differently from a weakly ionized gas in this regard. In a weakly ionized gas, we still have $\mathbf{j} = -n_e e \mathbf{u}_e$, where n_e is the density of charge carriers, i.e. electrons, but $\mathbf{u}_e$ will now be inversely proportional to the neutral density, n_n, if the principal contribution to resisting the electron flow comes from collisions with neutrals. In this case the current is proportional to n_e / n_n.

Our expression for η also shows that the resistivity of a fully ionized plasma varies inversely with $T_e^{3/2}$. As the temperature of a plasma is raised, its resistivity drops rapidly. Plasmas at very high temperatures are most likely to be 'perfectly conducting' or 'collisionless', meaning that their resistivity is negligible. We have seen in Chapter 8 the consequences of 'perfect conductivity' for the 'freezing' of plasma to magnetic field lines, which is now seen to be a concept that is particularly appropriate for high-temperature plasmas. However, the decrease in resistivity with increasing temperature has a severe disadvantage for one simple method of heating a plasma—namely, passing a current through it to dissipate some energy in heat ('ohmic' heating). The rate by which a plasma is heated by this method is ηj^2 per unit volume (equivalent to $I^2 R$ heating in electrical wires), which is simply due to the fact that the rate of energy transfer to electrons from the electric field is $-n_e e \mathbf{u}_e \cdot \mathbf{E} = \mathbf{j} \cdot \mathbf{E} = \eta j^2$. For fixed j, the heating rate drops as the temperature rises—so much so that ohmic heating is usually considered impractical, for example, at fusion temperatures.

The numerical value for the resistivity of a hydrogen plasma, after correcting equation (11.30) by the factor two, in SI units is

$$\eta = 5 \times 10^{-5} \ln\Lambda / T_e^{3/2} \ \Omega\,\mathrm{m} \tag{11.31}$$

where T_e is in eV. A plasma with $T_e = 100\,\mathrm{eV}$ has about the same resistivity as stainless steel ($7 \times 10^{-7}\,\Omega\,\mathrm{m}$), whereas a plasma at $T_e = 1\,\mathrm{keV}$ has as low a resistivity as copper ($2 \times 10^{-8}\,\Omega\,\mathrm{m}$).

11.4 ENERGY TRANSFER

Another collisional process that can be considered at this point is that of collisional energy transfer between a hotter electron component and a colder ion component. Specifically, we consider the temperature equilibration of a plasma in which $T_e \gg T_i$. This situation is possible (and common), as we will see, because the electron–ion energy transfer time, or 'equilibration time' τ_{eq}, is much longer than the characteristic times for the electrons and ions separately to come to thermodynamic equilibrium among themselves, which are ν_{ee}^{-1} and ν_{ii}^{-1} respectively.

When a light particle of mass m and initial velocity v_0 collides with a heavy particle of mass M initially at rest, the maximum energy and momentum transfer to the heavy particle occur for 180°-scattering (i.e. a 'head-on' collision). For this case of exactly 180°-scattering, the conservation of momentum and energy gives

$$mv_0 + mv_1 = MV \tag{11.32}$$
$$\tfrac{1}{2}mv_0^2 - \tfrac{1}{2}mv_1^2 = \tfrac{1}{2}MV^2 \tag{11.33}$$

where v_1 (backward) and V (forward) are the final velocities of the light and heavy particle, respectively. Combining these two equations yields $v_1 \approx v_0$ and

$$\tfrac{1}{2}MV^2 \approx \left(\frac{4m}{M}\right)\frac{mv_0^2}{2}. \tag{11.34}$$

Thus, only a fraction $\sim 4m/M$ of the energy of the light particle is transferred to the heavy particle. This simple result carries over to the physically more appropriate case of multiple small-angle collisions, as we will now see.

In the same spirit, we can calculate the rate at which energy is transferred from 'hot' electrons of mass m to 'cold' ions of mass M in a plasma. The change $\Delta \mathbf{v}$ in the velocity of an electron as a result of a Coulomb collision with an ion initially at rest can be related by momentum conservation to the velocity $\Delta \mathbf{V}$ acquired by the ion:

$$m\Delta \mathbf{v} = -M\Delta \mathbf{V}. \tag{11.35}$$

Referring to Figure 11.1, it is apparent that the scattering ion gains some amount of momentum, $M\Delta \mathbf{V}$, at the expense of a corresponding loss of the impinging electron's momentum, $m\Delta \mathbf{v}$, resulting from the deflection of the electron away from its initial trajectory. However, averaged over many such colliding electrons, each of them deflected in a different direction, there can be no net gain in ion *momentum*, provided of course that the electrons have an isotropic, e.g. Maxwellian, distribution with zero mean velocity. If the scattering ions are all initially at rest, however, each collision also results in a small gain in the ion *energy*. These increases in ion energy will accumulate, i.e. each colliding electron will contribute something, and the contributions from different electrons will not cancel out, as they do in the case of the (vector) momentum. From equation (11.35), the increase in ion energy occurring in a single collision is given by

$$\tfrac{1}{2}M|\Delta \mathbf{V}|^2 = \frac{m^2}{2M}|\Delta \mathbf{v}|^2. \tag{11.36}$$

We have seen that the change $\Delta \mathbf{v}$ in the velocity of the electron shown in Figure 11.1 is mainly in a direction *perpendicular* to its initial velocity vector, and the magnitude of this velocity change has been denoted $\Delta v_\perp$. The contribution from $\Delta v_\parallel$ is smaller, since $v\Delta v_\parallel \sim (\Delta v_\perp)^2$, as we saw from lowest-order energy conservation for the colliding electron. Accordingly, we can write

$$\tfrac{1}{2}M|\Delta \mathbf{V}|^2 = \frac{m^2}{2M}(\Delta v_\perp)^2 \tag{11.37}$$

where the quantity $(\Delta v_\perp)^2$ for a single electron colliding with a single ion has been given in equation (11.9). Equation (11.37) indicates that the electron energy transferred to the ion in this particular collision is $(m^2/2M)(\Delta v_\perp)^2$.

Now consider the case where there are many electrons colliding with many ions, as in a plasma. We allow the electron and ion densities, n_e and n_i, respectively, to be unequal, as they must be for example in a plasma with $Z \neq 1$. An average electron with velocity $\mathbf{v}$ is deflected by its many encounters with ions according to equation (11.11), i.e.

$$\frac{d\langle (\Delta v_\perp)^2 \rangle}{dt} = \frac{n_i Z^2 e^4 \ln \Lambda}{2\pi \epsilon_0^2 m^2 v}. \tag{11.38}$$

Integrating over a Maxwellian distribution of electrons

$$f_e(v) = n_e \left(\frac{m}{2\pi T_e} \right)^{3/2} \exp \left(-\frac{mv^2}{2T_e} \right) \tag{11.39}$$

we obtain a total rate of energy loss from the electrons by collisional transfer to ions, namely

$$\frac{dW_e}{dt} = -\frac{m^2}{2M} \int \frac{d\langle(\Delta v_\perp)^2\rangle}{dt} f_e(v) d^3 v \tag{11.40}$$

where $W_e = \frac{3}{2} n_e T_e$ is the energy density in the electrons. In each collision, the energy is transferred to a single ion, but this energy must then be shared among the entire population of ions, which are also assumed to be Maxwellian. The increase of the energy density in the ions, $W_i = \frac{3}{2} n_i T_i$, must balance the energy loss from the electrons, i.e.

$$\frac{dW_i}{dt} = -\frac{dW_e}{dt}. \tag{11.41}$$

Since only the average energies, and therefore the temperatures T_i and T_e, but not the densities are changed by elastic Coulomb collisions of this sort, we obtain an expression for the rate of increase of ion temperature:

$$\frac{dT_i}{dt} = \frac{m^2}{3Mn_i} \int \frac{d\langle(\Delta v_\perp)^2\rangle}{dt} f_e(v) d^3 v \tag{11.42}$$

$$= \frac{Z^2 e^4 \ln\Lambda}{6\pi \epsilon_0^2 M} \int \frac{f_e(v)}{v} d^3 v. \tag{11.43}$$

For a Maxwellian f_e, the integral in equation (11.43) is straightforward to evaluate (see equation (11.21)):

$$\int \frac{f_e(v)}{v} d^3 v = \left(\frac{2}{\pi}\right)^{1/2} \frac{n_e m^{1/2}}{T_e^{1/2}} \tag{11.44}$$

giving our final result

$$\frac{dT_i}{dt} = \frac{T_e}{\tau_{eq}} \tag{11.45}$$

where

$$\tau_{eq}^{-1} = \frac{n_e Z^2 e^4 m^{1/2} \ln\Lambda}{3\pi (2\pi)^{1/2} \epsilon_0^2 M T_e^{3/2}}. \tag{11.46}$$

Comparing this 'temperature equilibration rate' to the average electron–ion collision frequency defined earlier in this Chapter, we see that

$$\tau_{eq}^{-1} \approx 2(m/M)\langle v_{ei}\rangle. \tag{11.47}$$

Since we assumed that the ion scatterers were all initially at rest, these results are only valid for the initial increase of ion temperature, starting from $T_i \approx 0$. If the ions have a finite temperature, individual collisions will sometimes transfer energy from electrons to ions, and sometimes from ions to electrons.

From thermodynamic arguments, we know that, averaging over many collisions, the net transfer of energy must be from electrons to ions if $T_i < T_e$ and from ions to electrons if $T_e < T_i$. Thus, our analysis, which neglected the ion temperature entirely, must only be valid in the case $T_i \ll T_e$. However, if the ion scatterers are given some initial temperature T_i, our results could be generalized to show that the values of T_e and T_i approach each other at a rate given by

$$\frac{\mathrm{d}T_i}{\mathrm{d}t} = \frac{T_e - T_i}{\tau_{eq}}$$

$$\frac{\mathrm{d}T_e}{\mathrm{d}t} = \frac{n_i}{n_e} \frac{T_i - T_e}{\tau_{eq}}$$

(11.48)

the latter following from energy conservation. Although equation (11.48) is a thermodynamically plausible generalization of equation (11.45) for the case of finite T_i, its rigorous derivation requires a more complete treatment of electron–ion Coulomb collisions than can be given here. A more complete treatment of Coulomb collisions, including a derivation of equation (11.48), may be found, for example, in the classic monograph by L Spitzer (1962 *Physics of Fully Ionized Gases* 2nd edn, New York: Interscience).

Equation (11.48) implies that temperature equilibration (i.e. between electron and ion temperature) is a relatively slow process in a plasma. The rate at which electrons and ions exchange energy by collisions is not as rapid as the electron–ion or electron–electron collision frequencies, ν_{ei} and ν_{ee}, respectively, but is smaller by a factor of order m/M. The rate of energy exchange is also smaller that the ion–ion collision frequency, ν_{ii}, by a factor of order $(m/M)^{1/2}$.

Problem 11.4: Consider a hydrogen plasma in which the initial velocity distributions of electrons and protons are entirely arbitrary, i.e. non-Maxwellian. (We may assume, however, that the mean kinetic energies of electrons and protons are of the same order of magnitude.) Eventually, the plasma must come to thermodynamic equilibrium in which the electron and ion velocity distributions are both Maxwellian with $T_e = T_i$. Describe qualitatively the various stages of the approach to thermodynamic equilibrium. What happens first, what happens next, and so on?

11.5 BREMSSTRAHLUNG*

The Coulomb interaction of an electron with an ion results in acceleration of the electron as it is attracted towards the ion and executes the orbit shown in Figure 11.1. We know from electromagnetic theory that an accelerating charge produces electromagnetic radiation. The radiation of this sort produced by the electrons in a plasma making Coulomb collisions with the ions is called 'bremsstrahlung', or sometimes 'free–free bremsstrahlung'. ('Bremsstrahlung' is a German word for 'stopping radiation'.) It should be noted that collisions of electrons with other electrons produce no radiation of this sort in lowest order, because the accelerations of the two electrons are equal and opposite: since no net electron current is produced, there can be no radiation in the dipole approximation. Accordingly, we consider only electron–ion collisions.

Although an accurate treatment of bremsstrahlung for plasma parameters of interest generally requires a quantum mechanical calculation, we can obtain a reasonable approximation to the exact result by a classical calculation into which an appropriate quantum mechanical correction is introduced. From classical electromagnetic theory, we know that the power $\dot{W}$ radiated by a non-relativistic electron with acceleration a is given by Larmor's formula:

$$\dot{W} = \frac{e^2 a^2}{6\pi \epsilon_0 c^3}. \tag{11.49}$$

The acceleration can be expressed in terms of the Coulomb force of attraction by means of the equation of motion:

$$ma = Ze^2/4\pi \epsilon_0 r^2 \tag{11.50}$$

where r is the distance of the electron from the ion at this particular point of its orbit. Substituting equation (11.50) into equation (11.49), we obtain

$$\dot{W} = \frac{2Z^2 e^6}{3(4\pi \epsilon_0)^3 m^2 c^3 r^4}. \tag{11.51}$$

The total energy radiated as a result of this single collision is obtained by integrating equation (11.51) with respect to time along the electron's orbit. Denoting the distance along the orbit by s, the differential time element may be written $dt = ds/v$, where v is the instantaneous velocity of the electron. We obtain

$$W_{rad} = \frac{2Z^2 e^6}{3(4\pi \epsilon_0)^3 m^2 c^3} \int \frac{ds}{r^4 v}. \tag{11.52}$$

Since energy conservation provides a relation between the electron's kinetic energy, $mv^2/2$, and its potential energy in the Coulomb field of the ion, $-Ze^2/r$, we can express the instantaneous velocity v in terms of r, for a given value of the electron's initial velocity, i.e. its velocity far from the ion. Since we also know the geometry (a hyperbola) of the electron's orbit, we could in principle attempt to calculate the integral in equation (11.52) exactly. This would lead to an exact expression for the bremsstrahlung in the strictly classical case.

For present purposes, recognizing that our calculation must necessarily be approximate in its treatment of quantum mechanical effects, it is sufficient to evaluate the integral in equation (11.52) approximately, by integrating along the *unperturbed* electron orbit. This approximation treats all collisions as if they result in only small-angle deflections of the electron from a straight-line orbit. In this approximation, $v \approx$ constant and, measuring s from the mid-point of the orbit (i.e. the point at which the electron is closest to the ion), we have $r^2 = s^2 + b^2$, where b is the usual impact parameter. The integral then reduces to

$$\int \frac{ds}{r^4} = \int_{-\infty}^{\infty} \frac{ds}{(s^2 + b^2)^2} = \frac{\pi}{2b^3} \tag{11.53}$$

where the last step has been accomplished with the substitution $s = b \tan \alpha$. Our final expression for the energy radiated in this single collision becomes

$$W_{\text{rad}} \approx \frac{\pi Z^2 e^6}{3(4\pi\epsilon_0)^3 m^2 c^3 v b^3} \tag{11.54}$$

where the approximate equality sign is in recognition of the approximate treatment of the integral along the orbit.

In a differential time element dt, the number of ions with which the electron collides at impact parameters in a differential range db is given by multiplying the ion density, n_i, by the volume element $2\pi b \, db \, v \, dt$. The power (i.e. energy per second) radiated per unit volume of a plasma is obtained by multiplying equation (11.54) by $2\pi n_i v b \, db$, integrating over all b, and finally multiplying by the number of electrons in a unit volume, i.e. the electron density n_e. We obtain

$$
\begin{aligned}
P_{\text{br}} &\approx \frac{2\pi^2 n_i n_e Z^2 e^6}{3(4\pi\epsilon_0)^3 m^2 c^3} \int_{b_{\min}}^{\infty} \frac{db}{b^2} \\
&\approx \frac{2\pi^2 n_i n_e Z^2 e^6}{3(4\pi\epsilon_0)^3 m^2 c^3 b_{\min}}.
\end{aligned}
\tag{11.55}
$$

We note that it has not been necessary, in this calculation, to introduce an upper cut-off to the permitted values of the impact parameter, for example at the Debye length λ_D. This is because the integral over impact parameters does not diverge at large b: physically, large-angle collisions contribute about as much to bremsstrahlung as do multiple small-angle collisions. However, it has been necessary to introduce a *lower* cut-off to the permitted values of the impact parameter, which we have denoted $b_{\min}$. For the strictly classical case, we could estimate $b_{\min} \sim b_0$, where b_0 is the impact parameter for 90°-scattering, defined in equation (11.3) or equation (11.17). However, a more satisfactory procedure would be to calculate the integral along the electron orbit *exactly*, in which case the lower cut-off at approximately b_0 appears naturally.

We have found (see Problem 11.2) that quantum mechanical effects determine the minimum impact parameter in plasmas with $T_e > 10\,\text{eV}$. Moreover, bremsstrahlung is of interest mainly in this higher-temperature regime, since it is generally exceeded by other forms of radiation, for example line radiation, at lower electron temperatures. Even for a pure hydrogen plasma, recombination radiation, together with line radiation arising from transitions among excited recombined states, generally exceeds free–free bremsstrahlung at $T_e < 10\,\text{eV}$ (see Figure 10.11). Thus bremsstrahlung is of interest mainly in the quantum mechanical case where the minimum impact parameter is the de Broglie wavelength, i.e.

$$b_{\min} \approx \frac{\hbar}{mv} \tag{11.56}$$

with $v = (3T_e/m)^{1/2}$. Substituting equation (11.56) into equation (11.55), we obtain the power radiated by bremsstrahlung per unit volume in the quantum mechanical case, namely

$$P_{\text{br}} \approx \frac{2\pi^2 n_i n_e Z^2 e^6 T_e^{1/2}}{3^{1/2}(4\pi\epsilon_0)^3 m^{3/2} c^3 \hbar}. \tag{11.57}$$

Equation (11.57) is only approximate, since an exact calculation must be explicitly quantum mechanical, rather than classical with an *ad hoc* quantum mechanical cut-off. In addition, a proper averaging over a Maxwellian distribution of electrons is needed. Nonetheless, equation (11.57) is high relative to the exact quantum mechanical result (in which a so-called 'Gaunt factor' appropriate to electron temperatures in the keV range is used) by only about 34%. (The Gaunt factor is used to correct the original classical calculation for relativistic, as well as quantum mechanical, effects.) Including an additional factor of 0.75 on the right-hand side of equation (11.57) to correct this deficiency and substituting numerical values for the various physical constants which appear in equation (11.57), we obtain a final expression for the power radiated by bremsstrahlung:

$$P_{\text{br}} = 1.7 \times 10^{-38} Z^2 n_e n_i T_e^{1/2} (\text{W m}^{-3}) \tag{11.58}$$

where n_e and n_i are in m^{-3} and T_e is in eV. If the plasma contains several different ions with different charge numbers Z_i, then $n_i Z^2$ in equation (11.58) must be replaced by $\Sigma_i n_i Z_i^2$, where the summation is over the types of ions present. The reader interested in the exact quantum mechanical derivation of equation (11.58) is referred to W Heitler (1954 *Quantum Theory of Radiation* 3rd edn, Oxford: Oxford University Press).

Problem 11.5: Fusion reactions between deuterons and tritons produce charged helium ions ('alpha particles') with energy $E_\alpha = 3.5\,\text{MeV}$. If these ions remain confined, they will provide an internally generated plasma heating power of

$$n_D n_T \langle \sigma v \rangle_{DT} E_\alpha$$

per unit volume. In the temperature range 3–10 keV, the fusion reactivity, $\langle \sigma v \rangle_{DT}$, averaged over Maxwellian deuterons and tritons at the same temperature T_i, can be written $\langle \sigma v \rangle_{DT} (\mathrm{m^3\,s^{-1}}) \approx 10^{-34} T_i^3$ where T_i is in eV. (Beware that this formula is *not* good at temperatures above 10 keV.) Show that the alpha-particle heating power exceeds the power radiated by bremsstrahlung in a pure deuterium–tritium plasma with $n_D = n_T = n_e/2$ and $T_e = T_i = T$ only if $T > 4.3$ keV.

Chapter 12

Diffusion in plasmas

When a charged particle in a plasma collides with another particle, its velocity vector undergoes a small but abrupt change, causing the particle to move from one collisionless orbit to another. After a sufficient number of such collisions, the particle will have wandered a significant distance away from its original trajectory. In a non-uniform plasma, the result of this will be a net migration of particles from the highest-density region of the plasma to the lowest-density region, thereby tending to flatten-out the density gradient. This net migration is called 'diffusion'.

We will consider diffusion in weakly ionized gases, where collisions of charged particles with the much-more-abundant neutral atoms are more frequent than collisions with other charged particles, in addition to the case of diffusion in fully ionized plasmas. The case of the weakly ionized gas is of conceptual interest in the context of diffusion, not only because it serves to illustrate some of the physical ideas without too much algebraic complexity, but also because it differs importantly from the case of the fully ionized plasma in the mechanisms by which the charge neutrality of the ionized gas is maintained during the diffusion process. The weakly ionized gas is, of course, of practical interest in its own right, for example in high-pressure arcs and process plasmas.

We will limit ourselves in this Chapter to singly charged (e.g. hydrogen) ions. The generalization to ions with a multiple charge Ze is straightforward.

12.1 DIFFUSION AS A RANDOM WALK

It will be useful, first, to develop a heuristic understanding of the coefficient of diffusion (or 'diffusivity') of charged particles in a plasma. To do this, we need to introduce the concept of a 'random walk'.

Consider a group of particles moving along a straight line (the x axis) beginning at $x = 0$. The particles take one step at a time, each step of magnitude

Δx, and these steps are random in the sense that a step to the left is just as likely as a step to the right. Steps are taken at equal intervals of time Δt. On average, i.e. if very many similar particles are followed, all of them beginning at $x = 0$, an average particle will not move at all, since steps to the right are compensated by an approximately equal number of steps to the left. Thus, the 'average' position of the particles, denoted $\langle x \rangle$, is at all times given by

$$\langle x \rangle = 0. \tag{12.1}$$

However, after a sufficient length of time, the particles will have 'spread out' relative to their initial position, and a few of them will have succeeded in migrating quite far to the left or to the right. The root-mean-square spread in the particles' positions can be denoted $\langle x^2 \rangle^{1/2}$, and we will show that the quantity $\langle x^2 \rangle$ increases in time according to the relation

$$\frac{\mathrm{d}\langle x^2 \rangle}{\mathrm{d}t} = \frac{(\Delta x)^2}{\Delta t} \tag{12.2}$$

which can be integrated to give $\langle x^2 \rangle = (\Delta x)^2 t / \Delta t$. Thus, the 'spread' $\langle x^2 \rangle^{1/2}$ increases as the square root of the time.

In Section 12.3, we will derive the equivalent of equation (12.2) by solving a diffusion equation for the density $n(x, t)$ of particles on the line. The diffusion equation is applicable in the limit where Δx and Δt are both infinitesimally small.

12.2 PROBABILITY THEORY FOR THE RANDOM WALK*

Before introducing the diffusion equation, it may be of interest to some readers to derive equation (12.2) by analyzing the random walk using the methods of probability theory. This derivation is valid even in the case where Δx and Δt are not small. (Readers who are uninterested in such a derivation may omit this entire Section.)

The derivation of equation (12.2) from probability theory goes as follows. Let us consider a total of n steps and define $P_n(r)$ to be the probability that r of these steps are to the right, so that $n - r$ of the steps must be to the left. The probability of one particular prescribed ordered sequence of such steps is $2^{-r} \times 2^{-(n-r)} = 2^{-n}$, just as the probability of a prescribed sequence of heads and tails in n throws of a dice will be 2^{-n}. To obtain $P_n(r)$, where it does not matter in which of the n steps the r rightward ones occur, we must multiply by the number of ways of choosing r indistinguishable items from a total of n items, which is $n!/[r!(n-r)!]$. Thus

$$P_n(r) = \frac{n!}{r!(n-r)!} \frac{1}{2^n}.$$

After these n steps, which occur in a time $t = n\Delta t$, the particle has progressed a net distance to the right of $r\Delta x - (n-r)\Delta x = (2r - n)\Delta x$, and its mean-square position is given by

$$\langle x^2 \rangle = 4(\Delta x)^2 \sum_{r=0}^{n} \left(r - \frac{n}{2} \right)^2 P_n(r).$$

To make progress at this point we will employ a subtle trick that at first seems off the track. We write down the binomial expansion of a function $F_n(y)$ defined as follows:

$$F_n(y) \equiv \frac{(1 + y)^n}{2^n y^{n/2}} = \frac{1}{2^n} \sum_{r=0}^{n} \frac{n!}{r!(n-r)!} y^{(r-n/2)}$$

so that

$$y\frac{\mathrm{d}}{\mathrm{d}y} \left(y\frac{\mathrm{d}F_n(y)}{\mathrm{d}y} \right) = \frac{1}{2^n} \sum_{r=0}^{n} \frac{n!}{r!(n-r)!} \left(r - \frac{n}{2} \right)^2 y^{(r-n/2)}$$

$$= \sum_{r=0}^{n} \left(r - \frac{n}{2} \right)^2 P_n(r) y^{(r-n/2)}.$$

Substituting this into our expression for $\langle x^2 \rangle$, we obtain

$$\langle x^2 \rangle = 4(\Delta x)^2 \left[y\frac{\mathrm{d}}{\mathrm{d}y} \left(y\frac{\mathrm{d}F_n(y)}{\mathrm{d}y} \right) \right]_{y=1}$$

and now it is a simple matter to carry out these differentiations of the function $F_n(y)$ and then to substitute $y = 1$, which gives a value $n/4$ for the expression in square brackets. Thus

$$\langle x^2 \rangle = n(\Delta x)^2 = t(\Delta x)^2 / \Delta t$$

thereby proving equation (12.2). We want now to go to the limit in which the number of steps becomes extremely large, while the length of each step, Δx, and the time interval between steps, Δt, become infinitesimally small. To do this in a way that keeps the mean-square distance $\langle x^2 \rangle$ finite for finite time t requires that Δx and Δt approach zero with a particular relationship between each other: specifically, it is necessary that Δx and Δt approach zero in such a manner that $(\Delta x)^2 \propto \Delta t$.

12.3   THE DIFFUSION EQUATION

Returning now to the promised derivation of equation (12.2) by formulating and then solving a diffusion equation, which correctly models the random walk in the limit of infinitesimal Δx and Δt, we begin by defining a 'number density' of particles $n(x, t)$, such that the number of particles in an element of length dx at location x at time t is given by $n(x, t)dx$. We will show that this 'number density' (like a probability density) satisfies a 'diffusion equation' of the form

$$\frac{\partial n}{\partial t} = -\frac{\partial \Gamma}{\partial x} = \frac{\partial}{\partial x}\left(D\frac{\partial n}{\partial x}\right). \tag{12.3}$$

This equation, which is in the form of a continuity equation for our random walkers, is correct as long as the assumption that the flux Γ of particles in the x direction is proportional to the density gradient (with constant of proportionality $-D$) is valid. In the limit of small individual steps Δx, we can obtain this flux directly from the random walk picture of particles moving along the x axis. Evaluating the flux across $x = x_0$, we note that the positively directed flux arises from the particles in a line segment of length Δx, located immediately to the left of $x = x_0$, 'emptying out' in the x direction as a result of positive steps Δx in a time interval Δt. Noting that only half of these particles make rightward steps, the other half making leftward steps, we see that the positively directed flux is

$$\begin{aligned}
\Gamma_+ &= \frac{1}{2\Delta t}\int_{x_0-\Delta x}^{x_0} n(x)dx \\
&\approx \frac{1}{2\Delta t}\int_{x_0-\Delta x}^{x_0}\left(n(x_0) + (x - x_0)\frac{dn}{dx}\bigg|_{x_0}\right)dx \\
&\approx \frac{1}{2\Delta t}\left(n\Delta x - \frac{(\Delta x)^2}{2}\frac{dn}{dx}\right).
\end{aligned}$$

Similarly, the negatively directed flux from particles in a similar line segment to the right of $x = x_0$ making leftward steps is

$$\begin{aligned}
\Gamma_- &= -\frac{1}{2\Delta t}\int_{x_0}^{x_0+\Delta x} n(x)dx \\
&\approx -\frac{1}{2\Delta t}\left(n\Delta x + \frac{(\Delta x)^2}{2}\frac{dn}{dx}\right).
\end{aligned}$$

The net flux is

$$\Gamma = \Gamma_+ + \Gamma_- = -\frac{(\Delta x)^2}{2\Delta t}\frac{dn}{dx} \tag{12.4}$$

which corresponds to a diffusion coefficient in equation (12.3) given by

$$D = \frac{(\Delta x)^2}{2\Delta t}. \tag{12.5}$$

Thus the diffusion equation, equation (12.3), gives a correct fluid description of the flows that arise when many particles, with non-uniform density, execute random walks in the limit of a large number of steps with infinitesimal Δx and Δt.

We can find an exact solution of the diffusion equation for the case where all of the particles begin at $t = 0$ at $x = 0$, namely

$$n(x, t) = \frac{N}{(4\pi Dt)^{1/2}} \exp\left(-\frac{x^2}{4Dt}\right) \tag{12.6}$$

where N is the total number of particles, i.e. $N = \int n(x, t) \mathrm{d}x$. While this is an exact solution of equation (12.3), we should note that the diffusion equation is valid only in the limit of $\Delta t \ll t$ and $\Delta x \ll x$.

Problem 12.1: Show by direct substitution that equation (12.6) is indeed a solution of equation (12.3). By finding a counter-example where equation (12.6) (with equation (12.5) for D) cannot be valid, show that this solution is not correct for all x and t in the case where Δx and Δt are *finite*.

At time $t = 0$, the particles are distributed in a δ-function at $x = 0$. This formulation allows us to derive the result for the root-mean-square spread in particle positions that was implied by equation (12.2), namely

$$\langle x^2 \rangle = N^{-1} \int x^2 n(x, t) \mathrm{d}x = 2Dt = (\Delta x)^2 t / \Delta t.$$

The three-dimensional generalization of the diffusion equation, equation (12.3), is clearly given by

$$\frac{\partial n}{\partial t} = \nabla \cdot (D \nabla n). \tag{12.7}$$

In a diffusion process, there is on balance a net migration of particles in the direction opposite to ∇n. Since more particles are located in higher-density regions than in lower-density regions, when both spread out, there is a net flux in the $-\nabla n$ direction.

For a first example of diffusion in plasmas, we consider the case of a plasma not containing any significant electric or magnetic fields. In this case, a charged particle will move on a straight-line trajectory until it encounters, i.e. collides with, another particle—either another charged particle or a neutral atom. The diffusion coefficient D may be estimated based on our heuristic model as follows. The step size in the random walk is the mean-free path λ_{mfp}. The interval between steps is the inverse of the collision frequency, i.e. $\tau \approx \nu^{-1}$. Thus, according to the simple random-walk analysis discussed above, the diffusion coefficient is given by

$$D \sim \nu \lambda_{\text{mfp}}^2. \tag{12.8}$$

Equation (12.8) and subsequent estimates like it can be regarded as valid only in an order-of-magnitude sense; accordingly, we omit the factor 1/2 that appeared in our expression for D given in equation (12.5). Recalling that the mean-free path is obtained by dividing the particle velocity, which we take to be the thermal velocity of the diffusing species v_t, by the collision frequency ν, we can express the diffusion coefficient

$$D \sim \frac{v_t^2}{\nu} \sim \frac{T}{m\nu}. \tag{12.9}$$

Consider next the case of a plasma containing a strong magnetic field **B**. As illustrated in Figure 12.1, charged particles will move freely *along* the **B** field, unimpeded except by collisions with other particles. If the density of particles is non-uniform along the field, these non-uniformities will smooth themselves out by diffusion, in accordance with the relationship derived above for the case with no magnetic field, since **B** itself does not affect motion in the direction parallel to **B**. Accordingly, we can define a 'parallel' diffusion coefficient, $D_{\parallel}$, which takes the form given in equation (12.9).

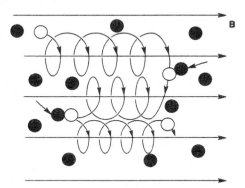

Figure 12.1. Diffusion of a charged particle (open circle) in a magnetic field due to collisions with other particles, either neutral or charged (full circles). Two collisions are shown, each of which contributes to diffusion along the field and, by changing the phase angle of the Larmor gyration, to diffusion across the field.

Note that in both of these cases, i.e. an unmagnetized plasma and diffusion parallel to a magnetic field, collisions actually *reduce* transport. In the case of diffusion perpendicular to a magnetic field, collisions *enhance* transport. Indeed, if there are no collisions, particles will not migrate at all in the *perpendicular* direction—they will continue to gyrate indefinitely about the same field line. There are, of course, particle *drifts* across **B**, due to field gradients, curvature or electric fields perpendicular to **B**, but these are often arranged to form closed drift orbits within a bounded plasma. A particular example is the cylindrical plasma column, where the electric field and the gradients are all in the radial direction, so that the drifts are azimuthal. In such cases, the particle drifts do not carry particles out of the plasma.

However, when there are collisions, particles migrate across **B** by a random walk process. When a charged particle collides with another particle, the direction of its velocity vector is turned through some finite angle. The particle continues to gyrate about the magnetic field in the same sense, but the phase of its gyration is changed discontinuously, thereby changing the location of the gyro-center. The radius of gyration (Larmor radius) may also change, but this is not essential to the process, and for now we can suppose for simplicity that the charged particle does not gain or lose perpendicular energy in the collision, so that the gyration radius is unchanged.

Because of the change of phase, the center of gyration ('guiding center') shifts position as a result of the collision and, if there is a succession of such collisions, the center of gyration undergoes a random walk, as shown in Figure 12.1. We assume here that the collision frequency is much less than the gyration frequency, so that most Larmor orbits are completed. The step size in the random walk is no longer the mean-free path λ_{mfp}, as in the magnetic-field-free case, but has instead the magnitude of the Larmor radius r_L. The interval between steps is again the inverse of the collision frequency, i.e. $\tau = \nu^{-1}$, giving a 'perpendicular' diffusion coefficient

$$D_\perp \sim \nu r_L^2. \tag{12.10}$$

Although the estimate given in equation (12.10) is based on the pictorial representation of perpendicular diffusion shown in Figure 12.1, it is important to remember that Coulomb collisions act mainly through the cumulative effect of many small-angle scatterings, rather than the relatively infrequent 90° scatterings depicted in Figure 12.1. A typical particle experiences collisional scattering of its perpendicular velocity vector through an angle of order $\Delta\theta \sim (2\pi\nu/\omega_c)^{1/2}$ in one gyro-period ($\Delta t \sim 2\pi/\omega_c$). (Remember that collisional scattering is itself a diffusive process in velocity space, so the angle of scattering $\Delta\theta$ is given by $(\Delta\theta)^2 \sim \nu\Delta t$.) If the perpendicular velocity vector is turned through an angle $\Delta\theta$, the particle's gyro-center moves by a distance $\Delta x \sim r_L\Delta\theta$, which gives rise to a spatial diffusion coefficient $D \sim (\Delta x)^2/\Delta t \sim \nu r_L^2$. Thus, a more correct approach based on multiple small-angle scattering events gives the same result as the heuristic derivation based on large-angle scatters, i.e. equation (12.10).

12.4 DIFFUSION IN WEAKLY IONIZED GASES

In a weakly ionized gas with a small enough ionization fraction, charged particles will interact primarily (by means of elastic collisions) with neutral atoms rather than with other charged particles. In this case, the applicable diffusion coefficients may be obtained from equations (12.9) and (12.10) by simply using the appropriate electron–neutral or ion–neutral collision frequency. Recalling that $\nu = n_n \langle \sigma_n v \rangle$ and noting that a neutral atom displays approximately the same cross section σ_n to a colliding electron as to a colliding ion, we see that the electron–neutral and ion–neutral collision frequencies are related to each other according to

$$\nu_{en}/\nu_{in} \sim v_{t,e}/v_{t,i} \sim (M/m)^{1/2} \gg 1. \tag{12.11}$$

Here, $v_{t,e}$ and $v_{t,i}$ are the thermal velocities of electrons and ions and, in relating them to the electron and ion masses, we have assumed that the electron and ion temperatures are roughly equal.

It follows that the electron and ion diffusion coefficients in the absence of a magnetic field (or along the field, if one is present) are related according to

$$D_{\|e}/D_{\|i} \sim (M/m)^{1/2} \gg 1. \tag{12.12}$$

The diffusion coefficients perpendicular to a strong magnetic field are related according to

$$D_{\perp e}/D_{\perp i} \sim (m/M)^{1/2} \ll 1 \tag{12.13}$$

because of the square-root dependence of the Larmor radius on the mass.

In a plasma, which must remain charge neutral to a very high degree of approximation, net motion of electrons and ions at separate rates will not occur. If the plasma is to remain neutral, the fluxes of electrons and ions will somehow adjust themselves so that the two species leave the plasma at the same rate. Not surprisingly, the process of adjustment of the two loss rates involves the electric field that arises as soon as a slight charge imbalance occurs. In the case without a magnetic field (or diffusion *along* the field, if one is present), the electrons have the larger diffusivity and tend to leave the ions behind. A very small positive charge is left in the region of highest plasma density, sufficient to create an outwardly directed electric field of such a size that the preferential loss of electrons is eliminated, although the loss of ions will tend to be increased somewhat in the process. In the case of diffusion *across* a magnetic field, the electric field will be inwardly directed, i.e. toward the region of highest plasma density, to eliminate the preferential loss of ions.

The results given in equations (12.9) and (12.10), which we have obtained from a heuristic single-particle picture, can be derived formally (but actually at the same level of approximation) using the fluid equations for a weakly ionized gas. This approach will also lead to a quantitative determination of the electric field. The fluid equation of motion, including collisions with neutral atoms, for either species of a hydrogenic plasma, is

$$mn\frac{d\mathbf{u}}{dt} = qn\mathbf{E} - \nabla p - mn\nu\mathbf{u} \tag{12.14}$$

where, as usual, q is the signed charge (i.e. $\pm e$ for ions/electrons). We will assume that the collision frequency ν has been averaged appropriately over the distribution of particle velocities and is a constant, i.e. independent of the fluid velocity. If we consider a steady state in which $\partial\mathbf{u}/\partial t = 0$, and in which the fluid element does not move very far in a collision time (i.e. $u/\nu \ll L$, where L is the characteristic dimension of the plasma), so that $(\mathbf{u} \cdot \nabla)\mathbf{u}$ is negligible, then inertia and acceleration may be neglected, and we obtain

$$\mathbf{u} = \frac{q\mathbf{E}}{m\nu} - \frac{T}{m\nu}\frac{\nabla n}{n}. \tag{12.15}$$

Here, we have also assumed an isothermal plasma, so that $\nabla p = T\nabla n$. The constant of proportionality between the flux $n\mathbf{u}$ and ∇n is the same as was obtained heuristically from the single-particle picture, see equation (12.9).

Let us now consider equation (12.15) for ions ($m \to M, q \to e$) and electrons ($m \to m, q \to -e$) separately. Equation (12.15) shows that electrons not only diffuse more rapidly than ions in the presence of a density gradient, but they also respond more readily to an electric field. (The coefficient of the electric field term in equation (12.15) is called the 'mobility'.) For the diffusion of ions and electrons to be at the same rate—sometimes called 'ambipolar diffusion'—the electric field must adjust itself so that the electron flow is reduced by a large factor of order $(M/m)^{1/2}$. More precisely, we can determine the electric field from equation—(12.15) by setting $\mathbf{u}_e = \mathbf{u}_i$, to obtain

$$\mathbf{E} = -\frac{T_e M\nu_{in} - T_i m\nu_{en}}{M\nu_{in} + m\nu_{en}}\frac{\nabla n}{ne}$$
$$\approx -\frac{T_e}{ne}\nabla n. \tag{12.16}$$

This is called the 'ambipolar electric field'. Substituting it back into equation (12.15), we find electron and ion fluxes

$$n\mathbf{u}_e = n\mathbf{u}_i = -D_a\nabla n \tag{12.17}$$

where D_a is the ambipolar diffusion coefficient, given by

$$D_a \approx \frac{T_e + T_i}{M\nu_{in}}. \tag{12.18}$$

For the case where $T_e \approx T_i$, the effect of the ambipolar electric field is to approximately double the diffusion coefficient of the ions; the electron pressure acts through the electric field effectively to push on the ions. The diffusion rate for the two species together is thus controlled primarily by the species that diffuses more slowly—in this case, the ions. Even if electron transport were infinitely faster, it would still only double the net particle transport.

It is interesting to note that in the opposite limit of $\lambda_{mfp} \sim u/\nu \gg L$, inertia and acceleration become dominant, collisions become unimportant, and the discussion in Chapter 7 of pressure balance parallel to a magnetic field in a collisionless plasma may be invoked. There too, the electron pressure is effectively added to the ion pressure through the action of the electric field, in that case driving ion acceleration rather than diffusion. Acceleration and diffusion are handled simultaneously by the fluid equations of motion when the collisional friction term is included. Thus, in the fluid picture, diffusive flow is a consequence of collisional friction between particles of different species becoming more important than acceleration in determining the fluid velocity that is produced as a response to pressure gradients. Diffusion appears naturally from the fluid equations including collisional friction and is *absolutely not* to be introduced separately, for example as an extra $\nabla \cdot (D\nabla n)$ term in the continuity equation.

Consider next the case of a weakly ionized gas in a magnetic field. As before, we start with the fluid equation of motion for either species including collisions with neutral atoms. Since in the parallel direction the equation of motion is the same as for an unmagnetized plasma, we focus on the perpendicular components:

$$mn\frac{d\mathbf{u}_\perp}{dt} = qn(\mathbf{E} + \mathbf{u}_\perp \times \mathbf{B}) - T\nabla n - mn\nu\mathbf{u}. \qquad (12.19)$$

We suppose that the magnetic field is in the z direction and that the density non-uniformity is in the x direction. The ambipolar electric field will also be in the x direction. Again assuming a steady state and neglecting inertia, we can solve the two perpendicular components of equation (12.19) for the two perpendicular components of the fluid velocity $\mathbf{u}$. Because of the $\mathbf{u}_\perp \times \mathbf{B}$ term, the two components of equation (12.19) are coupled, but it is a matter of simple algebra to solve the two equations simultaneously, to obtain

$$u_x = \frac{1}{1 + \omega_c^2\tau^2}\left(\frac{qE_x}{m\nu} - \frac{T}{m\nu}\frac{1}{n}\frac{dn}{dx}\right) \qquad (12.20)$$

$$u_y = -\frac{\omega_c^2\tau^2}{1 + \omega_c^2\tau^2}\left(\frac{E_x}{B} - \frac{T}{qB}\frac{1}{n}\frac{dn}{dx}\right) \qquad (12.21)$$

where τ is the mean collision time, ν^{-1}, and ω_c is the cyclotron frequency, eB/m.

Equation (12.20) compared with equation (12.15) shows that the effect of the magnetic field ($\omega_c \neq 0$) is to reduce the mobility and diffusivity relative to an unmagnetized plasma by a factor $1 + \omega_c^2 \tau^2$. Equation (12.21) shows that the effect of collisions ($\tau \neq \infty$) is to reduce the $\mathbf{E} \times \mathbf{B}$ and diamagnetic drifts relative to a collisionless plasma by a factor $\omega_c^2 \tau^2 / (1 + \omega_c^2 \tau^2)$. Since the density gradient is in the x direction, the flow-velocity component u_y does not contribute to diffusion, although this component is the larger in a collisionless plasma. The diffusive flows that carry particles from high-density to low-density regions of the plasma are contained in the component u_x.

When $\omega_c^2 \tau^2 \ll 1$, the magnetic field has little effect on diffusion. On the other hand, in the case where $\omega_c^2 \tau^2 \gg 1$, the magnetic field significantly retards the rate of diffusion across $\mathbf{B}$. In the limit $\omega_c^2 \tau^2 \gg 1$, we have

$$u_x = \frac{\nu}{m\omega_c^2}\left(qE_x - \frac{T}{n}\frac{\mathrm{d}n}{\mathrm{d}x} \right). \tag{12.22}$$

Again the constant of proportionality between the flux $n\mathbf{u}$ and ∇n, i.e. the diffusion coefficient $D_\perp$, is the same as was obtained heuristically from the single-particle picture, equation (12.10).

Compared with diffusion along a magnetic field (or in the magnetic-field-free case), we see that the role of the collision frequency has been reversed, as we found in the heuristic derivation. In diffusion *along* $\mathbf{B}$, the diffusion rate varies inversely with ν, since collisions impede the motion. In diffusion *perpendicular* to $\mathbf{B}$, the diffusion rate is proportional to ν, since collisions are needed for cross-field migration. The dependence on the particle's mass m has also been reversed. Keeping in mind that the collision frequency for charged particles colliding with neutral atoms is proportioned to $m^{-1/2}$ and that ω_c varies inversely with m, we see that the diffusion rate along $\mathbf{B}$ varies as $m^{-1/2}$, whereas the diffusion rate perpendicular to $\mathbf{B}$ varies as $m^{1/2}$. In diffusion along $\mathbf{B}$, electrons move faster than ions because of their higher thermal velocities; in perpendicular diffusion, ions migrate more rapidly than electrons because of their larger Larmor radii.

Let us now consider equation (12.22) for ions ($m \to M, q \to e$) and electrons ($m \to m, q \to -e$) separately. Because the diffusion coefficients are anisotropic in the presence of a magnetic field, the problem of ambipolar diffusion is not as straightforward as in the magnetic-field-free case. As we have just seen, in diffusion *perpendicular* to a magnetic field, the ion flux tends to exceed the electron flux. Ordinarily, a transverse electric field will then be set up so as to aid electron diffusion and retard ion diffusion. The electric field will be that needed to reduce the ion flow by a large factor of order $(M/m)^{1/2}$ which, from equation (12.22) for ions, can be seen to be given by

$$\mathbf{E}_\perp \approx \frac{T_\mathrm{i}}{ne}\nabla_\perp n. \tag{12.23}$$

The ambipolar electron and ion fluxes are then obtained from equation (12.22) for electrons:

$$n\mathbf{u}_{e\perp} = n\mathbf{u}_{i\perp} = -D_a \nabla_\perp n \tag{12.24}$$

where

$$D_a \approx \frac{\nu_{en}(T_e + T_i)}{m\omega_{ce}^2}$$

$$\approx \nu_{en}\langle r_{Le}^2 \rangle \left(1 + \frac{T_i}{T_e}\right) \tag{12.25}$$

where $\langle r_{Le}^2 \rangle = T_e/(m\omega_{ce}^2) = mT_e/(e^2 B^2)$ is the mean-square Larmor radius of the electrons. The diffusion coefficient D_a is seen to be inversely proportional to B^2. Clearly, our result agrees—at least in some sense—with the heuristic result given in equation (12.10), except that ambipolar diffusion is at the slower rate similar in order-of-magnitude to that given by equation (12.10) for *electrons*.

However, the electric field required for ambipolar diffusion perpendicular to a magnetic field can sometimes be short-circuited by an imbalance in fluxes along **B**. Specifically, the negative charge resulting from the net perpendicular outflux of ions can be dissipated by electrons escaping along field lines. Although the *total* diffusion must be ambipolar, the perpendicular part of the losses need not be ambipolar; the ions can diffuse *across* the field, while the electrons are lost primarily *along* the field.

Whether or not this occurs depends on the geometry of the particular magnetic configuration and on experimental conditions. In a mirror-trapped plasma on open field lines, the losses of *electrons* along the field generally far exceed the ion cross-field losses, so the plasma tends to become positively charged, in accordance with the requirement for ambipolar diffusion along (or without) a magnetic field. In the opposite case of a 'closed' plasma configuration, in which the field lines close back on themselves so that there is no possibility of escape along the field, the cross-field losses of *ions* are dominant, and the plasma tends to become negatively charged, in accordance with the requirement for ambipolar diffusion across a magnetic field. In a cylindrical plasma column with the field lines terminating on conducting end-plates, the ambipolar electric field is short-circuited out; each species is then able to diffuse radially at a different rate, provided there is sufficiently rapid compensating diffusion of net charge in the parallel direction to the end-plates.

12.5 DIFFUSION IN FULLY IONIZED PLASMAS

We will consider next the diffusion perpendicular to a magnetic field in a fully ionized plasma, where Coulomb collisions dominate over collisions with neutral atoms.

As in the case of the weakly ionized gas, the formal treatment of diffusion proceeds from the fluid equations of motion for the two species, including collisions. For a fully ionized plasma, these are effectively the single-fluid equation of motion and the plasma Ohm's law:

$$\rho \frac{d\mathbf{u}}{dt} = -\nabla p + \mathbf{j} \times \mathbf{B} \qquad (12.26)$$

$$\mathbf{E} + \mathbf{u} \times \mathbf{B} + \frac{\nabla p_e}{ne} - \frac{\mathbf{j} \times \mathbf{B}}{ne} = \eta \mathbf{j}. \qquad (12.27)$$

Since we have seen that diffusion flow-velocities tend to be less than or of order of the diamagnetic speed, we use here the so-called 'generalized Ohm's law'. This is equivalent to a two-fluid picture with electron inertia neglected. Collisions between the two species, electrons and ions, appear through the resistivity term in the Ohm's law. As in the case of the weakly ionized gas, we may assume that diffusion is a sufficiently slow process that the plasma is always in a state of equilibrium. We also assume that the diffusion velocity is much less than sonic, so that $\rho(\mathbf{u} \cdot \nabla)\mathbf{u}$ can be neglected compared with ∇p. These assumptions allow us to neglect the inertia of the ions, as well as that of the electrons, and to replace the equation of motion by the force balance equation:

$$\mathbf{j} \times \mathbf{B} = \nabla p. \qquad (12.28)$$

We again suppose that the magnetic field is in the z direction, and that the pressure non-uniformity is in the x direction. The ambipolar electric field will also be in the x direction, since there is no variation of any of the quantities in the y direction. (The generalization to the case of a cylinder of plasma in which the non-uniformity and electric field are both in the radial direction and there is no azimuthal variation is straightforward.) The force balance equation tells us that

$$j_x = 0 \qquad j_y = \frac{1}{B} \frac{dp}{dx}. \qquad (12.29)$$

Since there is no variation in the y direction, these currents automatically satisfy the quasi-neutrality condition $\nabla \cdot \mathbf{j} = 0$.

The generalized Ohm's law may be solved for the perpendicular components of the plasma fluid velocity:

$$u_y = -\frac{E_x}{B} + \frac{1}{neB} \frac{dp_i}{dx} \qquad u_x = -\frac{\eta}{B^2} \frac{dp}{dx} \qquad (12.30)$$

where we have substituted for j_x and j_y from equation (12.29). It is evident that the perpendicular velocity of the plasma is composed of two parts. First, there are the usual electric and diamagnetic drifts, perpendicular to the electric field and pressure gradient, respectively, and therefore not leading directly to any loss of plasma from high-density to low-density regions. (These drifts are in the y direction in our case of a plasma slab that is non-uniform in x, and they would be in the θ direction in the case of a cylinder of plasma that is non-uniform in r). Second, there is a resistively driven flow that is anti-parallel to the pressure gradient and that *does*, therefore, lead to particle loss. More generally, the resistive flow can be written

$$\mathbf{u}_\perp = -\frac{\eta}{B^2}\boldsymbol{\nabla}_\perp p. \tag{12.31}$$

Substituting this into the mass continuity equation gives

$$\frac{\partial \rho}{\partial t} = \boldsymbol{\nabla}_\perp \cdot \left(\frac{\rho\eta}{B^2}\boldsymbol{\nabla}_\perp p\right)$$
$$= \boldsymbol{\nabla}_\perp \cdot \left(\frac{\eta p}{B^2}\boldsymbol{\nabla}_\perp \rho\right) \tag{12.32}$$

where, in the second form, we have made the further assumption for simplicity that the plasma can be assumed to be isothermal, i.e. $p/\rho = nT/nM = T/M = $ constant. Equation (12.32) is a diffusion equation for the mass density. The diffusion coefficient is

$$D_\perp = \frac{\eta p}{B^2} \tag{12.33}$$

which is usually called the 'classical diffusion coefficient' for a fully ionized plasma.

The classical diffusion coefficient is seen to be inversely proportional to B^2, just as in the case of weakly ionized gases. This dependence can be traced back to the nature of diffusion as a random-walk process: for a random walk across a magnetic field, the step size must be the Larmor radius r_L. Indeed, writing $\eta \approx m\nu_{ei}/ne^2$ and $p = n(T_e + T_i)$ we obtain

$$D_\perp \sim \frac{\nu_{ei}m(T_e + T_i)}{e^2 B^2} \sim \nu_{ei}\langle r_{Le}^2\rangle\left(1 + \frac{T_i}{T_e}\right). \tag{12.34}$$

Thus cross-field diffusion in a fully ionized plasma can be described by a random walk of *electrons*, step size r_{Le} and frequency of steps ν_{ei}.

The classical diffusion coefficient is also seen to vary inversely with $T_e^{1/2}$. This is because the temperature variation of ν_{ei}, namely $T_e^{-3/2}$, outweighs the temperature dependence of the Larmor radius r_{Le}, namely $T_e^{1/2}$. Thus, the diffusion coefficient decreases as the electron temperature is raised. The reason is, of course, the velocity dependence of the Coulomb cross section. The implied improvement in magnetic confinement as a plasma is heated would lead to very optimistic projections of confinement in controlled fusion devices—were it not for the effect of turbulent processes arising from plasma collective effects, which, in practice, give rise to diffusion substantially exceeding the 'classical' processes described here.

Nonetheless, let us discuss some properties of classical diffusion that are different in a fully ionized plasma from those in a weakly ionized gas and that were not evident in the heuristic derivation. Classical diffusion in fully ionized plasmas is *intrinsically ambipolar*. Since we have satisfied the quasi-neutrality condition $\nabla \cdot \mathbf{j} = 0$ from the outset, it follows that *both electrons and ions diffuse out of the plasma at the same rate*. Moreover, it is not necessary to have some particular electric field $\mathbf{E}$ to equalize the electron and ion loss rates. These features can be illustrated by the case of a cylindrical plasma equilibrium with a $\mathbf{B}$ field entirely in the z direction. In this case, the current density is entirely azimuthal, given by $Bj_\theta = dp/dr$ and $j_r = 0$; thus there is no tendency for preferential ion or electron radial loss. Moreover, a radial electric field E_r merely produces a plasma rotation $u_\theta = -E_r/B$, changing neither the electron nor the ion diffusion. The surprising property of *intrinsic ambipolarity* turns out to be a consequence of conservation of total momentum in electron–ion collisions: as we will see in the next Section, if two particles of equal and opposite charge gyrating in a magnetic field are given equal and opposite increments of momentum, their gyration centers are moved an exactly equal distance in the same direction across the magnetic field. Thus, electrons and ions tend to diffuse *together* across a magnetic field.

Since the classical diffusion coefficient D is due to plasma resistivity, it arises from *electron–ion collisions*, and not from electron–electron nor ion–ion collisions. This, again, is somewhat surprising. Naively, we might have expected to obtain a diffusion coefficient due to ion–ion collisions of order $\nu_{ii}r_{Li}^2$—a factor $(M/m)^{1/2}$ times larger than the actual diffusion coefficient, which is of order $\nu_{ei}r_{Le}^2$. The reason this is absent is again related to the conservation of total momentum in collisions: if two ions gyrating in a magnetic field are given equal and opposite increments of momentum, their gyration centers are moved exactly equal and opposite distances across the magnetic field. As we will see in more detail in the next Section, this simple consequence of momentum conservation implies that, to the order at which a diffusive flux usually appears, there can be no net diffusion of particles due to like-particle (e.g. ion–ion) collisions.

In the two-fluid version of the equations of motion (see for example Chapter 6), the resistivity appears in the frictional force between particles of the two species. It is important to emphasize again, as was noted in our discussion of diffusion in weakly ionized gases, that it is by *solving* the fluid equations in the presence of this frictional force that the diffusive flows appear. Diffusion appears naturally from solving the fluid equations and is not to be introduced separately in the continuity equation.

It is worth noting that there are other collisional forces in the full fluid equations that are not being considered here; typically, these tend to be relatively small because they are higher order in kr_L. For example, gradients in flow velocities give rise to viscous forces, and gradients in temperature give rise to 'thermal forces'. These appear in the fluid equation of motion as extra terms in the pressure tensor, whose divergence represents a force density giving rise to additional effects which must be considered in a more complete treatment.

12.6 DIFFUSION DUE TO LIKE AND UNLIKE CHARGED-PARTICLE COLLISIONS

Here, we will use the single-particle picture to understand more deeply the simple fluid results of the preceding Section, that like-particle collisions (e.g. ion–ion collisions) do not lead to any cross-field diffusion, and that transport from unlike-particle collisions (e.g. electron–ion collisions) is intrinsically ambipolar.

Consider a magnetic field in the z direction and a density gradient in the x direction. We will focus on like-particle collisions involving *ions*, although our analysis is equally applicable to electron–electron collisions. Ions gyrate about the field, making circles in the (x, y) plane. For such circular orbits, the x coordinate of the guiding center's position, x_{gc}, is related to the actual particle position x by the formula

$$x_{gc} = x + v_y/\omega_c = x + Mv_y/eB. \qquad (12.35)$$

When two ions collide by arriving momentarily at the same location x, their velocity vectors are suddenly changed, as are the positions of their guiding centers. However, conservation of momentum in the y direction, i.e.

$$\left(\Sigma Mv_y\right)_{\text{initial}} = \left(\Sigma Mv_y\right)_{\text{final}} \qquad (12.36)$$

where the summation is over the two ions, implies by equation (12.35) that

$$(x_{gc}^{(1)} + x_{gc}^{(2)})_{\text{initial}} = (x_{gc}^{(1)} + x_{gc}^{(2)})_{\text{final}} \qquad (12.37)$$

where (1) and (2) denote the two ions. Thus, *conservation of momentum in the y direction assures that the center of mass of the two guiding centers along the direction of the density gradient, x, is unchanged.* Figure 12.2 shows the special case of a 90° collision, with ions approaching each other on 'initial' Larmor orbits with velocity vectors in the $\pm y$ direction, where the upper sign corresponds to orbit (1) and the lower sign to orbit (2). After such a collision, the momentum in the y direction is completely destroyed, and the two ions move off in the $\pm x$ direction, thereafter executing the 'final' Larmor orbits shown in the figure. Since these like-particle collisions cannot produce any net movement of the ions in any one direction, *even in a non-uniform plasma*, it could be argued that there is no continuous net diffusion, although there is certainly some

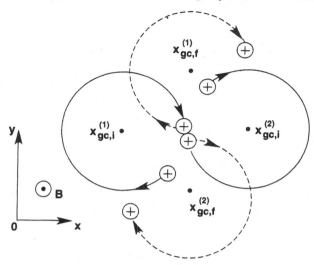

Figure 12.2. Initial (full lines) and final (broken lines) Larmor orbits of two ions making a 90° collision. The initial positions of the two guiding centers are $x_{gc,i}^{(1)}$ and $x_{gc,i}^{(2)}$; the final positions are $x_{gc,f}^{(1)}$ and $x_{gc,f}^{(2)}$.

small-scale rearrangement of the guiding centers on a scale-size of order the Larmor radius.

When we view diffusion as a 'spreading' of the particles' guiding centers or, in the case of a non-uniform plasma, as a net flux of guiding centers opposite to ∇n, this result of vanishing diffusion may still however seem paradoxical. Indeed, if we consider a different collision event, which is the exact reverse of the collision event illustrated in Figure 12.2, i.e. it corresponds to $f \to i$ (rather than $i \to f$), we see that the two guiding centers at $x = x_{gc,f}$, which start with zero spread in the x direction, have moved to the two locations $x = x_{gc,i}$, which have a spread of approximately two Larmor radii. Thus, at least from this one collision, there is a non-zero spreading of the guiding centers. Moreover, if we evaluate the flux of guiding centers across some surface $x = x_0$ drawn to the right of the location of the collision in Figure 12.2, we see that one guiding center has moved rightward across this line in the reversed collision, i.e. a flux has arisen from this one collision.

The resolution of this paradox comes from noting (i) that the frequency of collisions is proportional to the *product of the densities* at the locations of the two guiding centers, which are slightly different, and (ii) that the net flux involves an averaging over all collision events that cause a guiding center to cross the flux-evaluation surface, including that *pair of collisions* formed from any one collision event plus the reversed version of this event. We then find that the net flux of guiding centers due to like-particle collisions to lowest significant order, i.e. the flux that is proportional to ∇n, vanishes.

We can see this rather easily by examining the net flux of guiding centers from the pair of collisions illustrated in Figure 12.2, i.e. considering both 'forward' ($i \rightarrow f$) and 'reverse' ($f \rightarrow i$) collision events. If we again place an observation surface at $x = x_0$ just to the right of the location of the collision, we see that the forward collision event ($i \rightarrow f$) results in the leftward movement of one guiding center across this surface, whereas the reverse collision event ($f \rightarrow i$) results in the rightward movement of one guiding center across this same surface. The expected result of zero net flux can now be established by noting that, for this particular case, the frequencies of these two collision events are equal because, to first order in the size of the Larmor radius, the product of the guiding-center densities at $x_{\mathrm{gc},i}^{(1)}$ and $x_{\mathrm{gc},i}^{(2)}$ is equal to the product of the guiding-center densities at $x_{\mathrm{gc},f}^{(1)}$ and $x_{\mathrm{gc},f}^{(2)}$, *even allowing for the presence of a non-zero density gradient in the x direction*. Specifically, relative to a guiding-center density n_{gc} at $x = x_{\mathrm{gc},f}^{(1)} = x_{\mathrm{gc},f}^{(2)}$, the guiding-center densities at the two locations $x_{\mathrm{gc},i}^{(1)}$ and $x_{\mathrm{gc},i}^{(2)}$ are $n_{\mathrm{gc}} \pm r_{\mathrm{L}}(\mathrm{d}n_{\mathrm{gc}}/\mathrm{d}x)$, whose product is just n_{gc}^2, to first order in r_{L}. The constancy, i.e. before and after the collision, of the position of the center-of-mass of the two guiding centers, which we found to be a consequence of momentum conservation, enters this argument via the geometrical constraint embodied in Figure 12.2 that $x_{\mathrm{gc},i}^{(1)}$ and $x_{\mathrm{gc},i}^{(2)}$ are equidistant to the left and right of $x = x_{\mathrm{gc},f}^{(1)} = x_{\mathrm{gc},f}^{(2)}$. However, it is immediately apparent from this argument that *the net flux vanishes only to lowest significant order in* r_{L}: we can expect there to be non-zero fluxes which are higher order in r_{L} and which involve higher-order derivatives of the guiding-center density, i.e. $\mathrm{d}^2 n_{\mathrm{gc}}/\mathrm{d}x^2$, etc.

This same argument applies equally well to the more general case in which the colliding particles have unequal Larmor radii and undergo collisional scattering of their velocity vectors through some general angle, not necessarily 90°. To see this, consider Figure 12.3, which shows the guiding centers, $x_{\mathrm{gc}}^{(1)}$ and $x_{\mathrm{gc}}^{(2)}$, of two general particles that undergo a collision at $x = x_{\mathrm{c}}$, which causes the guiding centers to be displaced from their 'initial' locations, $x_{\mathrm{gc},i}$, to their 'final' locations, $x_{\mathrm{gc},f}$. We suppose that the collision results in a rightward displacement Δx of $x_{\mathrm{gc}}^{(1)}$, together with an equal and opposite (by equation (12.37)) leftward displacement Δx of $x_{\mathrm{gc}}^{(2)}$. We place a flux-evaluation surface at a location $x = x_0$ that is crossed by $x_{\mathrm{gc}}^{(1)}$, as shown in Figure 12.3.

If we consider *all* crossings of this flux evaluation surface by guiding centers, we must include both the 'forward' ($i \rightarrow f$) collision, in which the guiding center of particle (1) moves from $x_{\mathrm{gc},i}^{(1)}$ to $x_{\mathrm{gc},f}^{(1)}$, thereby crossing the flux

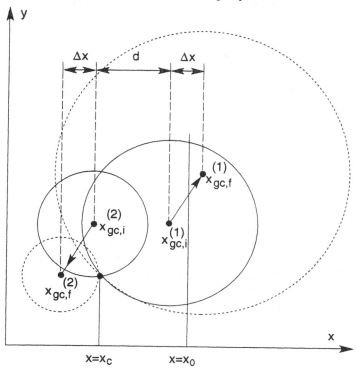

Figure 12.3. Illustration of two guiding centers at locations $x_{gc,i}$ which are displaced to locations $x_{gc,f}$ as a result of a collision occurring at $x = x_c$. The initial and final Larmor orbits are shown by full and broken lines, respectively. The flux is evaluated at $x = x_0$.

evaluation surface in the rightward direction, as shown in Figure 12.3, and the 'reverse' ($f \rightarrow i$) collision, in which the guiding center of particle (1) crosses the flux evaluation surface in the leftward direction. The frequency of each type of collision is proportional to the product of the distribution functions of guiding centers for the relevant velocities and at the respective, i.e. 'initial' or 'final', locations. The collision frequency is also proportional to $\sigma(v_{rel}, \theta)v_{rel}$, where the cross section σ for this particular collision depends on the magnitude of the relative velocity of the colliding particles, v_{rel}, and on the scattering angle in the center-of-mass frame, θ. We assume that the guiding-center distribution function of particles with velocity $\mathbf{v}$ is given by $f(x_{gc}, \mathbf{v}) \equiv n(x_{gc})f_M(v)$, where $f_M(v)$ is a Maxwellian distribution of velocities and where we allow a spatial gradient only in the density and not in the temperature.

For the particular collision under consideration, the probability of the 'forward' collision is proportional to

$$\sigma v_{\text{rel}} n(x_{\text{gc},i}^{(1)}) n(x_{\text{gc},i}^{(2)}) f_M(v_i^{(1)}) f_M(v_i^{(2)})$$

$$= \sigma v_{\text{rel}} \left(n(x_0) + (x_{\text{gc},i}^{(1)} - x_0) \frac{dn}{dx}\bigg|_{x_0} \right)$$

$$\times \left(n(x_0) + (x_{\text{gc},i}^{(2)} - x_0) \frac{dn}{dx}\bigg|_{x_0} \right) f_M(v_i^{(1)}) f_M(v_i^{(2)})$$

$$\approx \sigma v_{\text{rel}} \left(n^2(x_0) + (x_{\text{gc},i}^{(1)} + x_{\text{gc},i}^{(2)} - 2x_0) n(x_0) \frac{dn}{dx}\bigg|_{x_0} \right) f_M(v_i^{(1)}) f_M(v_i^{(2)}).$$

$$(12.38)$$

The total frequency of these 'forward' collisions is obtained by multiplying equation (12.38) by the product of the volume elements in velocity space of the two colliding particles, i.e. $d^3 v_i^{(1)} d^3 v_i^{(2)}$, and then integrating over the two velocity spaces. For like particles, for which the collision is essentially the same if particles (1) and (2) are interchanged, we should then divide by 2 to avoid counting the same collision twice.

In the same way, the probability of the 'reverse' collision is proportional to

$$\sigma v_{\text{rel}} n(x_{\text{gc},f}^{(1)}) n(x_{\text{gc},f}^{(2)}) f_M(v_f^{(1)}) f_M(v_f^{(2)})$$

$$\approx \sigma v_{\text{rel}} \left(n^2(x_0) + (x_{\text{gc},f}^{(1)} + x_{\text{gc},f}^{(2)} - 2x_0) n(x_0) \frac{dn}{dx}\bigg|_{x_0} \right) f_M(v_f^{(1)}) f_M(v_f^{(2)})$$

$$(12.39)$$

where to obtain the total frequency of these 'reverse' collisions, we must again multiply equation (12.39) by $d^3 v_f^{(1)} d^3 v_f^{(2)}$ and integrate over the two velocity spaces. Again we should then divide by 2 to avoid counting the same collision twice.

However, since each collision produces equal and opposite displacements Δx of the two guiding centers, we can again invoke equation (12.37), which shows that the terms in the large brackets of equations (12.38) and (12.39) are equal. Moreover, for Maxwellian distributions with the same temperature, the conservation of energy in a collision implies that the products of the two f_M in equations (12.38) and (12.39) are equal to each other. Also, the relative velocities, v_{rel}, for the forward and reverse collisions are the same, as are the scattering angles, θ, in the center-of-mass frame; thus the quantities σv_{rel} in equations (12.38) and (12.39) are the same. Finally, we note that the products of the two infinitesimal velocity-space volumes for the forward and reverse collisions are equal, i.e. $d^3v_i^{(1)}d^3v_i^{(2)} = d^3v_f^{(1)}d^3v_f^{(2)}$, where these infinitesimals are defined by the scattering dynamics of particles with initial velocities in the neighborhood of $\mathbf{v}_i^{(1)}$ and $\mathbf{v}_i^{(2)}$. This result can be proved formally by showing that both of these velocity-space-volume products are equal to $d^3V d^3v_{rel}$, where $\mathbf{V}$ is the mass velocity and $\mathbf{v}_{rel}$ is the relative velocity.

Problem 12.2: Prove this last statement, where the mass velocity $\mathbf{V}$ and relative velocity $\mathbf{v}_{rel}$ are defined by

$$(m_1 + m_2)\mathbf{V} = m_1\mathbf{v}^{(1)} + m_2\mathbf{v}^{(2)}$$
$$\mathbf{v}_{rel} = \mathbf{v}^{(1)} - \mathbf{v}^{(2)}.$$

(Hint: consider these relations as defining a transformation of coordinates, component by component. The transformation from $(v_x^{(1)}, v_x^{(2)})$ to $(V_x, v_{rel,x})$ defines a Jacobian J. Show that the determinant of the Jacobian, J, equals -1, which means that $dv_x^{(1)}dv_x^{(2)} = dV_x dv_{rel,x}$. The other components can be handled similarly.)

Thus when we integrate over the velocity spaces for every possible 'forward' collision, we find that the integral necessarily includes an exactly compensating 'reverse' collision, because the forward and reverse collisions are equi-probable. Thus, the flux of particles in one direction across the flux evaluation surface due to forward collisions is exactly compensated by a flux in the opposite direction due to reverse collisions.

Although on the basis of the simple random-walk picture presented at the beginning of this Chapter we might have expected a net second-order flux $\Gamma_x \sim -\nu(\Delta x)^2(dn/dx)$, in fact we have found that the net flux vanishes at this order; the rightward flux of guiding centers across $x = x_0$ in Figure 12.3 is exactly cancelled by a leftward flux of guiding centers, even in the presence of a finite density gradient, dn/dx. To take a particular example, in the case of a negative density gradient, $dn/dx < 0$, in Figure 12.3, the rightward flux of particles $(i \to f)$ across $x = x_0$, which is proportional to the product of the 'intermediate' densities at the two locations $x_{gc,i}^{(1)}$ and $x_{gc,i}^{(2)}$, is cancelled by the leftward flux $(f \to i)$, which is proportional to the product of the 'lowest' density at $x_{gc,f}^{(1)}$ and the 'highest' density at $x_{gc,f}^{(2)}$. It is apparent, however, that the *exact* flux to *all orders* in small quantities of order of the Larmor radius, need not vanish. Indeed, more detailed calculations show that there *is* a small non-vanishing higher-order flux, but this is not simply *diffusion*, since it is proportional to second and higher-order derivatives of $n(x)$.

Let us now apply the present type of analysis to the case of unlike-particle collisions. For unlike-particle collisions (e.g. electron–ion collisions), conservation of momentum implies that

$$(x_{\text{gc}}^{(1)} - x_{\text{gc}}^{(2)})_{\text{initial}} = (x_{\text{gc}}^{(1)} - x_{\text{gc}}^{(2)})_{\text{final}} \qquad (12.40)$$

or equivalently

$$x_{\text{gc, final}}^{(2)} - x_{\text{gc, initial}}^{(2)} = x_{\text{gc, final}}^{(1)} - x_{\text{gc, initial}}^{(1)}. \qquad (12.41)$$

Both guiding centers make a step in the same direction. For some collisions, the steps will both be in one direction, and for other collisions the steps will both be in the other direction. For a non-uniform plasma, there will be a preponderance of steps in the direction opposite to the density gradient. If we consider a surface at fixed $x = x_0$, if $\mathrm{d}n/\mathrm{d}x > 0$ there will be more guiding centers at $x > x_0$ to provide a source of negative flux across $x = x_0$ than guiding centers at $x < x_0$ to provide positive flux. Hence, diffusion *does* occur. Moreover, since the displacements of the two guiding centers are equal for each collision, the diffusion is *intrinsically ambipolar*, i.e. the same for electrons and ions.

Problem 12.3: A fully ionized plasma contains two different types of ions, with different masses M and different charge numbers Z. Consider the diffusion across a magnetic field that arises from collisions of ions of one type with ions of the other type. By generalizing the discussion of diffusion due to like and unlike charged-particle collisions given here, show that, whereas the two types of ions may diffuse relative to each other, there can be no net movement of ion *charge* in any one direction across the magnetic field. Describe qualitatively how diffusion of the two individual ion types can occur while satisfying this constraint. Do the electrons diffuse? Is diffusion still intrinsically ambipolar?

12.7   DIFFUSION AS STOCHASTIC MOTION*

The vanishing of the lowest-order diffusive flux from like-particle collisions was first shown by C L Longmire and M N Rosenbluth (1956 *Phys. Rev.* **103** 507), who also give a calculation of the non-zero flux proportional to the second and third derivatives of $n(x)$. The Longmire and Rosenbluth analysis is based on solving the Fokker–Planck equation to obtain the particle flux due to Coulomb collisions for small spatial non-uniformities in a Maxwellian plasma. The underlying concept of a more general type of 'stochastic motion' is introduced in this Section and used to provide a simplified version of the Longmire and Rosenbluth proof of the vanishing of the second-order flux due to like-particle collisions in one special case.

The simplified version of the Longmire and Rosenbluth approach described in this Section, applied to the case of unlike-particle collisions, also demonstrates in a more formal way the intrinsic ambipolarity of the second-order particle fluxes.

In order to provide this alternative demonstration both of the vanishing of the second-order diffusive flux due to like-particle collisions and of the intrinsic ambipolarity of diffusion due to unlike-particle collisions, we must first extend our concept of a random walk to include more general types of motion involving random steps, usually called 'stochastic motion'. In the simple random walk of particles on a straight line (the x axis), as discussed earlier in this Chapter, all of the particles take steps of exactly equal magnitude, Δx, and steps to the left are of exactly equal probability as steps to the right. We can generalize this concept in two ways. First, we can suppose that there is small difference in the average magnitudes of leftward and rightward steps. Redefining Δx now as the (signed) step in the positive-x direction, this means that there is a non-zero average net displacement $\langle \Delta x \rangle$ of the particles in the time interval Δt. The spreading of the particles' positions is now described by the mean square displacement $\langle (\Delta x)^2 \rangle$ in the same time interval Δt. The second generalization of the simple random walk is to allow the typical magnitude and degree of leftward/rightward imbalance of the steps to vary with location along the x axis, so that the quantities $\langle \Delta x \rangle$ and $\langle (\Delta x)^2 \rangle$ become functions of x.

To evaluate the net x-directed flux Γ across a surface located at $x = x_0$, we consider again the 'emptying out' of a line segment located immediately to the left of $x = x_0$ as a result of positive steps Δx in a time interval Δt. Allowing for x-dependent steps Δx, the length of the line-segment that empties out in the positive-x direction in the time interval Δt is

$$(\Delta x)_- \equiv \Delta x \big|_{x_0 - \Delta x} \approx \Delta x \big|_{x_0} - \Delta x \left. \frac{d\Delta x}{dx} \right|_{x_0} \tag{12.42}$$

The x-directed flux from the emptying out of this line-segment is

$$\begin{aligned}
\Gamma &= \frac{1}{\Delta t} \int_{x_0 - (\Delta x)_-}^{x_0} n(x) dx \\
&\approx \frac{1}{\Delta t} \int_{x_0 - (\Delta x)_-}^{x_0} \left(n(x_0) + (x - x_0) \left. \frac{dn}{dx} \right|_{x_0} \right) dx \\
&\approx \frac{1}{\Delta t} \left(n\Delta x - n\Delta x \frac{d\Delta x}{dx} - \frac{(\Delta x)^2}{2} \frac{dn}{dx} \right) \\
&\approx \frac{1}{\Delta t} \left(n\Delta x - \frac{1}{2} \frac{d}{dx} \left[n(\Delta x)^2 \right] \right).
\end{aligned} \tag{12.43}$$

Although we have derived this expression by considering a line-segment located to the left of $x = x_0$ emptying out by means of positive steps Δx, the analysis applies equally well to the leftward emptying out of a line-segment located to the right of $x = x_0$, provided Δx is taken to be a negative quantity. Thus we may average equation (12.43) over both rightward and leftward steps, to obtain our final expression for the average net flux:

$$\Gamma = \frac{1}{\Delta t}\left(n\langle\Delta x\rangle - \frac{1}{2}\frac{d}{dx}\left[n\langle(\Delta x)^2\rangle\right]\right). \tag{12.44}$$

Equation (12.44) is the generalization of equation (12.4) that applies to the present situation where there are x-dependent steps Δx and a mean step that is not necessarily zero. It is clear from our derivation that equation (12.44) is valid only up to second order in the step size Δx. Moreover, equation (12.44) is of interest mainly in situations where the rightward and leftward steps are *almost* in balance, so that the term $\langle\Delta x\rangle$, while apparently of first order in Δx, is in fact as small as the second-order term. Clearly, however, even a *small* leftward/rightward imbalance is sufficient to produce a contribution to the flux that is as important as the contribution from diffusive spreading. Equation (12.44) describes just this situation.

Problem 12.4: This simple derivation of equation (12.44) has assumed fixed (but slightly unequal) rightward and leftward steps, Δx, at each location x. It has also implicitly assumed that particles do not jump over each other in making these steps. Generalize this derivation by considering the (more physical) case where there is a distribution of possible steps with varying magnitude, so that the quantities $\langle\Delta x\rangle$ and $\langle(\Delta x)^2\rangle$ now refer to the mean step and mean spreading averaged over this distribution. (Hint: begin by defining a probability, $P(x, \Delta x)$, for each step Δx (positive or negative) at location x. Write down an expression for the flux of particles across a reference point x_0, and expand $n(x)$ around $x = x_0$ keeping only the terms in $n(x_0)$ and $dn/dx|_{x_0}$. The quantities $\langle\Delta x\rangle$ and $\langle(\Delta x)^2\rangle$ are defined by

$$\langle\Delta x\rangle = \int_{-\infty}^{\infty} \Delta x\, P(x, \Delta x)d(\Delta x)$$

$$\langle(\Delta x)^2\rangle = \int_{-\infty}^{\infty} (\Delta x)^2 P(x, \Delta x)d(\Delta x).$$

Your final result should be exactly the same as equation (12.44).)

Let us return now to the topic of the flux due to like-particle collisions. For the collision illustrated in Figure 12.3, assuming a negative density gradient,

$dn/dx < 0$, we have seen that the 'forward' collision ($i \to f$) produces a rightward flux of particles across $x = x_0$, which has the character of diffusion in the sense that it is in the opposite direction to the density gradient. However, we have seen that the 'reverse' collision ($f \to i$) produces a compensating leftward flux, which is in the *same direction* as the density gradient.

Expressed in the notation that we have introduced for describing general stochastic motion, the contribution to the flux that is in the same direction as the density gradient arises from a non-vanishing mean displacement $\langle \Delta x \rangle$ per time interval Δt, whereas the contribution to the flux that is opposite to the density gradient arises from the diffusive spreading $\langle (\Delta x)^2 \rangle$ in the same time interval. In this example of stochastic motion, the non-vanishing $\langle \Delta x \rangle$ arises from the non-uniformity in the density of the *scattering* particles, as distinct from that of the *scattered* particles. Indeed the flux from the term in $\langle \Delta x \rangle$, i.e. the first term in the expression on the right-hand side of equation (12.44), is found to be proportional to the density gradient for scattering particles. The diffusive flux from the term in $\langle (\Delta x)^2 \rangle$, i.e. the second term in the expression on the right-hand side of equation (12.44), has a term proportional to the density gradient for scattered particles, which appears explicitly in equation (12.44), and a term proportional to the density gradient for scattering particles, which appears implicitly through the dependence of $\langle (\Delta x)^2 \rangle$ on the density of scattering particles. When both scattered and scattering particles are the same, e.g. ions, there is the possibility of exact cancellation of these two contributions to the net flux, and this is what actually occurs.

We can see this most easily by considering a very simple case, somewhat like the case illustrated in Figures 12.2 and 12.3, in which the particles have velocities only perpendicular to the magnetic field, i.e. velocity components (v_x, v_y). We suppose that the collisions are such as to preserve this situation, i.e. no parallel velocities v_z are acquired as a result of collisions, so our analysis can be strictly two-dimensional in velocity space. We suppose that the velocity distribution function is Maxwellian in $v_\perp = \sqrt{(v_x^2 + v_y^2)}$, and that the temperature T is spatially uniform. Since we want to demonstrate both the vanishing of the particle fluxes in the case of like-particle collisions and the intrinsic ambipolarity of these fluxes in the case of unlike-particle collisions, we consider the general case of collisions between particles of two different types, i.e. types (1) and (2), with charges q_1 and q_2, masses m_1 and m_2, and densities $n_1(x)$ and $n_2(x)$, respectively. The two temperatures are assumed to be the same.

In treating a collision between a particle of type (1) and a particle of type (2), it is convenient to work in the 'center-of-mass' frame, in which the total momentum of the two colliding particles is zero. In this moving frame, as shown in Figure 12.4, the 'initial' velocity vectors satisfy $m_2 \mathbf{v}_i^{(2)} = -m_1 \mathbf{v}_i^{(1)}$, and since momentum is conserved in the collision, the 'final' velocity vectors likewise must satisfy $m_2 \mathbf{v}_f^{(2)} = -m_1 \mathbf{v}_f^{(1)}$. When center-of-mass velocity variables are used, the sum of the kinetic energies of the two particles in the rest frame, $W_1 + W_2$, can be written as the sum of the kinetic energy of the combined mass, $M = m_1 + m_2$, moving at the mass velocity $\mathbf{V}$, defined by $M\mathbf{V} = m_1\mathbf{v}^{(1)} + m_2\mathbf{v}^{(2)}$, and the kinetic energy of the 'reduced mass', $m = m_1 m_2/(m_1 + m_2)$, moving at the relative velocity $\mathbf{v}_{\text{rel}} = \mathbf{v}^{(1)} - \mathbf{v}^{(2)}$. In the center-of-mass frame, moving with velocity $\mathbf{V}$, the total kinetic energy is simply $m v_{\text{rel}}^2/2$. Since energy is conserved, we see that v_{rel} must be the same before and after the collision.

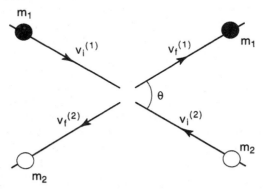

Figure 12.4. A collision between a particle of type (1), mass m_1, and a particle of type (2), mass m_2, shown in the center-of-mass frame in which the total momentum of the two colliding particles is zero. The relative speed of the colliding particles $v_{\text{rel}} = |\mathbf{v}^{(1)} - \mathbf{v}^{(2)}|$ is the same before and after the collision. The scattering angle is θ.

A particle of type (1) collides with particles of type (2) with frequency $\nu_{12} = n_2 \langle \sigma_{12}(v_{\text{rel}}, \theta) v_{\text{rel}} \rangle$, where the cross section σ_{12} is generally a function both of v_{rel} and of the scattering angle in the center-of-mass frame, θ (as shown in Figure 12.4). For the simple case considered here, we assume that the cross section is *independent* of the scattering angle θ, i.e. in the center-of-mass frame all scattering angles are equi-probable. We also assume that the quantity $\sigma_{12} v_{\text{rel}}$ may be treated as a constant, i.e. not dependent upon the relative velocity v_{rel}. (The more correct case for Coulomb collisions, where $\sigma_{12} v_{\text{rel}}$ depends on both v_{rel} and θ, is treated in the previously cited paper by Longmire and Rosenbluth; no fundamentally different effects are introduced, but the various velocity-space averages become more complicated.)

Let us focus our attention on a particle of type (1) and the average step $\langle \Delta x \rangle$ that its guiding center takes as a result of collisions with particles of type (2). The frequency of such collisions is proportional to the density of particles (2) at the location of the collision, x_c, which is the same as the density of guiding centers at

$$x_{\text{gc},i}^{(2)} = x_c + v_{y,i}^{(2)}/\omega_{c2} = x_{\text{gc},i}^{(1)} - v_{y,i}^{(1)}/\omega_{c1} + v_{y,i}^{(2)}/\omega_{c2}. \tag{12.45}$$

Here we have used equation (12.35) twice—once to relate $x_{gc,i}^{(2)}$ to x_c and once to relate $x_{gc,i}^{(1)}$ to x_c. We have also introduced the gyro-frequencies, ω_{c1} and ω_{c2}, of the two types of particles.

It is convenient at this point to transform from the initial individual particle velocities to the mass velocity, $\mathbf{V}$, which is the same before and after the collision, and the initial relative velocity, $\mathbf{v}_{rel,i} = \mathbf{v}_i^{(1)} - \mathbf{v}_i^{(2)}$. In terms of these velocities, the individual particle velocities can be written

$$\mathbf{v}_i^{(1)} = \mathbf{V} + \frac{m_2}{m_1 + m_2} \mathbf{v}_{rel,i}$$

$$\mathbf{v}_i^{(2)} = \mathbf{V} - \frac{m_1}{m_1 + m_2} \mathbf{v}_{rel,i}.$$

Substituting the y components of these relations into equation (12.45), we obtain

$$x_{gc,i}^{(2)} = x_{gc,i}^{(1)} - \left(\frac{1}{\omega_{c1}} + \frac{1}{\omega_{c2}} \right) V_y - \frac{m}{q_1 B} \left(1 + \frac{q_1}{q_2} \right) v_{rel,y,i} \qquad (12.46)$$

where we have again introduced the reduced mass, $m = m_1 m_2 / (m_1 + m_2)$.

In the collision being considered, the guiding center of particle (1) takes a step

$$\begin{aligned} \Delta x &= x_{gc,f}^{(1)} - x_{gc,i}^{(1)} \\ &= x_c + v_{y,f}^{(1)} / \omega_{c1} - x_{gc,i}^{(1)} \\ &= v_{y,f}^{(1)} / \omega_{c1} - v_{y,i}^{(1)} / \omega_{c1}. \end{aligned} \qquad (12.47)$$

We again transform from the individual particle velocities to the mass velocity and the relative velocity, now also defining a final relative velocity, $\mathbf{v}_{rel,f} = \mathbf{v}_f^{(1)} - \mathbf{v}_f^{(2)}$. Equation (12.47) then becomes

$$\begin{aligned} \Delta x &= \frac{1}{\omega_{c1}} \frac{m_2}{m_1 + m_2} \left(v_{rel,y,f} - v_{rel,y,i} \right) \\ &= \frac{m}{q_1 B} \left(v_{rel,y,f} - v_{rel,y,i} \right) \end{aligned} \qquad (12.48)$$

with the terms in V_y cancelling each other.

Because of our assumption that the cross section is independent of scattering angle, the angle through which the perpendicular velocity vector of particle (1) is scattered in the center-of-mass frame depicted in Figure 12.4 is entirely random—all angles θ in the range 0–2π are equi-probable. Thus the y component of the relative velocity just after the collision, $v_{rel,y,f}$, takes on positive and negative values with equal probability and averages to zero. Thus in a first averaging process, which averages over all possible 'final' relative velocity vectors while keeping the 'initial' relative velocity vector fixed, we can write

$$\langle \Delta x \rangle_f = -\frac{m}{q_1 B} v_{\text{rel},y,i}. \tag{12.49}$$

Next, we must carry out a second averaging process, which averages over all possible initial relative velocity vectors. This averaging process must take into account the proportionality of the collision frequency to the density of guiding centers (2) at the location $x_{\text{gc},i}^{(2)}$ given in equation (12.46), which varies with the initial velocity vectors. Denoting the collision frequency by $v_{12} = n_2 \sigma_{12} v_{\text{rel}}$, this second averaging involves a weighted average of the right-hand side of equation (12.49) with weights proportional to the guiding-center density for particles (2) at $x_{\text{gc},i}^{(2)}$. This gives an expression for the average displacement per unit time, namely

$$\frac{\langle \Delta x \rangle}{\Delta t} = -\sigma_{12} v_{\text{rel}} \left\langle n_{\text{gc}}^{(2)} (x_{\text{gc},i}^{(2)}) \frac{m v_{\text{rel},y,i}}{q_1 B} \right\rangle$$

$$= -\sigma_{12} v_{\text{rel}} \left\langle \left(n_{\text{gc}}^{(2)} (x_{\text{gc},i}^{(1)}) - (x_{\text{gc},i}^{(2)} - x_{\text{gc},i}^{(1)}) \left. \frac{\mathrm{d} n_{\text{gc}}^{(2)}}{\mathrm{d}x} \right|_{x_{\text{gc},i}^{(1)}} \right) \frac{m v_{\text{rel},y,i}}{q_1 B} \right\rangle$$

$$= \sigma_{12} v_{\text{rel}} \left(1 + \frac{q_1}{q_2} \right) \frac{m^2}{q_1^2 B^2} \langle v_{\text{rel},y,i}^2 \rangle \frac{\mathrm{d} n_{\text{gc}}^{(2)}}{\mathrm{d}x}. \tag{12.50}$$

In the last step in equation (12.50), we have substituted equation (12.46) for the quantity $x_{\text{gc},i}^{(2)} - x_{\text{gc},i}^{(1)}$, and we have then noted that only the term that is second order in $v_{\text{rel},y,i}$ survives in the averaging process. This is because positive and negative values of $v_{\text{rel},y,i}$ occur with equal weight in the averaging. In particular, the term in V_y from equation (12.46), which in the averaging in equation (12.50) is multiplied by a first-order term $v_{\text{rel},y,i}$, vanishes in the averaging process. Thus the mass velocity, **V**, disappears at this point from our calculation.

The final averaging indicated in equation (12.50) is to be carried out over all values of $v_{\text{rel},y,i}$ that occur in the velocity distributions of the colliding particles. We have assumed that the type-(1) and type-(2) particles both have Maxwellian distributions with the same temperature, T. We have also noted that the sum of the kinetic energies of two particles, one of each type, $W_1 + W_2$, can be written as the sum of the kinetic energy of the combined mass, $M = m_1 + m_2$, moving at the mass velocity, **V**, and the kinetic energy of the reduced mass, $m = m_1 m_2 / (m_1 + m_2)$, moving at the relative velocity, $\mathbf{v}_{\text{rel}}$, i.e.

$$W_1 + W_2 \equiv m_1 (v_i^{(1)})^2 / 2 + m_2 (v_i^{(2)})^2 / 2$$
$$= MV^2 / 2 + m v_{\text{rel}}^2 / 2$$

which means that the distribution of mass velocities and the distribution of relative velocities are also both Maxwellian, with the same temperature, T, and with masses M and m, respectively.

Problem 12.5: Verify this last statement formally by carrying out a transformation from the velocities $v_i^{(1)}$ and $v_i^{(2)}$ to V and v_{rel}.

It then follows that

$$\langle v_{rel,y,i}^2 \rangle = T/m. \tag{12.51}$$

Substituting this into equation (12.50), we obtain our final result for the average displacement per unit time:

$$\frac{\langle \Delta x \rangle}{\Delta t} = \sigma_{12} v_{rel} \left(1 + \frac{q_1}{q_2} \right) \frac{mT}{q_1^2 B^2} \frac{dn_2}{dx}. \tag{12.52}$$

In equation (12.52), there are two terms in $\langle \Delta x \rangle/\Delta t$ proportional to the density gradient of scattering particles, dn_2/dx, arising respectively from the 'unity' and the 'q_1/q_2' terms inside the parenthesis. The first of these terms gives the effect of the higher frequency of collisions on that side of the Larmor orbit of particle (1) where the density of particles (2) (i.e. scattering particles) is higher. In all cases, this produces a flux of scattered particles in the direction of the density gradient of scattering particles which, in the case of like particles, is *up* the density gradient. The second of these terms, i.e. the term in q_1/q_2, gives the effect of the non-zero y-directed diamagnetic drift of particles (2), which exerts a y-directed frictional force on particle (1) thereby producing an x-directed drift. Specifically, the y-directed diamagnetic drift of particles (2) is $(T/q_2 Bn_2)dn_2/dx$, which is to be multiplied by the reduced mass, m, and by the collision frequency, $n_2 \sigma_{12} v_{rel}$, to give the rate of momentum transfer, i.e. the frictional force on particle (1), $F_y = \sigma_{12} v_{rel}(mT/q_2 B)dn_2/dx$. This produces a drift, $(F \times B)/(q_1 B^2)$, of particle (1), which in this case is an x-directed drift of magnitude $\sigma_{12} v_{rel}(mT/q_1 q_2 B^2)dn_2/dx$, i.e. exactly the term in q_1/q_2 in equation (12.52). For like particles, this term produces another flux of scattered particles *up* the density gradient. However for unlike particles, in particular those with charges of opposite sign, the flux from this term is in the opposite direction. This is because the direction of the $F \times B$ drift depends on the sign of the charge of the drifting particle. For collisions between electrons and protons, it is apparent from equation (12.52) that these two terms cancel each other, so there is no net average displacement per unit time, $\langle \Delta x \rangle/\Delta t$, in this case.

Next consider the diffusive spreading of the guiding centers of particles (1) as a result of collisions with particles (2). In the collision being considered, the guiding center of particle (1) takes a step Δx, which is given in equation (12.47) and has been expressed in terms of the initial and final relative velocities in equation (12.48). Thus the average spreading of the guiding center positions *per collision* is given by

$$
\begin{aligned}
\langle (\Delta x)^2 \rangle &= \frac{m^2}{q_1^2 B^2} \langle (v_{\text{rel},y,f} - v_{\text{rel},y,i})^2 \rangle \\
&= \frac{m^2}{q_1^2 B^2} \left[\langle v_{\text{rel},y,f}^2 \rangle + \langle v_{\text{rel},y,i}^2 \rangle \right] \\
&= 2mT/q_1^2 B^2.
\end{aligned}
\tag{12.53}
$$

Here we have carried out two averagings essentially simultaneously, i.e. the averaging over the relative velocities just after the collision, $v_{\text{rel},y,f}$, and the averaging over the relative velocities just before the collision, $v_{\text{rel},y,i}$. Because of our assumption that the cross section is independent of scattering angle in the center-of-mass frame, these averagings are independent, so the cross term, $\langle v_{\text{rel},y,f} v_{\text{rel},y,i} \rangle$ vanishes. In the last step, in equation (12.53), we have made use of the fact that the relative velocities, both before and after the collision, have Maxwellian distributions in terms of the relative mass m and with temperature T, i.e. we have used both equation (12.51) and the similar relation for the relative velocities just *after* the collision. For collision frequency $v_{12} = n_2 \sigma_{12} v_{\text{rel}}$, the average spreading per unit time is now given by

$$
\frac{\langle (\Delta x)^2 \rangle}{\Delta t} = \sigma_{12} v_{\text{rel}} \frac{2mT n_2}{q_1^2 B^2}.
\tag{12.54}
$$

We can now substitute equations (12.52) and (12.54) into our general expression for the particle flux in stochastic motion, i.e. equation (12.44), to obtain

$$
\Gamma_{12} = \sigma_{12} v_{\text{rel}} \frac{mT}{q_1^2 B^2} \left[\left(1 + \frac{q_1}{q_2} \right) n_1 \frac{dn_2}{dx} - \frac{d(n_1 n_2)}{dx} \right].
\tag{12.55}
$$

Equation (12.55) gives the flux of particles of type (1) due to collisions with particles of type (2). For the case of like-particle collisions, where $q_1 = q_2$ and $n_1(x) \equiv n_2(x)$, the flux vanishes. Specifically, for this case of like-particle collisions, the flux arising from the non-zero average displacement $\langle \Delta x \rangle$ exactly cancels the flux from the 'spreading' term, $\langle (\Delta x)^2 \rangle$.

The calculation presented here has made use of the simplifying assumptions that the collision cross section is independent of the scattering angle in the center-of-mass frame and that the quantity $\sigma_{12} v_{\text{rel}}$ is independent of the relative velocity, v_{rel}. As a result, this quantity appears as just a multiplicative constant in equation (12.55). In more general (and more physical) cases, the quantity

$\sigma_{12} v_{rel}$ must be included in the various averagings that are carried out over velocity space.

The paper by Longmire and Rosenbluth cited at the beginning of this Section derives a result of the same general form as equation (12.55) for the more physical case of three velocity dimensions and the correct Coulomb scattering cross section. To obtain a non-zero net flux from like-particle collisions, it is necessary to retain terms of order $(\Delta x)^3$ and $(\Delta x)^4$ in the analysis.

Now consider equation (12.55) for the case of unlike-particle collisions. Noting a cancellation between the first part of the first term in the square bracket in equation (12.55) and that part of the second term containing dn_2/dx, we can write the flux of electrical charge of particles (1) due to collisions with particles (2) as

$$q_1 \Gamma_{12} = \sigma_{12} v_{rel} \frac{mT}{B^2} \left(\frac{n_1}{q_2} \frac{dn_2}{dx} - \frac{n_2}{q_1} \frac{dn_1}{dx} \right). \qquad (12.56)$$

Clearly, this is exactly equal and opposite to the flux of charge of particles (2) due to collisions with particles (1). Thus

$$q_1 \Gamma_{12} + q_2 \Gamma_{21} = 0 \qquad (12.57)$$

and we have demonstrated the intrinsic ambipolarity of the fluxes at this order due to unlike-particle collisions. In particular, in a hydrogen plasma, the fluxes of electrons and ions are exactly equal, which is what is generally meant by the statement that 'diffusion is ambipolar'.

12.8 DIFFUSION OF ENERGY (HEAT CONDUCTION)

A systematic treatment of heat conduction in plasmas is beyond our scope here. Indeed, in applications of the plasma fluid equations, we have so far considered only two limiting situations: first, the adiabatic equation of state, $d(p/\rho^\gamma)/dt = 0$, corresponding to the case where heat conduction is negligible on the time-scales of interest; second, the isothermal equation of state, $T =$ constant, corresponding to the case where heat conduction is very rapid. In practice, heat conduction in a fully ionized plasma is very anisotropic, i.e. it is extremely rapid along the magnetic field, but quite slow across the field.

In our systematic development of the fluid equations for a plasma, we have chosen to stop at the equation for momentum transfer, which led to the fluid equation of motion, and *not* to proceed further to the energy transfer equation, which would describe heat transport by convection and conduction as well as various heat sources and sinks. To proceed to the energy transfer equation would require consideration of the energy flows into and out of a differential volume element, as well as the deposition of energy into this volume, for example by ohmic heating ($\mathbf{j} \cdot \mathbf{E}$). Just as the pressure tensor arises in describing the momentum flow into a volume element, a 'heat-flux vector' enters into the energy transfer equation; in index notation, the heat-flux vector for an individual species is $Q_i = (m/2)\langle(v_i - u_i)(v_j - u_j)(v_j - u_j)\rangle$. (Note that a vector quantity is sufficient to describe the flux of heat, which is itself a scalar quantity, i.e. $(m/2)\langle(v_j - u_j)(v_j - u_j)\rangle$, whereas a tensor pressure is required to describe the flux of vector momentum.)

Often, the dominant effect needed to evaluate the heat flux vector is the heat *conduction*, which is a heat flux driven by a temperature gradient, although other terms are generally present also, e.g. convection of heat. When conduction dominates, the energy transfer equation, allowing different processes of heat conduction perpendicular and parallel to a magnetic field, can be written

$$\frac{3}{2}n\frac{\partial T}{\partial t} = -\nabla \cdot \mathbf{Q} = \nabla_\perp \cdot (\kappa_\perp \nabla_\perp T) + \nabla_\parallel(\kappa_\parallel \nabla_\parallel T). \tag{12.58}$$

The quantities $\kappa_\perp$ and $\kappa_\parallel$ are the 'thermal conductivities' perpendicular and parallel to the magnetic field, respectively. For each κ, the quantity κ/n, which has the dimensions of a diffusion coefficient, is sometimes called the 'thermal diffusivity'. It must be emphasized, however, that there are many processes of heat transport and heat generation or loss that have been omitted from equation (12.58). For example, in a plasma heat is convected at the fluid velocity **u**, and heat is generated by ohmic heating and lost by radiation. However, since heat *transport* is often dominated by thermal conduction, it is useful to examine the orders of magnitude of $\kappa_\perp$ and $\kappa_\parallel$.

Thermal conduction *along* the magnetic field arises mainly from electrons rather than ions. From our now-familiar random-walk argument, the two parallel diffusivities will take the form v_t^2/ν, which is larger for electrons than for ions by a factor $(M/m)^{1/2}$. Thus

$$\frac{\kappa_\parallel}{n} \sim \frac{v_{t,e}^2}{\nu_e} \tag{12.59}$$

where ν_e represents some combination of electron–electron and electron–ion collision frequencies, both of which will contribute.

Thermal conduction *across* the magnetic field arises mainly from ions rather than electrons. Ions have relatively large Larmor orbits and, when these are perturbed by the collision of two ions, a quantity of energy is exchanged between the two ions. In addition, the two guiding centers are displaced from their original positions by an amount of order an ion Larmor radius. (Note that, unlike the case of particle diffusion, there is no conservation law that constrains the combined energy from taking a 'step' in one direction or the other.) Thus, the 'energy' has made a random walk with a step size of about a Larmor radius r_{Li} in a characteristic time ν_{ii}^{-1}. Thus, the cross-field thermal diffusivity is approximately

$$\kappa_\perp/n \sim \nu_{ii}r_{Li}^2. \tag{12.60}$$

Comparing this result with the similar one for particle diffusion D, we see that *the cross-field thermal diffusivity arises mainly from ion–ion collisions, whereas cross-field particle diffusion arises only from electron–ion collisions.* Moreover, cross-field thermal diffusivity is larger than particle diffusivity by a factor $(M/m)^{1/2} \sim 40$. The conservation of total momentum in collisions prevents a lowest-order contribution from ion–ion collisions to particle diffusion, but it does not prevent a contribution to thermal diffusion. The theory of plasma transport across a magnetic field, including both density and temperature gradients, was developed first by M N Rosenbluth and A N Kaufman (1958 *Phys. Rev.* **109** 1), who give expressions for the cross-field electron and ion thermal diffusivities.

Problem 12.6: Our hydrogenic plasma is replaced by a plasma with multiply charged ions, each with charge Ze. The plasma is fully ionized and charge neutral, so that $n_e = Z n_i$. Give the dependence on Z of (i) the particle diffusivity perpendicular to a magnetic field, $D_\perp$, (ii) the electron thermal diffusivity parallel to a magnetic field, $\kappa_{\parallel e}/n_e$, and (iii) the ion thermal diffusivity perpendicular to a magnetic field, $\kappa_{\perp i}/n_i$. Where density appears in your formulae, be careful to distinguish between n_e and n_i.

Chapter 13

The Fokker–Planck equation for Coulomb collisions*

As we have seen already, collisional effects in fully ionized plasmas are predominantly due to the cumulative effect of many small-angle deflections, rather than to the effect of a few close collisions. In Chapter 11, we obtained estimates for the effective collision frequencies and for the plasma resistivity, but we have not yet provided a rigorous formalism for describing the effects of multiple small-angle Coulomb collisions on the distribution function, $f(v)$.

For the case of large-angle close collisions, a rigorous formalism is provided by the Boltzmann equation, which is discussed in standard textbooks on non-equilibrium statistical mechanics (see, for example, F Reif (1965 *Fundamentals of Statistical and Thermal Physics* New York: McGraw-Hill)) and which applies whenever the interparticle forces are of short range. The Fokker–Planck equation is the version of the Boltzmann equation applicable to the case of long-range interparticle forces. It can be derived from the Boltzmann equation by going to the limit of very-long-range interparticle forces (see, for example, D C Montgomery and D A Tidman (1964 *Plasma Kinetic Theory* New York: McGraw-Hill)), but our approach here is to derive the Fokker–Planck equation directly, by considering the effect of multiple small-angle Coulomb collisions on the distribution of velocities in a plasma.

Viewed in this way, the Fokker–Planck equation provides a general formulation for treating changes in a distribution function that result from a succession of collision 'events', each of which produces only a small change in the velocity of a particle. The equation was formulated in the period 1914–17 by A D Fokker and M Planck to treat Brownian motion (see, for example, S Chandrasekhar (1943 *Rev. Mod. Phys.* **15** 1).

13.1   THE FOKKER–PLANCK EQUATION: GENERAL FORM

Since collisional processes change the distribution of particle velocities, it is necessary to use the 'velocity distribution function' $f(\mathbf{v})$, introduced in Chapter 1, which is the number-density of particles in phase space, i.e. the number of particles per unit volume of physical space and per unit volume of velocity space. The density in physical space is given in terms of $f(\mathbf{v})$ by

$$n = \int f(\mathbf{v})\mathrm{d}^3 v. \tag{13.1}$$

The Fokker–Planck equation describes the evolution in time due to collisions of the function $f(\mathbf{v})$. Since collisional effects depend only on the *local* properties of f, the spatial variation of f can be ignored for present purposes.

We also define a function $\phi(\mathbf{v}, \Delta\mathbf{v})$, which is the probability that a particle with velocity $\mathbf{v}$ acquires an increment of velocity $\Delta\mathbf{v}$ in a time interval Δt. We will assume that collisions occur randomly enough that ϕ is independent of the history of the particle. From the definition of ϕ, it follows that the velocity distribution function at time t can be expressed in terms of the distribution function at a slightly earlier time, i.e.

$$f(\mathbf{v}, t) = \int f(\mathbf{v} - \Delta\mathbf{v}, t - \Delta t)\phi(\mathbf{v} - \Delta\mathbf{v}, \Delta\mathbf{v})\mathrm{d}^3 \Delta v \tag{13.2}$$

where the integral is over all possible velocity increments $\Delta\mathbf{v}$. Since the sum of the probabilities of all possible velocity increments must be unity, we have

$$\int \phi(\mathbf{v}, \Delta\mathbf{v})\mathrm{d}^3 \Delta v = 1. \tag{13.3}$$

Since the effects of Coulomb interactions can be described in terms of a sequence of small-angle deflections, i.e. a sequence of small velocity increments $\Delta\mathbf{v}$, we may expand the integrand $f\phi$ of equation (13.2) in powers of $\Delta\mathbf{v}$. In the case of the factor $\phi(\mathbf{v} - \Delta\mathbf{v}, \Delta\mathbf{v})$ appearing in this integrand, we can however only expand the *first* argument, in which $\Delta\mathbf{v}$ can be treated as small compared with $\mathbf{v}$, and we must leave ϕ unexpanded in regard to the second argument, which describes a strong variation of ϕ with respect to $\Delta\mathbf{v}$. Specifically, keeping terms up to second order in the expansions, we have

$$f(\mathbf{v} - \Delta\mathbf{v}, t - \Delta t) = f(\mathbf{v}, t - \Delta t)$$
$$- \Delta\mathbf{v} \cdot \frac{\partial}{\partial\mathbf{v}} f(\mathbf{v}, t - \Delta t) + \frac{1}{2}\Delta\mathbf{v}\Delta\mathbf{v} : \frac{\partial^2}{\partial\mathbf{v}\partial\mathbf{v}} f(\mathbf{v}, t - \Delta t)$$
$$\phi(\mathbf{v} - \Delta\mathbf{v}, \Delta\mathbf{v}) = \phi(\mathbf{v}, \Delta\mathbf{v})$$
$$- \Delta\mathbf{v} \cdot \frac{\partial}{\partial\mathbf{v}}\phi(\mathbf{v}, \Delta\mathbf{v}) + \frac{1}{2}\Delta\mathbf{v}\Delta\mathbf{v} : \frac{\partial^2}{\partial\mathbf{v}\partial\mathbf{v}}\phi(\mathbf{v}, \Delta\mathbf{v}).$$

The meaning of the somewhat unusual notation $\Delta \mathbf{v} \Delta \mathbf{v}$: and $(\partial^2/\partial \mathbf{v} \partial \mathbf{v})$: should be reasonably obvious—each of these quantities is a dyadic from which a scalar is to be formed by taking the double-dot product (:) of it and a similar dyadic. For example, in index notation the quantity $\Delta \mathbf{v} \Delta \mathbf{v} : \partial^2 f/\partial \mathbf{v} \partial \mathbf{v}$ is defined to mean $\Delta v_i \Delta v_j (\partial^2 f/\partial v_i \partial v_j)$, where repeated suffices are to be summed.

We substitute these into equation (13.2) and retain only terms up to second order in $\Delta \mathbf{v}$ in the product $f\phi$. Using equation (13.3) and assuming Δt is small, we obtain

$$
f(\mathbf{v}, t) - f(\mathbf{v}, t - \Delta t)
$$

$$
= -\int \Delta \mathbf{v} \cdot \left(\frac{\partial f}{\partial \mathbf{v}} \phi + \frac{\partial \phi}{\partial \mathbf{v}} f \right) \mathrm{d}^3 \Delta v
$$

$$
+ \frac{1}{2} \int \Delta \mathbf{v} \Delta \mathbf{v} : \left(\frac{\partial^2 f}{\partial \mathbf{v} \partial \mathbf{v}} \phi + 2 \frac{\partial f}{\partial \mathbf{v}} \frac{\partial \phi}{\partial \mathbf{v}} + \frac{\partial^2 \phi}{\partial \mathbf{v} \partial \mathbf{v}} f \right) \mathrm{d}^3 \Delta v
$$

$$
= -\frac{\partial}{\partial \mathbf{v}} \cdot \int f \phi \Delta \mathbf{v} \mathrm{d}^3 \Delta v + \frac{1}{2} \frac{\partial^2}{\partial \mathbf{v} \partial \mathbf{v}} : \int f \phi \Delta \mathbf{v} \Delta \mathbf{v} \mathrm{d}^3 \Delta v \qquad (13.4)
$$

where, to this order of approximation, f and ϕ appearing on the right-hand side are to be understood to mean $f(\mathbf{v}, t)$ and $\phi(\mathbf{v}, \Delta \mathbf{v})$. The rate of change of f due to collisions can now be written

$$
\left(\frac{\partial f}{\partial t} \right)_{\mathrm{coll}} = \frac{f(\mathbf{v}, t) - f(\mathbf{v}, t - \Delta t)}{\Delta t}
$$

$$
= -\frac{\partial}{\partial \mathbf{v}} \cdot \left(\frac{\mathrm{d} \langle \Delta \mathbf{v} \rangle}{\mathrm{d}t} f \right) + \frac{1}{2} \frac{\partial^2}{\partial \mathbf{v} \partial \mathbf{v}} : \left(\frac{\mathrm{d} \langle \Delta \mathbf{v} \Delta \mathbf{v} \rangle}{\mathrm{d}t} f \right) \qquad (13.5)
$$

since $f = f(\mathbf{v}, t)$ is independent of $\Delta \mathbf{v}$, where

$$
\frac{\mathrm{d} \langle \Delta \mathbf{v} \rangle}{\mathrm{d}t} = \frac{1}{\Delta t} \int \phi \Delta \mathbf{v} \mathrm{d}^3 \Delta v
$$

$$
\frac{\mathrm{d} \langle \Delta \mathbf{v} \Delta \mathbf{v} \rangle}{\mathrm{d}t} = \frac{1}{\Delta t} \int \phi \Delta \mathbf{v} \Delta \mathbf{v} \mathrm{d}^3 \Delta v. \qquad (13.6)
$$

Equation (13.5) for $(\partial f/\partial t)_{\mathrm{coll}}$ is called the 'Fokker–Planck equation'.

The quantity $\mathrm{d} \langle \Delta \mathbf{v} \rangle/\mathrm{d}t$ is the average rate of change of the particle's mean directed velocity due to Coulomb collisions. In an isotropic plasma, there can be no preferred direction for momentum acquired in collisions, nor is there any preferred direction toward which the particle's velocity vector can be deflected as it loses momentum. Thus the quantity $\mathrm{d} \langle \Delta \mathbf{v} \rangle/\mathrm{d}t$ will generally be in a direction exactly opposite to $\mathbf{v}$. Its magnitude is called the 'dynamical friction'. It gives rise to a slowing-down of the directed motion of the particle.

The quantities $\mathrm{d} \langle \Delta \mathbf{v} \Delta \mathbf{v} \rangle/\mathrm{d}t$ are 'velocity diffusion coefficients', since they have the effect of spreading the particle velocities over a wider region of velocity space. Often, velocity diffusion results in particular groups of particles gaining energy on average, e.g. a group of sub-thermal particles in a plasma with a Maxwellian distribution of velocities. Indeed it is the competition between, and eventual balance of, dynamical friction and velocity diffusion that gives rise to the Maxwellian distribution in steady state.

13.2   THE FOKKER–PLANCK EQUATION FOR ELECTRON–ION COLLISIONS

In Chapter 11, we analyzed the kinematics of a sequence of small-angle Coulomb collisions and produced quantities that are closely related to the dynamical friction and velocity diffusion coefficients. We did this for the case of electrons of mass m colliding with much heavier ions of charge Ze.

We obtained an expression for $\mathrm{d}(\Delta v_\parallel)/\mathrm{d}t$, given in equation (11.14), which is identical to the dynamical friction, i.e.

$$\frac{\mathrm{d}\langle \Delta \mathbf{v} \rangle}{\mathrm{d}t} = -\frac{n_\mathrm{i} Z^2 e^4 \ln \Lambda}{4\pi \epsilon_0^2 m^2 v^3} \mathbf{v}. \tag{13.7}$$

To obtain the velocity diffusion coefficients, we start by supposing that the particle is travelling in the z direction. The tensor $\mathrm{d}\langle \Delta \mathbf{v} \Delta \mathbf{v} \rangle/\mathrm{d}t$ will have xx and yy components given by

$$\frac{\mathrm{d}\langle (\Delta v_x)^2 \rangle}{\mathrm{d}t} = \frac{\mathrm{d}\langle (\Delta v_y)^2 \rangle}{\mathrm{d}t} = \frac{1}{2} \frac{\mathrm{d}\langle (\Delta v_\perp)^2 \rangle}{\mathrm{d}t} \tag{13.8}$$

and no other components. The vanishing of the other components of the velocity diffusion tensor can be explained as follows. All components such as $\mathrm{d}\langle \Delta v_x \Delta v_z \rangle/\mathrm{d}t$ must vanish because of the absence of any preferred direction for Δv_x. A similar argument shows that $\mathrm{d}\langle \Delta v_x \Delta v_y \rangle/\mathrm{d}t$ must vanish. The component $\mathrm{d}\langle (\Delta v_z)^2 \rangle/\mathrm{d}t$ is, strictly speaking, non-vanishing, but it is of higher order than the components that have been retained, since conservation of energy in collisions with infinitely massive ions gives $\Delta v_z \sim (\Delta v_\perp)^2/2v$, implying that $(\Delta v_z)^2 \sim (\Delta v_\perp)^4$, i.e. fourth-order in $\Delta v_\perp$.

Using our expression for $\mathrm{d}\langle (\Delta v_\perp)^2 \rangle/\mathrm{d}t$ from Chapter 11, i.e. equation (11.11), we can write

$$\frac{\mathrm{d}\langle \Delta \mathbf{v} \Delta \mathbf{v} \rangle}{\mathrm{d}t} = -\frac{n_\mathrm{i} Z^2 e^4 \ln \Lambda}{4\pi \epsilon_0^2 m^2 v^3} (I v^2 - \mathbf{v}\mathbf{v}) \tag{13.9}$$

where I denotes the unit tensor, and the final expression is independent of the original choice of $\mathbf{v}$ in the z direction.

Substituting these expressions for the dynamical friction and velocity diffusion coefficients into the Fokker–Planck equation, we obtain

$$\left(\frac{\partial f_e}{\partial t}\right)_{\text{coll}} = \frac{n_i Z^2 e^4 \ln\Lambda}{4\pi \epsilon_0^2 m^2} \left[\frac{\partial}{\partial \mathbf{v}} \cdot \left(\frac{\mathbf{v} f_e}{v^3}\right) + \frac{1}{2}\frac{\partial^2}{\partial \mathbf{v}\partial \mathbf{v}} : \left(\frac{I v^2 - \mathbf{v}\mathbf{v}}{v^3} f_e\right)\right]. \quad (13.10)$$

To cast equation (13.10) in a simpler form, it is convenient to use the identity

$$\frac{\partial}{\partial \mathbf{v}} \cdot \left(\frac{I v^2 - \mathbf{v}\mathbf{v}}{v^3}\right) = -\frac{2\mathbf{v}}{v^3} \quad (13.11)$$

which is most easily proved using index notation with the summation convention. The expression on the left-hand side of equation (13.11) represents a vector whose ith component is

$$\frac{\partial}{\partial v_j}\left(\frac{\delta_{ij}v^2 - v_i v_j}{v^3}\right) = -\frac{1}{v^2}\frac{\partial v}{\partial v_i} - \frac{\delta_{ij}v_j + 3v_i}{v^3} + \frac{3v_i v_j}{v^4}\frac{\partial v}{\partial v_j}$$

$$= -\frac{v_i}{v^3} - \frac{4v_i}{v^3} + \frac{3v_i}{v^3}$$

$$= -\frac{2v_i}{v^3}$$

where we have used $\partial v_i / \partial v_j = \delta_{ij}$, $\partial v_j / \partial v_j = 3$ and $v^2 = v_i v_i$ so that $\partial v / \partial v_i = v_i / v$.

Using equation (13.11), we obtain our final result for the Fokker–Planck equation:

$$\left(\frac{\partial f_e}{\partial t}\right)_{\text{coll}} = \frac{n_i Z^2 e^4 \ln\Lambda}{8\pi \epsilon_0^2 m^2}\frac{\partial}{\partial \mathbf{v}} \cdot \left(\frac{I v^2 - \mathbf{v}\mathbf{v}}{v^3} \cdot \frac{\partial f_e}{\partial \mathbf{v}}\right). \quad (13.12)$$

This form of the Fokker–Planck equation describes the evolution of the *electron* velocity distribution function $f_e(\mathbf{v}, t)$ due to collisions with fixed infinitely massive ions. Although the Fokker–Planck equation in this simple form applies only to electrons colliding with ions, a more general form of the Fokker–Planck equation can be derived that applies to electron–electron and ion–ion collisions as well. In all cases, the *structure* of the equation is preserved, i.e. there are dynamical friction and velocity diffusion coefficients, which appear exactly as in equation (13.5) and are derived by calculating the collisional effects on the velocities of individual particles of the species which the Fokker–Planck equation is to describe. The Fokker–Planck equation for a plasma was first derived in its complete form by M N Rosenbluth, W MacDonald and D Judd (1957 *Phys. Rev.* **107** 1).

13.3   THE 'LORENTZ-GAS' APPROXIMATION

The relatively simple form of the Fokker–Planck equation derived above describes electrons in the 'Lorentz-gas' approximation. A Lorentz gas is a plasma in which the electrons are supposed to collide only with (fixed) ions and not with other electrons. In practice, of course, in a $Z = 1$ plasma electron–electron collisions are about as frequent as electron–ion collisions. Nonetheless, the Lorentz-gas approximation is useful for many applications, especially since the resulting simple form of the Fokker–Planck equation is reasonably analytically tractable. The Lorentz-gas approximation will be quite accurate for a plasma composed of multiply charged ions of charge Ze, since for this case electron–ion collisions will be more frequent than electron–electron collisions by a factor of order $n_i Z^2 / n_e = Z$.

If we were to substitute a Maxwellian distribution $f_e \propto \exp(-mv^2/2T)$ into the preceding Fokker–Planck equation, the right-hand side of the equation would vanish. This must be true, of course, for *any* expression describing the effects of collisions, since the Maxwellian distribution implies thermodynamic equilibrium among the particles. However, the right-hand side of the Lorentz-gas form of the Fokker–Planck equation vanishes for any f_e that is *isotropic* in velocity space, i.e. any f_e that depends on v alone, because in this case

$$(\boldsymbol{I}v^2 - \mathbf{vv}) \cdot \frac{\partial f}{\partial \mathbf{v}} = (\boldsymbol{I}v^2 - \mathbf{vv}) \cdot \mathbf{v} \frac{\partial f}{v \partial v} = 0. \tag{13.13}$$

This property of the Lorentz-gas approximation arises from the fact that electron–ion collisions do not (to lowest order in m/M calculated here) change the magnitude of the electron velocity vectors; they only produce a scattering of the directions of the electron velocity vectors.

A somewhat simpler form of the Lorentz-gas Fokker–Planck equation is obtained by transforming to spherical coordinates in velocity space. Choosing some convenient direction for z, and writing $v_z = v\cos\theta$, $v_x = v\sin\theta\cos\phi$ and $v_y = v\sin\theta\sin\phi$, we can use the standard expressions for the gradient and divergence operators in spherical coordinates (see Appendix E, applied here to velocity space) to transform the Fokker–Planck equation to the form

$$\left(\frac{\partial f_e}{\partial t}\right)_{\text{coll}} = \frac{n_i Z^2 e^4 \ln\Lambda}{8\pi \epsilon_0^2 m^2 v^3} \left[\frac{1}{\sin\theta} \frac{\partial}{\partial \theta} \left(\sin\theta \frac{\partial f_e}{\partial \theta} \right) + \frac{1}{\sin^2\theta} \frac{\partial^2 f_e}{\partial \phi^2} \right] \tag{13.14}$$

The absence of terms in $\partial/\partial v$ is a further manifestation of the constancy of the velocity magnitude in electron–ion collisions.

Problem 13.1: Derive equation (13.14), beginning with equation (13.12) and using the formulae in Appendix E as appropriate.

13.4 PLASMA RESISTIVITY IN THE LORENTZ-GAS APPROXIMATION

As an application of the use of the Fokker–Planck equation, we will derive an exact expression for the *plasma electrical resistivity* in the Lorentz-gas approximation.

Suppose that the electron distribution is approximately Maxwellian:

$$f_{e0} = n_e \left(\frac{m_e}{2\pi T_e} \right)^{3/2} \exp\left(-\frac{m_e v^2}{2T_e} \right) \tag{13.15}$$

but that this equilibrium is slightly perturbed by the application of a small electric field in the z direction. The electric field will cause electrons to accelerate at a rate $-eE/m$, and so the velocity distribution function at time t can be expressed in terms of the distribution function at a slightly earlier time, $t - \Delta t$, by

$$f_e(\mathbf{v}, t) = f_e(\mathbf{v} + e\mathbf{E}\Delta t/m, t - \Delta t). \tag{13.16}$$

For small Δt, we may expand as follows:

$$f_e(\mathbf{v}, t) - f_e(\mathbf{v}, t - \Delta t) = \frac{e\mathbf{E}}{m} \cdot \frac{\partial f_e}{\partial \mathbf{v}} \Delta t$$
$$\left(\frac{\partial f_e}{\partial t} \right)_E = \frac{e\mathbf{E}}{m} \cdot \frac{\partial f_{e0}}{\partial \mathbf{v}}. \tag{13.17}$$

Here, the subscript E indicates a rate of change of f_e due to the effect of the E field alone. We have also assumed that the electric field $\mathbf{E}$ gives rise to only a small perturbation of the velocity distribution, so that $f_e \approx f_{e0}$ may be substituted in the term containing $\mathbf{E}$. Equation (13.17) constitutes a step towards the full 'Vlasov equation', which treats the evolution of $f(\mathbf{x}, \mathbf{v}, t)$ in a general force field. The Vlasov equation will be introduced in Chapter 22.

When the electrons reach a steady-state in which the accelerating force of the electric field is balanced by the collisional drag from the ions, we must have

$$0 = \frac{\partial f_e}{\partial t} = \left(\frac{\partial f_e}{\partial t} \right)_E + \left(\frac{\partial f_e}{\partial t} \right)_{coll} \tag{13.18}$$

i.e.

$$-\frac{e\mathbf{E}}{m} \cdot \frac{\partial f_{e0}}{\partial \mathbf{v}} = \left(\frac{\partial f_e}{\partial t} \right)_{coll}. \tag{13.19}$$

This equation must be solved for the non-Maxwellian part of f_e, which we will denote f_{e1}. The Fokker–Planck expression given in equation (13.14) will be used for the collision term on the right in equation (13.19). This expression contains only the non-Maxwellian part of the distribution function, i.e. f_{e1}, since we know that collisions can have no effect on a Maxwellian f_{e0}. For our present calculation, which uses the Lorentz-gas form of the Fokker–Planck equation, the isotropic property of the Maxwellian distribution is all that is needed to establish that only f_{e1}, and not f_{e0}, enters into the Fokker–Planck expression.

The distribution f_{e1} will be symmetric with respect to the azimuthal velocity angle about the z direction, i.e. there can be no dependence on ϕ, since $\mathbf{E}$ is in the z direction and the equation itself is symmetric in ϕ. Substituting the Maxwellian for f_{e0}, the equation to be solved is

$$\frac{eEv f_{e0}}{T_e}\cos\theta = \frac{n_i Z^2 e^4 \ln\Lambda}{8\pi \epsilon_0^2 m^2 v^3}\frac{1}{\sin\theta}\frac{\partial}{\partial\theta}\left(\sin\theta\frac{\partial f_{e1}}{\partial\theta}\right) \tag{13.20}$$

which has the solution

$$f_{e1} = -\frac{4\pi \epsilon_0^2 m^2 E v^4 f_{e0}\cos\theta}{n_i Z^2 e^3 T_e \ln\Lambda}. \tag{13.21}$$

The total electron distribution function obtained by adding the θ-independent Maxwellian distribution f_{e0} to the θ-dependent perturbation f_{e1} is a slightly asymmetric (in θ) distribution in which there are more electrons with $\pi > \theta > \pi/2$ than with $\pi/2 > \theta > 0$. In terms of cartesian coordinates, there are slightly more electrons with $v_z < 0$ than with $v_z > 0$. This is what would be expected for an electric field in the z direction, which accelerates negatively charged electrons in the negative-z direction.

We next calculate the current density in the z direction:

$$\begin{aligned}
j_z &= -e\int f_{e1}v\cos\theta\,\mathrm{d}^3 v \\
&= \frac{8\pi^2\epsilon_0^2 m^2 E}{n_i Z^2 e^2 T_e \ln\Lambda}\int_0^\infty v^7 f_{e0}\mathrm{d}v\int_0^\pi \cos^2\theta\sin\theta\,\mathrm{d}\theta \\
&= \frac{32\pi^{1/2}\epsilon_0^2 E(2T_e)^{3/2}}{m^{1/2}Ze^2\ln\Lambda}
\end{aligned} \tag{13.22}$$

where we have substituted $\mathrm{d}^3 v = 2\pi v^2\sin\theta\,\mathrm{d}\theta\,\mathrm{d}v$ and have also made use of charge neutrality, i.e. $n_e = Zn_i$. The integrals in equation (13.22) are straightforward to carry out: the integral over θ is done by writing $\sin\theta\,\mathrm{d}\theta = -\mathrm{d}(\cos\theta)$; the integral over v is done by first writing $v^7\mathrm{d}v = v^6\mathrm{d}(v^2/2)$, then noting that $f_{e0}\sim\exp(-v^2/2v_r^2)$ so that the integral over $v^2/2$ can be done by repeated integrations by parts. Thus, we obtain the plasma resistivity

$$\eta = \frac{m^{1/2}Ze^2\ln\Lambda}{32\pi^{1/2}\epsilon_0^2(2T_e)^{3/2}} \tag{13.23}$$

in the Lorentz-gas approximation.

Comparing this with the simple estimate for resistivity obtained in Chapter 11, i.e. equation (11.30), we see that the Lorentz-gas resistivity is smaller by a factor 3.4 than the simple estimate. The lower resistivity arises from the dominant role of higher-velocity electrons in carrying current in the Lorentz-gas approximation.

To obtain the *true* resistivity, we must include electron–electron collisions, and this calculation can only be done numerically. The resulting resistivity for a hydrogen plasma, first obtained by L Spitzer and R Harm (1953 *Phys. Rev.* **89** 977), is about 1.7 times larger than the Lorentz-gas resistivity and about 2.0 times smaller than the simple estimate obtained in Chapter 11, as already noted there. The role of electron–electron collisions is not to contribute *directly* to resistivity—which they do not, since they cannot affect the total momentum of the electron population—but rather to modify the electron distribution in such a way as to increase the total drag on electrons due to collisions with ions. The reason why the resistivity is *increased* by electron–electron collisions is obvious: in the Lorentz-gas approximation it is the suprathermal electrons that tend to carry most of the current, since the electron–ion collision frequency ($\sim v^{-3}$) decreases with increasing electron velocity. When electron–electron collisions are included, these suprathermal electrons are more strongly coupled to, and slowed down by, the thermal electrons, thereby indirectly increasing their collisional coupling to the ions.

Problem 13.2: Consider a neutral plasma composed of electrons and a single type of multiply charged ions, each with charge Ze. By considering the relative magnitude of electron–ion and electron–electron collisions in this case, give a formula for the plasma resistivity with a numerical coefficient that should be accurate in the limit of large Z, even when electron–electron collisions are included.

Problem 13.3: Describing electron–ion collisions by the Fokker–Planck equation, evaluate the plasma resistivity η in the case where *electron–electron collisions are imagined to be infinitely frequent compared with electron–ion collisions*. Obviously, this is the opposite limit from the Lorentz gas model. (Hint: remember that electron–electron collisions cannot cause resistivity on their own, but they can affect the resistivity by modifying the electron distribution function. In particular, they tend to lead to a Maxwellian distribution that is shifted about some non-zero mean electron velocity. This will arise from an electron–electron collision term in the Fokker–Planck equation that is dominant in determining the electron distribution function, but which is of a form that conserves the total momentum of the electron population. The momentum transfer between electrons and ions is still determined by a Lorentz-gas Fokker–Planck expression, but one in which the shape of the electron velocity-space distribution is effectively known, i.e. a shifted Maxwellian. Since the electric field can be assumed to be small, the shift in the Maxwellian distribution can be taken to be small compared with a thermal velocity.) Express your result for η in terms of the average electron–ion collision frequency, $\langle \nu_{ei} \rangle$, given in equation (11.22).

Chapter 14

Collisions of fast ions in a plasma*

A situation that arises in many naturally occurring plasmas, as well as fusion plasmas, is that of a 'beam' of fast ions moving through a plasma. The energy of the beam ions is typically much larger than the temperature of the background plasma, i.e. the beam-ion velocities are considerably 'suprathermal' relative to the background ions. The beam-ion velocities may be greater than, or less than, the thermal velocity of the background electrons. However, for the former case to apply, assuming a proton beam in a hydrogen plasma with $T_i \approx T_e$, the beam-ion velocity must exceed $(M/m)^{1/2} \sim 43$ times the background-ion thermal velocity (and the beam-ion energy must exceed 1800 times the plasma temperature); this does not often occur, at least in laboratory plasmas. More usually, the beam-ion velocity is much less than the background electron thermal velocity. The beam ions may be of the same type (i.e. same mass M and charge-number Z) as the background plasma ions, or they may be of some different type. Before its interaction with the plasma, the ion beam may be almost mono-energetic and unidirectional, or it may already have a substantial 'spread' in velocity magnitudes and directions.

14.1 FAST IONS IN FUSION PLASMAS

A case of particular interest in fusion research is that of a plasma self-heated by the energetic ions produced by the fusion reactions themselves. In particular, the deuterium–tritium reaction produces an energetic helium ion, or 'alpha particle' ($Z = 2$, atomic mass = 4), with energy approximately 3.5 MeV, which is about 200 times the temperature of the background plasma that will be typically needed in a fusion reactor. These alpha particles are born with an isotropic distribution of velocities, i.e. there is no preferred direction for their initial velocity vectors.

Experimental fusion plasmas are also often heated by energetic beams of ions, injected initially as neutral atoms, becoming ionized as they penetrate into the high-temperature plasma. Beam-ion energies of about 100 keV are presently used for this purpose, which is typically about 10–20 times the temperature of the background experimental plasma. For reactor-scale plasmas, beam energies of about 1 MeV will be needed. Such beams are usually highly directional, because their velocity vectors continue to point in the direction in which the beam was initially injected. Another commonly used heating technique is to accelerate a minority species of ions in the plasma to very high energies by radio-frequency waves, using a frequency equal to the cyclotron frequency of the minority ions. This produces a 'beam' of energetic ions with velocity vectors mainly perpendicular to the direction of the magnetic field.

Energetic beam ions will *thermalize* with the background plasma particles as a result of multiple Coulomb collisions. In this Chapter, we will describe this thermalization process, using the various results on Coulomb collisions that were derived in Chapter 11. We will also construct a Fokker–Planck equation for beam ions, which is somewhat more complex than the Lorentz-gas Fokker–Planck equation for electrons derived in Chapter 13.

The background plasma is assumed to be composed of Maxwellian ions and electrons. We will suppose that the density n_b of the beam ions is much less than the density n_i of the background plasma ions. Accordingly, the background plasma will itself be approximately charge-neutral, i.e. $n_e \approx Z n_i$, without any significant contribution from the beam ions to the charge density. The beam ions are supposed to have velocities V_b (in this Chapter, we will consistently use upper case for beam-ion velocities and lower-case for background plasma particle velocities) that are very much greater than the thermal velocity of the background plasma ions, $v_{t,i}$, but much less than the thermal velocity of the background electrons, $v_{t,e}$, i.e.

$$v_{t,i} \ll V_b \ll v_{t,e}. \qquad (14.1)$$

For maximum generality, we will allow the beam ions to be of a different type than the background plasma ions. As usual, the background-ion mass and charge-number will be denoted M and Z, respectively. For the beam ions, we will denote these quantities M_b and Z_b.

The beam ions will undergo Coulomb collisions with background ions and electrons. The result of these collisions will be frictional drag on the background ions and electrons, which will cause the beam ions to slow down, and angular scattering on the background ions, which will cause the beam ions to be deflected from their original direction.

14.2 SLOWING-DOWN OF BEAM IONS DUE TO COLLISIONS WITH ELECTRONS

First let us consider the collisions of beam ions with Maxwellian background plasma *electrons*. If we transform to the frame co-moving with the beam ions, we find a collisional situation similar to that considered in Chapter 11, i.e. electrons colliding with relatively massive and essentially stationary ions, and we see that the plasma electrons can transfer *momentum* to the beam ions, but not much *energy*. On average, the beam ions will gain directed momentum since in this frame the plasma electrons have non-zero average directed momentum because of the transformation of the Maxwellian distribution to the moving frame. In this frame, the momentum gained by the beam will be in the direction of the average electron momentum; that is, it will be exactly opposite to the velocity of the beam ions, i.e. the velocity of the moving frame in relation to the original background-plasma frame. The beam ions will thus lose directed momentum and slow down due to collisions with the electrons, but they will not be deflected significantly from their original direction of motion.

Indeed, the change in the beam-ion's energy is almost entirely due to its loss of energy of directed motion, rather than to any gain of energy associated with random motion, either along the original direction of motion of the ion, or perpendicular to it. This can be seen by considering, in the original background-plasma frame, a typical momentum-transfer collision between a beam ion and an electron. If the beam-ion's velocity changes by an amount ΔV, then conservation of momentum tells us that the electron must acquire a velocity $-(M_b/m)\Delta V$. By conservation of energy, the change in the beam-ion energy, i.e. $\Delta W_b = (M_b/2)(|V + \Delta V|^2 - V^2) \approx M_b V \cdot \Delta V$, must be equal and opposite to the change in electron energy, which is

$$\Delta W_e = (m/2)(M_b/m)^2 |\Delta V|^2 = (M_b^2/2m)|\Delta V|^2. \qquad (14.2)$$

Writing $\Delta V_\parallel$ for the component of ΔV in the direction of V (the quantity $\Delta V_\parallel$ will be negative), the energy conservation equation, i.e. $\Delta W_b = -\Delta W_e$, can be written

$$-M_b V \Delta V_\parallel = \frac{M_b^2}{2m}|\Delta V|^2 = \frac{M_b^2}{2m}[(\Delta V_\parallel)^2 + (\Delta V_\perp)^2] \qquad (14.3)$$

where $\Delta V_\perp$ is the increment of the beam-ion's velocity perpendicular to its original velocity vector V. Two conclusions follow from this energy conservation equation. First, since $M_b(\Delta V_\perp)^2/2 < mV|\Delta V_\parallel| \ll M_b V|\Delta V_\parallel|$, we see that the energy in the beam-ion perpendicular velocity components arising from the deflection of the beam-ion's velocity vector away from its original direction is much less than the energy decrement arising from slowing down without change of direction. Second, since equation (14.3) requires that $|\Delta V_\parallel| < (2m/M_b)V$, we see that the collision results in the beam ion losing a fraction of order m/M_b of its momentum, corresponding to the loss of a fraction of order m/M_b of its energy, i.e. an energy loss of order mV_b^2. Combining these two inequalities, we also see that $\Delta V_\perp < (2m/M_b)V$.

Essentially, the collision causes the beam-ion's velocity vector to be deflected through an angle at most of order m/M_b; the energy associated with this perpendicular motion is a fraction of order $(m/M_b)^2$ of the beam-ion's initial energy, to be compared with a fraction of order m/M_b that is lost by slowing down without change of direction. If the electron were also to transfer to the beam ion the maximum possible fraction of its thermal energy, which we saw in Chapter 11 to be of order $(m/M_b)T_e$, the gain in energy of random motion from this effect would also be only a small fraction, i.e. $(m/M_b)T_e/(mV_b^2) = T_e/(M_b V_b^2) \sim v_{t,i}^2/V_b^2 \ll 1$, of the loss of energy of directed motion. (We have assumed here that T_e and T_i are of similar order.) Thus, the force of the background electrons on the beam ions is mostly in the nature of a frictional drag, i.e. it directly opposes the motion of the beam, but does not cause any significant scattering of the beam.

The magnitude of this frictional drag can be calculated as follows. The increase $\Delta \mathbf{v}$ in the velocity of an electron as a result of a Coulomb collision with a beam ion can be related by momentum conservation to the velocity $\Delta \mathbf{V}$ lost by the beam ion:

$$m\Delta \mathbf{v} = -M_b \Delta \mathbf{V}. \tag{14.4}$$

Now suppose that the beam ions have density n_b and mean velocity $\langle \mathbf{V} \rangle$, and that the electrons have a Maxwellian distribution

$$f_e(\mathbf{v}) = n_e \left(\frac{m}{2\pi T_e} \right)^{3/2} \exp \left(-\frac{mv^2}{2T_e} \right) \tag{14.5}$$

with number density n_e. As a result of many collisions between beam ions and electrons, the beam-ion momentum decreases, and the electron momentum increases correspondingly:

$$n_b M_b \frac{d\langle \mathbf{V} \rangle}{dt} = -m \int \frac{d\langle \mathbf{v} \rangle}{dt} f_e(v) d^3 v. \tag{14.6}$$

The rate of change of the electron's directed velocity due to collisions with beam ions can be obtained by applying the formula obtained in Chapter 11, i.e. equation (11.15), appropriately adjusted to the laboratory frame, in which the ions have velocity $\mathbf{V}$:

$$\frac{d\langle \mathbf{v} \rangle}{dt} = -\nu_{eb}(\mathbf{v} - \mathbf{V}) \tag{14.7}$$

where

$$\nu_{eb} = \frac{n_b Z_b^2 e^4 \ln \Lambda}{4\pi \epsilon_0^2 m^2 |\mathbf{v} - \mathbf{V}|^3}. \tag{14.8}$$

Substituting this into the above expression for beam-ion slowing down, we obtain

$$\frac{d\langle \mathbf{V} \rangle}{dt} = \frac{Z_b^2 e^4 \ln \Lambda}{4\pi \epsilon_0^2 m M_b} \int \frac{\mathbf{v} - \mathbf{V}}{|\mathbf{v} - \mathbf{V}|^3} f_e d^3 v. \tag{14.9}$$

The evaluation of this integral over electron velocities requires first some vector calculus and then transformation to spherical coordinates in velocity space. The first step is to write

$$\frac{\mathbf{v} - \mathbf{V}}{|\mathbf{v} - \mathbf{V}|^3} = \frac{\partial}{\partial \mathbf{V}}(|\mathbf{v} - \mathbf{V}|^2)^{-1/2} = \frac{\partial}{\partial \mathbf{V}}\frac{1}{|\mathbf{v} - \mathbf{V}|}$$

which amounts to noting that the velocity-space 'force-field' $(\mathbf{v} - \mathbf{V})/|\mathbf{v} - \mathbf{V}|^3$ is derivable from a scalar potential field $|\mathbf{v} - \mathbf{V}|^{-1}$.

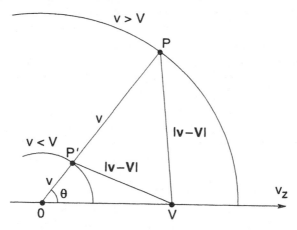

Figure 14.1. Spherical coordinate system for calculating the integral in equation (14.11), which is analogous to the 'gravitational potential' at a point V on the axis due to 'mass' distributed uniformly over spherical shells with density $f_e(v)$. The integral includes shells with $v > V$ and shells with $v < V$. Shells of both types are shown in the figure: P and P' are typical points on shells with $v > V$ and $v < V$, respectively. The magnitudes of the vectors $\mathbf{v} - \mathbf{V}$ are indicated in both cases.

We can then write

$$\frac{d\langle\mathbf{V}\rangle}{dt} = -\frac{Z_b^2 e^4 \ln\Lambda}{4\pi\epsilon_0^2 m M_b}\frac{\partial I}{\partial \mathbf{V}} \tag{14.10}$$

where I is the integral given by

$$I(\mathbf{V}) = -\int\frac{f_e d^3v}{|\mathbf{v} - \mathbf{V}|}. \tag{14.11}$$

Although we are assuming a Maxwellian distribution for $f_e(\mathbf{v})$, it is instructive to evaluate the integral $I(\mathbf{V})$ for a slightly more general class of distributions, namely those which are isotropic in velocity space, i.e. $f_e \equiv f_e(v)$. Examining equations (14.10) and (14.11), an analogy with inverse-square-law force fields, e.g. gravitation, is immediately apparent. Specifically, thinking of our velocity space as physical space, the vector $-\partial I/\partial \mathbf{V}$ is the gravitational force, described by a potential $I(\mathbf{V})$ and acting at a point whose position is given by the vector $\mathbf{V}$, due to the gravitational attraction of matter distributed in spherical shells with mass density $f_e(v)$.

The integral $I(\mathbf{V})$ can be evaluated for all values of the ratio of the beam-ion's speed V to the electron thermal velocity, $v_{t,e}$, as follows. First, we transform to a spherical coordinate system (v, θ, ϕ) in velocity space, with the $\theta = 0$ axis of the coordinate system lying in the direction of the vector $\mathbf{V}$. As indicated in Figure 14.1, we must distinguish between those shells in velocity space that lie outside the point $\mathbf{V}$, i.e. $v > V$, and those that lie inside this point, i.e. $v < V$. In both cases, the distance between the point given by the vector $\mathbf{V}$ and a general point P or P' (see Figure 14.1) on one of the shells is

$$|\mathbf{v} - \mathbf{V}| = (v^2 + V^2 - 2vV\cos\theta)^{1/2}$$

independent of the azimuthal coordinate φ. Thus we may integrate over φ from 0 to 2π immediately, and then proceed as follows:

$$
\begin{aligned}
I &= -2\pi \int \frac{f_e(v)v^2\sin\theta\,\mathrm{d}\theta\,\mathrm{d}v}{(v^2 + V^2 - 2vV\cos\theta)^{1/2}} \\
&= -2\pi \int_0^\infty v^2 f_e(v)\mathrm{d}v \int_{\cos\theta=-1}^{\cos\theta=1} \frac{\mathrm{d}(\cos\theta)}{(v^2 + V^2 - 2vV\cos\theta)^{1/2}} \\
&= -2\pi \int_0^\infty v^2 f_e(v)\mathrm{d}v \left[-\frac{1}{vV}(v^2 + V^2 - 2vV\cos\theta)^{1/2} \right]_{\cos\theta=-1}^{\cos\theta=1} \\
&= -\frac{2\pi}{V} \int_0^\infty v f_e(v)\mathrm{d}v\,(-|v - V| + v + V).
\end{aligned}
$$

At this point, we must distinguish between the shells that lie outside V, on which $|v - V| = v - V$, and those that lie inside V, on which $|v - V| = V - v$. Evaluating the two contributions to the integral separately and then adding these together, we obtain

$$I = -4\pi \int_V^\infty v f_e(v)\mathrm{d}v - \frac{4\pi}{V} \int_0^V v^2 f_e(v)\mathrm{d}v.$$

As is obvious from symmetry considerations, the quantity I depends only on the magnitude of $\mathbf{V}$, and not on its direction. Differentiating this 'potential' to obtain the 'force field', we obtain

$$-\frac{\partial I}{\partial \mathbf{V}} = -\frac{4\pi \mathbf{V}}{V^3} \int_0^V v^2 f_e(v)\mathrm{d}v$$

noting that the contributions from differentiating the limits of integration of the two integrals cancel each other. Thus, in terms of the gravitational analogy, the gravitational force is the same as would arise if all of the spherical shells of matter lying outside V were absent, and all of the matter in the spherical shells lying inside V were concentrated at $v = 0$.

For a Maxwellian $f_e(v)$ and for $V \ll v_{t,e}$, the remaining integral can be calculated explicitly, i.e.

$$\int_0^V v^2 f_e(v) dv = \frac{n_e}{(2\pi)^{3/2} v_{t,e}^3} \int_0^V v^2 \exp\left(-\frac{v^2}{2v_{t,e}^2}\right) dv$$
$$\approx \frac{n_e V^3}{3(2\pi)^{3/2} v_{t,e}^3}.$$

Putting all this together, we obtain our final result:

$$\frac{d\langle V \rangle}{dt} = -\frac{2^{1/2} n_e Z_b^2 e^4 m^{1/2} \ln\Lambda}{12\pi^{3/2} \epsilon_0^2 M_b T_e^{3/2}} V. \tag{14.12}$$

We note that the characteristic time for slowing down of the beam ions due to collisions with electrons, sometimes called the 'slowing down time', does not depend on the beam velocity, but it does depend inversely on the electron density and as the 3/2 power of the electron temperature. The higher the electron temperature, the lower the electron frictional drag on the beam ions. Taking the scalar product of equation (14.12) with $M_b V$, we obtain an expression for the rate by which the beam-ion kinetic energy W_b is reduced:

$$\frac{dW_b}{dt} = -\frac{2^{1/2} n_e Z_b^2 e^4 m^{1/2} \ln\Lambda}{6\pi^{3/2} \epsilon_0^2 M_b T_e^{3/2}} W_b. \tag{14.13}$$

Again, the characteristic time for beam-ion energy loss does not depend on beam energy, but it is strongly dependent on electron density and temperature.

14.3 SLOWING-DOWN OF BEAM IONS DUE TO COLLISIONS WITH BACKGROUND IONS

Next let us consider the collisions of beam ions with background plasma *ions*. We recall that the beam ions are assumed to have directed velocities greatly exceeding the background-ion thermal velocity. In such cases, there will be two processes that will reduce the directed velocity of the beam ions on roughly comparable time scales—namely, deflection of the beam-ion velocity vectors by background ions, and energy transfer from beam ions to background ions.

We will consider in turn the two limiting cases for the ratio of the beam-ion mass to the background-ion mass, i.e. the cases $M_b/M \gg 1$ and $M_b/M \ll 1$.

If the beam ions have *larger* mass than the background ions, then transfer of parallel energy to the background ions will be the dominant process by which directed momentum is lost, just as it was in the case of collisions of beam ions with electrons. This is because a heavier beam ion is only able to transfer a fraction of its energy to a stationary, lighter background ion, consistent with conservation of momentum. As we saw in our discussion of beam-ion collisions with background electrons (simply making the substitution $m \rightarrow M$ to treat collisions with background ions, mass $M \ll M_b$), the increment $\Delta V_{\perp}$ of the beam-ion's velocity perpendicular to its original velocity vector $\mathbf{V}$ is limited by $M_b(\Delta V_{\perp})^2/2 < MV|\Delta V_{\|}| \ll M_b V|\Delta V_{\|}|$, whereas the velocity decrement $\Delta V_{\|}$ in the direction of $\mathbf{V}$ arising from a typical collision is $|\Delta V_{\|}| < (2M/M_b)V$. Combining these two inequalities, we see that $\Delta V_{\perp} < (2M/M_b)V$. As in the case of beam ions colliding with electrons, the fraction of directed beam-ion momentum that is lost due to scattering at constant energy is of order $(M/M_b)^2$, versus a larger loss fraction of order M/M_b due to the frictional drag opposing the beam-ion's directed motion. For $M_b \gg M$, this results in energy and momentum loss without scattering. Thus, transfer of energy to background ions, without much scattering, will be the dominant process by which heavier beam ions lose momentum to lighter background ions.

In this case of a heavier beam ion, we can proceed as before to relate the change $\Delta \mathbf{v}$ in the velocity of a background ion to the change $\Delta \mathbf{V}$ in the velocity of a beam ion:

$$M\Delta \mathbf{v} = -M_b \Delta \mathbf{V}. \tag{14.14}$$

Equation (14.14) expresses momentum conservation in a single collision. We now proceed exactly as we did in the case of slowing down due to collisions with electrons. Suppose that the background ions have a Maxwellian distribution

$$f_i(v) = n_i \left(\frac{M}{2\pi T_i} \right)^{3/2} \exp \left(-\frac{Mv^2}{2T_i} \right) \tag{14.15}$$

with spatial density n_i. As a result of many collisions between beam and background ions, the average velocity of the beam ions decreases according to the relation

$$n_b M_b \frac{d\langle \mathbf{V} \rangle}{dt} = -M \int \frac{d\langle \mathbf{v} \rangle}{dt} f_i(v) d^3 v. \tag{14.16}$$

The rate of change of the background-ion's velocity due to collisions with beam ions can again be obtained by applying the formula obtained in Chapter 11, i.e. equation (11.15), adjusted to the laboratory frame:

$$\frac{d\langle \mathbf{v} \rangle}{dt} = -\nu_{ib}(\mathbf{v} - \mathbf{V}) \tag{14.17}$$

where

$$\nu_{ib} = \frac{n_b Z^2 Z_b^2 e^4 \ln \Lambda}{4\pi \epsilon_0^2 M^2 |\mathbf{v} - \mathbf{V}|^3}. \tag{14.18}$$

It should be remembered that this formula applies only to the case of heavier beam ions colliding with relatively light background ions. For the beam ions compared with thermal ions, as indicated by equation (14.1), we can still use the approximation $V \gg v$. Combining equations (14.16), (14.17) and (14.18) in this way, we obtain

$$\frac{d\langle \mathbf{V} \rangle}{dt} = -\frac{n_i Z^2 Z_b^2 e^4 \ln \Lambda}{4\pi \epsilon_0^2 M M_b V^3} \mathbf{V}. \tag{14.19}$$

In the approximation used to obtain equation (14.19), the right-hand side of equation (14.17) becomes simply $\nu_{ib}\mathbf{V}$, and the factor $|\mathbf{v} - \mathbf{V}|^3$ in the denominator of equation (14.18) becomes simply $|\mathbf{V}|^3$. The quantity $d\langle \mathbf{v} \rangle/dt$ is then independent of the background-ion velocity $\mathbf{v}$, so that the integral $d\langle \mathbf{v} \rangle/dt$ over the background-ion distribution in equation (14.16) becomes trivial, simply introducing a factor n_i.

We note that the characteristic time for slowing down of the beam ions, the 'slowing down time', does not depend on the background-ion temperature, but it does depend, as V^3, on the beam-ion velocity. Less energetic beams slow down more rapidly. The rate at which the beam-ion kinetic energy W_b decreases is obtained by taking the scalar product of equation (14.19) with $M_b\mathbf{V}$, giving

$$\frac{dW_b}{dt} = -\frac{2^{1/2} n_i Z^2 Z_b^2 e^4 M_b^{1/2} \ln \Lambda}{8\pi \epsilon_0^2 M W_b^{1/2}} \tag{14.20}$$

For this case of a heavier beam ion, the dominant effect of collisions with background ions is 'pure' slowing down, i.e. loss of directed momentum, as described by equation (14.19), without significant deflection of the beam-ion's velocity vector from its initial direction.

If the beam ions have *smaller* mass than the background ions, deflection of their velocity vectors will be the dominant process by which beam ions lose directed momentum. This is because the lighter beam ions can relatively easily transfer their momentum to the heavier background ions, without the latter gaining much of the beam-ions' energy. Even if a lighter beam ion loses all of its directed momentum $M_b\mathbf{V}$ in a collision with a heavier background plasma ion, so that the background-ion's velocity jumps to $M_b\mathbf{V}/M$, there would result an energy transfer of only $M|M_b\mathbf{V}/M|^2/2 = (M_b/M)M_b V^2/2$, i.e. a fraction M_b/M of the beam-ion's initial energy. The deflection of the beam-ion's velocity vector, a process usually termed 'pitch-angle scattering', in its pure form will not result in any change in the beam-ion's energy. In this case of a lighter beam ion, equation (14.19) will not describe the dominant process—but it is still relevant to ask at what rate the beam ion loses its *energy* to background ions, even if this occurs relatively slowly compared with pitch-angle scattering. In fact, it will turn out that equation (14.20) remains true for the case of a lighter beam ion, and this result applies equally well for all relative magnitudes of beam- and background-ion masses.

The case of a lighter beam ion can be analyzed as follows. Begin again with the relationship between the changes Δv and ΔV in the background- and beam-ion velocities, respectively:

$$M \Delta \mathbf{v} = -M_b \Delta \mathbf{V}. \tag{14.21}$$

The energy acquired by the background ion in a collision is

$$\frac{M}{2}|\Delta \mathbf{v}|^2 = \frac{M_b^2}{2M}|\Delta \mathbf{V}|^2 \tag{14.22}$$

which must be the same as the energy lost by the beam ion, so that $\Delta W_b = -(M_b^2/2M)|\Delta \mathbf{V}|^2$. For small-angle collisions of interest to us here, as discussed in Chapter 11, the deflections $\Delta \mathbf{V}$ in the beam-ion's velocity are mainly perpendicular to its initial velocity vector, i.e. $|\Delta \mathbf{V}|^2 \approx (\Delta V_\perp)^2$, and can be obtained from the analysis in Chapter 11 (for the corresponding case of electrons colliding with ions). The result of this analysis was expressed in equation (11.11) which, when applied to the present case, gives

$$\frac{d(\Delta V_\perp)^2}{dt} = \frac{n_i Z^2 Z_b^2 e^4 \ln \Lambda}{2\pi \epsilon_0^2 M_b^2 V_b}. \tag{14.23}$$

Thus, the beam-ion's energy decreases according to

$$\frac{dW_b}{dt} = -\frac{M_b^2}{2M}\frac{d(\Delta V_\perp)^2}{dt} = -\frac{n_i Z^2 Z_b^2 e^4 \ln \Lambda}{4\pi \epsilon_0^2 M V_b} = -\frac{2^{1/2} n_i Z^2 Z_b^2 e^4 M_b^{1/2} \ln \Lambda}{8\pi \epsilon_0^2 M W_b^{1/2}} \tag{14.24}$$

i.e. the same as equation (14.20).

Although we have derived the result for dW_b/dt given in equations (14.20) and (14.24) only for the two limiting cases $M \ll M_b$ and $M_b \ll M$, we will assume (as is indeed the case) that this result applies for all ratios of beam-ion mass to background-ion mass.

14.4 'CRITICAL' BEAM-ION ENERGY

If we combine our two expressions for the rates of beam-ion slowing down due to electron collisions and due to ion collisions, i.e. equations (14.13) and (14.20), we have

$$\frac{dW_b}{dt} = -\frac{2^{1/2} n_e Z_b^2 e^4 m^{1/2} \ln \Lambda}{6\pi^{3/2} \epsilon_0^2 M_b}\left(\frac{W_b}{T_e^{3/2}} + \frac{C}{W_b^{1/2}}\right) \tag{14.25}$$

where

$$C = \frac{3\pi^{1/2} Z M_b^{3/2}}{4m^{1/2} M} \approx 57 \tag{14.26}$$

the latter numerical value being for a case where both beam and background ions are protons. We see that above some 'critical' beam-ion energy $W_{b,\text{crit}}$, the collisions with electrons dominate the slowing down process. On the other hand, for $W_b < W_{b,\text{crit}}$, the slowing down is mainly due to collisions with background ions. The critical beam energy (at which the slowing down rates on electrons and ions are exactly equal) is given by

$$\frac{W_{b,\text{crit}}}{T_e} = C^{2/3} \approx 15 \tag{14.27}$$

the latter for the case where the beam and background ions are both protons.

As beam ions slow down in a plasma, they give up their energy increasingly to background ions, rather than to background electrons. Although the two contributions to the instantaneous slowing down rate are exactly equal at $W_b = W_{b,\text{crit}}$, the slowing down on background ions begins to dominate as soon as the beam-ion energy drops below $W_{b,\text{crit}}$.

Problem 14.1: Suppose that it is desired that the beam ions contribute exactly equal amounts of energy to background ions and electrons over the entire slowing down process. For a mono-energetic injected beam, obtain an estimate of the required injection energy in terms of $W_{b,\text{crit}}$. (Note: to do this, you may choose to carry out a simple integration numerically. A high degree of accuracy is not required; any simple numerical integration technique will suffice.)

14.5 THE FOKKER–PLANCK EQUATION FOR ENERGETIC IONS

Equations (14.12) and (14.19) give the two contributions to the dynamical friction for beam ions slowing down in a background plasma, and these may now be used to obtain a Fokker–Planck equation for the beam ions. As we have seen, the main effect of collisions with background plasma electrons is to slow down the motion of the beam ions, rather than to deflect the beam from its original direction. We have seen that collisions with background plasma ions also slow down the motion of the beam ions, although in this case, especially if the beam ions are lighter than the background plasma ions, there is also a significant pitch-angle scattering effect. However, if we choose for the present to ignore pitch angle scattering, as would be appropriate if our 'beam' originates from an isotropic source, such as fusion reactions, we can examine the effects on the distribution of beam-ion energies of 'pure' slowing down due to collisions with background electrons and ions, as described by equations (14.12) and (14.19), respectively. Thus, returning to the general form of the Fokker–Planck equation given in equation (13.5), we can neglect the velocity diffusion coefficients relative to the dynamical friction, writing simply

$$\left(\frac{\partial f}{\partial t}\right)_{\text{coll}} = -\frac{\partial}{\partial \mathbf{V}} \cdot \left(\frac{\mathrm{d}\langle\Delta \mathbf{V}\rangle}{\mathrm{d}t} f\right). \tag{14.28}$$

Substituting our expressions for the two contributions to the dynamical friction, i.e. equations (14.12) and (14.19), into equation (14.28), we obtain a Fokker–Planck equation for the beam-ion distribution function $f_b(\mathbf{V})$:

$$\frac{\partial f_b}{\partial t} = \frac{n_e Z Z_b^2 e^4 \ln\Lambda}{4\pi \epsilon_0^2 M_b M} \frac{\partial}{\partial \mathbf{V}} \cdot \left[\frac{\mathbf{V}}{V^3}\left(1 + \frac{V^3}{V_{\text{crit}}^3}\right) f_b\right] \tag{14.29}$$

where

$$V_{\text{crit}} = (2W_{b,\text{crit}}/M_b)^{1/2} = 3^{1/3} Z^{1/3} (\pi/2)^{1/6} [T_e/(m^{1/3} M^{2/3})]^{1/2}$$

i.e. the beam-ion velocity at the 'critical' energy $W_{b,\text{crit}}$.

If we are not interested in the direction of the beam velocities, but only in the magnitudes of the velocities, or if we have an energetic-ion population that is isotropic in velocity space, then it is more convenient to transform to spherical coordinates in velocity space. This will give an equation for $f_b(V)$, where $V = |\mathbf{V}|$. Using the divergence operator in spherical coordinates (see Appendix E), we can transform equation (14.29) to

$$\frac{\partial f_b}{\partial t} = \frac{n_e Z Z_b^2 e^4 \ln\Lambda}{4\pi \epsilon_0^2 M_b M} \frac{1}{V^2} \frac{\partial}{\partial V}\left[\left(1 + \frac{V^3}{V_{\text{crit}}^3}\right) f_b\right]. \tag{14.30}$$

Since our result for the rate of decrease of beam-ion energy applied equally well to lighter and heavier beam ions, similarly equation (14.30) applies for all ratios of beam-ion to background-ion mass.

We can apply equation (14.30) to a variety of situations involving a population of energetic or 'fast' ions in a plasma. Although we will continue to refer to these particles as a 'beam' and denote their distribution function f_b, this nomenclature and notation can refer also to the important practical case where the energetic ions are approximately isotropic in velocity space, either because they are injected isotropically or because they are born in the plasma itself with an isotropic distribution. For such cases, equation (14.28) gives an essentially complete description of the energetic ion distribution which results from collisions with background plasma particles.

Suppose, for example, that the energetic ions are injected into the plasma all at the same initial velocity V_0. Then a source term must be added to the right-hand side of equation (14.30), and in this case the source term will be in the form of a δ-function in velocity space, centered at $V = V_0$. Specifically, if S is the rate of injection of particles per second and if the particles are injected isotropically in velocity space, then the source term will be given by

$$\left(\frac{\partial f_b}{\partial t}\right)_{\text{source}} = \frac{S\,\delta(V - V_0)}{4\pi V^2}. \tag{14.31}$$

(Integrating this equation over all velocity space with a velocity volume element in spherical coordinates of $4\pi V^2 dV$ gives dn/dt equal to the source rate S.) Adding such a term to the right-hand side of equation (14.28) allows a steady-state solution to be found, in which the injected ions slow down until they are lost in a 'sink' at $V = 0$. In practice, this 'sink' does not require that the energetic ions be actually lost, only that there is a denser Maxwellian background-ion distribution into which they may be absorbed. (The inclusion of additional terms in the Fokker–Planck equation that become important when $V_b \sim v_{t,i}$ will of course result in the complete 'Maxwellianization' of the slowed-down beam-ion distribution.)

The steady-state distribution function may be obtained from equation (14.30), with the source-term added, taking care to apply a 'boundary condition' at $V = V_0$ obtained by integrating across the δ-function. The first step is to note that, away from the source at $V = V_0$, the right-hand side of equation (14.30) must in steady state be set to zero, giving

$$\left(1 + \frac{V^3}{V_{\text{crit}}^3}\right) f_b = C \tag{14.32}$$

for $V < V_0$, where C is a constant, as yet undetermined. For $V > V_0$, we must have the trivial solution $f_b = 0$, since the source at $V = V_0$ cannot supply particles to larger velocities. In our model, the beam particles are only slowed down by their interactions with the background plasma, not accelerated. The constant C in equation (14.32) is then obtained by including the source term in equation (14.30), assuming steady state, substituting our solution for f_b, i.e. equation (14.32), multiplying by V^2, and integrating just across $V = V_0$, giving

$$-C\frac{n_e Z Z_b^2 e^4 \ln\Lambda}{4\pi \epsilon_0^2 M_b M} + \frac{S}{4\pi} = 0 \tag{14.33}$$

thereby determining the constant C in terms of the source S. Finally, then, the beam distribution function is given by

$$f_b(V) = \frac{S\epsilon_0^2 M M_b}{n_e Z Z_b^2 e^4 \ln\Lambda}\left(\frac{1}{1 + V^3/V_{\text{crit}}^3}\right) \qquad V < V_0 \tag{14.34}$$

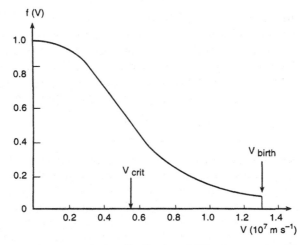

Figure 14.2. Steady-state velocity distribution $f(V)$ of energetic alpha particles in a deuterium–tritium plasma with $T_e = 20\,\text{keV}$. The vertical scale is arbitrary.

with $f_b(V) = 0$ for $V > V_0$. Thus, we have obtained an explicit solution for an isotropic distribution function of slowing down beam ions, usually termed the 'slowing down distribution'.

Equation (14.34) has an immediate application to the slowing down of alpha particles in a deuterium–tritium (D–T) fusion plasma, viewing the alpha particles as a 'beam' of energetic ions. Alpha particles, which are energetic helium ions (charge $Z = 2$, atomic mass $= 4$), are created continuously by the D–T fusion reaction at a 'source rate' S given by

$$S = n_\text{D} n_\text{T} \langle \sigma v \rangle_\text{DT} \tag{14.35}$$

where n_D and n_T are the deuterium and tritium ion densities (usually about equal to each other, and about half the electron density), and $\langle \sigma v \rangle_\text{DT}$ is the product of the D–T fusion cross section σ and the ion velocity averaged over a Maxwellian distribution of reacting ions. This quantity has a strong dependence on plasma temperature, but is given by $\langle \sigma v \rangle_\text{DT} \approx 4.2 \times 10^{-22}\,\text{m}^3\,\text{s}^{-1}$ at $T_i = 20\,\text{keV}$. The alpha particles are born with energy approximately 3.5 MeV, i.e. velocity $1.3 \times 10^7\,\text{m s}^{-1}$, and their birth distribution is isotropic in velocity direction. Thus, equation (14.34) describes the velocity distribution of energetic alpha particles in a D–T fusion plasma. The 'critical energy' $W_{b,\text{crit}}$ in this case (i.e. helium ions slowing down in a D–T plasma) is about $30T_e$, i.e. about 600 keV assuming $T_e = 20\,\text{keV}$. Accordingly, slowing down due to collisions with electrons is the dominant process by which alpha particles give up their energy to the background plasma, until they have slowed down to energies of about 600 keV. Figure 14.2 shows the alpha particle distribution in this case. For purposes of these calculations, we have treated the background D–T plasma as if it were composed of a single species of $Z = 1$ ions with mass number 2.5.

Problem 14.2: By integrating the appropriate slowing down distribution function over all velocities (note: this can be done analytically), find the total density of energetic alpha particles and the average alpha particle energy in a deuterium–tritium fusion plasma. The background plasma should have an electron density $n_e = 10^{20}$ m^{-3}, equal deuteron and triton densities, $n_D = n_T = n_e/2$, and temperatures $T_i = T_e = 20$ keV; it can be treated as if it had a single species of ions, mass-number 2.5. At these temperatures, we may take $\langle \sigma v \rangle_{DT} \approx 4.2 \times 10^{-22}$ m^3 s^{-1}. Express the alpha particle density and pressure as fractions of the background-plasma density and pressure. Are these fractions dependent on the background plasma density?

14.6   PITCH-ANGLE SCATTERING OF BEAM IONS

As we have seen, in the case of relatively light beam ions the dominant effect of collisions with heavier background ions will be deflection of the beam-ion's velocity vector from its initial direction. The equivalent problem of scattering of electrons by ions was considered in Chapter 11, and the corresponding Fokker–Planck equation was derived in Chapter 13. Applied to the present case, the beam ion acquires velocity increments $\Delta V_\perp$ perpendicular to its initial direction, at a rate given by equation (14.23). This occurs at approximately constant beam-ion energy, so that $(\Delta V_\perp)^2 + 2V\Delta V_\parallel = 0$, with the result that the directed momentum of the beam ion is reduced according to

$$\frac{dV_\parallel}{dt} = -\frac{n_i Z^2 Z_b^2 e^4 \ln \Lambda}{4\pi \epsilon_0^2 M_b^2 V_b^2} \tag{14.36}$$

(see equation (11.14) for the equivalent case of electrons colliding with ions).

Consider next the case of a distribution of beam ions, produced for example by injection of a directed beam into a plasma. If the energetic ions are not injected isotropically, and it is necessary to follow their distribution in velocity directions as well as velocity magnitudes, then a velocity-angle scattering term of the type given in equation (13.14) must be added to the beam-ion Fokker–Planck equation. This will describe the angle-scattering of energetic beam ions by collisions with background ions.

Strictly, equation (13.14) was derived only for the case where the colliding particles have much smaller mass than the scattering particles, e.g. electrons colliding with ions. Applied here, a velocity-angle scattering term of this type will be strictly valid only in the case where the scattered beam-ion is lighter than the scattering ion. For this case—or assuming that the same expression is at least a reasonable approximation even in the case where the two ion masses are comparable, which turns out in fact to be true—we can write

$$\left(\frac{\partial f_b}{\partial t}\right)_{scatt} = \frac{n_i Z^2 Z_b^2 e^4 \ln\Lambda}{8\pi\epsilon_0^2 M_b^2 V^3} \frac{1}{\sin\theta} \frac{\partial}{\partial\theta}\left(\sin\theta \frac{\partial f_b}{\partial\theta}\right). \tag{14.37}$$

Here, we have assumed that there is some direction, taken to be the z direction, about which the beam-ion distribution is symmetric in azimuthal velocity angle. Specifically, in spherical velocity coordinates, we have assumed that $f_b(\mathbf{V})$ is a function only of V and θ, but not of ϕ. In this case, the second term on the right-hand side in equation (13.14), which describes scattering in ϕ, can be dropped. Frequently, the presence of a strong magnetic field, taken to be in the z direction, ensures this kind of symmetry, because of the rapid Larmor gyration of the beam ions about the magnetic field. When the gyration frequency is much larger than any collision frequency, this Larmor gyration will rapidly average-out the azimuthal velocity phase-angles ϕ, so that f_b becomes effectively a function only of V and θ. The polar coordinate θ, which is given by

$$\sin\theta = V_\perp / V \tag{14.38}$$

is often called the 'pitch angle' of the particle.

Adding the 'pitch-angle scattering' term given in equation (14.37) to the 'slowing down' term given in equation (14.30), we obtain a final combined Fokker–Planck equation for the beam ions:

$$\frac{\partial f_b}{\partial t} = \frac{n_e Z Z_b^2 e^4 \ln\Lambda}{4\pi\epsilon_0^2 M_b M}\left\{ \frac{M}{2M_b} \frac{1}{V^3} \frac{1}{\sin\theta} \frac{\partial}{\partial\theta}\left(\sin\theta \frac{\partial f_b}{\partial\theta}\right) \right.$$
$$\left. + \frac{1}{V^2} \frac{\partial}{\partial V}\left[\left(1 + \frac{V^3}{V_{crit}^3}\right) f_b\right]\right\}. \tag{14.39}$$

Physically, this equation describes a combination of slowing down in velocity-magnitude V and spreading in pitch-angle θ. For example, if beam ions are injected at $V = V_0$ and all in a single direction, say $\theta = 0$, they will progressively spread over a wider range of θ values as they slow down to speeds below V_0.

Problem 14.3: A unidirectional beam of energetic ions, density n_b, mass M_b, charge-number Z_b, is continuously injected into a charge-neutral background plasma composed of electrons and ions with charge-number Z. The density of the beam can be considered to be very small compared with the density of the background plasma. The momentum injected via the beam is balanced by 'friction' on the beam ions arising from collisions with background-plasma electrons and ions. The electrons thereby acquire a finite mean velocity in the direction of the beam. Collisional 'friction' between the electrons and the background-plasma ions must then arise. For simplicity, the background plasma ions should be taken to be infinitely massive, so that they acquire no directed velocity, but can ultimately absorb the injected momentum, allowing an equilibrium to arise. By writing down simple expressions for the various collisional 'frictional' forces that arise in the direction of the beam, calculate the magnitude and direction of the net electrical current in this equilibrium, by adding together the current carried in the beam ions themselves and the current carried by the electrons. Show that the electrons tend to 'cancel out' the beam-ion current, but that this cancellation is inexact if $Z \neq Z_b$. (Hint: you do not need to use the Fokker–Planck equation to solve this problem, but you do need implicitly to include pitch-angle scattering, where it contributes importantly to momentum loss, as part of the collisional 'friction'.)

14.7 'TWO-COMPONENT' FUSION REACTIONS

Our analysis of fast-ion slowing down in plasmas has another immediate and interesting application to fusion—namely, the idea of injecting a beam of *reacting* ions into a fusion plasma.

For example, suppose a beam of deuterium ions is injected into a pure-tritium background plasma in order to generate fusion reactions. Since the peak of the D–T fusion cross section $\sigma_{DT}(v)$ occurs at an energy of about 120 keV, the injected beam should be somewhat more energetic than this, so that it will pass through the region of peak reactivity as it slows down. The frictional drag on the beam from collisions with background ions will be irreducible: for a given beam energy, it will depend linearly on the background-ion density, as will the reaction rate also, with the result that the density dependence effectively cancels out. However, the frictional drag from collisions with background electrons can be reduced by raising the electron temperature; for a 140 keV deuterium beam, the electron temperature must be raised to 10 keV to reduce the electron drag to equal the tritium ion drag at the injection speed.

Figure 14.3 shows the slowing down of a 180 keV beam deuteron injected into a pure tritium plasma with an electron temperature of 5 keV. The time-scale is normalized by plasma density, so that the figure applies for all plasma densities. As the deuteron's energy W_D (full line) decreases, cumulative energy increments ΔW (also full lines) are transferred to background electrons and

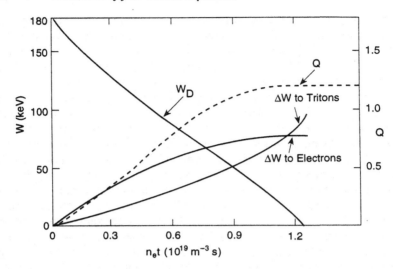

Figure 14.3. Slowing down of a 180 keV deuteron injected into a tritium plasma with $T_e = 5$ keV. The energy of the deuteron is W_D, and energy increments ΔW are given to plasma tritons and electrons. An amount of fusion energy is produced, expressed as a fraction Q of the deuteron's initial energy.

tritons, as shown. At $t \to \infty$, the sum of the two ΔW will equal the initial deuteron energy $W_D(0)$. As the deuteron moves through the plasma, it undergoes fusion reactions with instantaneous probability proportional to $\sigma_{DT}(v)v$, where v is the deuteron's velocity. An amount of thermonuclear energy is released, which is expressed in Figure 14.3 as a fraction Q (broken line) of the initial 180 keV energy of the deuteron. When the deuteron has completely slowed down, the Q value has reached about 1.15—indicating that about 200 keV of fusion energy has been produced. For higher electron temperatures, the Q value can be somewhat higher, since slowing down by collisions with electrons is reduced further. Because of the inefficiencies of converting fusion energy to electricity, the Q value in a practical reactor must be very much larger than this (~ 20)—implying that reactions among the Maxwellian background ions themselves must play the major role, rather than beam–plasma reactions.

The use of 'two-component' fusion reactions of this type to produce Q values of about unity and significant levels of fusion power density in an experimental fusion reactor was first proposed by J M Dawson, H P Furth and F H Tenney (1971 *Phys. Rev. Lett.* **26** 1156). Reactions of this sort typically contribute about a half of the fusion power produced in present-day beam-heated deuterium–tritium tokamaks.

UNIT 4

WAVES IN A FLUID PLASMA

We will now begin the study of the propagation of waves in a plasma. The natural dynamical motion of plasmas often generates waves, and laboratory plasmas are often heated or probed using waves. Thus the study of wave propagation—linear and nonlinear, electromagnetic and electrostatic—is a very important part of plasma physics.

In this Unit we will treat the plasma in the fluid approximation. (Wave–particle interactions are introduced in Chapters 23 and 24.) We will also assume that the oscillating plasma and electromagnetic quantities in the wave are small enough that we can 'linearize' the equations, i.e. ignore terms in the fluid equation that are second order in these so-called 'perturbed' quantities. This will make it convenient to Fourier analyze the time-dependent behavior. We will also assume that the plasma is uniform on space scales much greater than a wavelength, permitting Fourier analysis in space as well. The formalism we will use to describe the waves is presented in Chapter 15.

There is a rich variety of waves that can propagate in a plasma, spanning a wide range in frequency, ω, and wave-number, k. We will start in Chapter 16 by considering waves in an unmagnetized plasma, where we will find distinct electromagnetic and electrostatic waves. In Chapter 17 we will see that magnetic fields introduce a range of interesting effects at frequencies of order the electron–cyclotron and plasma frequencies. Finally in Chapter 18, we will examine plasma behavior at lower frequencies, where ion motion becomes important. We will finally show how all these waves can be derived by treating a magnetized plasma as an anisotropic medium with complex tensor electrical conductivity.

Chapter 15

Basic concepts of small-amplitude waves in anisotropic dispersive media

Systems of linear differential equations can often be studied conveniently using Fourier analysis. If any one quantity oscillates sinusoidally at a particular frequency, ω, then all the others must oscillate at the same frequency (or not at all), and the problem becomes one of finding the relative amplitudes and phases of the various oscillating quantities. The fluid plasma equations do *not* constitute a set of linear differential equations, so we cannot in general assume that nonlinear coupling between frequencies will be absent. However, if we consider only situations where the oscillations are small enough, then the equations can be 'linearized'. This means that the fluid equations are solved to zeroth order with no waves present. In the simplest case, considered here, that solution is the trivial one—a uniform isotropic plasma immersed in a steady (or even zero) magnetic field. Next we consider a first-order expansion of the equations in terms of small wave-like perturbations, neglecting second- and higher-order terms. This means that whenever we see two oscillating quantities multiplied together, since they are both small, we consider this to be a higher-order term and we neglect it. For any real situation, we then have to go back and verify that this neglect is justified: are the amplitudes we calculate in our real situation small enough that the nonlinear terms are actually negligible compared to the linear ones? For now, however, we will consider just the idealized small-amplitude limit.

15.1 EXPONENTIAL NOTATION

In the linear regime, all oscillating quantities can be represented with 'exponential notation'. For example, the density perturbation could be

$$n_1 = \bar{n}_1 \exp[i(\mathbf{k} \cdot \mathbf{x} - \omega t + \delta_n)] \tag{15.1}$$

where the overbar on the $\bar{n}_1$ indicates that it is serving as a real wave amplitude, rather than an oscillating quantity (note that the overbar does *not* indicate a time average.) The quantity **k** is the vector wave-number, or 'wave-vector', and λ, the wavelength, is $2\pi/k$. The vector **k** can have components in all directions. In an anisotropic medium like a magnetized plasma, the direction as well as the magnitude of **k** plays a crucial role in the wave dynamics. Along directions in which the component of **k** is large, the wavelength is short, so quantities vary rapidly in space; along directions in which the component of **k** is small, the wavelength is long, and so quantities vary slowly in space. Of course, the fact that we have small-amplitude perturbations does not imply that this plane-wave spatial variation necessarily gives the best description of the oscillations. Indeed, planar geometry is too simple to treat a cylindrical or otherwise specially shaped real situation, if the size of the plasma is not much greater than a wavelength. Then only the $\exp(-i\omega t + i\delta_n)$ time dependence applies, and a different spatial dependence is appropriate.

For now, we will deal with idealized plane waves only. In the particularly simple case where the plane wave-fronts align with surfaces of constant x, we can write

$$n_1 = \bar{n}_1 \exp[i(k_x x - \omega t + \delta_n)]. \qquad (15.2)$$

For definiteness, we can take δ_n to be 0 (i.e. no phase shift, an assumption that does not sacrifice generality since we can choose to measure the phase shift of *everything else* relative to n_1). If we choose the standard convention that the measurable part of n_1 is its real part, we have

$$n_1 = \bar{n}_1 \cos[(k_x x - \omega t)]. \qquad (15.3)$$

This represents a wave traveling with a phase velocity $v_p \equiv \omega/k_x$.

In the case of a vector wave-number, we define a vector phase velocity

$$\mathbf{v}_p \equiv \omega \mathbf{k}/k^2 = (\omega k_x/k^2)\hat{\mathbf{x}} + (\omega k_y/k^2)\hat{\mathbf{y}} + (\omega k_z/k^2)\hat{\mathbf{z}}.$$

An observer traveling at speed ω/k in the direction of propagation of the wave, $(\mathbf{k}/k)$, stays at a constant wave phase. We can see this by supposing that **x** varies as $\mathbf{v}_p t$, in which case the argument of the exponential, $i(\mathbf{k}\cdot\mathbf{x} - \omega t + \delta_n)$, is independent of time. In this Unit we will *always* consider $\text{Re}(\omega)$ to be positive, since a negative $\text{Re}(\omega)$ corresponds to a wave propagating in the opposite direction from **k**; we will handle such a case with $\mathbf{k} \to -\mathbf{k}$. The quantity $\text{Im}(\omega)$ represents damping ($\text{Im}(\omega) < 0$) or growth ($\text{Im}(\omega) > 0$) of the wave in time. Similarly, $\text{Im}(\mathbf{k})$ represents growth or damping in space.

Other quantities such as flow velocities and electric and magnetic fields will have the same character of spatial and temporal variation, i.e. $\exp[i(\mathbf{k}\cdot\mathbf{x} - \omega t)]$, but will have different phases and amplitudes. Indeed, each vector component of each quantity has its own phase and amplitude. For example, we can write the electric field as

$$\begin{aligned}
\mathbf{E}_1 &= \bar{E}_{x1}\hat{\mathbf{x}}\cos[(\mathbf{k}\cdot\mathbf{x}-\omega t+\delta_{Ex})] + \bar{E}_{y1}\hat{\mathbf{y}}\cos[(\mathbf{k}\cdot\mathbf{x}-\omega t+\delta_{Ey})] \\
&\quad + \bar{E}_{z1}\hat{\mathbf{z}}\cos[(\mathbf{k}\cdot\mathbf{x}-\omega t+\delta_{Ez})] \\
&= \mathrm{Re}\{\bar{E}_{x1}\hat{\mathbf{x}}\exp[i(\mathbf{k}\cdot\mathbf{x}-\omega t+\delta_{Ex})] + \bar{E}_{y1}\hat{\mathbf{y}}\exp[i(\mathbf{k}\cdot\mathbf{x}-\omega t+\delta_{Ey})] \\
&\quad + \bar{E}_{z1}\hat{\mathbf{z}}\exp[i(\mathbf{k}\cdot\mathbf{x}-\omega t+\delta_{Ez})]\} \\
&= \mathrm{Re}\{\bar{E}_{x1}\exp(i\delta_{Ex})\hat{\mathbf{x}}\exp[i(\mathbf{k}\cdot\mathbf{x}-\omega t)] + \bar{E}_{y1}\exp(i\delta_{Ey})\hat{\mathbf{y}} \\
&\quad \times \exp[i(\mathbf{k}\cdot\mathbf{x}-\omega t)] + \bar{E}_{z1}\exp(i\delta_{Ez})\hat{\mathbf{z}}\exp[i(\mathbf{k}\cdot\mathbf{x}-\omega t)]\} \\
&= \mathrm{Re}\{[\bar{E}_{x1}\exp(i\delta_{Ex})\hat{\mathbf{x}} + \bar{E}_{y1}\exp(i\delta_{Ey})\hat{\mathbf{y}} + \bar{E}_{z1}\exp(i\delta_{Ez})\hat{\mathbf{z}}] \\
&\quad \times \exp[i(\mathbf{k}\cdot\mathbf{x}-\omega t)]\}
\end{aligned} \tag{15.4}$$

where δ_{Ex}, δ_{Ey} and δ_{Ez} are real phase delays between E_{x1}, E_{y1}, E_{z1} and n_1, and all the amplitude factors (the quantities with the overbars) are again taken to be real. This is a painfully non-compact form for $\mathbf{E}_1$. The same information can be written as

$$\mathbf{E}_1 = \mathrm{Re}\{\underline{E}_1\exp[i(\mathbf{k}\cdot\mathbf{x}-\omega t)]\} \tag{15.5}$$

where the *underlined italic* $\underline{E}_1$ is now a complex vector (i.e. it has six scalars associated with it), but it is independent of time and space. To translate between these two notations recognize that, for example,

$$\tan\delta_{Ex} = \mathrm{Im}(\underline{E}_1\cdot\hat{\mathbf{x}})/\mathrm{Re}(\underline{E}_1\cdot\hat{\mathbf{x}}) \tag{15.6}$$

and

$$\bar{E}_{x1} = |\underline{E}_1\cdot\hat{\mathbf{x}}| = [(\underline{E}_1\cdot\hat{\mathbf{x}})(\underline{E}_1\cdot\hat{\mathbf{x}})^*]^{1/2} \tag{15.7}$$

where the asterisk indicates a complex conjugate. In equations (15.6) and (15.7), the terms on the far left-hand side are the real phase delay and the real amplitude, while the other terms are built from the complex wave amplitudes.

As we proceed to use this notation, we will take even more advantage of its compactness. All of the first-order terms in our equations (and therefore one multiplier in every additive term in the first-order equations) will contain the same exponential factor. Therefore we can simply drop the exponential factor without difficulty, so long as we are always clear about which are the first-order multiplicative terms. (For example, we will often find terms like $\underline{E}_1 \times \mathbf{B}_0$, and it is important to remember which one is the perturbed quantity.) Finally, in the interest of further conciseness of notation, we will drop the underlined italics which indicates a complex wave amplitude: *all the first-order terms will be complex wave amplitudes*, so that we may return to using a simple bold-faced vector such as $\mathbf{E}_1$, *with the understanding that the exponential factor is implicit and that the physical vector quantity is the real part*. We will, however, retain the subscripts indicating order everywhere in this Unit, as well as the distinction of boldface versus plain to show vector versus scalar quantities.

There is one pitfall in this more-or-less standard approach. Sometimes we find ourselves multiplying together two first-order quantities to evaluate some second-order quantity, and often then time-averaging this second-order quantity. For example, suppose we want the time average of $A_1 \cdot B_1$; the proper answer is $\frac{1}{2}\mathrm{Re}[A_1 \cdot B_1^*]$.

Problem 15.1: Show that the time average of the dot product of two physical vector fields, A_1 and B_1, is $\langle A_1 \cdot B_1 \rangle = \frac{1}{2}\mathrm{Re}[A_1 \cdot B_1^*]$. The left-hand side of this equation represents the time-average of the *physical* fields, while the right-hand side evaluates this time-average in terms of the complex wave amplitudes. Allow arbitrary phase differences between A_1 and B_1.

15.2   GROUP VELOCITIES

We have already discussed the phase velocity of a wave—the speed at which a point of constant phase propagates forward along k/k. If we make up a wave-packet of fast oscillations grouped together in time and space, as shown in Figure 15.1, this is the speed at which individual crests within the packet travel. However, these crests need not travel at the speed that the overall packet moves; the crests within the packet can slide forward or backward relative to the bundle of energy and information that constitutes the wave-packet. Indeed this frequently must be the case, since we will find that phase velocities in a plasma often exceed the speed of light, but the velocity of the group of waves (the 'group velocity') must be less than this, from fundamental considerations of special relativity.

Figure 15.1 shows a packet of oscillations with a Gaussian envelope. The amplitude $A(x)$ is given by

$$A(x) = \mathrm{Re}[\exp(-x^2/2\sigma^2)\exp(ik_0x)] \tag{15.8}$$

where we have chosen $k_0\sigma \gg 1$, so that there are many oscillations within the packet. The question we would like to investigate is: how does this wave-packet propagate in a dispersive medium where ω depends on k? Without deriving the principles of Fourier analysis, let us assert and later prove that the same $A(x)$ given in equation (15.8) can also be written

$$A(x) = \mathrm{Re}\left(\frac{\sigma}{\sqrt{2\pi}} \int_{-\infty}^{\infty} \exp(ikx)\exp[-\sigma^2(k-k_0)^2/2]dk\right). \tag{15.9}$$

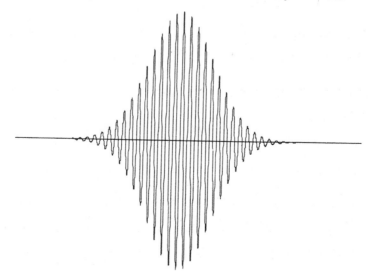

Figure 15.1. Wave-packet with a Gaussian envelope, constructed such that $k_0\sigma \ll 1$.

Equation (15.9) says that a wave-packet localized in space, x, can be considered to have been constructed of an integral over plane waves localized in wave-number, k.

Problem 15.2: Prove that the two forms of $A(x)$ given in equations (15.8) and (15.9) are equivalent. (A few tricks: transform $k' = k - k_0$; use the technique of completing the square in the exponent to transform the integral into an integral over a simple Gaussian; finally, use the facts that there are no poles in the complex plane for the resulting integrand, and that it goes to zero exponentially as Re $k \rightarrow \pm\infty$, so that any integral along a contour parallel to the real axis will give the same result.)

Equation (15.9) (and Figure 15.1) can be viewed as $t = 0$ freeze-frames of a set of propagating waves. The time evolution of this system is then just

$$A(x, t) = \text{Re}\left(\frac{\sigma}{\sqrt{2\pi}} \int_{-\infty}^{\infty} \exp\{i[kx - \omega(k)t]\}\exp[-\sigma^2(k - k_0)^2/2]dk\right)$$

(15.10)

where we have explicitly denoted the k dependence of ω by using $\omega(k)$. For a narrow enough wave-packet in k space (which means a large σ, i.e. wide in physical space), we can approximate $\omega(k) \approx \omega(k_0) + (\partial\omega/\partial k)_{k_0}(k - k_0)$.

We further assume that the medium is dispersive, but not too dispersive, by neglecting quadratic terms in the expansion of ω in $(k - k_0)$. So, proceeding for our moderately dispersive medium, we obtain

$$
A(x, t) = \text{Re}\bigg(\exp\{i[k_0(\partial\omega/\partial k)_{k_0} - \omega(k_0)]t\}
$$
$$
\times \frac{\sigma}{\sqrt{2\pi}} \int_{-\infty}^{\infty} \exp\{i[kx - k(\partial\omega/\partial k)_{k_0}t]\}\exp[-\sigma^2(k - k_0)^2/2]dk \bigg).
$$
(15.11)

Now the factor beginning with $\sigma/\sqrt{2\pi}$ is *exactly* $A(x - (\partial\omega/\partial k)_{k_0}t, 0)$—in other words, the original $t = 0$ freeze-frame, but translating at velocity $(\partial\omega/\partial k)_{k_0}$. This is just what we were looking for: the velocity of our wave-packet. So what is the factor on the first line? It is an overall space-independent time oscillation corresponding to the fact that the wave fronts are propagating at the phase velocity, ω/k, while the wave-packet moves at the group velocity, $\partial\omega/\partial k$, not equal to ω/k.

15.3   RAY-TRACING EQUATIONS

In an inhomogeneous plasma the trajectory of a wave-packet will be curved, responding to gradients in the plasma properties. We can derive the ray-tracing equations for the propagation of localized wave energy in a plasma simply from the considerations above. Consider a wave-packet localized not only in the longitudinal direction (parallel to $\mathbf{k}_0$), but also in the transverse direction (perpendicular to $\mathbf{k}_0$). For simplicity (but without loss of generality) let us assume $\mathbf{k}_0 \parallel \hat{\mathbf{x}}$, giving $\mathbf{k}_0 = k_0\hat{\mathbf{x}}$. Then the wave amplitude we desire can be expressed as

$$
A(\mathbf{x}) = \text{Re}[\exp(-x^2/2\sigma_x^2 - y^2/2\sigma_y^2 - z^2/2\sigma_z^2)\exp(ik_0x)]. \tag{15.12}
$$

By analogy with equation (15.9), we can re-express $A(\mathbf{x})$ in terms of its Fourier transform:

$$
A(\mathbf{x}) = \text{Re}\bigg(\frac{\sigma_x\sigma_y\sigma_z}{(2\pi)^{3/2}} \int_{-\infty}^{\infty} \exp(i\mathbf{k} \cdot \mathbf{x})
$$
$$
\times \exp[-\sigma_x^2(k_x - k_0)^2/2 - \sigma_y^2 k_y^2/2 - \sigma_z^2 k_z^2/2]d^3\mathbf{k} \bigg). \tag{15.13}
$$

As before, we now consider this as a 'freeze-frame' picture at $t = 0$, and include a factor $\exp(-i\omega t)$, acknowledging that $\omega = \omega(\mathbf{k})$, where $\mathbf{k}$ is a vector quantity in our anisotropic medium. Carrying through a Taylor expansion as before, we approximate

$$\omega \simeq \omega(\mathbf{k}_0) + (\mathbf{k} - \mathbf{k}_0) \cdot \nabla_{\mathbf{k}}\omega|_{k_0} \qquad (15.14)$$

where the meaning of $\nabla_{\mathbf{k}}\omega|_{k_0}$ is given by

$$\nabla_{\mathbf{k}}\omega \equiv \hat{\mathbf{x}}\frac{\partial\omega}{\partial k_x} + \hat{\mathbf{y}}\frac{\partial\omega}{\partial k_y} + \hat{\mathbf{z}}\frac{\partial\omega}{\partial k_z} = \frac{\partial\omega}{\partial \mathbf{k}} \qquad (15.15)$$

evaluated at $\mathbf{k} = \mathbf{k}_0$. If we carry through the same analysis as equations (15.9)–(15.11), but in three dimensions, we will find our 'freeze-frame' $A(\mathbf{x})$ translating at a vector group velocity given by

$$\mathbf{v}_g = \frac{\partial\omega}{\partial \mathbf{k}} \qquad (15.16)$$

with an overall time-dependent oscillation superimposed, as before. Note that $\mathbf{v}_g$ may not only have a different magnitude from $\mathbf{v}_p$, but even a different *direction*.

Problem 15.3: Prove Equation (15.16), following the derivation given in one dimension in equations (15.9)–(15.11).

We are assuming that the plasma medium is inhomogeneous so, based on our experience with light rays and lenses, there is no reason to expect the location of the peak of the $\mathbf{k}$ spectrum, $\mathbf{k}_0$, to be preserved. On the other hand, since the background medium is by hypothesis linear and time-independent, $\omega(\mathbf{k}_0)$ should be constant. This means that the total derivative of ω, moving with the wave-packet, must vanish. Assuming we know $\omega = \omega(\mathbf{x}, \mathbf{k})$ for our medium, the total derivative of ω can be expressed in terms of its partial derivatives by

$$\frac{d\omega}{dt} = \mathbf{v}_g \cdot \frac{\partial\omega}{\partial\mathbf{x}}\bigg|_{\mathbf{k}} + \frac{d\mathbf{k}_0}{dt} \cdot \frac{\partial\omega}{\partial\mathbf{k}}\bigg|_{\mathbf{x}} = 0. \qquad (15.17)$$

The partial derivative with respect to $\mathbf{x}$ is at fixed $\mathbf{k}$, and *vice versa*. Thus we have, in general, 'equations of motion' or 'ray-tracing equations' for our wave-packet:

$$\frac{d\mathbf{k}_0}{dt} = -\frac{\partial\omega}{\partial\mathbf{x}}\bigg|_{\mathbf{k}} \qquad \frac{d\mathbf{x}_0}{dt} = \frac{\partial\omega}{\partial\mathbf{k}}\bigg|_{\mathbf{x}}. \qquad (15.18)$$

As the wave-packet propagates it maintains the peak of its frequency spectrum, but its wave-number spectrum transforms. To trace out a 'ray' one must integrate forward in time the packet's position in both $\mathbf{x}$- and $\mathbf{k}$-space, since the future propagation depends on *both* $\mathbf{x}_0$ and $\mathbf{k}_0$.

The analogy to Hamiltonian mechanics is evident, as is the parallel with quantum mechanics, where $\hbar\omega$ is identified as the energy of a photon and $\hbar\mathbf{k}$ as its momentum. The ray-tracing equations are only valid in the limit of so-called 'geometrical optics', where the wave-packet is also well localized in physical space such that $\delta\mathbf{x} \cdot \partial\omega/\partial\mathbf{x} \ll \omega$, where $\delta\mathbf{x} = \sigma_x\hat{\mathbf{x}} + \sigma_y\hat{\mathbf{y}} + \sigma_z\hat{\mathbf{z}}$, and is well localized in $\mathbf{k}$-space such that $\delta\mathbf{k} \cdot \partial\omega/\partial\mathbf{k} \ll \omega$, where $\delta\mathbf{k} = \hat{\mathbf{x}}/\sigma_x + \hat{\mathbf{y}}/\sigma_y + \hat{\mathbf{z}}/\sigma_z$.

In this same limit of geometrical optics, we can use the Wentzel–Kramers–Brillouin (WKB) approximation to determine the wave phase at any location along the ray trajectory. In this approach we note that $\mathbf{k}_0(t)$ is implicitly a function of $\mathbf{x}_0(t)$ along the ray, since both are explicitly functions of t. If we imagine sending out a steady beam of radiation, rather than a wave-packet, the energy will still propagate along the group velocity vector. Along this ray-trajectory, now, the continuous spatial derivative of the wave phase will be $\mathbf{k}_0$, while the time-derivative of the phase will continue to be $-\omega_0$ (which does not vary in time or space). Thus the phase difference at fixed time between two points $\mathbf{x}_0$ and $\mathbf{x}_1$ along the ray path, $\mathbf{l}$, is given by

$$\Delta\phi = \int_{\mathbf{x}_0}^{\mathbf{x}_1} \mathbf{k}_0 \cdot d\mathbf{l}.$$

Chapter 16

Waves in an unmagnetized plasma

For simplicity we will begin by considering waves in an unmagnetized plasma—a simple homogeneous isotropic system. Such systems are somewhat unusual, since plasmas tend to be confined by magnetic fields, and they also tend to generate magnetic fields due to their own internal currents. Nonetheless this is an interesting situation to analyze. Furthermore, some plasma oscillations behave as if there were no magnetic field, even if one is present. For example at high enough frequencies, far above the electron cyclotron frequency, the particle trajectories cannot trace out any fraction of a cyclotron orbit before the wave fields reverse sign. (In the fluid equations this means the inertial, pressure, and/or electric-field terms dominate the $\mathbf{j} \times \mathbf{B}$ term.) There are also waves whose electric fields are polarized along the equilibrium magnetic field, $\mathbf{B}_0$, with the result that the driven particle motion never interacts with the magnetic field.

16.1 LANGMUIR WAVES AND OSCILLATIONS

If electrons in an unmagnetized plasma are displaced from their equilibrium positions as an initial condition, leaving the ions unmoved, the electric field that is created will act as a restoring force, pulling the electrons back towards exactly neutralizing the ion charge. The energy initially stored in the electric field will be converted into electron kinetic energy, however, and when the electrons arrive at their 'home' positions they will have kinetic energy, and as a result will overshoot, and build up a new out-of-equilibrium density distribution on the other side. This process, called a Langmuir oscillation, is illustrated in Figure 16.1.

As we will see, the period of this oscillation is very short and in this short time the ions have too much inertia to respond. Thus we can consider the ions to be a stationary background to the calculation. On the other hand, the whole process depends on the electron inertia (which is what gives rise to

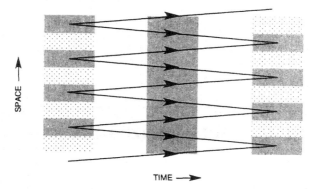

Figure 16.1. Schematic diagram of a Langmuir oscillation. The dots represent electron density; time flows to the right.

the overshoot), so we have to include $mn_e\dot{\mathbf{u}}_e$ in the electron-fluid equation of motion: the Boltzmann distribution will not describe the electron dynamics on this rapid time scale. Notice that we are finding a hierarchy of approaches to the equation of motion:

(i)   Very fast time scale—assume the species does not move at all.
(ii) Medium time scale—include inertial effects.
(iii) Very slow time scale—Boltzmann distribution.

Having dispensed with the ion force balance equation by legislating that the ions remain at rest, let us consider the electrons. We will consider the case with no **B** field, but finite scalar electron pressure, so that the equation of motion of the electron fluid is

$$mn_e[\dot{\mathbf{u}}_e + (\mathbf{u}_e \cdot \nabla)\mathbf{u}_e] = -en_e\mathbf{E} - \nabla p_e. \tag{16.1}$$

We will also make use of the electron continuity equation:

$$\dot{n}_e + \nabla \cdot (n_e\mathbf{u}_e) = 0. \tag{16.2}$$

With our assumptions, the ion continuity equation is uninteresting. Furthermore, since we are not going to circumvent Poisson's equation via the Boltzmann relation, we need it also: assuming ions with $Z = 1$, it is

$$\epsilon_0\nabla \cdot \mathbf{E} = e(n_i - n_e). \tag{16.3}$$

In this analysis, we will only consider the case where the electrons move in the direction of propagation of a plane wave, and the electric field points in this direction as well. This is not the only possible physical situation, by any means, but we will start with this case—the so-called 'Langmuir wave', or 'plasma wave'. Because of this restricted choice of motion, we will find that the displacement current ($\epsilon_0 \dot{E}_1$) is equal and opposite to the current carried by the electrons, so there is no first-order **B** field, and these are totally electrostatic (as opposed to electromagnetic) waves. For notational simplicity, we will assume that the wave is propagating in the x direction (we do not have a special direction such as a magnetic field in this case). The ∇ operator simply becomes $\hat{x}\partial/\partial x$. If we take all of our first-order quantities to vary as $\exp[i(kx - \omega t)]$, then $\partial/\partial x$ is further simplified to ik, and $\partial/\partial t$ becomes simply $-i\omega$. Dropping the subscript 'e', since we are only considering the electrons here, linearizing equation (16.1), and dotting with $\hat{x}$ we obtain

$$-i\omega m n_0 u_1 = -e n_0 E_1 - ik p_1 \tag{16.4}$$

where we had to recognize that $u_0 = 0$ and $E_0 = 0$. We have dropped terms quadratic in u_1, consistent with our linearization scheme.

The pressure perturbation, p_1, needs to be related to n_1 from the equation of state. If we assume that the compression of the electrons happens one-dimensionally, and adiabatically (faster than thermal conduction), we have $p \propto n^\gamma$, $\gamma = 3$, so that p_1 can be derived as follows:

$$
\begin{aligned}
p &= C n^\gamma \\
dp/dn &= \gamma C n^{\gamma-1} = \gamma p/n = \gamma T \\
dp &= \gamma T \, dn \\
p_1 &= \gamma T n_1.
\end{aligned}
\tag{16.5}
$$

Thus we have, for the equation of motion of the electron fluid,

$$i\omega m n_0 u_1 = e n_0 E_1 + 3ik T n_1. \tag{16.6}$$

The continuity equation linearizes to

$$-i\omega n_1 + ik n_0 u_1 = 0. \tag{16.7}$$

(We can see here that interesting physics is probably contained in terms like $n_1 u_1$, which must become important as the waves grow in amplitude, and something new could arise from $u_0 n_1$ in a moving plasma, i.e. non-zero u_0, but we leave these nonlinear physics aspects aside for now.) Poisson's equation is just

$$ik\epsilon_0 E_1 = -e n_1 \tag{16.8}$$

where the ion contribution to the charge density has served only to neutralize the equilibrium electron contribution, and has no perturbed component. Since there is no zeroth-order electron flow, the first-order current carried by the electrons is given by

$$j_1 = -en_0u_1 = -e(\omega/k)n_1 = i\omega\epsilon_0 E_1 = -\epsilon_0 \dot{E}_1 \qquad (16.9)$$

which may be substituted into the Maxwell equation

$$\nabla \times B_1 = \mu_0 j_1 + \mu_0\epsilon_0 \dot{E}_1 \qquad (16.10)$$

to show that there is no perturbed magnetic field: this 'longitudinal' wave ('longitudinal' means $E_1 \parallel k$) is indeed electrostatic, as we claimed earlier. Another way to see that longitudinal waves in general must be electrostatic is to recognize that $\nabla \times X_1$ for any perturbed quantity X_1 varying as $\exp[i(k \cdot x - \omega t)]$, is the same as $ik \times X_1$. If E_1 is parallel to k, then $\nabla \times E_1 = 0$, which tells us that $i\omega B_1 = 0$.

Problem 16.1: Prove that $\nabla \times X_1 = ik \times X_1$ for quantities X_1 that vary as $\exp[i(k \cdot x - \omega t)]$.

Now let us solve equation (16.7) for n_0u_1

$$n_0u_1 = (\omega/k)n_1 \qquad (16.11)$$

and equation (16.8) for E_1

$$E_1 = -en_1/ik\epsilon_0 \qquad (16.12)$$

and then substitute into equation (16.6) to arrive at an equation with n_1 as the only first-order quantity. We obtain

$$\frac{i\omega^2 mn_1}{k} = \frac{-e^2 n_0 n_1}{ik\epsilon_0} + 3iTn_1. \qquad (16.13)$$

Multiplying through by $-ik/mn_1$, assuming we are not allowing the trivial solution $n_1 = 0$, we obtain the 'Bohm–Gross dispersion relation', first derived by D Bohm and E P Gross (1949 *Phys. Rev.* **75** 1851):

$$\omega^2 = \omega_{pe}^2 + 3k^2 T/m = \omega_{pe}^2 + 3k^2 v_{t,e}^2 \qquad (16.14)$$

where ω_{pe}, the 'electron plasma frequency', is given by

$$\omega_{pe}^2 \equiv n_e e^2/\epsilon_0 m_e \qquad (16.15)$$

and $v_{t,e} = (T/m)^{1/2}$ is the usual electron thermal velocity. There is also an ion plasma frequency, with all ion quantities in its definition, but it is less commonly encountered, so ω_p without a species subscript generally refers to ω_{pe}.

Equation (16.14) can be cast in the form $\omega = \omega(k)$—this is referred to as a 'dispersion relation', in this case for the electrostatic plasma wave, or Langmuir wave. It is useful to plot this Bohm–Gross dispersion relation on dimensionless axes, by dividing both sides by ω_p, as shown in Figure 16.2.

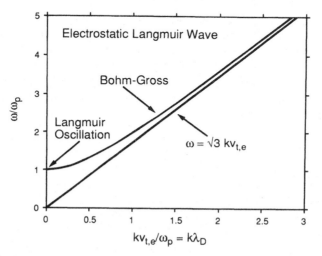

Figure 16.2. Bohm–Gross dispersion relation for the high-frequency electrostatic Langmuir wave in an unmagnetized plasma.

First of all, we note that there are no Langmuir waves at all with $\omega < \omega_p$. Furthermore, waves with $\omega > \omega_p$ only occur as a result of the finite-temperature effect. At long wavelength (low k), or low temperature, the wave phase velocity ω/k (which is proportional to the slope of a line from the origin to the dispersion curve) can become arbitrarily large, much greater than the electron thermal velocity, and even greater than c. This certainly justifies our approximation of adiabaticity, that thermal conduction cannot keep up with the moving wave front. By contrast, the group velocity $\partial\omega/\partial k$ (which is proportional to the slope of a line tangent to the dispersion curve) goes to zero in this vicinity, so no information or energy propagates. This non-propagating wiggle at low k is sometimes referred to as a 'plasma oscillation', since it was the first oscillation observed in this new state of matter.

At large k (short wavelength), or high temperature, the Bohm–Gross dispersion relation begins to look rather like an electron sound wave. The group and phase velocities both converge to $\sqrt{3}v_{t,e}$, and the wave propagates forward in the manner of a sound wave. In contrast to a sound wave in a gas, the dynamics are mediated both by the electric field and by ∇p_e. The greatest differences come, however, when we include collisionless kinetic effects associated with the class of particles that move at velocities close to the wave phase velocity. These effects, called Landau damping, will be discussed in Chapter 24.

16.2 ION SOUND WAVES

Let us now look at another longitudinal ($\mathbf{k} \parallel \mathbf{E}_1$), and therefore electrostatic, wave in an unmagnetized plasma. In this case we will assume (and later verify) that the frequency is low enough that the ions can participate in the motion, but the electrons are able to establish nearly exact force balance (i.e. a Boltzmann distribution) on the oscillation time scale. We will continue to take $\mathbf{B}_0 = 0$, and since we will take $\mathbf{k} \parallel \mathbf{E}_1$, we have $\mathbf{B}_1 = 0$ as well. The ion fluid equation, for scalar pressure, is then

$$M n_i [\dot{\mathbf{u}}_i + (\mathbf{u}_i \cdot \nabla) \mathbf{u}_i] = e n_i \mathbf{E} - \nabla p_i \tag{16.16}$$

where the upper-case M indicates an ion mass, and we have again assumed $Z = 1$. We now (as usual) linearize this equation, taking advantage of the electrostatic nature of the oscillation to write $\mathbf{E}_1$ as the gradient of a potential ($\nabla \times \mathbf{E}_1 = 0$), and using the equation of state to relate p_{i1} to n_{i1}. We make our usual plane wave and sinusoidal assumptions and note that, in this unmagnetized longitudinal wave, the fluid motion has no reason to be in any direction other than $\mathbf{k}$, so we treat u_{i1} as a scalar, i.e. the component of $\mathbf{u}_{i1}$ in the $\mathbf{k}$ direction. Equation (16.16) becomes

$$-i\omega M n_{i0} u_{i1} = -e n_{i0} i k \phi_1 - \gamma_i T_i i k n_{i1}. \tag{16.17}$$

For the electrons, we assume a Boltzmann distribution:

$$n_e = n_{e0} \exp(e\phi_1 / T_e) \approx n_{e0}(1 + e\phi_1 / T_e)$$
$$n_{e1} = n_{e0}(e\phi_1 / T_e). \tag{16.18}$$

Next we use Poisson's equation. (If we were treating only the limit of small k, i.e. $k\lambda_D \ll 1$, our knowledge of Debye shielding would tell us that we could instead use quasi-neutrality: $n_{i1} = n_{e1}$). We obtain

$$\epsilon_0 \nabla \cdot \mathbf{E}_1 = \epsilon_0 k^2 \phi_1 = e(n_{i1} - n_{e1}) = e[n_{i1} - n_{e0}(e\phi_1 / T_e)] \tag{16.19}$$

allowing us to solve for n_{i1} as a function of ϕ_1, i.e.

$$n_{i1} = [n_{i0}(e/T_e) + \epsilon_0 k^2 / e]\phi_1 \tag{16.20}$$

where we have assumed $n_{i0} = n_{e0}$, required for our case of $Z = 1$. We also need the linearized ion continuity equation:

$$i\omega n_{i1} = n_{i0}iku_{i1}. \tag{16.21}$$

We can now cast all the first-order terms in equation (16.17) in terms of n_{i1}, using equations (16.20) and (16.21). We obtain (multiplying throughout by i)

$$\omega M n_{i0}\frac{\omega n_{i1}}{kn_{i0}} = \frac{en_{i0}kn_{i1}}{n_{i0}e/T_e + \epsilon_0 k^2/e} + \gamma_i T_i kn_{i1}. \tag{16.22}$$

Dividing throughout by Mkn_{i1} (again assuming we are not looking for the trivial solution $n_{i1} = 0$), we obtain

$$(\omega/k)^2 = \frac{T_e/M}{1 + k^2\lambda_D^2} + \gamma_i T_i/M \tag{16.23}$$

where we define $\lambda_D^2 \equiv \epsilon_0 T_e/n_e e^2 = v_{t,e}^2/\omega_p^2$, as is common in the literature.

In the long-wavelength ($k \to 0$) limit, this is very similar to a normal sound wave, where we note that the electrons and the ions both contribute pressure, but the ions contribute essentially all of the mass. The appearance, effectively, of $\gamma_e = 1$ is consistent with our assumption of a Boltzmann distribution, and so *isothermal* electrons. The phase velocity of this wave is of the order of the *ion* sound speed, so the electrons generally have plenty of time to free-stream and equilibrate their temperature ahead of the wave propagation. This is not the case for the ions. In the limit of sufficient collisions to prevent ion thermal diffusion at speeds close to the sound speed, we should take γ_i to be the usual adiabatic isotropic 5/3. In the absence of collisions, but with $T_i \ll T_e$ so ion thermal motion cannot keep up with the wave, we can assume a one-dimensional adiabatic compression for the ions, equivalent to $\gamma_i = 3$. Note that since many laboratory plasmas designed for wave studies have $T_i \ll T_e$, the 'ion sound speed', C_s, is usually defined as $(T_e/M)^{1/2}$.

For large wavelengths (small k), the ion sound wave is a constant phase velocity and constant group velocity wave. At short wavelengths (large k), i.e. less than a Debye length (where this λ_D is defined without the T_i term shown in equation (1.36)), the ion sound wave turns into a constant-frequency wave, at the ion plasma frequency $\Omega_p \equiv (m/M)^{1/2}\omega_p$ (where the upper-case Ω here indicates ions).

There is an interesting complementarity between the ion and electron longitudinal ($\mathbf{k} \parallel \mathbf{E}_1$) electrostatic waves in an unmagnetized plasma. The electron waves have constant frequency ω_p at $kv_{t,e}/\omega_p = k\lambda_D \ll 1$, but travel at a constant phase velocity of $\sqrt{3}v_{t,e}$ at shorter wavelengths (larger k). The ion waves, by contrast, travel at constant phase velocity C_s for $k\lambda_D \ll 1$, but become constant frequency Ω_p for $k\lambda_D \gg 1$. The electron waves look like electron sound waves at short wavelengths, and the ion waves look like ion plasma oscillations at short wavelengths. In Figure 16.3 we plot the ion sound dispersion relation, for $T_i = 0$, on appropriate dimensionless axes. In collisionless plasmas, unless $T_e \gg T_i$, ion sound waves are subject to ion Landau damping, analogous to the electron effects for Langmuir waves, as will be discussed in Chapter 24.

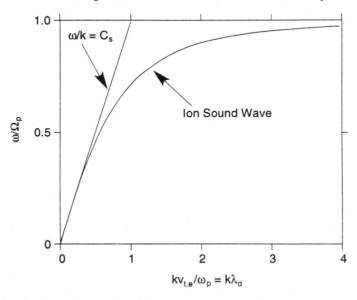

Figure 16.3. Dispersion relation for the ion sound wave in an unmagnetized plasma.

Both the Langmuir wave and the ion sound wave dispersion relations also apply in a magnetized plasma, for the special case of $\mathbf{k}||\mathbf{E}_1||\mathbf{B}_0$, since the Lorentz force will not come into play in such geometry.

16.3   HIGH-FREQUENCY ELECTROMAGNETIC WAVES IN AN UNMAGNETIZED PLASMA

So far we have found two electrostatic waves in an unmagnetized plasma, the 'plasma wave' with $\omega \geq \omega_p$, and the ion sound wave with $\omega \leq \Omega_p$. There is also a high-frequency electromagnetic wave in an unmagnetized plasma, which we will now study. To do this, we will need more of Maxwell's equations.

Given our assumption of sinusoidal plane waves, we can write

$$i\mathbf{k} \times \mathbf{B}_1 = \mu_0 \mathbf{j}_1 - i\omega \mathbf{E}_1/c^2 \tag{16.24}$$

$$i\mathbf{k} \times \mathbf{E}_1 = i\omega \mathbf{B}_1. \tag{16.25}$$

Taking the cross-product of equation (16.25) with **k**, we obtain

$$i\mathbf{k} \times \mathbf{k} \times \mathbf{E}_1 = \omega(\mu_0 \mathbf{j}_1 - i\omega \mathbf{E}_1/c^2) = (\omega^2/c^2)[\mathbf{j}_1/(\epsilon_0 \omega) - i\mathbf{E}_1] \qquad (16.26)$$

or using a vector identity to simplify the left-hand side (see Appendix D) and multiplying throughout by i, we find

$$k^2\mathbf{E}_1 - \mathbf{k}(\mathbf{k} \cdot \mathbf{E}_1) = (\omega^2/c^2)[\mathbf{E}_1 + i\mathbf{j}_1/(\epsilon_0 \omega)]. \qquad (16.27)$$

The first terms on each side give rise to the usual result for electromagnetic waves propagating in vacuum. This equation was not needed in the electrostatic case. For $\mathbf{k} \parallel \mathbf{E}_1$, the left-hand side is zero, and we see that the displacement current and the real current cancel each other, an effect we noted before. All the interesting physics came from the continuity equation and the electric fields associated with σ, the charge density—as would be expected in an *electrostatic* wave. Now, however, we will take $\mathbf{E}_1$ to be 'transverse', the opposite of longitudinal, i.e. $\mathbf{k} \cdot \mathbf{E}_1 = 0$. As we will show in Problem 16.2, there is no wave in an *unmagnetized* plasma with **E** at an oblique angle to **k**; the components of **E** longitudinal and transverse to **k** simply propagate separately as an electrostatic and an electromagnetic wave, respectively, in an unmagnetized plasma. (Notice that we are not using the words 'parallel' and 'perpendicular' relative to **k**; we reserve those terms for use relative to $\mathbf{B}_0$, when we introduce a zero-order magnetic field in the next Chapter.) Since $\mathbf{k} \cdot \mathbf{E}_1 = 0$ (i.e. $\nabla \cdot \mathbf{E} = 0$), we have $\sigma = 0$ at all times for the waves we are looking at here. Thus we do not need to consider the continuity equation in this calculation.

We are working in the high-frequency regime where we can consider the ions to be stationary, so we write

$$\mathbf{j}_1 = -n_0 e\mathbf{u}_1 \qquad (16.28)$$

where we have dropped the subscript 'e' since the ions are not of interest in this calculation. The relevant linearized fluid equation of motion for the electrons, for this case, is simply

$$-i\omega m\mathbf{u}_1 = -e\mathbf{E}_1 \qquad (16.29)$$

so that

$$\mathbf{j}_1 = -n_0 e^2 \mathbf{E}_1/i\omega m. \qquad (16.30)$$

One might ask about the absence of ∇p_e from the fluid equation of motion. This is because if $\sigma = 0$ due to $\mathbf{k} \cdot \mathbf{E} = 0$, and the ions are not moving, then there is no n_{e1}, and so no p_{e1}, no matter what equation of state we use. We say 'the fluid motion is incompressible', meaning that this particular wave does not compress the fluid. Equation (16.27) becomes

$$(c^2k^2 - \omega^2)\mathbf{E}_1 = i\omega \mathbf{j}_1/\epsilon_0 = (-n_0 e^2/m\epsilon_0)\mathbf{E}_1 \qquad (16.31)$$

or

$$\omega^2 = c^2 k^2 + \omega_p^2 \qquad (16.32)$$

so

$$\omega = (c^2 k^2 + \omega_p^2)^{1/2} = ck(1 + \omega_p^2/c^2 k^2)^{1/2}. \qquad (16.33)$$

This is the dispersion relation for an electromagnetic wave propagating in an unmagnetized plasma. (This dispersion relation also holds for high-frequency electromagnetic waves in a weakly magnetized plasma, where $\omega \gg \omega_c$. Furthermore, it is also correct for high-frequency electromagnetic waves with $\mathbf{E}_1 \| \mathbf{B}_0$, since the Lorentz force will not affect such a wave.) This is the classic example of a wave in a dispersive medium. Figure 16.4 shows this dispersion relation on appropriate dimensionless axes.

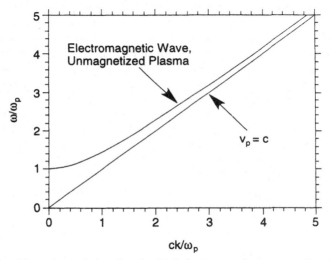

Figure 16.4. Dispersion relation for the high-frequency electromagnetic wave in an unmagnetized plasma.

Problem 16.2: Start with the electromagnetic wave equation, equation (16.27), and substitute for $\mathbf{j}_1$ in terms of $\mathbf{E}_1$ by using the electron fluid equation of motion ((equation (16.1), including the electron pressure) and then Poisson's equation. By separately dotting and crossing the resulting equation with $\mathbf{k}$, show how to generate the dispersion relations for longitudinal plasma waves and for high-frequency electromagnetic waves, and show also that one dispersion relation $\omega(\mathbf{k})$ must hold if $\mathbf{k} \cdot \mathbf{E}_1 \neq 0$, and the other must hold if $\mathbf{k} \times \mathbf{E}_1 \neq 0$. This implies that there is no class of waves that propagates with $\mathbf{k}$ at an intermediate angle to $\mathbf{E}$.

From the last step in equation (16.33), we see that

$$v_p \equiv \omega/k = c(1 + \omega_p^2/c^2k^2)^{1/2} > c. \tag{16.34}$$

The phase velocity is greater than the speed of light, and varies with k (or, equivalently, with ω). Of course the group velocity cannot be greater than c. From the first step of equation (16.33), we have

$$v_g \equiv \partial\omega/\partial k = c^2k/(c^2k^2 + \omega_p^2)^{1/2}$$
$$= c/(1 + \omega_p^2/c^2k^2)^{1/2} < c. \tag{16.35}$$

At low k (long wavelength, $ck \ll \omega_p$), these are constant-frequency waves at $\omega = \omega_p$, while as the frequency increases and the wavelength decreases (k increases), they turn into vacuum electromagnetic waves propagating at the speed of light. Eventually the frequency becomes so high that the electrons' inertia keeps them from responding significantly.

An interesting feature of this wave is that it cannot propagate in a plasma with $\omega_p > \omega$. There is a 'cutoff density' given by

$$n_c = m_e \epsilon_0 \omega^2 / e^2 \tag{16.36}$$

above which a wave impinging on a plasma is reflected back. This is the means by which low-frequency electromagnetic waves are reflected from the ionosphere and propagate around the Earth, allowing short-wave buffs in the Northern hemisphere to talk with their friends in Australia. Fortunately for the Australians, high-frequency waves such as those used for high-bandwidth broadcasting, like television, are not reflected by the ionosphere, and so the Australians are not directly afflicted with American television (and *vice versa*).

Problem 16.3: Based on the above observations estimate a lower and an upper bound for the electron density in the ionosphere.

It is interesting to calculate how deeply an electromagnetic wave penetrates into an overdense plasma (one where $\omega < \omega_p$, or equivalently $n > n_c$, so the wave cannot propagate). The dispersion relation can be solved for real ω but imaginary k in this case:

$$k = (\omega^2 - \omega_p^2)^{1/2}/c = \pm i(\omega_p^2 - \omega^2)^{1/2}/c. \tag{16.37}$$

For our 'plane wave' geometry, which admits of evanescent solutions, we have

$$\exp(ikx) = \exp[-x(\omega_p^2 - \omega^2)^{1/2}/c] \tag{16.38}$$

where the sign of k is chosen to match the physical situation. The penetration depth is then $c/(\omega_p^2 - \omega^2)^{1/2}$. The solution at $\omega = 0$, namely c/ω_p, is sometimes referred to as the 'collisionless skin depth'. Note that this evanescence is not caused by dissipation of the wave energy—the wave and its energy are simply reflected. In the presence of some dissipation, such as collisions, the penetration depth is finite (i.e. k has an imaginary part) even for $\omega > \omega_p$.

Problem 16.4: A plasma with $n \leq n_c$ can bend electromagnetic radiation considerably. Consider a cylindrical plasma with a hollow $n(r)$ profile, $n(r) = n_c r^2/a^2$. Show that the ray-tracing equations for a wave-packet of electromagnetic radiation can form a circle of radius $r = a/\sqrt{2}$. (You can think of the wave-fronts as 'steering' around the circle because λ is greater in the higher-density plasma on the outside than in the lower-density plasma further inside.) Hint: the mathematics is simplified if you work in terms of $n/n_c = \omega_p^2/\omega^2$.

Having referred to the electromagnetic waves we have just studied as 'high-frequency' waves, we might appropriately ask whether there are any 'low-frequency' electromagnetic waves in an unmagnetized plasma. In fact, there are none. The electron motion 'shorts' them out, and no waves propagate with $\omega < \omega_p$. When we include a magnetic field (in the next two Chapters), the electrons are prevented from shorting out certain waves, and as a result whole classes of new waves becomes possible at lower frequencies. The high-frequency waves also become more interesting—in particular, the electromagnetic waves are no longer purely transverse and incompressible.

Chapter 17

High-frequency waves in a magnetized plasma

In this Chapter, we will introduce a magnetic field, B_0, into the background plasma equilibrium, and begin to investigate propagation in the resulting anisotropic medium. The direction of propagation of the wave, k/k, now affects the dynamics, as well as the polarization of the wave electric field relative to the equilibrium magnetic field. In this Chapter, we will assume that the wave frequency is high enough that the ions can be considered as stationary. The new dynamical effect (i.e. Larmor gyration) that is introduced by the magnetic field creates the possibilities of wave 'resonances' as well as 'cutoffs' such as we just derived for the unmagnetized electromagnetic wave at $\omega = \omega_p$. We will take this opportunity to 'compare and contrast' cutoffs and resonances. In this Chapter we will only examine waves propagating perfectly perpendicular and perfectly parallel to B_0. In Chapter 18 we will develop the formalism for arbitrary angles of propagation.

17.1 HIGH-FREQUENCY ELECTROMAGNETIC WAVES PROPAGATING PERPENDICULAR TO THE MAGNETIC FIELD

We now treat the case of high-frequency electromagnetic waves in the presence of a zeroth-order magnetic field, B_0. We will start with waves propagating perpendicular to B_0 (i.e. $k \perp B_0$—'perpendicular propagation' as opposed to 'parallel propagation', $k \parallel B_0$, which we will study in the next Section. Note that we use the nomenclature 'perpendicular' and 'parallel' to describe the orientation of k or E relative to B_0.) For perpendicular waves we find another division of wave types, the 'ordinary' and 'extraordinary' waves. The 'ordinary' waves (sometimes abbreviated to 'O-waves') are just that—ordinary. They arise where we have a wave propagating perpendicular to $\mathbf{B}_0$, but the wave's electric field is oriented along $\mathbf{B}_0$. This means that the magnetic field plays no role in the wave dynamics, so we can apply the previous results for high-frequency electromagnetic waves in an unmagnetized plasma with $\mathbf{u}_1 \parallel \mathbf{E}_1$. The Lorentz force $\mathbf{u}_1 \times \mathbf{B}_0$ is null, i.e. the ordinary mode never notices $\mathbf{B}_0$. Thus the previous calculation and its resulting dispersion relation go through as before for the ordinary mode. Remember that the high-frequency electromagnetic wave in an unmagnetized plasma is purely transverse ($\mathbf{E}_1 \perp \mathbf{k}$); for the ordinary wave in a magnetized plasma with $\mathbf{k} \perp \mathbf{B}_0$ ('perpendicular waves'), this is just $\mathbf{E}_1 \perp \mathbf{k}$ and $\mathbf{E}_1 \parallel \mathbf{B}_0$.

Problem 17.1: An O-wave of angular frequency ω_0 propagating through a plasma with $n_e < n_c$ has a longer wavelength than a wave at the same frequency propagating in vacuum. For a wave propagating in the x direction, with a slowly varying spatially non-uniform wave-number, $k(x)$, the formula for the phase factor of the wave at location x_1 and time t_1 is given by the WKB approximation

$$\exp\left(i \int_{x_0}^{x_1} k(x)\mathrm{d}x - i\omega_0(t - t_0) \right)$$

if we take the phase to be zero at $t = t_0$ and $x = x_0$. Find the difference in phase at location x_1 and time t_1 between an O-wave propagating between x_0 and x_1 in a plasma, and a wave propagating the same distance through vacuum, given $n_e(x)$ along the path of propagation in the plasma. Assume n_e/n_c to be small, but do the calculation to second order in n_e/n_c.

The other possible orientation of $\mathbf{E}_1$, i.e. $\mathbf{E}_1 \perp \mathbf{B}_0$, has some extraordinary properties, and it is referred to, appropriately, as the 'extraordinary' wave (sometimes abbreviated to 'X-wave'). It has both transverse and longitudinal components, depending on the frequency ω. When ω is very close to the 'upper-hybrid' resonance (which we will define shortly), it is purely longitudinal ($\mathbf{E}_1 \parallel \mathbf{k}$), but elsewhere it has a transverse component ($\mathbf{E}_1 \perp \mathbf{k}$). In general, the electric field of this wave has a component along $\mathbf{k}$ ($\perp$ to $\mathbf{B}_0$) and also a component perpendicular to both $\mathbf{k}$ and $\mathbf{B}_0$. If we choose $\mathbf{B}_0$ to lie in the z direction, and $\mathbf{k}$ to lie in the x direction, then $\mathbf{E}_1$ may have components in both the x and y directions.

We will take the ions to be stationary, since their inertia is too large for them to respond to a high-frequency wave, and we will neglect the electron pressure—which actually can matter here since this wave is *not* incompressible. This approximation is sometimes called 'cold plasma' theory; it is equivalent to assuming $T_e = T_i = 0$.

With these assumptions, the linearized fluid equation of motion for the electrons is

$$-i\omega m u_{x1} = -e(E_{x1} + u_{y1} B_0)$$
$$-i\omega m u_{y1} = -e(E_{y1} - u_{x1} B_0)$$

(17.1)

which can be straightforwardly solved for u_{x1} and u_{y1}. We may use the method of determinants to solve this system of linear equations, expressed in the form

$$eE_{x1}/m = i\omega u_{x1} - \omega_c u_{y1}$$
$$eE_{y1}/m = \omega_c u_{x1} + i\omega u_{y1}.$$

(17.2)

The determinant is $\omega_c^2 - \omega^2$, and the solutions are

$$u_{x1} = \frac{(e/m)(i\omega E_{x1} + \omega_c E_{y1})}{(\omega_c^2 - \omega^2)}$$
$$u_{y1} = \frac{(e/m)(i\omega E_{y1} - \omega_c E_{x1})}{(\omega_c^2 - \omega^2)}.$$

(17.3)

We proceed to substitute these electron fluid velocities into the wave equation, i.e. equation (16.27):

$$k^2 \mathbf{E}_1 - \mathbf{k}(\mathbf{k} \cdot \mathbf{E}_1) = (\omega^2/c^2)[\mathbf{E}_1 + i\mathbf{j}_1/(\epsilon_0 \omega)]$$

(17.4)

with $\mathbf{j}_1 = -n_0 e \mathbf{u}_1$. First we break this vector equation into its two components in the x and y directions. The vector $\mathbf{E}_1$ has no component in the z direction, since this would correspond to the ordinary mode we have already treated. Remember that we have chosen $\mathbf{k}$ to point in the x direction, so the left-hand side of equation (17.4) has no x component. From the x component of the right-hand side of equation (17.4), we obtain

$$E_{x1} = \frac{i(n_0 e^2/\epsilon_0 m)(i\omega E_{x1} + \omega_c E_{y1})}{\omega(\omega_c^2 - \omega^2)}$$

(17.5)

and from the y component, multiplying through by c^2/ω^2, we get

$$(1 - c^2 k^2/\omega^2)E_{y1} = \frac{i(n_0 e^2/\epsilon_0 m)(i\omega E_{y1} - \omega_c E_{x1})}{\omega(\omega_c^2 - \omega^2)}.$$

(17.6)

Noting the presence of ω_p^2 in these equations, and observing also that we have here two linear equations in two unknowns, we multiply through by $(\omega_c^2 - \omega^2)$

and rewrite the equations as

$$(\omega_c^2 - \omega^2 + \omega_p^2)E_{x1} - i(\omega_p^2\omega_c/\omega)E_{y1} = 0$$
$$i(\omega_p^2\omega_c/\omega)E_{x1} + [(1 - c^2k^2/\omega^2)(\omega_c^2 - \omega^2) + \omega_p^2]E_{y1} = 0. \tag{17.7}$$

Once again we can solve these equations by the method of determinants, but we do not obtain the amplitude of $\mathbf{E}_1$ from these equations: when the right-hand-side of this matrix equation vanishes then the solution is degenerate, and instead we obtain a criterion on the coefficients, that their determinant should vanish. This gives us

$$(\omega_c^2 - \omega^2 + \omega_p^2)[(1 - c^2k^2/\omega^2)(\omega_c^2 - \omega^2) + \omega_p^2] - (\omega_p^2\omega_c/\omega)^2 = 0 \tag{17.8}$$

which is effectively the dispersion relation $[\omega = \omega(k)]$ that we are seeking.

We define the 'upper-hybrid' frequency as

$$\omega_h^2 \equiv \omega_p^2 + \omega_c^2. \tag{17.9}$$

The dispersion relation may now be written

$$(1 - c^2k^2/\omega^2)(\omega_c^2 - \omega^2) + \omega_p^2 = (\omega_p^2\omega_c/\omega)^2/(\omega_h^2 - \omega^2). \tag{17.10}$$

Looking at this equation, we see clearly that something interesting will happen at $\omega = \omega_h$, the upper-hybrid 'resonance'. It *looks* as if something may happen as well at $\omega = \omega_c$, but this is an illusion. If we substitute $\omega = \omega_c$ into equation (17.10), we obtain the null result that $\omega_p^2 = \omega_p^2$, and the k value is then not defined by this equation. For ω *arbitrarily close* to ω_c, however, the k value is perfectly well defined, and the same on either side. Thus it is helpful to remove this spurious apparent sign of activity, since it is not a physical result, and a more transparent form of the dispersion relation may exist. (We should, however, note that in the full kinetic treatment—the so-called 'hot plasma' theory—some new physics arises at $\omega = \omega_c$, and indeed at $\omega = n\omega_c$ for all n; new modes, called 'electron Bernstein modes' appear, named for their discoverer; see I B Bernstein (1958 *Phys. Rev.* **109** 10).) In any event, to simplify the cold-plasma dispersion relation, we proceed as follows:

$$(1 - c^2k^2/\omega^2)(\omega_c^2 - \omega^2) = \frac{-\omega_p^2(\omega_p^2 + \omega_c^2 - \omega^2) + (\omega_p^2\omega_c/\omega)^2}{(\omega_h^2 - \omega^2)}$$
$$= \frac{-\omega_p^2(\omega_c^2 - \omega^2) + (\omega_p^4/\omega^2)(\omega_c^2 - \omega^2)}{(\omega_h^2 - \omega^2)}. \tag{17.11}$$

Then dividing through by $(\omega_c^2 - \omega^2)$ and rearranging terms we obtain

$$\frac{c^2k^2}{\omega^2} = \frac{c^2}{v_p^2} = 1 - \frac{\omega_p^2(\omega^2 - \omega_p^2)}{\omega^2(\omega^2 - \omega_h^2)}. \tag{17.12}$$

Here we see the result that at the upper-hybrid resonance, $k \to \infty$, i.e. the wavelength goes to zero. This is what is meant by a *resonance*. When $k \to \infty$, the phase velocity goes to zero, and the wave-fronts 'pile up'. We can see from the first part of equation (17.7), that when $\omega = \omega_h$, then $E_{y1}/E_{x1} \to 0$. Since k is in the x direction, this means that the resonance has $\mathbf{k} \parallel \mathbf{E}_1$, implying that the upper-hybrid resonance is purely electrostatic.

The dispersion relation also has two *cutoffs*, defined as where $k \to 0$, i.e. the wavelength goes to infinity. These can be found by setting

$$\omega^2(\omega^2 - \omega_c^2 - \omega_p^2) = \omega_p^2(\omega^2 - \omega_p^2). \tag{17.13}$$

This is a quadratic equation for ω^2; dividing both sides by $\omega^2(\omega^2 - \omega_p^2)$, we obtain

$$1 - \omega_c^2/(\omega^2 - \omega_p^2) = \omega_p^2/\omega^2 \tag{17.14}$$

and, continuing the algebraic manipulation, we have

$$(1 - \omega_p^2/\omega^2) = (\omega_c^2/\omega^2)/(1 - \omega_p^2/\omega^2) \tag{17.15}$$

from which we take the square root, to obtain

$$(1 - \omega_p^2/\omega^2) = \pm(\omega_c/\omega) \tag{17.16}$$

which gives us two quadratic equations, with (presumably) a total of four solutions. The quadratic equations can be rewritten

$$\omega^2 \pm \omega\omega_c - \omega_p^2 = 0 \tag{17.17}$$

and the (four) solutions contain two *independent* $\pm$ symbols:

$$\omega = \tfrac{1}{2}[\pm\omega_c \pm (\omega_c^2 + 4\omega_p^2)^{1/2}]. \tag{17.18}$$

However, a negative ω is meaningless—by convention $\omega > 0$, and $\mathbf{k}$ is a signed vector to give us the direction of propagation—so two of the solutions are not useful, and we obtain only two physically distinct cutoff frequencies:

$$\omega = [\pm\omega_c + (\omega_c^2 + 4\omega_p^2)^{1/2}]/2 \equiv \begin{cases} \omega_R \\ \omega_L \end{cases}. \tag{17.19}$$

The $+$ sign gives the 'right-hand' cutoff frequency, and the $-$ sign the 'left-hand' cutoff frequency, denoted ω_R and ω_L respectively. The reason for this nomenclature is that these same frequencies will appear as cutoffs for left-hand and right-hand circularly polarized waves propagating parallel to $\mathbf{B}_0$, which we will discuss in the next Section. *Cutoff frequencies do not vary with angle of propagation.* By contrast, the upper-hybrid *resonance* falls in frequency as the wave moves away from perpendicular. It is also worthwhile noting here that the upper-hybrid resonance and the right-hand cutoff are clearly in the high-frequency domain. However, if $\omega_p \ll \omega_c$ then the left-hand cutoff can appear at low frequencies, where ion dynamics can be important in the calculation of ω_L and in the wave dynamics in its vicinity. This is something we will investigate in the next Chapter.

The cutoffs and resonances are important in part because they define the pass and stop bands where waves can propagate in a plasma. This is clearly illustrated in the dispersion relation for the extraordinary, or 'X-wave', where in Figure 17.1 we have chosen plasma parameters such that $\omega_c^2 = 2\omega_p^2$. This shows that waves can propagate in the pass band regions of $\omega_L < \omega < \omega_h$ and $\omega > \omega_R$, but a stop band exists in the range $\omega_h < \omega < \omega_R$. The cutoffs are at the frequencies where the dispersion relation disappears into $k = 0$, and the resonances are where $k \to \infty$. Note the curiosity that, for the lower pass band of the X-wave, $v_p > c$ for $\omega < \omega_p$, while $v_p < c$ for $\omega > \omega_p$.

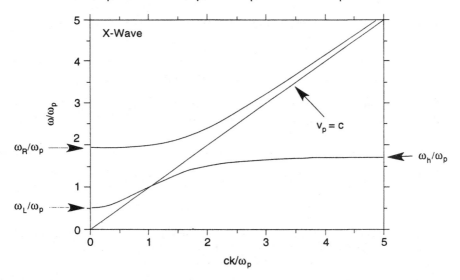

Figure 17.1. Dispersion relation for the extraordinary wave propagating perpendicular to **B** in a magnetized plasma, with ω_c^2 chosen to be equal to $2\omega_p^2$.

In an experimental situation, generally we find a fixed-frequency wave being driven by a radio-frequency (rf) generator, and we are interested in studying the propagation of this wave in a specified plasma. Thus for a wave to reach the upper-hybrid resonance it must propagate *down* a density gradient or magnetic-field gradient, so that $\omega < \omega_h \equiv (\omega_p^2 + \omega_c^2)^{1/2}$ in the propagation region, but then $\omega = \omega_h$ at the resonance. Propagating a wave down a density gradient is usually difficult to arrange, since the rf source is usually located outside of the plasma, so one has to rely instead on propagating down a magnetic-field gradient. To reach the ω_R cutoff, by contrast, requires propagating *up* a density or magnetic-field gradient, which is easier to arrange.

If we nonetheless imagine an X-wave propagating perpendicular to $\mathbf{B}_0$, and directly down a density or magnetic-field gradient until it reaches the upper hybrid 'layer', something interesting is clearly going to happen as the phase velocity goes to zero. In this situation, it is also interesting to look at the group velocity, $d\omega/dk$. In the vicinity of the upper hybrid resonance we can see graphically from Figure 17.1 that the group velocity is going to zero. Multiplying equation (17.12) by ω^2 and differentiating, we have, as $\omega \to \omega_h$:

$$2c^2k\,dk \approx 2\omega\,d\omega + \frac{2\omega_p^2\omega\,d\omega}{(\omega_h^2 - \omega^2)} + \frac{2\omega_p^2\omega_c^2\omega\,d\omega}{(\omega_h^2 - \omega^2)^2} \tag{17.20}$$

and the third term dominates near $\omega_h = \omega$. Using an approximation for v_p obtained by solving equation (17.12) near $\omega = \omega_h$, i.e.

$$v_p \equiv \omega/k \approx c\omega_h(\omega_h^2 - \omega^2)^{1/2}/\omega_p\omega_c \tag{17.21}$$

we have

$$\frac{d\omega}{dk} = v_g \approx \frac{(\omega_h^2 - \omega^2)^2 c^2}{\omega_p^2\omega_c^2 v_p} \approx \frac{(\omega_h^2 - \omega^2)^{3/2}c}{\omega_p\omega_c\omega_h}. \tag{17.22}$$

If we consider a wave-packet as a bundle of energy and v_g the velocity at which the bundle travels, then when v_g goes to zero, the bundle stops moving. Since the rf generator keeps 'sending in bundles', it is interesting to ask whether the energy stops at the resonance and builds up wave energy-density until some process outside of linear cold-plasma theory absorbs or otherwise converts the energy, or whether the wave energy is reflected back out of the plasma.

The concept of a discrete wave-packet begins to break down under these circumstances, but we can obtain a feel for the answer by examining the spatial dependence of the group velocity, v_g. From the equation of motion of a ball rolling in a well, it is clear that for wave energy to be reflected we need the *spatial derivative* of v_g^2 to be non-zero, in order for there to be finite acceleration at the top of the roll of the ball (or equivalently at the resonance, where v_g is zero). Just as ordinary kinematics tells us that acceleration can be written $(d/dx)(v^2/2) = v\,dv/dx = dv/dt$, the acceleration of a wave-packet is also given by $(d/dx)(v_g^2/2)$. If the spatial derivative of v_g *is* zero, the ball has found a shelf at the top of its roll, and stays there, as illustrated in Figure 17.2, and by analogy the wave energy in this case just steadily builds up at the resonance as more packets are sent in.

Near the resonance, v_g^2 varies as

$$v_g^2 \approx (\omega_h^2 - \omega^2)^3 c^2/(\omega_p\omega_c\omega_h)^2. \tag{17.23}$$

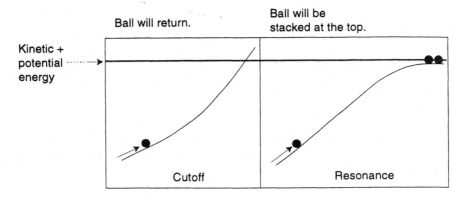

Figure 17.2. Mechanical analog to wave cutoffs and resonances.

Assuming that the spatial profiles of all the plasma parameters are smooth, it is clear that the spatial derivative of v_g^2 will be zero at just the place where v_g is zero, since a factor that vanishes at least as rapidly as $(\omega_h^2 - \omega)^2$ will remain after the right-hand side of equation (17.23) is differentiated with respect to a spatial coordinate. Thus, within the confines of linear cold-plasma theory and assuming purely perpendicular wave propagation, the wave amplitude must grow steadily at the upper-hybrid layer when we pump energy in from the outside—which is why this is called a resonance!

To see how cutoffs work—which is very different from resonances—it is simplest to consider the cutoff of the electromagnetic wave at $\omega = \omega_p$ in an unmagnetized plasma. (This has the same dispersion relation as the ordinary-mode ($\mathbf{E}_1 \parallel \mathbf{B}_0$) perpendicular ($\mathbf{k} \perp \mathbf{B}_0$) waves in a magnetized plasma.) There we had

$$\omega^2 = \omega_p^2 + k^2 c^2 \tag{17.24}$$

and

$$v_g = c/(1 + \omega_p^2/k^2c^2)^{1/2} \tag{17.25}$$

so

$$
\begin{aligned}
v_g^2 &= c^2/(1 + \omega_p^2/k^2c^2) \\
&= c^2/[1 + \omega_p^2/(\omega^2 - \omega_p^2)] \\
&= c^2/[\omega^2/(\omega^2 - \omega_p^2)] \\
&= c^2(1 - \omega_p^2/\omega^2)
\end{aligned}
\tag{17.26}
$$

which clearly has a non-zero spatial derivative (assuming, of course, that the density has a non-zero gradient) at the place where $n = n_c$, i.e. at $\omega = \omega_p$.

Thus there is a 'restoring force' at the cutoff, accelerating the wave energy back out of the plasma, so that it does not accumulate at the cutoff. Generally wave energy is essentially fully reflected at a cutoff, although refraction effects need to be taken into account for the specific geometry under consideration. In some geometries refraction will bend the ray trajectory such that it never even reaches the cutoff.

Problem 17.2: Show that wave energy does not in general accumulate at the ω_R and ω_L cutoffs of the X-wave.   Hint: first re-express equation (17.12) in the elegant form:

$$\frac{c^2 k^2}{\omega^2} = \frac{(\omega^2 - \omega_L^2)(\omega^2 - \omega_R^2)}{\omega^2(\omega^2 - \omega_h^2)}.$$

When you obtain the formula for the group velocity, note that it does not contain $(\omega^2 - \omega_R^2)$ or $(\omega^2 - \omega_L^2)$ to any power greater than unity, so the spatial gradient of the square of the group velocity at the cutoff need not be zero. (For some perverse choice of plasma parameters and gradients, the vanishing of this spatial gradient could be arranged, but it is not a fundamental feature of the equations, as at the upper-hybrid resonance.)

17.2 HIGH-FREQUENCY ELECTROMAGNETIC WAVES PROPAGATING PARALLEL TO THE MAGNETIC FIELD

In the previous Section, we treated high-frequency waves propagating perpendicular to $\mathbf{B}_0$ (i.e. $\mathbf{k} \perp \mathbf{B}_0$). Now we treat the case of parallel propagation, $\mathbf{k} \parallel \mathbf{B}_0$, again in the high-frequency limit where the ions can be considered stationary compared to the electrons.

As usual we will have $\mathbf{B}_0$ in the z direction, and so $\mathbf{k}$ is now also in the z direction. Once again, we use the wave equation

$$k^2 \mathbf{E}_1 - \mathbf{k}(\mathbf{k} \cdot \mathbf{E}_1) = (\omega^2/c^2)[\mathbf{E}_1 + i\mathbf{j}_1/(\epsilon_0 \omega)]. \tag{17.27}$$

There is a longitudinal mode ($\mathbf{E}_1 \parallel \mathbf{k}$) propagating parallel to $\mathbf{B}_0$, i.e. with $\mathbf{k} \parallel \mathbf{B}_0$, but it is just the electrostatic Langmuir wave which we have already studied for $\mathbf{B}_0 = 0$. In fact, in the cold-plasma limit we are using here, it is just the Langmuir oscillation at $\omega = \omega_p$, independent of k. To find a new electromagnetic mode, we will take $\mathbf{k} \cdot \mathbf{E}_1 = 0$. Remembering that we have taken $\mathbf{k}$ to be in the z direction, we have

$$\mathbf{E}_1 = E_{x1}\hat{\mathbf{x}} + E_{y1}\hat{\mathbf{y}}. \tag{17.28}$$

In our calculation of the extraordinary wave, we calculated u_{x1} and u_{y1} in terms of E_{x1} and E_{y1}; see equation (17.3). Since $\mathbf{B}_0$ was in the z direction in that calculation also, the results apply in our present case as well, i.e.

$$
\begin{aligned}
u_{x1} &= \frac{(e/m)(i\omega E_{x1} + \omega_c E_{y1})}{(\omega_c^2 - \omega^2)} \\
u_{y1} &= \frac{(e/m)(i\omega E_{y1} - \omega_c E_{x1})}{(\omega_c^2 - \omega^2)}.
\end{aligned}
\tag{17.29}
$$

Next we will substitute for $\mathbf{j}_1$ into equation (17.27) using $\mathbf{j}_1 = -n_e e \mathbf{u}_1$. The y component of the equation goes through just as for the extraordinary wave

$$
(1 - c^2 k^2/\omega^2)E_{y1} = \frac{i(n_0 e^2/\epsilon_0 m)(i\omega E_{y1} - \omega_c E_{x1})}{\omega(\omega_c^2 - \omega^2)}
\tag{17.30}
$$

and because $\mathbf{k}$ is now in the z direction, the x component this time looks very much like the y component, i.e.

$$
(1 - c^2 k^2/\omega^2)E_{x1} = \frac{i(n_0 e^2/\epsilon_0 m)(i\omega E_{x1} + \omega_c E_{y1})}{\omega(\omega_c^2 - \omega^2)}.
\tag{17.31}
$$

Casting these in the appropriate form for solution by matrix methods, we have

$$
\begin{aligned}
&i(\omega_p^2 \omega_c/\omega)E_{x1} + [(1 - c^2 k^2/\omega^2)(\omega_c^2 - \omega^2) + \omega_p^2]E_{y1} = 0 \\
&[(1 - c^2 k^2/\omega^2)(\omega_c^2 - \omega^2) + \omega_p^2]E_{x1} - i(\omega_p^2 \omega_c/\omega)E_{y1} = 0
\end{aligned}
\tag{17.32}
$$

and again we require the determinant to be zero, i.e.

$$
(\omega_p^2 \omega_c/\omega)^2 - [(1 - c^2 k^2/\omega^2)(\omega_c^2 - \omega^2) + \omega_p^2]^2 = 0
\tag{17.33}
$$

so we have two solutions:

$$
(\omega_p^2 \omega_c/\omega) = \pm[(1 - c^2 k^2/\omega^2)(\omega_c^2 - \omega^2) + \omega_p^2].
\tag{17.34}
$$

To solve for $\tilde{n}^2 \equiv c^2 k^2/\omega^2 = c^2/v_p^2$ (where $\tilde{n}$ is the index of refraction) we multiply through by ± 1 and divide through by $(\omega_c^2 - \omega^2)$ to obtain

$$
(1 - c^2 k^2/\omega^2) = \frac{\pm(\omega_p^2 \omega_c/\omega) - \omega_p^2}{(\omega_c^2 - \omega^2)}
\tag{17.35}
$$

or

$$
\tilde{n}^2 \equiv \frac{c^2 k^2}{\omega^2} = 1 + \frac{\omega_p^2(\omega \mp \omega_c)}{\omega(\omega_c^2 - \omega^2)} = 1 - \frac{\omega_p^2}{\omega(\omega \pm \omega_c)}.
$$

For the upper and lower signs, we call these the 'L-wave' and 'R-wave', respectively. Both of these solutions correspond to circularly polarized waves. This means that E_{x1} and E_{y1} oscillate $\pi/2$ out of phase with each other, but have the same amplitude. This can be seen from the first part of equation (17.32) coupled with equation (17.34). The solution corresponding to the upper sign has $E_{y1} = -iE_{x1}$, and the solution corresponding to the lower sign has $E_{y1} = iE_{x1}$. As it turns out, this implies that the upper sign corresponds to a wave rotating according to a left-hand rule: left thumb along $\mathbf{B}_0$, fingers showing the direction of rotation of the electric field vector. The lower sign follows the corresponding right-hand rule. To illustrate this let us take $x = 0$ and an overall phase delay of zero for E_{x1}. Then the time dependences of E_{x1} and E_{y1} for the upper sign (L-wave) are given by

$$E_{x1}(t) = \mathrm{Re}\{\bar{E}_{x1}[\cos(-\omega t) + \mathrm{i}\sin(-\omega t)]\} = \bar{E}_{x1}\cos(\omega t)$$
$$E_{y1}(t) = \mathrm{Re}\{-\mathrm{i}\bar{E}_{x1}[\cos(-\omega t) + \mathrm{i}\sin(-\omega t)]\} = -\bar{E}_{x1}\sin(\omega t) \qquad (17.36)$$

where the overbarred quantities are real wave amplitudes, and the quantities on the left-hand side are the physical fields. Figure 17.3 shows what happens as time progresses from 0 to $\pi/2\omega$. The E-field vector rotates according to the left-hand rule, since $\mathbf{B}_0$ is in the z direction.

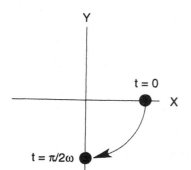

Figure 17.3. Time progression of the E-field vector for a left-hand circularly polarized wave with $\mathbf{B}_0$ along the z direction, out of the page.

   Physically, the direction of rotation of the wave has impact on the dispersion relation $\omega(k)$, because it is connected with the direction of rotation of the particles which carry $\mathbf{j}_1$. Both the L- and R-waves can propagate in either direction along $\mathbf{B}_0$; the dispersion relation only includes k^2. The L-wave by definition always has its electric field vector rotating with the left-hand rule, in relation to the direction of $\mathbf{B}_0$. Similarly, the R-wave will always rotate according to the right-hand rule along $\mathbf{B}_0$, which is of course the sense in which the electrons gyrate about $\mathbf{B}_0$. Thus it is not surprising that the R-wave has a resonance at ω_c. (Note that in plasma physics the 'handedness' of a circularly polarized wave is defined relative to $\mathbf{B}_0$, but in other research fields it is usually defined relative to $\mathbf{k}$.)

In the case of the ordinary (O) wave (propagating perpendicular to $\mathbf{B}_0$), the electric field vector is always parallel to $\mathbf{B}_0$, so it is a plane-polarized wave. The electric-field vector in the extraordinary (X) wave (also propagating perpendicular to $\mathbf{B}_0$) has both longitudinal and transverse components (but always perpendicular to $\mathbf{B}_0$) which are phased 90° apart, as can be seen from equation (17.7). However the amplitudes of the two components are not the same, so the wave is elliptically polarized. It becomes linearly polarized, with $\mathbf{E}_1 \parallel \mathbf{k}$ (longitudinal, electrostatic) at the ω_h resonance. Above and below the resonance, the direction of rotation of the electric field vector in the X-wave changes sign.

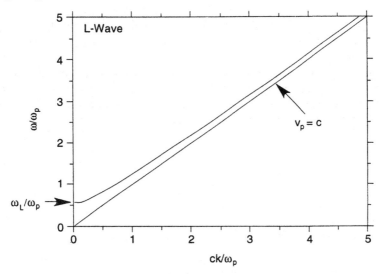

Figure 17.4. Left-hand circularly polarized electromagnetic wave propagating parallel to $\mathbf{B}_0$ in a magnetized plasma, with ω_c^2 chosen to equal $2\omega_p^2$.

It is interesting now to look at the ω versus k diagrams for the R- and L-waves. Let us begin with the simpler L-wave, shown in Figure 17.4, which corresponds to the solution of equation (17.36) with the + sign in the denominator. For the case plotted, we chose $\omega_c^2/\omega_p^2 = 2$, although this ratio can take on any value, depending on plasma parameters. This wave evidently has a simple dispersion curve. It propagates with $v_p > c$ and $v_g < c$ (necessarily) for any frequency above the cutoff frequency, ω_L. This is rather like the electromagnetic wave in the absence of a $\mathbf{B}_0$ field, which is, however, cut off at $\omega = \omega_p$. As in the case of the O-mode, wave energy is reflected from the cutoff. Setting $k = 0$ in the dispersion relation with the upper sign gives us a quadratic equation for the cutoff frequency ω_L:

$$\omega_L^2 + \omega_L\omega_c - \omega_p^2 = 0 \tag{17.37}$$

which has the solution

$$\omega_L = [-\omega_c + (\omega_c^2 + 4\omega_p^2)^{1/2}]/2 \tag{17.38}$$

which we found before as a cutoff of the extraordinary wave. (A $\pm$ sign is obtained before the square-root term from solving the quadratic equation, but since $\omega > 0$ by convention, we must take the $+$ sign.) Note that ω_c plays no special role in the L-wave. This is to be expected, since the electrons rotate around the magnetic field in the right-hand sense (in the nomenclature we are using here), so there is no resonance for the L-wave. However, for $\omega_p \ll \omega_c$, the cutoff $\omega_L \approx \omega_p(\omega_p/\omega_c)$ can be at a low enough frequency that the ion dynamics, which we have ignored so far, can become important.

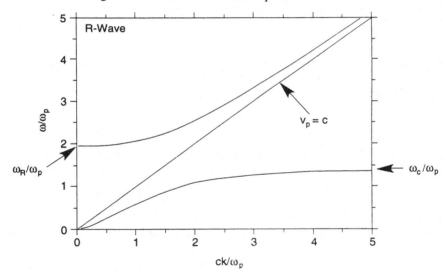

Figure 17.5. Right-hand circularly polarized electromagnetic wave propagating parallel to $\mathbf{B}_0$ in a magnetized plasma, with ω_c^2 chosen to equal $2\omega_p^2$.

The ω versus k plot for the R-wave is more complicated, as a result of the ω_c resonance, as shown in Figure 17.5. We have again chosen to plot the case $\omega_c^2/\omega_p^2 = 2$. We find a cutoff at ω_R, given by setting $k = 0$ in the dispersion relation with the lower sign, i.e.

$$\omega_R = [\omega_c + (\omega_c^2 + 4\omega_p^2)^{1/2}]/2 \tag{17.39}$$

which we have also found before as a cutoff of the extraordinary wave. Again the + sign before the square root term is forced by the requirement to have $\omega > 0$. We note that $\omega_R > \omega_c$ (unlike the case for ω_L) so ion dynamics cannot be important for this cutoff.

Note that now we obtain a resonance at $\omega = \omega_c$, a result that should not be too surprising. A resonance, as we can see here, has $\omega/k \to 0$, or $\tilde{n} = ck/\omega \to \infty$, because $k \to \infty$, $\lambda \to 0$, as we discussed with respect to the upper-hybrid resonance of the extraordinary wave propagating perpendicular to $\mathbf{B_0}$. In that case, the resonant frequency was $\omega_h^2 = \omega_p^2 + \omega_c^2$, and the plasma-frequency term entered because the wave compresses electrons, and so generates an electrostatic restoring force. The cyclotron-frequency term came from the Lorentz force, which causes this perpendicular resonance to be affected by the magnetic field, unlike the Langmuir oscillation. In the present case, since the transverse wave propagating parallel to $\mathbf{B_0}$ has $\mathbf{k} \cdot \mathbf{E} = 0$, there is no electron compression or charge-density buildup, and so no ω_p^2 in the resonance frequency. However, for $\mathbf{k} \parallel \mathbf{B_0}$, and right-circularly polarized waves, the electron cyclotron frequency enters strongly.

Problem 17.3: We showed that the resonance of the X-wave at the upper-hybrid frequency was purely electrostatic. Is the R-wave resonance at the electron-cyclotron frequency electrostatic or electromagnetic?

An especially interesting feature of the ω versus k plot for the R-wave is the presence of a new wave in the region of $\omega < \omega_c$. The low frequency part of the R-wave is called the 'whistler' wave. In the frequency range far below ω_c, visible in the lower left corner of Figure 17.5, the group velocity increases with frequency. This means that white radio noise generated in a burst in the ionosphere due to lightning flashes, and propagating as a whistler, will travel faster at high frequencies than at low. A ground-based receiver in the northern hemisphere will then hear a 'whistle' going from high frequencies to low due to lightning flashes in the southern hemisphere located along the same magnetic field lines. The ray trajectories of the whistler focus along $\mathbf{B_0}$, so one can use the properties of the received signal to deduce plasma conditions along individual field lines. When we include ion dynamics in Chapter 18, we will see that a low-frequency pass band also appears for the L-wave.

Another interesting feature of the L- and R-wave dispersion curves is that, if we choose a frequency in the upper band of the R-wave, it always has a higher phase velocity than the corresponding L-wave. This may be seen by examining Figures 17.4 and 17.5, in particular comparing the k values corresponding to the same ω value (choosing the upper frequency band of Figure 17.5). Because of this, when linearly polarized rf energy propagates parallel to $\mathbf{B}_0$, the angle of polarization of the wave rotates as it travels. This is called 'Faraday rotation'.

A linearly polarized wave can be viewed as the superposition of two counter-rotating (right and left) circularly polarized waves—and this is just how the plasma treats parallel-propagating high-frequency electromagnetic waves. Let us assume that at $z = 0$, we have a linearly polarized electromagnetic wave at frequency ω, with $\mathbf{E}_1$ in the x direction and both $\mathbf{k}$ and $\mathbf{B}_0$ in the z direction. We can decompose the electric field vector at this location into the sum of two circularly polarized field vectors as follows:

$$\mathbf{E}(z = 0) = \hat{\mathbf{x}} E_0 \mathrm{Re}[\exp(-i\omega t)] = \mathrm{Re}(\mathbf{E}_R + \mathbf{E}_L) \tag{17.40}$$

where

$$\mathbf{E}_R(z = 0) = \left(\frac{E_0}{2}\hat{\mathbf{x}} + i\frac{E_0}{2}\hat{\mathbf{y}} \right) \exp(-i\omega t)$$

$$\mathbf{E}_L(z = 0) = \left(\frac{E_0}{2}\hat{\mathbf{x}} - i\frac{E_0}{2}\hat{\mathbf{y}} \right) \exp(-i\omega t). \tag{17.41}$$

At a distance l away from $z = 0$, along $\mathbf{B}_0$, $\mathbf{E}_R$ and $\mathbf{E}_L$ evolve to

$$\mathbf{E}_R(z = l) = \left(\frac{E_0}{2}\hat{\mathbf{x}} + i\frac{E_0}{2}\hat{\mathbf{y}} \right) \exp(i\phi_R - i\omega t)$$

$$\mathbf{E}_L(z = l) = \left(\frac{E_0}{2}\hat{\mathbf{x}} - i\frac{E_0}{2}\hat{\mathbf{y}} \right) \exp(i\phi_L - i\omega t) \tag{17.42}$$

where

$$\phi_R = \int_0^l k_R(\omega)\mathrm{d}z \qquad \phi_L = \int_0^l k_L(\omega)\mathrm{d}z. \tag{17.43}$$

At a fixed ω, $k_L > k_R$, so ϕ_L will be greater than ϕ_R. The result is that a plane wave reconstructed at $z = l$ will have rotated from the incoming plane wave at $z = 0$:

$$\begin{aligned}
\mathbf{E}(z = l) &= \mathrm{Re}[\mathbf{E}_R(z = l) + \mathbf{E}_L(z = l)] \\
&= \hat{\mathbf{x}}\frac{E_0}{2}\mathrm{Re}\left\{\exp(-i\omega t)[\exp(i\phi_R) + \exp(i\phi_L)]\right\} \\
&\quad + \hat{\mathbf{y}}\frac{E_0}{2}\mathrm{Re}\left\{i\exp(-i\omega t)[\exp(i\phi_R) - \exp(i\phi_L)]\right\} \\
&= E_0\left[\hat{\mathbf{x}}\cos\left(\frac{\phi_L - \phi_R}{2}\right) + \hat{\mathbf{y}}\sin\left(\frac{\phi_L - \phi_R}{2}\right)\right] \\
&\quad \times \mathrm{Re}\left\{\exp[i(\phi_L + \phi_R)/2 - i\omega t]\right\}
\end{aligned} \tag{17.44}$$

(the last step will be proved in Problem 17.4).

For $\phi_L > \phi_R$ this corresponds to a linearly polarized field at $z = l$, rotated at an angle $(\phi_L - \phi_R)/2$ in the R direction relative to the polarization at $z = 0$. Measuring this Faraday rotation is a convenient way to determine the magnetic field in a plasma, if we can first determine the plasma density, and thereby the plasma frequency, ω_p, by other means (such as measuring the phase shift, using the same beam!). In astrophysical situations it is not possible to control the radiation source (perhaps a rapidly rotating neutron star), or provide a 'reference' beam, but much information can be gleaned about magnetic fields by studying Faraday rotation as a function of frequency. It is also possible to use pulse delay as a function of frequency to measure plasma densities.

Problem 17.4: Prove the last step in equation (17.44). This can be done in a number of different ways using trigonometric identities. Then in the spirit of Problem 17.1, calculate the Faraday rotation of a plane-polarized transverse wave propagating parallel to $\mathbf{B}_0$, given $n_e(z)$ and $B(z)$. In this case, assume $\omega \gg \omega_p$, ω_c and go only to first order in ω_R^2/ω^2, ω_c^2/ω^2, ω_p^2/ω^2.

Problem 17.5: How would you use pulse delay, as a function of frequency, to measure the average plasma density between the Earth and a radio pulsar?

Chapter 18

Low-frequency waves in a magnetized plasma

In this Chapter, we investigate classes of waves that become available at lower frequencies due to ion motion. These waves have considerable practical and theoretical interest. The equations, however, become considerably more complex with both ion and electron dynamics included, so it helps to introduce an overall formalism in which we consider the plasma to have either a complex tensor dispersive electrical conductivity or, more conventionally, a complex tensor dispersive dielectric response. This permits us to unify all the waves we have investigated: X- and O-waves in the perpendicular direction, electrostatic waves, and R- and L-waves in the parallel direction, using a single formalism valid for all angles of propagation.

18.1 A BROADER PERSPECTIVE—THE DIELECTRIC TENSOR

Before we begin the algebraically formidable task of analyzing wave propagation at low frequencies where ion motion has to be taken into account, it is useful to take an overview of what we have been doing in the process of calculating dispersion relations. In doing so, we can generalize our results to include ion motion, finite pressure and arbitrary angle of propagation. First let us write down the linearized fluid equation of motion, where we will not be specific yet as to ions or electrons:

$$mn_0\frac{\partial \mathbf{u}_1}{\partial t} = qn_0(\mathbf{E}_1 + \mathbf{u}_1 \times \mathbf{B}_0) - \gamma T \nabla n_1. \tag{18.1}$$

Without losing any generality in our plane-wave solutions, we have been taking $\mathbf{B}_0$ in the z direction, and the $\mathbf{k}$-vector to have components only in the x and z directions. Insofar as $\mathbf{k}$ forms an angle with $\mathbf{B}_0$, this shows up as k_x. Fourier analyzing the three components of this equation in the usual way, and dividing through by n_0, we obtain

$$-i\omega m u_{x1} = q(E_{x1} + u_{y1}B_0) - ik_x\gamma T n_1/n_0 \qquad (18.2)$$

$$-i\omega m u_{y1} = q(E_{y1} - u_{x1}B_0) \qquad (18.3)$$

$$-i\omega m u_{z1} = q E_{z1} - ik_z\gamma T n_1/n_0. \qquad (18.4)$$

The continuity equation is $\nabla \cdot (n_0\mathbf{u}_1) = -\partial n_1/\partial t$, which becomes

$$ik_x u_{x1} + ik_z u_{z1} = i\omega n_1/n_0. \qquad (18.5)$$

Let us define θ to be the angle between $\mathbf{k}$ and $\mathbf{B}_0$, so that $k_x = k\sin\theta$ and $k_z = k\cos\theta$. Then we have

$$n_1/n_0 = (k/\omega)(u_{x1}\sin\theta + u_{z1}\cos\theta). \qquad (18.6)$$

This can be used to substitute for n_1/n_0 in equations (18.2) and (18.4) above, so that we then have three equations for the unknown components of $\mathbf{u}_1$ in terms of the components of $\mathbf{E}_1$. Equations (18.2) and (18.4) become

$$-i\omega m u_{x1} = q(E_{x1} + u_{y1}B_0) - i(k^2/\omega)\gamma T(u_{x1}\sin^2\theta + u_{z1}\sin\theta\cos\theta) \quad (18.7)$$

$$-i\omega m u_{z1} = q E_{z1} - i(k^2/\omega)\gamma T(u_{x1}\sin\theta\cos\theta + u_{z1}\cos^2\theta). \qquad (18.8)$$

Equations (18.3), (18.7) and (18.8) form a set of linear equations for the components of $\mathbf{u}_1$. They can be solved to give each of these components as a linear combination of the components of $\mathbf{E}_1$. Indeed, we have done this in the two previous Chapters, but assuming either $\theta = 0$ or $\theta = \pi/2$, and $T = 0$. Combining the fluid velocities to form an electrical current, this result can be expressed as a complex frequency-dependent tensor electrical conductivity:

$$\mathbf{j}_1 = \sum n_0 q \mathbf{u}_1 = \underline{\sigma} \cdot \mathbf{E}_1 \qquad (18.9)$$

where the summation is over species and $\underline{\sigma}$ is a tensor quantity. (As before, we indicate a tensor by bold italics but, where Greek characters are used, as in this case, we add underlining for additional clarity.) This tensor conductivity can be substituted into the wave equation to construct a dispersion relation. Reiterating the wave equation:

$$k^2\mathbf{E}_1 - \mathbf{k}(\mathbf{k} \cdot \mathbf{E}_1) = (\omega^2/c^2)[\mathbf{E}_1 + i\mathbf{j}_1/(\epsilon_0\omega)] \qquad (18.10)$$

we obtain

$$k^2\mathbf{E}_1 - \mathbf{k}(\mathbf{k} \cdot \mathbf{E}_1) = (\omega^2/c^2)(I + i\underline{\sigma}/\epsilon_0\omega)\cdot\mathbf{E}_1 \qquad (18.11)$$

where I is just the identity tensor or, in index notation, the matrix with ones along the main diagonal and zeros elsewhere, δ_{ij}. It is more conventional to work in terms of a *dielectric tensor*, replacing the usual scalar dielectric constant in the equation for wave propagation in a non-dispersive, isotropic dielectric medium:

$$k^2 \mathbf{E}_1 - \mathbf{k}(\mathbf{k} \cdot \mathbf{E}_1) = \omega^2 \mu_0 \underline{\epsilon} \cdot \mathbf{E}_1. \tag{18.12}$$

Thus in the plasma case we have a dielectric tensor, denoted by $\underline{\epsilon}$, given by

$$\underline{\epsilon} = \epsilon_0 (I + i\underline{\sigma}/\epsilon_0 \omega) \tag{18.13}$$

where we have used $\epsilon_0 \mu_0 c^2 = 1$. In the low-frequency limit for a cold plasma, we will see later that the diagonal components of this dielectric tensor corresponding to the directions perpendicular to $\mathbf{B}_0$ are just the perpendicular plasma dielectric constant $\epsilon_\perp = \epsilon_0 + \rho/B^2$ that we have encountered before.

Since we are using tensor notation, we re-express the left-hand side of the wave equation in tensor notation:

$$k^2 X \cdot \mathbf{E}_1 \equiv [k^2 \mathbf{E}_1 - \mathbf{k}(\mathbf{k} \cdot \mathbf{E}_1)]$$

where X is the tensor defined by this equation, so $X = I - \mathbf{kk}/k^2$. Remembering that we have chosen $k_y = 0$, so $\mathbf{k} = k\sin\theta \hat{\mathbf{x}} + k\cos\theta \hat{\mathbf{z}}$, we can see that

$$\mathbf{kk}/k^2 = (\sin\theta \hat{\mathbf{x}} + \cos\theta \hat{\mathbf{z}})(\sin\theta \hat{\mathbf{x}} + \cos\theta \hat{\mathbf{z}})$$

so

$$
X \equiv
\begin{array}{ccccc}
\hat{\mathbf{x}}\hat{\mathbf{x}}\cos^2\theta & + & 0 & - & \hat{\mathbf{x}}\hat{\mathbf{z}}\sin\theta\cos\theta \\
+ \quad 0 & + & \hat{\mathbf{y}}\hat{\mathbf{y}} & + & 0 \\
- \quad \hat{\mathbf{z}}\hat{\mathbf{x}}\sin\theta\cos\theta & + & 0 & + & \hat{\mathbf{z}}\hat{\mathbf{z}}\sin^2\theta
\end{array}
\tag{18.14}
$$

where equation (18.14) is written in a manner so as to clearly display the corresponding matrix elements. Our wave equation is just

$$(\omega^2 \mu_0 \underline{\epsilon} - k^2 X) \cdot \mathbf{E}_1 = 0. \tag{18.15}$$

The dispersion relation is then derived from the requirement that the determinant of the tensor quantity in parentheses in equation (18.15) be zero.

For the equations of motion we have considered, including finite plasma pressure, this is sometimes called the 'warm' plasma dispersion relation. If we had taken $T = 0$ in these equations, we would have obtained the 'cold' plasma dispersion relation, which is a generalization of the dispersion relations we have been considering in the two previous Chapters, including ion motion and arbitrary angle of propagation. The nomenclature 'hot' is usually reserved for fully kinetic calculations, including the effects of classes of particles that move at velocities close to the wave phase velocity. Note, incidentally, by looking back at equations (18.3), (18.7) and (18.8), that $\underline{\sigma}$ (and therefore $\underline{\epsilon}$)

does *not* contain the wave-vector **k** anywhere except in the $T \neq 0$ terms. In the *cold* plasma dispersion relation, then, we see that the wave-vector **k** enters only through the k^2 term in equation (18.15), with its direction entering only through X. The additional **k** terms that enter for a warm plasma permit new solutions of the dispersion relation, such as the ion acoustic wave, which do not exist *at all* in a cold plasma, as well as modifications of waves by compressional motion and by finite Larmor radius effects, when kr_L is not small. The full 'hot' plasma dispersion relation effectively brings in terms to all higher orders in k.

18.2 THE COLD-PLASMA DISPERSION RELATION

One can straightforwardly calculate the matrix components of equation (18.15) for the case of a cold plasma. It is easier if we make the following conventional definitions:

$$\tilde{n} \equiv ck/\omega = c/v_p$$
$$c\mathbf{k}/\omega \equiv \tilde{n}\sin\theta\hat{\mathbf{x}} + \tilde{n}\cos\theta\hat{\mathbf{z}}$$
$$\omega_p, \Omega_p \equiv \text{electron and ion plasma frequencies}$$
$$\omega_c, \Omega_c \equiv \text{electron and ion cyclotron frequencies}$$
$$R \equiv 1 - (\omega_p^2/\omega)/(\omega - \omega_c) - (\Omega_p^2/\omega)/(\omega + \Omega_c)$$
$$L \equiv 1 - (\omega_p^2/\omega)/(\omega + \omega_c) - (\Omega_p^2/\omega)/(\omega - \Omega_c)$$
$$S \equiv (R + L)/2$$
$$D \equiv (R - L)/2$$
$$P \equiv 1 - \omega_p^2/\omega^2 - \Omega_p^2/\omega^2.$$

Equation (18.15), multiplied through by c^2/ω^2, becomes

$$[\hat{\mathbf{x}}\hat{\mathbf{x}}(S - \tilde{n}^2\cos^2\theta) \quad - \quad \hat{\mathbf{x}}\hat{\mathbf{y}}iD \quad + \quad \hat{\mathbf{x}}\hat{\mathbf{z}}\tilde{n}^2\sin\theta\cos\theta$$
$$+\hat{\mathbf{y}}\hat{\mathbf{x}}iD \quad + \quad \hat{\mathbf{y}}\hat{\mathbf{y}}(S - \tilde{n}^2) \quad + \quad 0$$
$$+\hat{\mathbf{z}}\hat{\mathbf{x}}\tilde{n}^2\sin\theta\cos\theta \quad + \quad 0 \quad + \quad \hat{\mathbf{z}}\hat{\mathbf{z}}(P - \tilde{n}^2\sin^2\theta)] \cdot \mathbf{E}_1 = 0$$

$$(18.16)$$

where again we have arranged the terms to display the matrix elements.

Problem 18.1: Derive equation (18.16) from equation (18.15), in the limit $T = 0$. (This involves a fair bit of algebra, but it is useful to do it once in order to be comfortable with the resulting cold-plasma dispersion relation.)

Setting the determinant to zero we obtain

$$(S-\tilde{n}^2\cos^2\theta)(S-\tilde{n}^2)(P-\tilde{n}^2\sin^2\theta)-\tilde{n}^4\sin^2\theta\cos^2\theta(S-\tilde{n}^2)-D^2(P-\tilde{n}^2\sin^2\theta) = 0.$$
(18.17)

At first glance this looks like a sixth-order equation for k, at any given θ and ω. Fortunately the $\tilde{n}^6$ terms cancel, and only $\tilde{n}^0$, $\tilde{n}^2$ and $\tilde{n}^4$ terms remain. This means that we have a quadratic in $\tilde{n}^2$, which we can solve relatively easily. Gathering together terms by their powers of $\tilde{n}$, we find

$$(S^2 P - D^2 P) - \tilde{n}^2(SP\cos^2\theta + SP + S^2\sin^2\theta - D^2\sin^2\theta)$$
$$+ \tilde{n}^4(P\cos^2\theta + S\sin^2\theta) = 0. \quad (18.18)$$

This is a quadratic equation for $\tilde{n}^2$. Thus for any value of ω, θ and plasma parameters, there are at most two real positive solutions for k, corresponding to the two 'branches' of the dispersion relation we have been studying for parallel propagation (R and L) and for perpendicular propagation (X and O). Interestingly, all the $\sin\theta$ and $\cos\theta$ terms can be simplified, for the cold-plasma dispersion relation. Replacing $\cos^2\theta$ with $1 - \sin^2\theta$, and using $S^2 - D^2 = RL$, we obtain

$$RLP - \tilde{n}^2[2SP + (RL - SP)\sin^2\theta] + \tilde{n}^4[P + (S - P)\sin^2\theta] = 0 \quad (18.19)$$

or

$$RLP - 2\tilde{n}^2 SP + \tilde{n}^4 P = \tilde{n}^2(RL - SP)\sin^2\theta - \tilde{n}^4(S - P)\sin^2\theta$$

or

$$\sin^2\theta = \frac{-P(\tilde{n}^4 - 2S\tilde{n}^2 + RL)}{\tilde{n}^4(S - P) + \tilde{n}^2(SP - RL)}. \quad (18.20)$$

To proceed further toward a useful form of the dispersion relation, we put $\sin^2\theta = 1 - \cos^2\theta$ in equation (18.19) and obtain

$$RLP - \tilde{n}^2[SP + RL + (SP - RL)\cos^2\theta] + \tilde{n}^4[S + (P - S)\cos^2\theta] = 0 \quad (18.21)$$

or

$$\cos^2\theta = \frac{S\tilde{n}^4 - (PS + RL)\tilde{n}^2 + PRL}{\tilde{n}^4(S - P) + \tilde{n}^2(PS - RL)}. \quad (18.22)$$

We can now divide equation (18.22) into equation (18.20) to obtain

$$\tan^2\theta = \frac{-P(\tilde{n}^4 - 2S\tilde{n}^2 + RL)}{S\tilde{n}^4 - (PS + RL)\tilde{n}^2 + PRL}. \quad (18.23)$$

Using the fact that $2S = (R + L)$ in the numerator, both the numerator and the denominator can now be factored to give

$$\tan^2\theta = \frac{-P(\tilde{n}^2 - R)(\tilde{n}^2 - L)}{(S\tilde{n}^2 - RL)(\tilde{n}^2 - P)}. \quad (18.24)$$

Equation (18.24) is a very useful form for the cold-plasma dispersion relation and it provides good physical insight. First, for parallel propagation ($\theta = 0$), we have two solutions, $\tilde{n}^2 = R$ and $\tilde{n}^2 = L$, which are the familiar right- and left-circularly polarized waves. For perpendicular propagation, we also have two roots, $\tilde{n}^2 = P$ (the ordinary wave) and $\tilde{n}^2 = RL/S$ (the extraordinary wave), now with the ion dynamics included automatically through the definitions of R, L, S and P. The resonances can be found from this equation by setting $\tilde{n} \to \infty$ ($k \to \infty, \lambda \to 0$); we then have $\tan^2\theta = -P/S$. Thus the resonance frequencies vary with the angle of propagation. For $\theta = 0$, they occur where $P = 0$, or $S \to \infty$. The case with $P = 0$ is the plasma resonance at ω_p. (Remember that the Langmuir oscillation has $\omega = \omega_p$, independent of k, for a cold plasma.) The case $S \to \infty$ can be arranged via either R or $L \to \infty$, which occur at the electron and ion cyclotron resonances respectively. At $\theta = \pi/2$, we need $P \to \infty$, which cannot occur for finite ω and ω_p, or we need $S \to 0$. This latter gives the upper- and lower-hybrid resonances, including ion dynamics. (We will learn about the lower-hybrid resonance later in this Chapter.)

To obtain the cutoffs from this equation is not as straightforward, since setting $\lambda \to \infty, \tilde{n} \to 0$ gives the curious result that $-PRL/PRL = \tan^2\theta$. Such a thing cannot occur for any real θ, unless $PRL = 0$, a result that can be obtained more explicitly by going back to equation (18.18), where we can see that $\tilde{n}^2 = 0$ implies $P(S^2 - D^2) = PRL = 0$. Note then that the cutoffs do not vary with θ, as we have observed before. Specifically, $P = 0$ is just the ω_p cutoff of the ordinary wave and the cutoff/resonance of the Langmuir oscillation. The cases $R = 0$ and $L = 0$ correspond to the ω_R and ω_L cutoffs, with the ion dynamics included.

18.3   COLDWAVE

In order to let you study the properties of the cold-plasma dispersion relation, we have provided a simple graphical program, COLDWAVE, which solves equation (18.18) for $\tilde{n}^2$. It works by stepping through a prescribed range in ω, for given plasma parameters and given θ, and finding the $\tilde{n}$ and therefore k which corresponds to each ω. The instructions for how to use this program are included in files called COLDWAVE.WRI on the IBM PC disk and README-COLDWAVE on the Mac[1] disk. (Computer source code is included as well.) Note that in this program frequencies are normalized to the electron plasma frequency, and all wave-numbers to ω_p/c.

The plot shown in Figure 18.1 has parameters $\omega_c/\omega_p = 1.414$, $Z_i = 1$, $A_i = 1$, $\theta = 7°$, $\omega_{min} = 0$, $\omega_{max} = 5\omega_p$, linear ω scale, $k_{min} = 0$, $k_{max} = 5$, and linear k scale. Each vertical pixel (ω axis) corresponds to five evaluations of $k(\omega)$. You

---

[1] Macintosh is a registered trademark of Apple Computer, Inc.

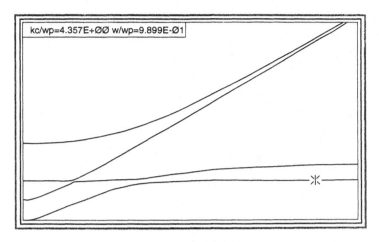

Figure 18.1. Dispersion relation plot from COLDWAVE. $\theta = 7°$, RE = 5, $\omega_c^2 = 2\omega_p^2$. kc/ω_p and ω/ω_p indicate location of pointer. $kc/\omega_p = 4.357$; $\omega/\omega_p = 0.9899$.

can see that various sections of three different curves are about to vanish into the Langmuir oscillation at $\omega = \omega_p$ when θ reaches 0. One of the four curves vanishes in a different way as θ goes to 90°, as you will see in Problem 18.2.

Problem 18.2: Use COLDWAVE to study the angular range 0–90° for high-frequency waves, in the cases where $\omega_c = \omega_p/2$ and where $\omega_c = 2\omega_p$. Describe qualitatively how the different waves transform.

Problem 18.3: Use COLDWAVE to explore the low-frequency range, $\omega \ll \Omega_c$, for $\theta = 0°$. Find the relation between the phase velocity as $\omega \to 0$ and the low-frequency perpendicular dielectric constant calculated in Chapter 4.

18.4 THE SHEAR ALFVÉN WAVE

Now we proceed to consider the low-frequency range of the R- and L-waves. As before, let us consider propagation purely parallel to $\mathbf{B}_0$, i.e. $\mathbf{k} \parallel \mathbf{B}_0$. We will further take $\mathbf{E}_1 \perp \mathbf{B}_0$, and thus $\mathbf{E}_1 \perp \mathbf{k}$. This is because the choice $\mathbf{E}_1 \parallel \mathbf{B}_0 \parallel \mathbf{k}$ at low frequency, with a warm plasma, just gives the ion acoustic wave we studied previously—the Lorentz force plays no role in that mode. Just as we saw that the electrostatic Langmuir wave is uncoupled from the high-frequency purely electromagnetic waves, so is the ion-acoustic wave decoupled from the low-frequency R- and L-waves, propagating parallel to $\mathbf{B}_0$ in the modes we are considering, and since $\mathbf{B}_0 \parallel \hat{\mathbf{z}}$, $E_{z1} = u_{z1} = 0$. Examining our equations of motion (equations (18.3), (18.7) and (18.8)), we see that T plays no role in these modes. As a result, our u_{x1} and u_{y1} for the electrons—even for finite pressure—are the same as those given in equation (17.3), when we first considered the case of $u_{x1} \neq 0, u_{y1} \neq 0$. For the ions, we simply use equation (17.3) with $e \rightarrow -e$, $\omega_c \rightarrow -\Omega_c$, and $m \rightarrow M$. (Remember that $\omega_c \equiv -q_e B/m_e$, while $\Omega_c \equiv +q_i B/m_i$). We then have for the ions

$$u_{xi1} = \frac{-(e/M)(i\omega E_{x1} - \Omega_c E_{y1})}{(\Omega_c^2 - \omega^2)}$$

$$u_{yi1} = \frac{-(e/M)(i\omega E_{y1} + \Omega_c E_{x1})}{(\Omega_c^2 - \omega^2)}. \tag{18.25}$$

For the electrons we will assume $\omega \lesssim \Omega_c \ll \omega_c$, in which case the electron fluid exhibits pure $\mathbf{E}_1 \times \mathbf{B}_0$ drift:

$$u_{xe1} = \frac{(e/m)E_{y1}}{\omega_c} = \frac{eE_{y1}}{M\Omega_c}$$

$$u_{ye1} = \frac{-(e/m)E_{x1}}{\omega_c} = \frac{-eE_{x1}}{M\Omega_c}. \tag{18.26}$$

The last step in the above equations was motivated by the desire to express frequencies in terms of ion quantities, and it conveniently brings a common factor of e/M in front of both the ion and the electron equations.

Our conductivity tensor is then given by

$$\mathbf{j}_1 = \frac{n_0 e^2}{M} \left[\frac{-i\omega}{(\Omega_c^2 - \omega^2)} \hat{\mathbf{x}}\hat{\mathbf{x}} + \left(\frac{\Omega_c}{(\Omega_c^2 - \omega^2)} - \frac{1}{\Omega_c} \right) \hat{\mathbf{x}}\hat{\mathbf{y}} \right.$$

$$\left. - \left(\frac{\Omega_c}{(\Omega_c^2 - \omega^2)} - \frac{1}{\Omega_c} \right) \hat{\mathbf{y}}\hat{\mathbf{x}} - \frac{i\omega}{(\Omega_c^2 - \omega^2)} \hat{\mathbf{y}}\hat{\mathbf{y}} \right] \cdot \mathbf{E}_1$$

$$= \underline{\boldsymbol{\sigma}} \cdot \mathbf{E}_1. \tag{18.27}$$

Note that the conductivity tensor is reduced to a 2×2 tensor for the shear Alfvén wave propagating parallel to $\mathbf{B}_0$, because we have taken $j_{z1} = E_{z1} = 0$. If we take $\omega \rightarrow 0$, the off-diagonal elements go to zero as ω^2, while the diagonal elements go to zero as ω, so we obtain a purely scalar (but imaginary) conductivity tensor. Using $\Omega_p^2 \equiv n_0 e^2/M\epsilon_0$, so the first factor on the right-hand side in equation (18.27) is just $\epsilon_0 \Omega_p^2$, this scalar conductivity becomes

$$\sigma = -i\omega n_0 e^2/M\Omega_c^2 = -i\omega\epsilon_0 \Omega_p^2/\Omega_c^2. \tag{18.28}$$

We can also view this as giving rise to a low-frequency dielectric response. Remember from equation (18.13) that $\underline{\epsilon} = \epsilon_0(I + i\underline{\sigma}/\epsilon_0\omega)$, so we obtain

$$\epsilon_\perp = \epsilon_0(1 + \Omega_p^2/\Omega_c^2) = \epsilon_0 + n_0 M/B^2 \qquad (18.29)$$

which is just the dielectric constant that we derived from the polarization drift in Chapter 4.

Returning to finite ω, in the general case the dispersion relation is given by

$$|| \omega^2 \mu_0 \underline{\epsilon} - k^2 X || = 0 \qquad (18.30)$$

where the double vertical bars indicate the determinant of the equivalent matrix. In terms of $\underline{\sigma}$, and multiplying through by c^2/ω^2, this is

$$||I - \tilde{n}^2 X + i\underline{\sigma}/\epsilon_0\omega|| = 0. \qquad (18.31)$$

For $\theta = 0$, we have $X = \hat{x}\hat{x} + \hat{y}\hat{y}$. Our calculations here did not find the $\hat{z}\hat{z}$ component of $\underline{\sigma}$, but this does not matter since the matrix we have now constructed for the case of $\mathbf{k} \parallel \mathbf{B}_0$ has non-zero elements in the upper left 2×2 area plus a single potentially non-zero component in the lower right corner. The other elements of $\underline{\sigma}$ ($\hat{z}$ currents driven by non-$\hat{z}$ fields, and non-$\hat{z}$ currents driven by $\hat{z}$ fields) are zero. Thus the upper left 2×2 matrix must have determinant zero (the dispersion relation we are looking for) *or* the element in the lower right hand corner, corresponding to purely $\mathbf{E} \parallel \mathbf{B}$ dynamics, must be zero (the Langmuir wave at high frequencies and the ion-acoustic wave at low, for warm plasmas) for the whole matrix to have determinant zero.

Thus for the case of interest here, which has $\mathbf{E}_1 \perp \mathbf{B}_0$, we obtain

$$\begin{vmatrix} 1 - \tilde{n}^2 + \Omega_p^2/(\Omega_c^2 - \omega^2) & i\Omega_p^2\omega/[\Omega_c(\Omega_c^2 - \omega^2)] \\ \\ -i\Omega_p^2\omega/[\Omega_c(\Omega_c^2 - \omega^2)] & 1 - \tilde{n}^2 + \Omega_p^2/(\Omega_c^2 - \omega^2) \end{vmatrix} = 0. \qquad (18.32)$$

The symmetry of this matrix is reminiscent of the high-frequency parallel-propagating R- and L-waves. We obtain

$$1 - \tilde{n}^2 + \frac{\Omega_p^2}{(\Omega_c^2 - \omega^2)} = \pm \frac{\Omega_p^2\omega}{\Omega_c(\Omega_c^2 - \omega^2)} \qquad (18.33)$$

and, once again, we have circularly polarized R- and L-waves. We can see this because E_{x1} and E_{y1} are equal in magnitude and $\pi/2$ out-of-phase, by the same arguments as before. To see this, we simply have to reconstruct the linear equations that come from dotting the matrix shown in equation (18.32) with $\mathbf{E}_1$, and setting the result equal to zero. Given equation (18.33), the terms multiplying E_{x1} and E_{y1} differ by a factor of $\pm i$. Proceeding to find a more compact dispersion relation, we obtain

$$\tilde{n}^2 = c^2 k^2/\omega^2 = 1 + \frac{\Omega_p^2 \Omega_c \mp \Omega_p^2 \omega}{\Omega_c(\Omega_c^2 - \omega^2)}$$

$$= 1 + \frac{\Omega_p^2}{\Omega_c(\Omega_c \pm \omega)}$$

$$= \frac{\Omega_c^2 + \Omega_p^2 \pm \Omega_c \omega}{\Omega_c(\Omega_c \pm \omega)}. \tag{18.34}$$

The upper signs go with right-hand polarization (R-wave), in the sense we defined in Chapter 17, while the lower signs go with left-hand polarization (L-wave).

Thus for the R-wave, dividing top and bottom by Ω_c, we have

$$\tilde{n}^2 = c^2 k^2/\omega^2 = \frac{\Omega_c + \Omega_p^2/\Omega_c + \omega}{(\Omega_c + \omega)}. \tag{18.35}$$

The shear-Alfvén R-wave has no cutoffs and no resonances in this low-frequency range, since neither the numerator nor the denominator can go to zero. This is not surprising, since ion motion is left-handed. As we go up in frequency, the shear-Alfvén R-wave smoothly goes over into the whistler, which has its resonance at $\omega = \omega_c$. At the low-frequency end, we have a 'simple' light wave propagating in a medium with a large scalar dielectric constant. As $\omega \to 0$, equation (18.35) gives an index of refraction

$$\tilde{n} = (1 + \Omega_p^2/\Omega_c^2)^{1/2} \tag{18.36}$$

and so a phase velocity

$$v_p = \omega/k = c/\tilde{n} = c(1 + \Omega_p^2/\Omega_c^2)^{-1/2}. \tag{18.37}$$

If we define an 'Alfvén speed', v_A, by

$$v_A \equiv c\Omega_c/\Omega_p = c(eB/M)/(ne^2/\epsilon_0 M)^{1/2} = cB/\sqrt{nM/\epsilon_0}$$

$$= B/\sqrt{\mu_0 n M} \tag{18.38}$$

then the phase velocity can be written

$$v_p = c/(1 + c^2/v_A^2)^{1/2}.$$

Multiplying top and bottom by v_A/c, and then taking $v_A/c = \Omega_c/\Omega_p \ll 1$ (which is correct for $\omega_p \sim \omega_c$), we have

$$v_p = v_A/(1 + v_A^2/c^2)^{1/2} \approx v_A. \tag{18.39}$$

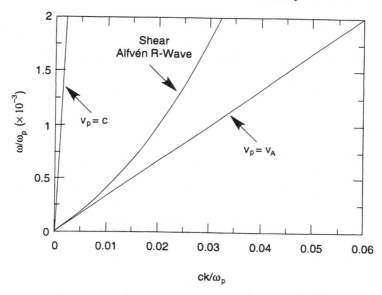

Figure 18.2. Dispersion relation for the shear Alfvén R-wave, with ω_c^2 chosen to equal $2\omega_p^2$.

Figure 18.2 shows an ω versus k diagram for the shear-Alfvén R-wave. As usual, we (arbitrarily) choose $\omega_c^2 = 2\omega_p^2$, and we will also choose $M/m = 1837$. This fixes the relevant quantities in equation (18.35). For example

$$v_A/c = \Omega_c/\Omega_p = (\Omega_c/\omega_c)(\omega_c/\omega_p)(\omega_p/\Omega_p) = (1/1837)(\sqrt{2})\sqrt{1837}$$
$$= \sqrt{2}/\sqrt{1837} = 0.033.$$

In order to allow comparison with our previous plots at high frequency, we use the same dimensionless axes, but in order to see these new results we must rescale the axes—and the two axes must be scaled differently from one another, since ω/k is now $\sim c/30$.

The left-handed shear Alfvén wave (L-wave) has the dispersion relation

$$\tilde{n}^2 = c^2 k^2/\omega^2 = \frac{\Omega_c + \Omega_p^2/\Omega_c - \omega}{(\Omega_c - \omega)} \tag{18.40}$$

which is shown in Figure 18.3. This has the same low-frequency behavior as the R-wave. At low frequencies, *plane-polarized* shear Alfvén waves exist, and do not undergo Faraday rotation. The L-wave, however, clearly has a resonance at $\omega = \Omega_c$, associated with the left-handed ion cyclotron motion. In addition,

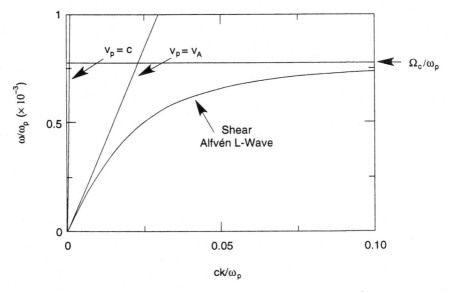

Figure 18.3. Dispersion relation for the shear Alfvén L-wave, with ω_c^2 chosen to equal $2\omega_p^2$.

it has a cutoff at $\omega = \omega_L = \Omega_c + \Omega_p^2/\Omega_c$ (Figure 18.3 does not go this high in frequency). This does not *appear* to be the 'same' ω_L as we encountered in our high-frequency calculation: in fact, however, it actually *is* the same L cutoff, just calculated with the assumption that ω is low, and so different terms can be ignored.

The full definitions of ω_L and ω_R, including both ion and electron effects— and not assuming anything *in advance* about their magnitude—can be derived from setting $L = 0$ or $R = 0$ respectively, using the general definitions for R and L given earlier in this Chapter. If we take $\Omega_c \ll \omega_c$ and $\Omega_p \ll \omega_p$, which are always justified for an ion–electron plasma, we obtain

$$\omega_R^2 - \omega_R\omega_c - \omega_c\Omega_c - \omega_p^2 = 0 \qquad (18.41)$$

and

$$\omega_L^2 + \omega_L\omega_c - \omega_c\Omega_c - \omega_p^2 = 0. \qquad (18.42)$$

In the high-frequency range, the third term is negligible (and we did not include it in equation (17.38)), and in the low-frequency range (the present calculation) the first term has been neglected. There is no positive low-frequency solution for ω_R, and so no low-frequency right-hand cutoff. For some combinations of plasma parameters, however, ω_L *may* fall in the low-frequency range, and then our shear Alfvén dispersion relation will give the cutoff in approximately the correct place. However this cutoff usually falls (for ω_p of order ω_c) in a frequency range where electron motion must be taken into account beyond just the $\mathbf{E}_1 \times \mathbf{B}_0$ drift, so the present calculation will be inaccurate. For example, for the parameters we have chosen in our plots, $\omega_L = 0.37\omega_c = 0.52\omega_p$, which violates our initial assumption of $\omega \ll \omega_c$, so the high-frequency calculation of Chapter 17 is closer to the correct cutoff given in equation (18.42). The neglected ion terms in the previous calculation are much smaller and its approximations were therefore satisfactory. It is important to be clear that, for a given density and magnetic field, there is only one ω_L and one ω_R cutoff frequency, which can be calculated from the full equations for ω_L and ω_R above.

So, finally, what *are* shear Alfvén waves? In the lowest-frequency range ($\omega \ll \Omega_c$), both the ions and the electrons are $\mathbf{E}_1 \times \mathbf{B}_0$ drifting, and the ions have a simple low-frequency polarization drift, which is small compared to their $\mathbf{E}_1 \times \mathbf{B}_0$ drift. The magnetic field lines themselves also 'move' with the same $\mathbf{v}_\perp = \mathbf{E}_1 \times \mathbf{B}_0/B^2$. As we learned in Chapter 8, this is described as the plasma being 'frozen' to the field lines, and we have encountered it before for low-frequency phenomena. In the present case, the field lines are twisting—moving circularly in the (x, y) plane—with different phases of rotation along z rather like the lines on a barber pole. Thus the name 'shear' (or sometimes 'torsional') Alfvén waves. The twisting of the field lines pulls the magnetic configuration away from its lowest energy state, and magnetic energy is stored in the 'twist'. The ions provide the inertia for this wave, causing the field-lines to continue to move circularly, rather than come to rest. The twisting motion of the shear Alfvén mode has $\nabla \cdot \mathbf{u}_1 = 0$, so there is no compression, no perturbed pressure, p_1, and hence no pressure effects on the waves.

Problem 18.4: For finite ω, the shear Alfvén wave exhibits Faraday rotation. In the spirit of Problem 17.4 calculate the Faraday rotation given $B(z)$ and $n_e(z)$ along the trajectory of the wave.

Problem 18.5: Derive ω_R and ω_L for an electron–positron plasma. You may start from R and L as defined for use in the cold-plasma dispersion relation (18.24).

Now we have completed our study of waves propagating parallel to $\mathbf{B}_0$, i.e. with $\mathbf{k} \parallel \mathbf{B}_0$. In Chapter 17, we had a curious asymmetry between the L-wave and the R-wave. The R-wave had two pass bands, with the whistler as the lower-frequency band, but the L-wave had only one pass band. Now that we have included ion motion, we find another pass band for the L-wave, below the *ion* cyclotron frequency. Remember that left-hand circular polarization resonates with *ion* Larmor motion. So the total picture for $\mathbf{k} \parallel \mathbf{B}_0, \theta = 0$ in a cold plasma is as follows:

R-Wave $(\mathbf{E}_1 \perp \mathbf{B}_0, \mathbf{k} \parallel \mathbf{B}_0)$ (2 pass bands)
$\omega > \omega_R$ high-frequency pass band
$\qquad$ $v_p \rightarrow c$ as $\omega \rightarrow \infty$
$\omega < \omega_c$ 'whistler' wave, becoming shear Alfvén R-wave at low frequency
$\qquad$ $v_p \rightarrow v_A$ as $\omega \rightarrow 0$

L-Wave $(\mathbf{E}_1 \perp \mathbf{B}_0, \mathbf{k} \parallel \mathbf{B}_0)$ (2 pass bands)
$\omega > \omega_L$ high-frequency pass band
$\qquad$ $v_p \rightarrow c$ as $\omega \rightarrow \infty$
$\omega < \Omega_c$ shear Alfvén L-wave
$\qquad$ $v_p \rightarrow v_A$ as $\omega \rightarrow 0$

Langmuir oscillation $(\mathbf{E}_1 \parallel \mathbf{B}_0, \mathbf{k} \parallel \mathbf{B}_0)$
$\omega = \omega_p$ zero group velocity Langmuir oscillation
$\qquad$ v_p undefined

For finite temperature $(\mathbf{E}_1 \parallel \mathbf{B}_0, \mathbf{k} \parallel \mathbf{B}_0)$
$\omega > \omega_p$ Langmuir wave
$\qquad$ $v_p \rightarrow \sqrt{3}v_{t,e}$ as $\omega \rightarrow \infty$
$\omega < \Omega_p$ ion sound wave
$\qquad$ $v_p \rightarrow C_s$ as $\omega \rightarrow 0$

The high-frequency pass bands of the R- and L-waves become simple vacuum light waves at very high frequency. In the high-frequency range, they have a difference in phase velocity that causes Faraday rotation of plane-polarized waves.

At all frequencies, the parallel $(\mathbf{k} \parallel \mathbf{B}_0)$ R- and L-waves are fully transverse, $(\mathbf{k} \perp \mathbf{E}_1)$, and so have neither flows nor electric fields along $\mathbf{B}_0$. They are completely compressionless, and do not give rise to any perturbed particle density nor charge density, and so they are not affected by finite pressure effects.

By contrast, the Langmuir oscillation is fully electrostatic—$\dot{\mathbf{B}}_1$ is exactly zero—so the physical effects come from differential ion and electron compression effects, giving rise to a perturbed charge directly, σ_1. In the cold-plasma limit, there is just one oscillation at ω_p. Warm-plasma effects give rise to the propagating Langmuir wave and the ion acoustic wave. If we take the warm-plasma calculation in the limit of $T_e, T_i \rightarrow 0$, the Langmuir wave collapses into the Langmuir oscillation at $\omega = \omega_p$, and the ion acoustic wave disappears into the horizontal axis at $\omega = 0$.

18.5 THE MAGNETOSONIC WAVE

The final class of waves we need to discuss are low-frequency waves propagating perpendicular to $\mathbf{B}_0$. These can be broken up into the two categories of extraordinary (X) and ordinary (O) waves. The X-wave has its electric field oriented everywhere perpendicular to $\mathbf{B}_0$, giving rise to 'extraordinary' phenomena due to the Lorentz force, while the O-wave is 'ordinary', with $\mathbf{E}_1 \parallel \mathbf{B}_0$. This categorization is academic, however, because the O-wave does not exist in this frequency range. It was cut off at $\omega = \omega_p$, and without the Lorentz force, the ion dynamics cannot bring it back at low frequency. Thus we have only the X-wave to analyze.

The X-wave is tricky when ion dynamics are included, however. There are interesting phenomena in the X-wave at a frequency of order $\sqrt{\omega_c \Omega_c}$ so we have to be careful in our ordering of ω relative to the two cyclotron frequencies if we want to recover all the important behavior. In particular, for this wave we must include the electron current in the direction of $\mathbf{E}_1$ (the polarization current) in the approximation $\omega \ll \omega_c$. For the shear Alfvén wave, we neglected the electron current along the diagonal of $\underline{\sigma}$ altogether relative to the ion current. At low frequencies this is valid (the ratio is M/m), but at higher frequencies we have to be careful. Looking ahead at equation (18.43) (where we have put in the electron polarization current), we can see that the ion polarization current's contribution along the diagonal for $\omega \gg \Omega_c$ is approximately $-\Omega_p^2/\omega^2$, while the electron polarization drift contribution for $\omega \ll \omega_c$ is ω_p^2/ω_c^2. Setting these to be comparable in magnitude gives $\omega^2 \sim \omega_c^2(\Omega_p^2/\omega_p^2)$, or $\omega \sim \sqrt{\omega_c \Omega_c}$. Thus, in this 'lower hybrid' frequency range (i.e. frequencies of order $\sqrt{\omega_c \Omega_c}$), we cannot neglect the electron polarization current as we did for the shear Alfvén wave. Other than adding this extra term, the determinant we need to solve is only changed from equations (18.31) and (18.32) by the fact that the X tensor, given in equation (18.14), now needs to be evaluated for $\theta = \pi/2$, giving $X = \hat{\mathbf{y}}\hat{\mathbf{y}} + \hat{\mathbf{z}}\hat{\mathbf{z}}$. We again need to evaluate only the upper left 2×2 part of $\underline{\sigma}$ because, for this geometry of wave propagation and electric field polarization, in the third row and column only the lower corner is potentially non-zero. The determinant is

$$\begin{vmatrix} 1 + \Omega_p^2/(\Omega_c^2 - \omega^2) + \omega_p^2/\omega_c^2 & i\Omega_p^2\omega/[\Omega_c(\Omega_c^2 - \omega^2)] \\ -i\Omega_p^2\omega/[\Omega_c(\Omega_c^2 - \omega^2)] & 1 - \tilde{n}^2 + \Omega_p^2/(\Omega_c^2 - \omega^2) + \omega_p^2/\omega_c^2 \end{vmatrix} = 0.$$

(18.43)

Equation (18.43) is very like equation (18.32). However, since $\tilde{n}^2$ does not appear twice here we will not obtain two waves, but only the one X-wave. We have retained enough electron dynamics to retrieve the ω_R and ω_L cutoffs of the X-wave in this determinant, but that is not the topic of interest here. Here we are interested in the low-frequency dynamics, and the lower-hybrid resonance which forms the bottom of the stop band between the extraordinary ion wave and the high-frequency X-wave. The dispersion relation can be written

$$\tilde{n}^2 \left(\frac{\Omega_c^2 - \omega^2 + \Omega_p^2}{\Omega_c^2 - \omega^2} + \frac{\omega_p^2}{\omega_c^2} \right) = \left(\frac{\Omega_c^2 - \omega^2 + \Omega_p^2}{\Omega_c^2 - \omega^2} + \frac{\omega_p^2}{\omega_c^2} \right)^2 - \left(\frac{\Omega_p^2 \omega}{\Omega_c (\Omega_c^2 - \omega^2)} \right)^2 .$$

(18.44)

Since a resonance is where $k \to \infty$, and $\tilde{n} \equiv ck/\omega$, the term in brackets on the left-hand side must go to zero at the resonance. (Nothing special happens in the cold-plasma limit at $\omega = \Omega_c$, but hot-plasma theory introduces *ion*-Bernstein waves at all harmonics of Ω_c.) The first term in brackets on the right-hand side will also be zero, but the second term should be well behaved. So, cross-multiplying, we obtain resonances where

$$\omega_p^2 (\Omega_c^2 - \omega^2) + \omega_c^2 (\Omega_c^2 - \omega^2 + \Omega_p^2) = 0 \qquad (18.45)$$

or

$$\omega^2 = \frac{\omega_p^2 \Omega_c^2 + \omega_c^2 \Omega_c^2 + \omega_c^2 \Omega_p^2}{\omega_p^2 + \omega_c^2} . \qquad (18.46)$$

The first and third terms in the numerator differ only due to their mass dependences. Specifically, the first term in the numerator is m/M times the third term, and thus is negligible compared to it. To conform to conventional notation, we multiply top and bottom by m/M, and obtain

$$\omega^2 = \omega_{lh}^2 = \frac{\omega_c \Omega_c (\Omega_c^2 + \Omega_p^2)}{\Omega_p^2 + \Omega_c \omega_c} \qquad (18.47)$$

or even more conventionally (taking $\Omega_p^2 \gg \Omega_c^2$, which is the case if $\omega_p^2 \approx \omega_c^2$):

$$\omega_{lh}^{-2} = \Omega_p^{-2} + (\Omega_c \omega_c)^{-1} . \qquad (18.48)$$

Here, ω_{lh} is called the 'lower hybrid' frequency. We plot the dispersion relation for the low-frequency X-wave in Figure 18.4 (for our usual case of $\omega_c^2 = 2\omega_p^2$). Note that in the limit of $\omega \to 0$, we recover the Alfvén wave dispersion relation:

$$\tilde{n}^2 = 1 + \Omega_p^2 / \Omega_c^2 + \omega_p^2 / \omega_c^2 \approx 1 + \Omega_p^2 / \Omega_c^2 \qquad (18.49)$$

which we can obtain from equation (18.44).

This low-frequency X-wave is generally referred to as the 'magnetosonic wave'. It is clear from our derivation why this is so. Unlike the shear or torsional Alfvén wave, this wave—sometimes also referred to as the 'compressional Alfvén wave'—does have a finite $\mathbf{k} \cdot \mathbf{u}_1$, and so 'compresses' the plasma. Again,

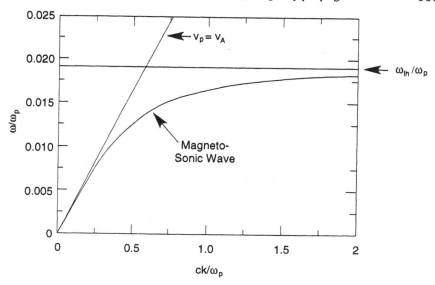

Figure 18.4. Dispersion relation for the magnetosonic wave in the cold plasma limit, with ω_c^2 chosen to equal $2\omega_p^2$.

since plasma is 'stuck' to the field lines at the lowest frequencies considered here ($\omega \ll \Omega_c$), the magnetic field is also compressed. The wave propagates across the magnetic field, alternately compressing and expanding it like the pressure in a sound wave, thus the name 'magnetosonic'. If the plasma has finite pressure, this wave is affected by 'warm' plasma terms, and its phase velocity increases. To understand this better, we will next calculate the dispersion relation for Alfvén waves in the very low-frequency limit, using the 'warm plasma' dielectric tensor. This will permit us to derive results for arbitrary angle of propagation, θ, so we will also get a look at the low-frequency limit of the shear Alfvén wave.

18.6 LOW-FREQUENCY ALFVÉN WAVES, FINITE T, ARBITRARY ANGLE OF PROPAGATION*

To study Alfvén waves in the low-frequency limit, ($\omega \ll \Omega_c$), including warm-plasma effects, we will follow the general prescription developed at the beginning of this Chapter, specialized for low frequencies. First we need to find the conductivity tensor $\underline{\sigma}$. To do this we recast equations (18.3), (18.7) and (18.8) as a set of linear equations for the fluid velocities, in a uniform format for solution by matrix methods:

$$E_{x1} = -i\left(\frac{\omega m}{q} - \frac{k^2\gamma T}{\omega q}\sin^2\theta\right)u_{x1} - B_0 u_{y1} + \left(\frac{ik^2}{\omega q}\gamma T\sin\theta\cos\theta\right)u_{z1}$$

$$E_{y1} = B_0 u_{x1} - \frac{i\omega m}{q}u_{y1} \tag{18.50}$$

$$E_{z1} = \left(\frac{ik^2}{\omega q}\gamma T\sin\theta\cos\theta\right)u_{x1} - i\left(\frac{\omega m}{q} - \frac{k^2\gamma T}{\omega q}\cos^2\theta\right)u_{z1}$$

where we have not yet specified species. To solve for $\underline{\sigma}$ we will need the determinant of the coefficients in equations (18.50), which we will denote Δ. We obtain

$$\Delta = i\left(\frac{\omega m}{q} - \frac{k^2\gamma T}{\omega q}\sin^2\theta\right)\left(\frac{\omega m}{q} - \frac{k^2\gamma T}{\omega q}\cos^2\theta\right)\left(\frac{\omega m}{q}\right)$$

$$- i\left(\frac{k^2}{\omega q}\gamma T\sin\theta\cos\theta\right)^2\frac{\omega m}{q} - iB_0^2\left(\frac{\omega m}{q} - \frac{k^2\gamma T}{\omega q}\cos^2\theta\right)$$

$$\simeq -iB_0^2\left(\frac{\omega m}{q} - \frac{k^2\gamma T}{\omega q}\cos^2\theta\right) \tag{18.51}$$

where the final simplification comes from assuming $\omega \ll \Omega_c$ and $kr_L \ll 1$ (the corresponding inequalities for electrons being even better satisfied). These approximations eliminate any differences between the ion and electron $\mathbf{E}_1 \times \mathbf{B}_0$ drifts, with the result that the associated currents precisely cancel. This makes the dispersion relation for the R and L waves identical, so that in the low-frequency limit Alfvén waves can be viewed as either linearly or circularly polarized at $\theta = 0$. As we will see, in the case where θ can take on any value, it will be advantageous to consider linear polarization.

Solving for $\mathbf{u}$, we obtain

$$u_{x1}\Delta = E_{x1}\left(\frac{-\omega m}{q}\right)\left(\frac{\omega m}{q} - \frac{k^2\gamma T}{\omega q}\cos^2\theta\right) - iE_{y1}B_0\left(\frac{\omega m}{q} - \frac{k^2\gamma T}{\omega q}\cos^2\theta\right)$$

$$- E_{z1}\left(\frac{\omega m}{q}\right)\left(\frac{k^2}{\omega q}\gamma T\sin\theta\cos\theta\right)$$

$$u_{y1}\Delta = iE_{x1}B_0\left(\frac{\omega m}{q} - \frac{k^2\gamma T}{\omega q}\cos^2\theta\right)$$

$$- E_{y1}\left[\left(\frac{\omega m}{q} - \frac{k^2\gamma T}{\omega q}\sin^2\theta\right)\left(\frac{\omega m}{q} - \frac{k^2\gamma T}{\omega q}cos^2\theta\right)\right.$$

$$\left. - \left(\frac{k^2\gamma T}{\omega q}\sin\theta\cos\theta\right)^2\right] + iE_{z1}B_0\left(\frac{k^2}{\omega q}\gamma T\sin\theta\cos\theta\right)$$

$$u_{z1}\Delta = -E_{x1}\left(\frac{\omega m}{q}\right)\left(\frac{k^2}{\omega q}\gamma T\sin\theta\cos\theta\right) - iE_{y1}B_0\left(\frac{k^2}{\omega q}\gamma T\sin\theta\cos\theta\right)$$

$$- E_{z1}\left[\left(\frac{\omega m}{q}\right)\left(\frac{\omega m}{q} - \frac{k^2\gamma T}{\omega q}\sin^2\theta\right) - B_0^2\right]. \tag{18.52}$$

Next we will use $\mathbf{j} = \Sigma nq\mathbf{u} = \underline{\sigma}\cdot\mathbf{E}$ and $\underline{\epsilon} = \epsilon_0 I + i\underline{\sigma}/\omega$ to obtain the dielectric tensor. However, at this point we will explicitly form the sum over species. From our previous work, we will assume that $\omega/k \sim v_A$. If we assume $\beta_i(\equiv n_iT_i/(B^2/2\mu_0)) \ll 1$, and $\beta_e(\equiv n_eT_e/(B^2/2\mu_0)) \gg m/M$, then we can deduce $v_{t,i} \ll v_A$ and $v_{t,e} \gg v_A$. We will also assume $\beta_e \ll 1$, resulting in $C_s \ll v_A$, $\beta_i \gg m/M$ and $\epsilon_\perp = \epsilon_0 + n_0M/B_0^2 \gg \epsilon_0$. These assumptions will simplify the mathematics.

Problem 18.6: Prove that if $\beta_i \ll 1$ and $\beta_e \gg m/M$, then $v_{t,i} \ll v_A$ and $v_{t,e} \gg v_A$.

We find the following components for the low-frequency warm-plasma dielectric tensor:

$$\epsilon_{xx} = \epsilon_0 + \sum_s \frac{n_0 m}{B_0^2} \simeq \epsilon_0 + \frac{n_{i0}M}{B_0^2} \simeq \frac{n_{i0}M}{B_0^2} \qquad \epsilon_{xy} = \epsilon_{yx} = 0$$

$$\epsilon_{xz} = \epsilon_{zx} = \sum_s \frac{n_0 m}{B_0^2}\frac{k^2\gamma T\sin\theta\cos\theta}{(\omega^2 m - k^2\gamma T\cos^2\theta)}$$

$$\simeq \frac{n_{i0}M}{B_0^2}\left(\frac{k^2\gamma_i T_i\sin\theta\cos\theta}{\omega^2 M}\right) - \frac{n_{e0}m}{B_0^2}\left(\frac{\sin\theta}{\cos\theta}\right) \simeq \frac{n_{i0}M}{B_0^2}\left(\frac{k^2\gamma_i v_{t,i}^2\sin\theta\cos\theta}{\omega^2}\right)$$

$$\epsilon_{yy} = \epsilon_0 + \sum_s \frac{n_0 m}{B_0^2}\frac{\omega^2 m - k^2\gamma T}{\omega^2 m - k^2\gamma T\cos^2\theta} \simeq \frac{n_{i0}M}{B_0^2}\left(1 - \frac{k^2\gamma_i v_{t,i}^2\sin^2\theta}{\omega^2}\right)$$

$$\epsilon_{yz} = -\epsilon_{zy} = -i\sum_s \frac{n_0 q}{\omega B_0}\frac{k^2\gamma T\sin\theta\cos\theta}{(\omega^2 m - k^2\gamma T\cos^2\theta)}$$

$$\simeq -i\frac{n_{e0}e}{\omega B_0}\left(\frac{\sin\theta}{\cos\theta}\right) - i\frac{n_{i0}e}{\omega B_0}\left(\frac{k^2\gamma_i v_{t,i}^2\sin\theta\cos\theta}{\omega^2}\right) - i\frac{n_{e0}e}{\omega B_0}\left(\frac{\omega^2\sin\theta}{k^2\gamma_e v_{t,e}^2\cos^3\theta}\right)$$

$$\epsilon_{zz} \simeq \epsilon_0 - \sum_s \frac{nq^2}{\omega^2 m - k^2\gamma T\cos^2\theta} = \epsilon_0 + \frac{n_e e^2}{k^2\gamma_e T_e\cos^2\theta} - \frac{n_i e^2}{\omega^2 M}$$

$$\simeq \frac{n_e e^2/m}{k^2\gamma_e v_{t,e}^2\cos^2\theta}. \tag{18.53}$$

To derive ϵ_{xx}, we used $M \gg m$. For $\epsilon_{xy} = \epsilon_{yx}$, the currents associated with the ion and electron $\mathbf{E}_1 \times \mathbf{B}_0$ drifts canceled. For $\epsilon_{xz} = \epsilon_{zx}$, we used $\beta_i \gg m/M$. For ϵ_{yy}, we used $M \gg m$. For ϵ_{zz}, we used $\beta_e \gg m/M$. In all cases, we have gone to first order in $(v_{t,i}/v_A)^2$ and in $(v_A/v_{t,e})^2$.

If we now put the wave equation, equation (18.30), into a dimensionless form

$$\| \omega^2 \mu_0 \underline{\epsilon}/k^2 - X \| = 0 \tag{18.54}$$

we can easily evaluate the order of each term in the matrix, assuming $\omega/k \sim v_A$, as follows:

xx term:
$O(\omega^2 \mu_0 \epsilon_{xx}/k^2 - \cos^2\theta) = 1$.

xz, zx terms:
$O(\omega^2 \mu_0 \epsilon_{xz}/k^2 + \sin\theta\cos\theta) = O(\omega^2 \mu_0 \epsilon_{zx}/k^2 + \sin\theta\cos\theta) = 1$ (with the plasma term of order $(v_{t,i}/v_A)^2$).

yy term:
$O(\omega^2 \mu_0 \epsilon_{yy}/k^2 - 1) = 1$ (with a correction term of order $(v_{t,i}/v_A)^2$).

yz, zy terms:
$O(\omega^2 \mu_0 \epsilon_{yz}/k^2) = O(\omega^2 \mu_0 \epsilon_{zy}/k^2) = \Omega_c/\omega$ (with correction terms of order $(\Omega_c/\omega)(v_{t,i}/v_A)^2$ and $(\Omega_c/\omega)(v_A/v_{t,e})^2$).

zz term:
$O(\omega^2 \mu_0 \epsilon_{zz}/k^2 - \sin^2\theta) = (\Omega_c/\omega)^2(v_A/C_s)^2$ (with a correction term of order unity).

Now we can evaluate the order of each of the terms that make up the determinant:

$O[(xx)(yy)(zz)] = (\Omega_c/\omega)^2(v_A/C_s)^2$ (with a correction term of order $(\Omega_c/\omega)^2$).

$O[-(xz)(yy)(zx)] = 1$ (with correction terms of order $(v_{t,i}/v_A)^2$).

$O[-(xx)(zy)(yz)] = (\Omega_c/\omega)^2$ (with correction terms of order $(\Omega_c/\omega)^2(v_{t,i}/v_A)^2$ and $(\Omega_c/\omega)^2(v_A/v_{t,e})^2$).

Clearly the second term in the determinant can be neglected compared with the others. Thus the dispersion relation we are seeking can be written

$$(\omega^2 \mu_0 \epsilon_{xx}/k^2 - \cos^2\theta)[(\omega^2 \mu_0 \epsilon_{yy}/k^2 - 1)(\omega^2 \mu_0 \epsilon_{zz}/k^2 - \sin^2\theta)$$
$$- (\omega^2 \mu_0/k^2)^2 \epsilon_{yz}\epsilon_{zy}] = 0. \tag{18.55}$$

Setting the first term in parentheses to zero gives us the linearly polarized shear Alfvén wave in the low-frequency limit:

$$\omega = kv_A\cos\theta = k_\parallel v_A. \qquad (18.56)$$

This also implies that the only non-zero electric field is in the x direction, giving rise to a divergence-free y-directed $\mathbf{E}_1 \times \mathbf{B}_0$ drift, since $\mathbf{k}$ points only in the x and z directions. Equation (18.56) is the generalization of the low-frequency shear Alfvén wave to arbitrary angle of propagation. At low frequency, the non-compressional nature of the wave is preserved in this linear polarization. With our approximation of $kr_L \ll 1$, no warm-plasma effects are observed.

Setting the term in square brackets in equation (18.55) to zero will allow finite E_y and E_z. The E_y field gives rise to compression through the x-directed $\mathbf{E}_1 \times \mathbf{B}_0$ drift. If there is any finite k_z, there will be a density gradient along $\mathbf{B}_0$. The nonzero E_z arises because $v_{t,e} \gg v_p \gg v_{t,i}$ so the ions cannot flow down the density gradient, and the electrons establish a Boltzmann distribution along $\mathbf{B}_0$, with a resulting E_z. Due to the large size of ϵ_{zz}, this electric field is much smaller than E_y. The dispersion relation for this compressional mode to lowest order is

$$\left[\frac{\omega^2\mu_0}{k^2}\frac{n_{i0}M}{B_0^2}\left(1 - \frac{k^2\gamma_i v_{t,i}^2\sin^2\theta}{\omega^2}\right) - 1\right]\left(\frac{\omega^2\mu_0}{k^2}\frac{n_{e0}e^2/m}{k^2\gamma_e v_{t,e}^2\cos^2\theta} - \sin^2\theta\right)$$
$$= \frac{\omega^2\mu_0^2}{k^4}\frac{n_{e0}^2 e^2}{B_0^2}\frac{\sin^2\theta}{\cos^2\theta}. \qquad (18.57)$$

If we neglect the final $\sin^2\theta$ on the left-hand side, since it is small by a factor of $(C_s/v_A)^2(\omega/\Omega_c)^2$, we can greatly simplify this expression:

$$\frac{\omega^2\mu_0}{k^2}\frac{n_{i0}M}{B_0^2}\left(1 - \frac{k^2\gamma_i v_{t,i}^2\sin^2\theta}{\omega^2}\right) - 1 \simeq \frac{\mu_0 n_{e0}m\gamma_e v_{t,e}^2\sin^2\theta}{B_0^2}$$

$$\frac{\omega^2}{k^2}\left(1 - \frac{k^2\gamma_i v_{t,i}^2\sin^2\theta}{\omega^2}\right) = \frac{B_0^2}{\mu_0 n_{i0}M} + \frac{m\gamma_e v_{t,e}^2\sin^2\theta}{M}$$

$$\frac{\omega^2}{k^2} = v_A^2 + \left(\gamma_i v_{t,i}^2 + \gamma_e C_s^2\right)\sin^2\theta.$$

Here $C_s^2 \equiv T_e/M$. When $\theta = 0$ there is no plasma compression, and finite temperature plays no role in this mode, as we observed previously when we examined the Alfvén waves propagating parallel to $\mathbf{B}_0$. Note that the phase velocity is independent of angle in a cold plasma, but increases as the wave points away from the magnetic field in a warm plasma. This is because bending the magnetic field lines does not also store energy in the plasma pressure the way compressing them does. It is even clearer now why this mode, the compressional Alfvén wave, is also called a 'magnetic + sound' = 'magnetosonic' wave. The proper values to use for γ_e and γ_i depend on subtleties like the precise angle of propagation of the wave relative to $\mathbf{B}_0$ (the slightest angle away from perpendicular allows the electrons to be isothermal, $\gamma_e = 1$, since the phase velocity is assumed to be small compared to the electron thermal speed). The wave frequency relative to the ion–ion and electron–electron collision frequencies also plays a role in determining γ_i and γ_e.

18.7 SLOW WAVES AND FAST WAVES

It is interesting to see how the shear Alfvén L- and R-waves transform into the magnetosonic wave as θ goes from 0 to $\pi/2$. The R-wave, with its resonance at ω_c, transforms into the magnetosonic wave, with its resonance at the lower hybrid frequency ω_{lh} (remember that resonances, unlike cutoffs, vary with θ); this branch is sometimes called the 'fast wave', except near resonance where $k \to \infty$ and its propagation slows. The L-wave, with its $\theta = 0$ resonance at Ω_c, maintains the non-compressional character of a shear Alfvén wave, and disappears into the $\omega = 0$ axis as θ goes from 0 to $\pi/2$, and so it is sometimes referred to as the 'slow wave'. In the lowest frequency range, $\omega \ll \Omega_c$, these waves are best classified as two linearly polarized modes, compressional and shear. In the cold-plasma limit, the compressional wave obeys $\omega = k v_A$ and the shear wave obeys $\omega = k_\parallel v_A$.

Problem 18.7: Show that in the limit $\omega \ll \Omega_c$, and $T = 0$, there are two linearly polarized Alfvén modes, one with $\omega = k v_A$ and the other with $\omega = k_\parallel v_A$. Show that the fluid flows give rise to plasma compression in the first case, while in the second case they do not. Work directly from the cold-plasma dielectric tensor, equation (18.16). Can you come up with a physical explanation for why the shear Alfvén wave propagates more slowly as $k_\parallel$ decreases?

In general, for a given wave frequency, ω, and a given set of plasma parameters, $(\omega_c, \omega_p, m/M)$, and angle of propagation, θ, there are at most two cold-plasma waves with distinct values of k. As θ varies, the values of k never coalesce, so a distinct branch with a smaller k can be identified as the 'fast' wave, while the branch with the larger k is always the 'slow' wave. At $\theta = 0$, these branches are the R- and L-waves, and at $\theta = \pi/2$ they are the O- and X-waves, which we have been studying. The identification of the R-, L-, O-

and X-waves as 'fast' and 'slow', and the continuous interconnection of the O- and X-waves at $\theta = \pi/2$ with the R- and L-waves at $\theta = 0$, as θ varies, depend both on ω and on the plasma parameters.

Finally, it is appropriate to remark on the practical importance of the low-frequency resonances we have just examined. We found earlier that the shear Alfvén L-wave has a resonance at $\omega = \Omega_c$, and we have now seen that the compressional Alfvén wave has a resonance at ω_{lh}. These resonances are especially important for fusion applications, because plasma heating is a necessary element in almost any fusion experiment. High-power microwave sources in the frequency range of tens to hundreds of gigahertz are difficult to produce, because the components of such sources generally need to be small but high-powered. On the other hand, high-power systems in the tens to hundreds of megahertz frequency range are readily available, and have been commercialized for communications applications. Thus heating at the Ω_c or ω_{lh} resonances is attractive for very practical reasons, even though the physics is more complex than resonance heating at ω_c or the upper hybrid frequency, ω_h. The propagation of the waves to the resonance region is a complex issue in real geometry, and the heating mechanism at the resonance can be delicately dependent on plasma parameters. This is an active area of theoretical and experimental plasma physics research.

Summarizing what we know about waves propagating in a cold plasma with $\mathbf{k} \perp \mathbf{B}_0$:

O-Mode $(\mathbf{E}_1 \perp \mathbf{B}_0, \mathbf{k} \parallel \mathbf{B}_0)$(1 pass band)
$\omega > \omega_p$ high-frequency pass band
$\qquad v_p \rightarrow c$ as $\omega \rightarrow \infty$

X-Mode $(\mathbf{E}_1 \perp \mathbf{B}_0, \mathbf{k} \parallel \mathbf{B}_0)$(3 pass bands)
$\omega > \omega_R$ high-pass region of X-mode
$\qquad v_p \rightarrow c$ as $\omega \rightarrow \infty$

$\omega < \omega_h$ mid-pass region of X-mode
$\omega > \omega_L$ $v_p = c$ at $\omega = \omega_p$

$\omega < \omega_{lh}$ compressional Alfvén wave, or magnetosonic wave
$\qquad v_p \rightarrow v_A$ as $\omega \rightarrow 0$

In the frequency range $\omega \ll \Omega_c$, the Alfvén wave divides conveniently into two linear polarizations which are preserved as the angle of propagation varies. The shear Alfvén wave propagates with $\omega = k_\parallel v_A$, and the compressional Alfvén wave propagates with $\omega = k v_A$, independent of angle. Warm-plasma effects accelerate the compressional wave when $\theta \neq 0$.

The cutoffs for $\mathbf{k} \parallel \mathbf{B}_0$ and $\mathbf{k} \perp \mathbf{B}_0$, in order from low frequency to high, are ω_L, ω_p, ω_R. Cutoffs do not vary with θ. The resonances at $\theta = 0$ or $\pi/2$ are Ω_c, ω_{lh}, ω_c, ω_h (and for the Langmuir oscillation, ω_p is both a cutoff and a resonance), but the resonances do vary with θ.

Problem 18.8: Show that the lower hybrid resonance of the X-wave is purely electrostatic.

Problem 18.9: Use COLDWAVE to explore the variations of the all the resonances with θ. Do both the case $\omega_c = 2\omega_p$ and also the case $\omega_c = \omega_p/2$. Plot each resonance frequency versus θ. (Be sure to explore the region near $\theta = 90°$ that connects to the ion-cyclotron resonance at $\theta = 0°$. This region is sometimes called the Alfvén resonance, since it is decoupled from the cyclotron motion.)

There are many interesting and practical aspects of waves in plasmas that are impossible to treat in an introductory text. For a more complete overview of this field, we recommend T H Stix (1992 *Waves in Plasmas* New York: American Institute of Physics).

UNIT 5

INSTABILITIES IN A FLUID PLASMA

In Unit 4, we considered a variety of small-amplitude wave-like perturbations of a fluid plasma. Since the plasma is in equilibrium without the perturbations, some kind of external disturbance is needed to excite one of these waves. For the simplest case, where the plasma is spatially uniform in all three directions, the possible perturbations can be Fourier analyzed into independent plane waves with frequencies ω and wave-vectors $\mathbf{k}$, connected by a 'dispersion relation'. For almost all cases considered in Unit 4, the dispersion relation $\omega \equiv \omega(\mathbf{k})$ gave only real values ω for real $\mathbf{k}$, implying that the amplitude of a wave with a single $\mathbf{k}$ value will remain constant in time.

For some fluid plasmas with strong spatial non-uniformities, there are modes of perturbation for which the dispersion relation gives *complex* ω values, some of which have positive imaginary parts, γ, and correspond to disturbances *growing in time* as $\exp(\gamma t)$. Such perturbations can be assumed to grow essentially *spontaneously*, since the infinitesimal disturbances needed to excite them will always be present. They correspond to 'instabilities' of the plasma.

In this Unit, we will consider three types of instability of a non-uniform fluid plasma, which differ from each other in the source of energy to drive the instability: the Rayleigh–Taylor or 'flute' instability is driven by gravitational potential energy or the thermal energy of a plasma confined by a curved magnetic field, the resistive tearing instability draws entirely upon magnetic energy of non-uniform $\mathbf{B}$ fields, and the resistive drift instability uses thermal energy associated with a pressure gradient to destabilize a wave traveling at the diamagnetic speed.

Chapter 19

The Rayleigh–Taylor and flute instabilities

In Chapter 9, we learned that magnetohydrodynamic plasma equilibria must be determined self-consistently, i.e. the presence of currents flowing in the plasma modifies the magnetic configuration in which the plasma rests. A static magnetohydrodynamic equilibrium (plasma fluid velocity $\mathbf{u} = 0$, hence electric field $\mathbf{E} = 0$) occurs when the plasma pressure gradients are balanced by magnetic ($\mathbf{j} \times \mathbf{B}$) forces.

However, even if a magnetohydrodynamic equilibrium exists in some particular case, the lack of *plasma stability* can lead to the spontaneous generation of $\mathbf{E}$ fields and associated plasma velocities $\mathbf{u}$. For if the plasma is disturbed slightly, its motion can deform the magnetic field in such a way as to produce magnetic forces that tend to amplify the original disturbance. This type of phenomenon is called a 'magnetohydrodynamic (MHD) instability'.

Because of the complexity of the magnetohydrodynamic equations, we are generally only able to treat analytically the case of *linear* stability, i.e. stability against *infinitesimally small* disturbances, in relatively simple geometries. For spatially uniform plasmas, infinitesimal perturbations will generally have a wave-like spatial structure. In such cases, as was discussed in Chapter 15, a plane wave with a single wave-vector $\mathbf{k}$ will generally have a single frequency ω. Thus, for a uniform plasma, this plane wave will be a 'normal mode'. For non-uniform plasmas, such as those considered in the present Chapter, it will be necessary to find the 'eigenfunctions', describing the spatial structure in the direction of non-uniformity, of the normal modes of perturbations, i.e. the modes which oscillate (or grow) with a single (possibly complex) frequency ω.

The theory of magnetohydrodynamic stability has been developed rigorously and applied analytically and numerically for a variety of plasmas using a variational principle, known as the 'MHD energy principle'. The MHD energy principle was formulated by I B Bernstein, E A Frieman, M D Kruskal and R M Kulsrud (1958 *Proc. R. Soc. (London)* A **744** 17). The energy principle lies outside the scope of this book, however. Rather, we will limit ourselves to a simple configuration for which the normal modes can be obtained explicitly, and we will then use general arguments to extend our results qualitatively to other configurations.

19.1 THE GRAVITATIONAL RAYLEIGH–TAYLOR INSTABILITY

Perhaps the most important MHD instability is the Rayleigh–Taylor (or 'gravitational') instability.   In ordinary hydrodynamics, a Rayleigh–Taylor instability arises when one attempts to support a heavy fluid on top of a light fluid: the interface becomes 'rippled', allowing the heavy fluid to fall through the light fluid. In plasmas, a Rayleigh–Taylor instability can occur when a dense plasma is supported against gravity by the pressure of a magnetic field.

The situation would not be of much interest or relevance in its own right, since actual gravitational forces are rarely of much importance in plasmas. However, in curved magnetic fields, the centrifugal force on the plasma due to particle motion along the curved field-lines acts like a 'gravitational' force. (Expressed differently, as we saw in Chapters 2 and 3, the electron and ion drifts due to magnetic-field gradient and curvature (∇B and curvature drifts) are similar to the particle drifts that arise from a gravitational field (gravitational drift).) For this reason, the analysis of the Rayleigh–Taylor instability provides useful insight as to the stability properties of plasmas in curved magnetic fields. Rayleigh–Taylor-like instabilities driven by actual field curvature are the most virulent type of MHD instability in non-uniform plasmas.

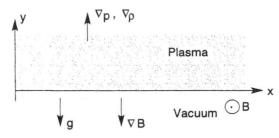

Figure 19.1. An equilibrium in which a plasma is supported against gravity by a magnetic field.

To treat the simplest case, we consider a plasma that is non-uniform in the y direction only and is immersed in a magnetic field in the z direction. To be specific, we suppose that the density gradient $\nabla\rho$ is in the y direction and that the gravitational field **g** is opposite to it, i.e. in the negative y direction. This corresponds to the case of a dense plasma supported against gravity by a magnetic field, as shown in Figure 19.1. Although Figure 19.1 suggests that there is a sharp boundary between the plasma and the vacuum, this is only one possible case and is used here primarily for illustration; the density 'profile' $\rho_0(y)$ may, in practice, be a smoothly increasing function of y. For the purposes of our present analysis, we will assume that the density has an exponential shape in y, i.e.

$$\rho_0(y) \propto \exp(y/s) \tag{19.1}$$

where s denotes the density-gradient 'scale length'. The plasma is bounded by conducting walls at $y = 0$ and $y = h$. This is illustrated in Figure 19.2.

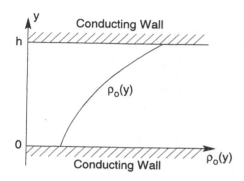

Figure 19.2. The profile of plasma mass density $\rho_0(y)$ between conducting walls at $y = 0, h$.

The equilibrium situation has $\mathbf{u}_0 = 0$, and p_0, B_0 and ρ_0 functions of y alone. (Here, the subscript '0' denotes an *equilibrium* quantity.) The pressure-balance condition (Chapter 9), including an additional gravitational force, requires that

$$\frac{\partial}{\partial y}\left(p_0 + \frac{B_0^2}{2\mu_0}\right) + \rho_0 g = 0 \tag{19.2}$$

where g is the magnitude of the gravitational acceleration, i.e. $\mathbf{g} = -g\hat{\mathbf{y}}$. From equation (19.2) and by referring to Figures 19.1 and 19.2, we see that the field strength B_0 must be larger in the 'vacuum' region than in the 'plasma' region, both to support the pressure gradient and to balance the gravitational force, implying that $\partial B_0/\partial y < 0$.

We now embark on a linearized small-amplitude stability analysis of this equilibrium. We suppose that the plasma equilibrium is perturbed in some way, so that all quantities (densities, fields, etc.) differ from their equilibrium values by infinitesimal but non-zero amounts. However, we neglect all products of two or more infinitesimal quantities (linearized analysis). Unlike the equilibrium, the perturbations will vary in time. For linearized equations, the three types of time dependence that can arise for a perturbation quantity ψ can all be expressed in the form $\psi \propto \exp(-i\omega t)$, where a real value of the 'frequency' ω will correspond to an oscillating perturbation, an ω value with a positive imaginary part will correspond to an exponentially growing perturbation (instability), and an ω value with a negative imaginary part will correspond to a damped perturbation.

For an equilibrium that is spatially uniform in some direction, say the x direction, the spatial eigenfunctions of the linearized system of equations will be sinusoidal in x, i.e. they can be expressed in the form $\psi \propto \exp(ikx)$, where k is the wave-number. If the equilibrium is not only uniform but also infinitely long in the x direction, then all real k values are allowed. Thus, stability problems of this kind are generally analyzed by assuming that perturbation quantities vary, for example, like

$$\psi \propto \hat{\psi}(y)\exp(ikx - i\omega t) \tag{19.3}$$

for some complex ω to be determined. If ω turns out to be imaginary (with a positive imaginary part), the system can be said to be 'unstable'.

Since the particular equilibrium under investigation here is uniform and infinitely long in the x direction, we adopt precisely the above form for all perturbation quantities. Moreover, the dynamics of the Rayleigh–Taylor instability is *purely two-dimensional*: there is no variation at all (equilibrium or perturbations) *along* the magnetic field (z direction). Thus, while a more general perturbation would have the form

$$\psi \propto \hat{\psi}(y)\exp(ik_x x + ik_z z - i\omega t) \tag{19.4}$$

we may take $k_z = 0$ in this particular problem. In all cases, the eigenfunctions $\hat{\psi}(y)$ are to be determined by finding solutions that correspond to normal modes, i.e. perturbations that have a single (complex) frequency ω.

Accordingly, we are to investigate perturbations of the equilibrium shown in Figures 19.1 and 19.2, in which all quantities (densities, pressures, fields and so on) are of the form

$$f = f_0(y) + f_1(y)\exp(ikx - i\omega t) \tag{19.5}$$

where the subscript '1' denotes small perturbations, and where we have suppressed the suffix in k_x, writing simply k for the x component of the **k**-vector. Such solutions represent wave-like perturbations of the plasma–vacuum interface, as illustrated in Figure 19.3. If the frequency ω is real, the wave-like perturbation travels in the x direction. The wave-like perturbation is created by the periodic upward and downward (i.e. in the y direction) motion of plasma elements: the plasma elements themselves do not need to move significantly in the x direction. (The situation is exactly analogous to propagating water waves, which are caused mainly by the upward and downward motion of the water, rather than by any lateral motion of the water, so long as the wavelength is short compared with the water depth.) If the ω value is purely imaginary, the wave-like perturbation grows in amplitude, but the wave pattern does not move in the x direction.

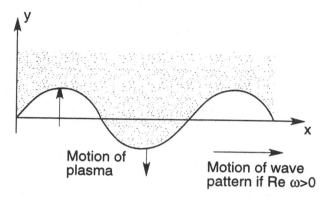

Figure 19.3. A wave-like perturbation of the plasma–vacuum interface shown in Figure 19.1.

An important simplification results from noting that, for this type of perturbation, the field lines *remain straight* even in the perturbed state. This is intuitively obvious from our general result that plasma elements initially on some given field line remain on the same field line in any 'ideal' (i.e. infinite-conductivity) magnetohydrodynamic motion. For, if plasma elements simply move up or down in a wave-like pattern that extends uniformly to infinity in the z direction, then there is no way in which the field lines can become bent. The same result may be obtained formally by examining each component of the linearized version of the usual combination of Faraday's law and the ideal MHD Ohm's law, namely

$$\frac{\partial \mathbf{B}_1}{\partial t} = \nabla \times (\mathbf{u}_1 \times \mathbf{B}_0) = (\mathbf{B}_0 \cdot \nabla)\mathbf{u}_1 - (\mathbf{u}_1 \cdot \nabla)\mathbf{B}_0 - \mathbf{B}_0(\nabla \cdot \mathbf{u}_1) \qquad (19.6)$$

where we have dropped a term in $\nabla \cdot \mathbf{B}_0$ from the right-hand side. (Note that, in this case, the plasma velocity $\mathbf{u}_0$ is zero in the equilibrium and has *only* a perturbed value, denoted by $\mathbf{u}_1$.) If we examine the x and y components of equation (19.6) we see that, in each case, all three terms on the right-hand side vanish identically. The first term on the right-hand side always vanishes since $\mathbf{B}_0 \cdot \nabla = B_0(\partial/\partial z) = 0$. The x and y components of the second and third terms vanish because $\mathbf{B}_0$ has only a z component. Thus, no components B_x or B_y can arise, and the field lines remain straight.

For straight field lines, the linearized perturbed fluid equation of motion is simply

$$\rho_0 \frac{\partial \mathbf{u}_1}{\partial t} = \rho_1 \mathbf{g} - \nabla\left(p_1 + \frac{B_0 B_{z1}}{\mu_0}\right). \qquad (19.7)$$

Here we have linearized the magnetic-pressure perturbation, i.e. $(B^2)_1 = 2B_0 B_{z1}$.

Both x and y components of this linearized equation of motion provide useful information. However, since we do not at present have much additional information about either p_1 or B_{z1}, it is convenient to eliminate these two quantities by taking the z component of the curl of the equation of motion, i.e. operating on both sides of equation (19.7) with the operator $\hat{z} \cdot \nabla \times$. This corresponds to taking $\partial/\partial x$ of the y component and subtracting $\partial/\partial y$ of the x component, eliminating the entire gradient term on the right-hand side, since the curl of a gradient vanishes. What remains is

$$-i\omega \left(ik\rho_0 u_y - \frac{\partial}{\partial y}(\rho_0 u_x) \right) = -ik\rho_1 g \tag{19.8}$$

where we have dropped the subscript '1' from the velocity components.

Let us, for the moment, suppose that the plasma motion is incompressible, i.e.

$$0 = \nabla \cdot \mathbf{u}_1 = iku_x + \frac{\partial u_y}{\partial y}$$
$$u_x = \frac{i}{k} \frac{\partial u_y}{\partial y}. \tag{19.9}$$

(This assumption replaces the adoption of an adiabatic or isothermal equation of state. Its validity is only approximate, but will be verified later after we have completed our calculation.) With this assumption, the density perturbation can be obtained from the continuity equation, as follows:

$$\frac{\partial \rho_1}{\partial t} + \mathbf{u}_1 \cdot \nabla \rho_0 = 0 \tag{19.10}$$

giving

$$-i\omega\rho_1 = -u_y \frac{\partial \rho_0}{\partial y} = -\frac{\rho_0 u_y}{s}$$
$$\rho_1 = \frac{\rho_0 u_y}{i\omega s} \tag{19.11}$$

the latter for our particular form of $\rho_0(y)$. Substituting from the continuity equation (19.11) for ρ_1, and the incompressibility relation (19.9) for u_x into the equation of motion (19.8), we obtain

$$\frac{1}{\rho_0} \frac{\partial}{\partial y} \left(\rho_0 \frac{\partial u_y}{\partial y} \right) - k^2 \left(1 + \frac{g}{s\omega^2} \right) u_y = 0. \tag{19.12}$$

This is a second-order differential equation for a single spatial variable, $u_y(y)$, as a function of an unknown scalar quantity ω, which can be solved once the appropriate boundary conditions are specified. Since the differential equation is homogeneous, it will be possible to satisfy *two* boundary conditions only for some discrete set of 'eigenvalues', which will determine the allowed set of values for ω. As we have already indicated in Figure 19.2, we suppose that the plasma is bounded above and below by conducting walls, taken to be at $y = 0$ and $y = h$. (A conducting wall cannot have any **E** field parallel to its surface, and thus the perpendicular component of the plasma velocity must also vanish. In this sense, the wall is a 'rigid' boundary in regard to fluid motion.) Thus, the boundary conditions are

$$u_y = 0 \qquad \text{at} \qquad y = 0, h. \qquad (19.13)$$

By design, we chose a form for $\rho_0(y)$ for which the differential equation can be solved analytically. By using an integrating factor $\exp(-y/2s)$, discrete solutions ('eigenfunctions') of equation (19.12) may be found of the form

$$u_y(y) = \sin\left(\frac{n\pi y}{h}\right) \exp\left(-\frac{y}{2s}\right) \qquad (19.14)$$

for all integer values of n. The 'eigenvalues', which for equation (19.12) will give the allowed values for the quantity $g/(s\omega^2)$, are given by the relation

$$k^2\left(1 + \frac{g}{s\omega^2}\right) = -\frac{1}{4s^2} - \frac{n^2\pi^2}{h^2}. \qquad (19.15)$$

Problem 19.1: Verify equation (19.15) by direct substitution of equation (19.14) into equation (19.12).

For the case where g and s are both positive, as they are for the configuration illustrated by Figures 19.1 and 19.2, we see immediately that there are no solutions unless ω^2 is negative, corresponding to ω being pure imaginary. Solving for ω, we obtain

$$\omega = \pm i \left(\frac{g}{s} \frac{h^2 k^2}{n^2\pi^2 + h^2 k^2 + h^2/4s^2}\right)^{1/2}. \qquad (19.16)$$

The solution for ω with a positive imaginary part represents an *exponentially growing perturbation*, i.e. an *instability*. The solution with a negative imaginary part represents a decaying perturbation that is of no interest.

The lowest mode that satisfies our boundary conditions has $n = 1$. This is the 'longest wavelength' mode in the y direction and is more rapidly growing than modes with $n > 1$. The fastest growing modes tend to be those with the shortest wavelengths in the x direction, however, i.e. large k values. Indeed, for all modes with wavelengths in the x direction that are shorter than both the density scale-length s and the geometric height of the plasma h, i.e. those with $hk \gg \pi$ and $ks \gg 1$, the growth rate γ (the imaginary part of ω for the growing $n = 1$ mode) is given by

$$\gamma = (g/s)^{1/2}. \tag{19.17}$$

The 'growth time' $\gamma^{-1} = (s/g)^{1/2}$ is just the time for 'free fall' over a distance s due to the gravitational acceleration g.

If the sign of either g or s is reversed, corresponding to the case of the plasma density increasing in the direction of the gravitational force $\mathbf{g}$, the solutions for ω are all real. This case is *stable*, and the eigenmodes are propagating wave-like disturbances.

19.2   ROLE OF INCOMPRESSIBILITY IN THE RAYLEIGH–TAYLOR INSTABILITY

In the discussion of the Rayleigh–Taylor instability given in the previous Section, we assumed the plasma flow to be incompressible, i.e.

$$\nabla \cdot \mathbf{u} = 0. \tag{19.18}$$

We will now verify the validity of this approximation.

Physically, incompressibility is a good approximation because the potential energy of the plasma in the gravitational field is usually insufficient to provide either the increase in thermal energy that occurs in compression of the plasma, or the increase in magnetic-field energy that occurs as the magnetic field is (necessarily) compressed along with the plasma. Let us consider this latter effect, since it is the more important in a plasma with a low β value ($p \ll B^2/2\mu_0$).

The geometrical configuration is the same as in the previous Section, as shown in Figure 19.1. As we saw before, the magnetic field lines remain straight, and no B_x or B_y components arise. The perturbation in the B_z component may be obtained by combining Faraday's and Ohm's laws in the usual manner:

$$\frac{\partial \mathbf{B}_1}{\partial t} = \nabla \times (\mathbf{u}_1 \times \mathbf{B}_0) = (\mathbf{B}_0 \cdot \nabla)\mathbf{u}_1 - (\mathbf{u}_1 \cdot \nabla)\mathbf{B}_0 - \mathbf{B}_0(\nabla \cdot \mathbf{u}_1). \tag{19.19}$$

Taking the z component gives

$$\frac{\partial B_{1z}}{\partial t} + (\mathbf{u}_1 \cdot \nabla)B_0 = -i\omega B_{z1} + u_y \frac{\partial B_0}{\partial y} = -B_0(\nabla \cdot \mathbf{u}_1). \tag{19.20}$$

This simply tells us that the magnetic field is convected and compressed along with the plasma. Henceforth, we again drop the subscript '1' from the velocity components.

To relate the energy needed to produce this amount of compression to the potential energy that is available, we consider one of the individual components of the equation of motion, say the x component:

$$\rho_0 \frac{\partial u_x}{\partial t} = -\frac{\partial}{\partial x}\left(p_1 + \frac{B_0 B_{z1}}{\mu_0}\right). \tag{19.21}$$

This equation balances the forces arising from compression of the plasma and magnetic field with the accelerating or decelerating flow that drives this compression. Recall that, in the previous Section, we conveniently eliminated both p_1 and B_{z1} by taking $\partial/\partial y$ of this x component of the equation of motion and subtracting $\partial/\partial x$ of the y component. The assumption of incompressibility allowed us to use this trick to avoid treating the effects of p_1 and B_{z1} directly. Here, we must retain these two quantities and use equation (19.21) in the form

$$-i\omega\rho_0 u_x \approx -ik(p_1 + B_0 B_{z1}/\mu_0). \tag{19.22}$$

We now use the adiabatic gas law to find the perturbation in the pressure, p_1. From $\mathrm{d}p/\mathrm{d}t = (\gamma p/\rho)\mathrm{d}\rho/\mathrm{d}t$, we obtain

$$\frac{\partial p_1}{\partial t} + (\mathbf{u}_1 \cdot \nabla)p_0 = -i\omega p_1 + u_y \frac{\partial p_0}{\partial y} = -\gamma p_0 (\nabla \cdot \mathbf{u}_1). \tag{19.23}$$

We may now substitute equation (19.20) for B_{z1} and equation (19.23) for p_1 into equation (19.22). After considerable rearranging of terms, equation (19.22) then becomes:

$$iku_x = \frac{k^2}{\omega^2}\left(\frac{\gamma p_0}{\rho_0} + \frac{B_0^2}{\rho_0 \mu_0}\right)\nabla \cdot \mathbf{u}_1 + \frac{k^2 u_y}{\omega^2 \rho_0}\frac{\partial}{\partial y}\left(p_0 + \frac{B_0^2}{2\mu_0}\right). \tag{19.24}$$

We may simplify the second term on the right-hand side of equation (19.24) by using the equilibrium relation (19.2). For the eigenfunctions and eigenvalues described by equations (19.14) and (19.16), respectively, it will then be seen that the second term on the right-hand side of equation (19.24) has the same order-of-magnitude as the term on the left-hand side. However, the coefficient of the first term on the right-hand side of equation (19.24) (for $p_0 \ll B_0^2/\mu_0$) is approximately $k^2 B_0^2/\omega^2 \rho_0 \mu_0 = k^2 v_A^2/\omega^2$. Thus, from equation (19.24), we obtain the order-of-magnitude relationship:

$$\frac{\nabla \cdot \mathbf{u}_1}{iku_x} \sim \frac{\omega^2}{k^2 v_A^2} \tag{19.25}$$

where $v_A = B_0/(\rho_0\mu_0)^{1/2}$ is the Alfvén speed. Noting that

$$\nabla \cdot \mathbf{u}_1 = iku_x + \frac{\partial u_y}{\partial y}$$

we see that equation (19.25) expresses the *neglected* quantity ($\nabla \cdot \mathbf{u}_1$) as a fraction of a *retained* quantity, in this case iku_x. This fraction clearly measures how good the incompressibility approximation is. If the fraction is very small, the two terms in $\nabla \cdot \mathbf{u}_1$ must almost cancel, i.e. to a good approximation we may assume that $\nabla \cdot \mathbf{u}_1 = 0$. Thus, the incompressibility approximation is valid whenever

$$|\omega^2| \ll k^2 v_A^2. \tag{19.26}$$

Conversely, a flow with finite compression, i.e. in which $\nabla \cdot \mathbf{u}_1$ is as large as either of its constituent parts, e.g. iku_x, would result in a higher-frequency wave, whose phase velocity perpendicular to the magnetic field would be comparable to the Alfvén speed. In the terminology of Chapter 18, this would be the 'compressional' Alfvén wave, or the 'magnetosonic' wave.

In the case of an instability, the magnitude of the growth rate will be a measure of the amount of potential energy available to drive the compression. For the Rayleigh–Taylor instability, which has a growth rate (see equation (19.16)) given by

$$|\omega^2| = |\gamma^2| = \frac{g}{s} \frac{h^2 k^2}{n^2 \pi^2 + h^2 k^2 + h^2/4s^2}$$

the incompressibility condition, equation (19.26), is valid whenever

$$gs \ll v_A^2 \left(\frac{n^2 \pi^2 s^2}{h^2} + k^2 s^2 + \frac{1}{4} \right). \tag{19.27}$$

Equation (19.27) is least easily satisfied for the longest wavelengths, i.e. the smallest values of n and ks. Even then, it is satisfied whenever

$$\rho g s \ll \rho v_A^2 \approx B^2/\mu_0 \tag{19.28}$$

i.e. whenever the gravitational potential energy is much less than the magnetic field energy. For shorter wavelengths, the approximation is even better.

This agrees with our initial intuitive observation: incompressibility should be a very good approximation whenever the potential energy that is available from the gravitational field is inadequate to provide the energy needed for compression of the magnetic field.

It must be emphasized that the approximate incompressibility of the plasma is the consequence, for the particularly simple geometry under consideration here, of the plasma's inability to compress the magnetic field due to the smallness of the available gravitational potential energy. Equivalently, the compressional Alfvén wave, or magnetosonic wave, cannot be excited: the instability arises, in effect, in the 'shear' Alfvén wave in the special case where $k_\parallel = 0$. For this wave, to minimize the effect of the magnetosonic branch, the perturbation quantities B_{z1} and p_1 are relatively small' (although non-zero) and are related to each other through the equation of motion, e.g. equation (19.21). They are also both described in terms of a combination of convection and a small amount of compression, as given in equations (19.20) and (19.23), respectively. Equation (19.20) expresses the conservation of magnetic flux in our assumed perfectly conducting plasma which is *exact*, in contrast to incompressibility, which is only *approximate*. We will see below that there are other geometries in which the Rayleigh–Taylor instability can be *driven* by expansion (i.e. negative compression) of the plasma. In these cases, the expansion is just that necessary to conserve magnetic flux in a plasma that is convecting into a region of reduced magnetic field. There is still little expansion/compression of the *magnetic field*, i.e. still little coupling to the magnetosonic wave.

19.3   PHYSICAL MECHANISMS OF THE RAYLEIGH–TAYLOR INSTABILITY

As a complement to the fluid picture developed above, the physical mechanism at work in the Rayleigh–Taylor instability can also be understood in terms of the *gravitational drifts* of ions and electrons.

From Chapter 2, we recall that an external force $\mathbf{F}$ (such as a gravitational force $\mathbf{F} = M\mathbf{g}$) perpendicular to a magnetic field $\mathbf{B}$ causes a charged particle (in particular, an ion with charge $+e$) to drift with a velocity

$$\mathbf{v}_d = \frac{\mathbf{F} \times \mathbf{B}}{eB^2} = \frac{M\mathbf{g} \times \mathbf{B}}{eB^2}. \qquad (19.29)$$

In our case (see Figure 19.1), this gravitational drift is in the negative-x direction, and has the magnitude $v_d = Mg/eB$. There is also an electron drift in the opposite direction, but this is much smaller because of the smaller electron mass.

Suppose a small wave-like ripple should develop on a 'plasma–vacuum interface', as shown in Figure 19.3. The gravitational drift of ions on the plasma side of the interface will cause positive charge to build up on one side of the ripple, as illustrated in Figure 19.4; the depletion of ions causes a negative charge to build up on the other side of the ripple. Due to this separation of charges, a small electric field $\mathbf{E}_1$ develops, and this electric field changes sign going from crest to trough of the perturbation, again as shown in Figure 19.4. It is apparent that the resulting $\mathbf{E}_1 \times \mathbf{B}_0$ drift is always upward in those regions where the interface has already moved upward, and downward in those regions where the interface has already moved downward. Thus the initial ripple grows larger, as a result of $\mathbf{E} \times \mathbf{B}$ drifts that are phased so as to amplify the initial perturbation.

The Rayleigh–Taylor instability can also be understood from an energy viewpoint, i.e. in terms of the lowering of the plasma's potential energy in the

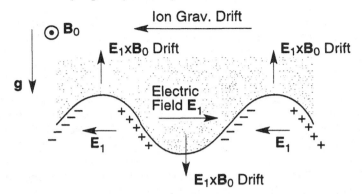

Figure 19.4. The mechanism of the Rayleigh–Taylor instability. The ion gravitational drift leads to charge separation on the plasma–vacuum interface, producing electric fields and $\mathbf{E} \times \mathbf{B}$ drifts that increase the amplitude of the perturbation.

gravitational field due to the growth of the instability. However, the change in potential energy is *second order* in the amplitude of the perturbations. For the simple case illustrated in Figure 19.3, this second-order change in the gravitational potential energy can be calculated explicitly. Suppose the plasma shown in Figure 19.3 has uniform density ρ and extends from the plasma–vacuum interface at $y = 0$ to some fixed upper boundary at $y = h$. Before the onset of the wave-like perturbation of the plasma's lower surface, the gravitational potential energy is simply

$$\int \rho g y \,\mathrm{d}x\mathrm{d}y = \rho g L h^2/2$$

where the integral over y has been taken from $y = 0$ to $y = h$ and the integral over x has been taken over some length L. Now add a sinusoidal perturbation of the plasma's lower surface, which may be assumed to take the shape $y = \xi \sin kx$, as shown in Figure 19.3. This perturbation satisfies the incompressibility constraint since the area of the plasma in the (x, y) plane is unchanged (see Figure 19.3). The plasma fills the region above this deformed lower boundary, still with uniform mass density, ρ. The gravitational potential energy is still $\int \rho g y \mathrm{d}x\mathrm{d}y$, but the integral over y must now be taken from $y = \xi \sin kx$ to $y = h$ and the integral over x may most conveniently be taken over the length of a full period, $L = 2\pi/k$; the gravitational potential energy becomes

$$\rho g \int (h^2 - \xi^2 \sin^2 kx)\mathrm{d}x/2 = \rho g L(h^2 - \xi^2/2)/2.$$

Thus the gravitational potential energy is *lowered* by an amount $\rho g L \xi^2/4$ (second order in the perturbation amplitude ξ) by the onset of the perturbation.

When potential energy can be lowered by such a perturbation, so that the energy released can go into kinetic energy of plasma motion, this can provide the energy necessary to drive an instability.

19.4 FLUTE INSTABILITY DUE TO FIELD CURVATURE

Real gravitational forces are generally totally negligible in laboratory plasma physics: plasmas are much too rarefied for gravity to compete with the strong pressure gradients and magnetic forces. The importance of the Rayleigh–Taylor instability lies in the close analogy between *gravitational drifts* and the ∇B and *curvature drifts* that arise in non-uniform magnetic fields.

In Chapter 3, we obtained the following expression for the combined ∇B and curvature drifts of an ion with charge e in a *vacuum* magnetic field (which should provide an adequate approximation to the actual magnetic field in a low-β plasma without strong field-aligned currents):

$$\mathbf{v_d} = \frac{M}{e} \left(\frac{v_\perp^2}{2} + v_\parallel^2 \right) \frac{\mathbf{R_c} \times \mathbf{B}}{R_c^2 B^2} \tag{19.30}$$

where $\mathbf{R_c}$ is the vector radius-of-curvature (a vector drawn from the local center-of-curvature to the field line, intersecting the field line normally and pointing away from the center-of-curvature). By comparing equation (19.30) with the expression for the gravitational drift given in equation (19.29), we see that the gravitational drift provides a good model for the drifts in a curved magnetic field, provided the vectors $\mathbf{g}$ and $\mathbf{R_c}$ are in the same direction, and the magnitude of g is defined by

$$g = \left(\frac{v_\perp^2}{2} + v_\parallel^2 \right) \frac{1}{R_c}. \tag{19.31}$$

If we average over a thermal distribution of particle velocities $v_\perp$ and $v_\parallel$, we can write $\langle v_\parallel^2 \rangle = \langle v_\perp^2/2 \rangle = T/M = p/\rho$, which shows that the magnitude of g should be related to the ion pressure p of a plasma in a curved magnetic field by

$$g = \frac{2p}{\rho R_c}. \tag{19.32}$$

Since the thermal velocities of electrons are much larger than those of ions, *both* particle species have comparable curvature and ∇B drifts, whereas the gravitational drift is important only for ions. The effect of this is that the total pressure, ions and electrons, should be used for p in equation (19.32).

Thus a plasma in a curved magnetic field can be viewed as having analogous particle drifts to a plasma in a gravitational field—and therefore a potential for charge build-up and unstable growth of perturbations. Since the Rayleigh–Taylor instability arises whenever the gravitational force is directed away from the region of maximum plasma density, the corresponding instability of a plasma in a curved field arises whenever the *radius-of-curvature vector is directed away from the region of maximum plasma pressure*, i.e. *whenever the plasma is confined by a magnetic field that is concave towards the plasma*.

The growth rate γ of the instability can be estimated by replacing g by $2p/\rho R_c$ in the expression for γ given in equation (19.17) and by equating the scale-length s to the pressure-gradient scale-length, i.e. $s^{-1} = |\nabla p|/p$. We obtain

$$\gamma \approx (2|\nabla p|/\rho R_c)^{1/2}. \tag{19.33}$$

We reiterate that this instability occurs only if the radius-of-curvature vector is directed away from the region of maximum plasma pressure, i.e. only if $\mathbf{R}_c$ and ∇p are oppositely directed.

This pressure-driven version of the Rayleigh–Taylor instability, which in the next Section we will learn to call the 'flute instability', is rapidly growing. The growth time (i.e. γ^{-1}) can be estimated by noting that $p/\rho \approx C_s^2$, where C_s is the sound speed in the plasma, giving

$$\gamma \sim C_s/(s R_c)^{1/2}. \tag{19.34}$$

Thus, the characteristic growth time is the time it takes a sound wave to traverse a distance that is the geometric mean of the pressure-gradient scale-length and the radius-of-curvature.

Problem 19.2: An annular cylindrical plasma, as shown in Figure 19.5, is *infinitely long* in the z direction. It has a purely azimuthal magnetic field $B_\theta(r)$, produced mainly by the current I in a central conductor at $r = 0$. The plasma pressure $p(r)$ falls to zero on both the inside of the annular cylinder, $r = r_1$, and on the outside, $r = r_2$, peaking somewhere between r_1 and r_2. Describe carefully by means of an illustration why you would expect this plasma to be subject to the Rayleigh–Taylor flute instability. For simplicity, you may suppose that $p \ll B_\theta^2/\mu_0$, so that the field is approximately the vacuum field, $B_\theta \propto r^{-1}$. Indicate in your illustration the particle drifts that give rise to this instability, and show the form that the unstable perturbations will take.

19.5 FLUTE INSTABILITY IN MAGNETIC MIRRORS

One configuration that is obviously susceptible to the pressure-driven version of the Rayleigh–Taylor instability is the *magnetic mirror*, in which a cylindrical plasma with an approximately axial magnetic field is constricted at both ends

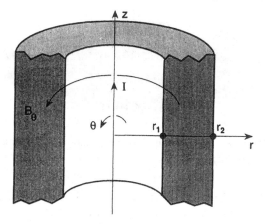

Figure 19.5. Annular cylindrical plasma, infinitely long in the z direction, has a purely azimuthal field $B_\theta(r)$ produced by the current I in a central conductor at $r = 0$. See Problem 19.2.

by regions of higher field strength, as shown in Figure 19.6. In this case, the curvature of the magnetic field is clearly concave toward the plasma in the central region. Approximating the plasma as a long cylinder, in which the pressure is considered to be a function of the radius r, the growth rate of the instability will be given by

$$\gamma \approx \left(-\frac{2p'(r)}{\rho R_c}\right)^{1/2} \tag{19.35}$$

where the prime denotes differentiation with respect to r.

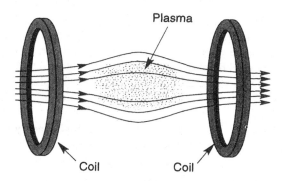

Figure 19.6. Plasma equilibrium in a 'magnetic mirror' configuration. Note that the magnetic field curvature is concave toward the plasma in the central region where the plasma pressure is largest.

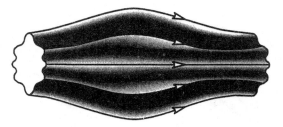

Figure 19.7. Flute-like perturbation of a magnetic-mirror plasma produced by the Rayleigh–Taylor instability.

This Rayleigh–Taylor instability will produce a rippling of the plasma surface in the azimuthal direction, and the ripples will extend uniformly along the length of the cylinder. The form of the perturbation is illustrated in Figure 19.7. The pressure-driven version of the Rayleigh–Taylor instability is called the 'flute instability' because of the resemblance of the perturbed surface of a quasi-cylindrical plasma such as this to a fluted Greek column.

Problem 19.3: Consider a cylindrical plasma with an axial field B_0 that is made flute-unstable by constricting the ends to form a magnetic-mirror configuration. Consider a flute instability with azimuthal mode number m, i.e. a mode in which the perturbations vary as $\exp(im\theta)$. Use the appropriate expression for the growth rate γ to show that the incompressibility approximation is valid whenever $\beta r / R_c \ll m^2$.

The basic energy reason for the flute instability in a curved magnetic field is very similar to the energy reason for the gravitational instability. Just as a fluid supported against gravity can lower its *potential* energy by perturbations that push downward in the direction of **g**, so the *thermal* energy of a flute-unstable plasma can be lowered by perturbations that push outward in the direction of R_c. That such perturbations produce a net expansion of the plasma, and thus release thermal energy, can be shown explicitly in the case of a low-β mirror-confined plasma, as follows.

We have already seen that there is not enough energy to compress the magnetic field, but in a low-β plasma an even stronger condition applies, namely that the magnetic field is essentially a vacuum field and remains approximately unchanged even when the plasma pushes outward across this field. However, the total magnetic flux contained within the plasma, i.e. the quantity $\int B\,dS$ integrated over the plasma cross section, must remain exactly constant, and so the only type of perturbation permitted is that illustrated in Figure 19.7, in which the surface of the plasma becomes rippled by 'filaments' of plasma moving outward, while compensating 'filaments of vacuum' move inward so as to conserve the total magnetic flux. The perturbations must be 'flutes', i.e. uniform along the entire length of the plasma, so as to avoid 'bending' the magnetic field, which would require additional energy. To the extent that special effects occur at the ends of the magnetic mirror which limit the allowed perturbations in this area (e.g. conducting plates could be placed at the ends of the mirror), then these effects will have a stabilizing influence; this topic is beyond the scope of the present discussion. Such effects are required, however, to explain the stability of the Earth's magnetosphere.

If the strength of the magnetic field decreases in the radially outward direction (as it does in the central region of the magnetic mirror, where the field gradient arises because the field is concave towards the plasma), the rippling perturbation of the plasma surface that conserves magnetic flux must result in a small (second-order) increase in the area of the plasma cross section. This is because the filaments of plasma which move outward are moving into a region of lower field, and so these cross section areas must increase, relative to the cross sectional areas of the 'vacuum filaments' of equal magnetic flux which move inward into a region of higher field. This increase in net cross sectional area results in a corresponding increase in plasma volume. The concave (towards the plasma) curvature of the magnetic field results in another (second-order) increase in the plasma volume, because the plasma filaments moving outward are lengthened slightly, relative to the vacuum filaments moving inward, which are shortened. For vacuum magnetic fields the gradient and curvature effects are always additive (corresponding to the ∇B and curvature drifts always being in the same direction). The increase in volume, due both to increased cross sectional area and increased field-line length, corresponds to *expansion* of the plasma and a lowering of its thermal energy, thereby making energy available for the unstable perturbation.

From a single-particle perspective, the drop in perpendicular and parallel particle kinetic energy associated with moving to lower B and higher R_c is invested in $j \cdot E$ work, as discussed in Section 3.5. This $j \cdot E$ work *drives* the instability to higher amplitudes.

Closer examination of the mirror field configuration, however, shows that there are regions of *favorable* curvature (convex toward the plasma) near the ends, in addition to the main region of unfavorable (concave) curvature at the center. In general, however, in axisymmetric mirror configurations the unfavorable curvature is dominant. However, *non-axisymmetric* mirror configurations have been designed for fusion applications in which current-carrying rods, first used by M C Ioffe (see Y B Gott *et al* 1962 *Nuclear Fusion Suppl.* p 1042), are placed outside the plasma, parallel to its axis, so as to create a B_θ field with favorable curvature, i.e. convex toward the plasma. In such cases, the combined curvature can be favorable everywhere; indeed the plasma is located in the region of an absolute minimum in the strength of the vacuum magnetic field.

The correct weighting of the favorable and unfavorable regions in a 'simple mirror' can be derived as follows. Take cylindrical coordinates (r, θ, z), with z along the axis of the mirror field. Overall stability will be determined by the average net angular drift of particles over their complete orbits along the mirror field from one end to the other. If the sign of this average net angular drift corresponds to field curvature that is concave toward the plasma, there will be a build-up of charges on the edges of the flutes which will give rise to azimuthal **E** fields that produce unstable growth in the amplitude of the flute-like perturbations. In the simple mirror geometry, the ∇B and curvature drifts are entirely azimuthal in direction, so that the angular drift speed of an individual particle is given by

$$r\frac{d\theta}{dt} = \frac{m}{eR_cB}\left(v_\parallel^2 + \frac{v_\perp^2}{2}\right). \tag{19.36}$$

In one complete orbit along the mirror field, the net angular drift of this particle is given by

$$\Delta\theta = \frac{m}{e}\int \frac{(v_\parallel^2 + v_\perp^2/2)d\ell}{rR_cBv_\parallel} \tag{19.37}$$

where we have written $dt = d\ell/v_\parallel$, where ℓ is a length coordinate along the field line. The particle's velocity components, $v_\parallel$ and $v_\perp$, change as the particle moves along the field line, i.e. are functions of ℓ in the integral in equation (19.37), and these changes will be such as to conserve the particle energy $W = mv^2/2$ and the magnetic moment $\mu = mv_\perp^2/2B$.

To obtain the net angular drift averaged over all particles in a filamentary 'flux tube', i.e. a thin tube which follows the magnetic field and contains a given number of magnetic field lines, it is simplest to return to equation (19.36) and average $d\theta/dt$ over the velocity-space distribution function, f, and over a flux tube containing a small amount of magnetic flux, $\Delta\Phi$. At any point along this flux tube, its cross sectional area is given by $\Delta A = \Delta\Phi/B$. The total number of particles contained in the flux tube is $\Delta N = \int n dA d\ell$. Dividing equation (19.36) by r, multiplying by the distribution function, f, and integrating both over velocity space and over the volume of the flux tube, we obtain

$$\Delta N\left\langle\frac{d\theta}{dt}\right\rangle = \frac{m}{e}\Delta\Phi\int\frac{v_\parallel^2 + v_\perp^2/2}{rR_cB^2}f d^3v d\ell \tag{19.38}$$

Equation (19.38) gives the average rate at which the entire population of particles of a given species in a given flux tube drifts azimuthally in θ to a neighboring flux tube. The direction of the drift is opposite for electrons and ions, as expected for gradient and curvature drifts, so the contributions from both species to the drift of *charge* are additive. Carrying out the local velocity-space integrals in equation (19.38) and omitting various positive multiplicative factors, we find that the average angular drift of charge is given by

$$\left\langle \frac{d\theta}{dt} \right\rangle \propto \int \frac{p_{\parallel} + p_{\perp}}{r R_c B^2} \, d\ell. \tag{19.39}$$

Adopting the convention that field lines that are concave toward the plasma have positive radii-of-curvature, while convex field lines have negative radii-of-curvature, the condition for flute instability is that the integral in equation (19.39) be positive, i.e. that the regions of positive R_c outweigh the regions of negative R_c. The point of inflection, which separates these two regions, has an infinite R_c and contributes negligibly to the integral in equation (19.39).

Unfortunately, the weighting due to $1/rB^2$ in the integrand of equation (19.39) is unfavorable, in that B is smallest where R_c is positive. In general, therefore, the simple mirror is unstable to flutes.

The flute instability in magnetic mirrors was analyzed first by M N Rosenbluth and C L Longmire (1957 *Ann. Phys.* **1** 120).

19.6   FLUTE INSTABILITY IN CLOSED FIELD LINE CONFIGURATIONS*

An even simpler stability criterion can be obtained for the case where the plasma pressure is isotropic, i.e. $p_{\parallel} = p_{\perp} = p$. In this case, the condition for equilibrium demands that the pressure be uniform along the field, i.e. $\mathbf{B} \cdot \nabla p = 0$. For a mirror-confined plasma, this condition can never be satisfied, or else the plasma would extend infinitely far along the field lines. However, it is possible to create certain 'closed field line' configurations in which each field line closes on itself, so that the plasma pressure can be exactly constant along field lines. An example of such a configuration is the 'toroidal quadrupole' shown in Figure 19.8. Here the plasma entirely surrounds the two coils that produce the magnetic field. (In a practical situation, the coils must either by supported and electrically fed by leads that pass through the plasma, or they must be superconducting and supported magnetically for the duration of the plasma pulse.) From Figure 19.8, it may be seen that some of the plasma lies on field lines that encircle only one coil, whereas the rest of the plasma lies on field lines that pass around both coils. On the inner sides of the plasma which face each single coil, the curvature of the magnetic field is convex toward the plasma, and this interface is stable to flutes. On the outer side of the plasma there are regions of both concave and convex curvature, and so the stability of this interface depends on the appropriate averaging of the favorable and unfavorable contributions, expressed in the form of a criterion that we will now derive. We will do this for isotropic pressure, and we will assume that the plasma (as in the simple mirror) is axisymmetric, i.e. that the configuration is symmetric to rotation in θ about the z axis in Figure 19.8. In such cases, the pressure can be brought outside the integral in equation (19.39), which then becomes

$$\left\langle \frac{d\theta}{dt} \right\rangle \propto \oint \frac{d\ell}{r R_c B^2} \tag{19.40}$$

with instability corresponding to the case where this integral is positive. (The integral is to be taken along the entire closed field line.)

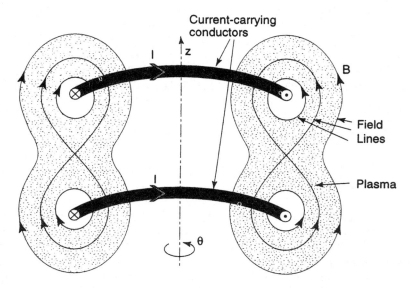

Figure 19.8. The toroidal quadrupole configuration. The plasma entirely surrounds the two current-carrying conductors that produce the magnetic field shown. The configuration is axisymmetric, i.e. symmetric to rotation in θ about the z axis.

In order to derive an even simpler stability criterion, consider two neighboring field lines in the same azimuthal plane (i.e. same θ value) of an axisymmetric configuration. Examine two infinitesimal elements of these neighboring field lines bounded by the same two radius-of-curvature vectors, as shown in Figure 19.9. The field strengths on these two elements are denoted B and $B + \delta B$ and the (infinitesimal) lengths of the elements are denoted $d\ell$ and $d\ell + \delta(d\ell)$. For a vacuum magnetic field, we can use Stokes' theorem to show that

$$\oint \mathbf{B} \cdot d\boldsymbol{\ell} = \int (\nabla \times \mathbf{B}) \cdot d\mathbf{S} = 0 \tag{19.41}$$

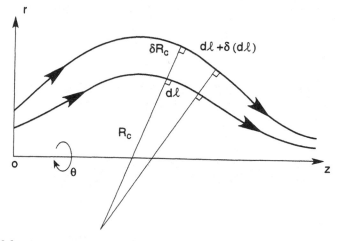

Figure 19.9. Two neighboring field lines in a mirror-like configuration with local radii-of-curvature R_c and $R_c + \delta R_c$. The configuration is axisymmetric, i.e. symmetric to rotation in θ about the z axis.

which, when applied to the infinitesimal closed contour shown in Figure 19.9, tells us that

$$B\,d\ell = (B + \delta B)[d\ell + \delta(d\ell)] \tag{19.42}$$

that is,

$$\frac{\delta B}{B} = -\frac{\delta(d\ell)}{d\ell} = -\frac{\delta R_c}{R_c}. \tag{19.43}$$

Here, in the final step, we have used simple geometry to relate $\delta(d\ell)$ to the perpendicular distance between the two field lines, δR_c. Since we want to apply equation (19.43) at all points along the two magnetic field lines, it is more convenient to define their separation not by the geometrical distance between them, δR_c, which varies along the field line, but by the magnetic flux between them, which is the same at all points along the field line. A convenient measure of this is the magnetic flux passing through an annular band obtained by rotating the element δR_c shown in Figure 19.9 by one revolution in θ about the axis. Specifically, this magnetic flux is

$$\delta\Phi = 2\pi r B \delta R_c \tag{19.44}$$

so that

$$\frac{\delta B}{B} = -\frac{\delta\Phi}{2\pi r R_c B}. \tag{19.45}$$

We can now write equation (19.40) as

$$\left\langle \frac{\mathrm{d}\theta}{\mathrm{d}t} \right\rangle \propto -\frac{1}{\delta\Phi} \oint \frac{\delta B}{B^2} \mathrm{d}\ell \qquad (19.46)$$

(omitting the factor 2π). Let us now consider the quantity $\oint \mathrm{d}\ell/B$ and its variation between neighboring field lines, such as those shown in Figure 19.9. We have

$$\delta \oint \frac{\mathrm{d}\ell}{B} = -\oint \frac{\delta B}{B^2} \mathrm{d}\ell + \oint \frac{\delta(\mathrm{d}\ell)}{B}. \qquad (19.47)$$

End-point variations do not need to be considered in this closed-loop integral. Using equation (19.43) to relate $\delta(\mathrm{d}\ell)$ to δB, we obtain

$$\delta \oint \frac{\mathrm{d}\ell}{B} = -2 \oint \frac{\delta B}{B^2} \mathrm{d}\ell. \qquad (19.48)$$

Thus, in the limit of vanishing differentials, equation (19.46) reduces to

$$\left\langle \frac{\mathrm{d}\theta}{\mathrm{d}t} \right\rangle \propto \frac{\mathrm{d}}{\mathrm{d}\Phi} \oint \frac{\mathrm{d}\ell}{B}. \qquad (19.49)$$

Thus, the condition for instability, which corresponds to a *positive* value of $\langle \mathrm{d}\theta/\mathrm{d}t \rangle$, is that the quantity $\oint \mathrm{d}\ell/B$ be *increasing outward*.

This is the simplest form of the stability condition for flute modes in closed field line configurations: in such configurations an isotropic-pressure plasma is stable or unstable depending on whether the quantity $\oint \mathrm{d}\ell/B$ decreases or increases away from the center of the plasma; the integral is to be taken completely around a closed field line. Quadrupole configurations, such as that shown in Figure 19.8, can be made flute-stable according to this criterion.

The criterion for instability derived here, namely that $\oint \mathrm{d}\ell/B$ must be increasing outward (i.e. in the direction opposite to that of the pressure-gradient vector), has applicability to a broader class of closed field line configurations than the axisymmetric (i.e. rotationally symmetric about the z axis) configurations discussed so far. Indeed, from the fluid viewpoint, this criterion could be obtained intuitively by considering whether a net expansion of the plasma occurs (thereby releasing kinetic energy) when flux tubes containing equal amounts of magnetic flux are interchanged. Consider a thin flux tube containing an amount $\delta\Phi$ of magnetic flux. At different points along this flux tube, its area δA is given by $\delta\Phi = B\delta A$, and so the volume of the entire flux tube is given by

$$\delta V = \oint \delta A \cdot \mathrm{d}\ell = \delta\Phi \oint \mathrm{d}\ell/B.$$

Now consider a 'rippling' perturbation of the plasma surface in which a plasma flux tube moves outward, while a 'vacuum flux tube' containing exactly the same amount of magnetic flux moves inward; we could call this the 'interchange' of these two flux tubes. If the quantity $\oint d\ell/B$ is increasing outward, the plasma flux tube will expand as it moves outward, while the vacuum flux tube will contract as it moves inward. The overall effect will be a net expansion of the plasma and a reduction in its thermal energy, which then provides the energy needed to drive the instability.

It is clear from this discussion that these unstable flute perturbations do not occur only at a plasma–vacuum interface, but can occur interior to the plasma, in which case a flux-tube containing *high-pressure* plasma is interchanged with a flux-tube containing *lower-pressure* plasma. In this case, instability will occur if the quantity $\oint d\ell/B$ is increasing in the direction of lower plasma pressure (the equivalent of 'outward' in the case of a plasma–vacuum interface). As in the case of the gravitational Rayleigh–Taylor instability, we note that the release of energy is again second order in a displacement vector $\boldsymbol{\xi}$, since it scales as $-(\boldsymbol{\xi} \cdot \nabla p)[\boldsymbol{\xi} \cdot \nabla(\oint d\ell/B)]$.

One possible method for stabilizing the flute instability would be to add some 'shear' to the magnetic field. A magnetic field is said to be 'sheared' if the direction of the field vector rotates as one moves from one constant-pressure surface to the next. For example, in the quadrupole configuration shown in Figure 19.8, the addition of a B_θ component (e.g. by placing a current-carrying conductor along the z axis) would provide magnetic shear. In a sheared magnetic field, the interchange of two flux tubes cannot occur without 'twisting' the field lines, thereby increasing the magnetic energy. In this case, the energy made available by plasma expansion must compete with the increase required in the magnetic energy; this will generally impose a lower limit on the plasma β value for the instability to be possible.

Even in configurations that are flute-stable according to the $\oint d\ell/B$ criterion, e.g. the quadrupole configuration shown in Figure 19.8, there are generally regions along each field line where the magnetic curvature is unfavorable, i.e. concave towards the plasma. Although the flute instabilities discussed in this Chapter all extend uniformly along the entire length of the field lines (hence their name 'flutes'), it is clearly possible, in principle, for instabilities with the same driving mechanism to arise that are localized to finite regions of unfavorable curvature. Such instabilities will cause the plasma to 'balloon' outward along these finite portions of field lines. Conservation of magnetic flux then requires that the field lines 'bend', and this bending will generally increase the magnetic energy. As in the case of a sheared field, the energy made available by plasma expansion must compete with this increase in magnetic energy, and the instabilities—called 'ballooning instabilities'—also arise only above some threshold β value.

19.7   FLUTE INSTABILITY OF THE PINCH

Another configuration that is obviously susceptible to flute instabilities is the cylindrical 'self-pinched plasma' (see Chapter 9). Here, the magnetic field is produced by an axial current flowing in the plasma. The magnetic field is azimuthal (B_θ) and its radius-of-curvature is simply the radial coordinate r. Clearly, the field-curvature is always unfavorable (concave towards the plasma). In this case, the flute perturbations are azimuthal, as shown in Figure 19.10. From the shape of the perturbed plasma, this instability is sometimes called the 'sausage instability'.

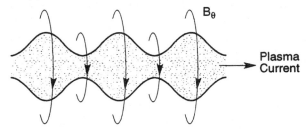

Figure 19.10. The flute, or 'sausage', instability of a self-pinched plasma.

The growth of the sausage instability is very rapid, since the radius-of-curvature of the field lines is effectively just the radius of the pinch column. From our previous formula, we estimate the growth rate to be

$$\gamma = \left(-\frac{2p'}{\rho r}\right)^{1/2} \tag{19.50}$$

where a prime denotes again a derivative with respect to the radial coordinate r.

19.8   MHD STABILITY OF THE TOKAMAK*

Before ending this Chapter, it may be useful to discuss very briefly the stability of the *tokamak* in the 'ideal MHD' model which has been used here to derive the Rayleigh–Taylor and flute instabilities. The tokamak configuration in the 'cylindrical approximation' was introduced in Chapter 9 and is illustrated in Figure 9.6. The actual tokamak geometry is toroidal, and the main magnetic field (corresponding to B_z in the cylindrical approximation) is toroidally directed, with the smaller magnetic field (B_θ in the cylindrical approximation) directed azimuthally the short way around the torus. The 'cylindrical tokamak' would clearly be vulnerable to flute instabilities, because the helical magnetic field produced by the combination of the B_z and B_θ fields has its curvature concave toward the plasma. On the other hand, the field also has considerable magnetic shear, which we have seen to be a stabilizing effect. In the actual toroidal geometry, however, it turns out that the effect of the additional curvature introduced by 'bending' the cylinder into a torus generally dominates over the effect of the helical curvature in regard to the stability of flute modes. For a torus with major radius R, the toroidal curvature is favorable (convex toward the plasma) on the small-R side of the plasma and unfavorable (concave toward the plasma) on the large-R side. When a calculation is carried out for the actual toroidal geometry, the 'weighting' of the small-R side turns out to be slightly greater than the weighting of the large-R side, so the net effect of the toroidal curvature is stabilizing. For the net favorable toroidal curvature to exceed the unfavorable helical curvature (in the case of a tokamak of approximately circular plasma cross section), it is necessary only that $q \equiv rB_z/RB_\theta > 1$. In practice, the q value in a tokamak typically rises from about unity at the center of the plasma ($r = 0$) to three or higher at the plasma edge ($r = a$). Thus, this condition is usually satisfied in the tokamak, so that pure flutes are stable.

Following any helical field line around the torus, it is clear that the field line will alternately lie on the small-R and large-R sides of the plasma. Thus, as in the case of the closed field line quadrupole configuration shown in Figure 19.8, there are regions of favorable curvature and regions of unfavorable curvature on each field line; as we saw, this gives rise to the possibility of 'ballooning' instabilities. Since the field line makes exactly q transits the long way around the torus for each transit the short way around, these regions of favorable and unfavorable curvature are a distance of order qR apart along a field line. For a displacement ξ, the energy released per unit volume by a flute-like instability is of order $p'\xi^2/R$, whereas the energy per unit volume needed to bend the magnetic field over a distance of order qR (field-line bending being unavoidable for a ballooning instability, as distinct from a pure flute) is of order $(B_z^2/2\mu_0)(\xi^2/q^2R^2)$. Thus, ballooning instabilities will arise in tokamaks only when $p'/R > B_z^2/2\mu_0q^2R^2$, i.e. only for $\beta > \beta_{crit} \approx a/q^2R$, where we have estimated $p' \sim p/a$. This result should be taken only as a rough order-of-magnitude estimate: in practical cases, tokamaks tend to be stable to ballooning instabilities up to β values in the range 3–6%.

The tokamak can, however, exhibit an entirely different type of MHD instability, which is driven by the magnetic energy that is available in the tokamak magnetic field, rather than by the thermal energy that is available from plasma expansion. This instability, which can arise also in the cylindrical tokamak approximation, is called the 'kink', and it takes the form of a helical displacement of the plasma cylinder. The instability arises whenever such a perturbation lowers the magnetic energy of the B_θ field—the field component that is produced by currents in the plasma itself. In practice, kink instabilities tend to arise only at relatively low q values. We will not pursue them further here, except to note that kinks are closely related (in regard to their source of energy) to a more slowly growing, but also more pervasive, instability that arises when resistivity is added to the MHD model. This instability, which occurs in many types of laboratory and naturally occurring plasmas in magnetic fields, is discussed in the next Chapter. For simplicity, we choose there to consider a simpler magnetic configuration (a plane current slab), which we find to be stable in the ideal MHD model.

The reader who is interested in pursuing further the topic of MHD instabilities in tokamaks is referred to J Wesson (1987 *Tokamaks* Oxford: Clarendon Press), or to R B White (1989 *Theory of Tokamak Plasmas* Amsterdam: North-Holland).

Chapter 20

The resistive tearing instability*

In the previous Chapter, we analyzed an important instability, the Rayleigh–Taylor (or flute) instability, which can arise in an ideal magnetohydrodynamic (MHD) plasma, i.e. a plasma in which the electrical resistivity is assumed to be zero and where the additional terms that enter in the 'generalized' Ohm's law are also negligible. For such cases, as we have seen, the plasma and the magnetic field are 'frozen' together. We found the flute instability to be very rapidly growing, with a growth time comparable to the time it takes a sound wave to travel a distance that is the geometric mean of the size of the plasma and the radius-of-curvature of the magnetic field. Since sound waves travel rapidly in high-temperature plasmas, such times are very short.

Even if a plasma is not subject to MHD instabilities, to be certain that it is *completely* stable we must also examine non-MHD perturbations that have the potential to grow at much slower rates. We have seen that the ideal MHD approximation breaks down for very long time-scales: eventually, the plasma will 'leak' across the magnetic field or, equivalently, the magnetic field will 'diffuse' into the plasma. Thus for slow plasma phenomena, non-zero resistivity must be included in the stability analysis, specifically in the plasma Ohm's law. Although resistivity often acts to damp out perturbations, there are important cases where resistivity is actually *destabilizing*. Indeed, there is an entirely new class of plasma instabilities, of which the most important is the 'resistive tearing instability' to be discussed here, that arise only in the presence of resistivity. The reason why resistivity can be destabilizing is that it frees the plasma from the constraint that it remain 'frozen' to the magnetic field, thereby allowing qualitatively different types of plasma perturbations. In particular, these 'resistive' perturbations can more effectively draw upon the magnetic energy generated by currents in the plasma itself, which is available to drive instabilities.

Intuitively, one might expect that 'resistive instabilities' would grow exceedingly slowly, specifically on time-scales comparable to the characteristic times for resistive diffusion of plasma across a magnetic field. If so, they would be of little interest, since most plasma equilibria are changing on such time-scales anyway, and the occurrence of a comparably slowly growing mode of instability might not make much difference in practice. However, some resistive instabilities, certainly including the tearing instabilities to be considered here, *grow much faster than this.* The reason is that the instability is able to take whatever form most efficiently releases the magnetic energy on which it feeds. Just as the flute instability was found to be driven by the non-uniformity of the plasma *pressure* (i.e. by the plasma *thermal* energy), the resistive tearing instability in its simplest form is driven by various types of non-uniformity of the *magnetic field* (i.e. by the ability of the *magnetic* energy to find a path to a lower energy state). It is this 'pent up' energy in the magnetic field, trying to find a way to relax to a lower energy state, that drives the tearing instability. The growth rate can be larger than one might intuitively expect because the resistive diffusion of plasma across the magnetic field occurs on a much shorter spatial scale-length than the plasma size and yet still can release significant amounts of magnetic energy; because of the shorter scale-length, the resistive diffusion can proceed quite quickly. The theory of resistive tearing instabilities, including their surprisingly large growth rates, was developed first in a paper by H P Furth, J Killeen and M N Rosenbluth (1963 *Phys. Fluids* **6** 459).

20.1 THE PLASMA CURRENT SLAB

We will analyze the resistive tearing instability for the simplest configuration in which it occurs, namely a 'plasma current slab'. Specifically, we consider an infinite plasma that contains a finite slab (or thick sheet) of current, directed parallel to the surface of the slab, namely

$$j_z = \begin{cases} j_{z0} & -a < x < a \\ 0 & |x| > a. \end{cases} \tag{20.1}$$

The plasma is uniform in the y and z directions. Solving Ampere's law, $\nabla \times \mathbf{B} = \mu_0 \mathbf{j}$, i.e. $dB_y/dx = \mu_0 j_z(x)$, we obtain

$$B_y(x) = \begin{cases} B'_{y0}x & -a < x < a \\ -B'_{y0}a & x < -a \\ B'_{y0}a & x > a \end{cases} \tag{20.2}$$

where $B'_{y0} = \mu_0 j_{z0}$. The functions $j_z(x)$ and $B_y(x)$ are sketched in Figure 20.1.

The magnetic field lines in the (x, y) plane are illustrated in Figure 20.2. Here, we have indicated the strength of the B_y field at different locations x by the density of field lines at x: the field is stronger where the field lines are

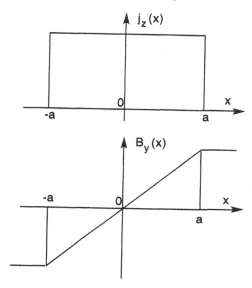

Figure 20.1. The 'plasma current sheet' equilibrium.

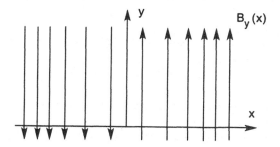

Figure 20.2. Magnetic field lines for the 'plasma current sheet' equilibrium. There is also a strong approximately uniform field B_z.

more crowded together. This plasma could possibly be subject to ideal MHD perturbations (although we will in fact find it to be ideal-MHD stable), but these would not change the basic configuration, since the magnetic flux through any plasma surface element in the (x, z) plane (i.e. the number of magnetic field lines of the B_y field crossing such a surface element) must remain fixed. However, the inclusion of plasma resistivity will allow the negative B_y field on the left of $x = 0$ to diffuse into the region of positive B_y field on the right of $x = 0$, thereby *annihilating* it. This 'annihilation' (or 'cancelling out') of the magnetic field will clearly occur most effectively in the vicinity of $x = 0$, which is where we will find the largest plasma flows in the resistive tearing instability.

It is easy to see that this annihilation of magnetic field is *energetically favored*. For example, if we consider the modification of $B_y(x)$ that would result from cancellation of the positive and negative B_y components in some small region $|x| < \delta$, it is clear that the magnetic energy, $\int (B_y^2/2)\mathrm{d}V$, would be reduced. The actual resistive tearing instability cannot annihilate magnetic field in such a neat and simple way: rather it involves wave-like perturbations of the entire plasma, well to the left and right of $x = 0$, which cause a wave-like 'break-up' of the magnetic topology near $x = 0$. Overall, however, the magnetic energy is lowered by this type of perturbation.

The current-slab configuration illustrated in Figure 20.2 may have an additional magnetic field in the z direction. If such a field is *not* present, the plasma can be in equilibrium only if it has a pressure $p(x)$ that varies in x in such a way as to balance the variation in magnetic pressure, i.e. to satisfy $p + B_y^2/2\mu_0 = $ constant. On the other hand, if a *large* B_z field is introduced, small variations of it will easily be sufficient to balance the pressure variations (assuming $p \ll B_z^2/2\mu_0$), and the functions $p(x)$ and $B_y(x)$ become essentially independent of each other. A strong B_z field will also play another role: as in the case of the Rayleigh–Taylor instability, it will constrain the plasma flow in the (x, y) plane to be incompressible, satisfying $\nabla \cdot \mathbf{u}_\perp = 0$. In the particular example analyzed in this Chapter, we will assume that a strong B_z field *is* in fact present. It should be emphasized that these assumptions are made largely for analytic simplicity. Resistive tearing instabilities can occur at a surface where $B_y(x) = 0$, if energetically favored, even with finite pressure in the equilibrium and a weak (or zero) B_z field, so that the flow becomes compressible.

Once the B_z field is introduced, it becomes clear that the configuration we are considering is simply one particular example of more general 'plane slab' configurations with field components $B_y(x)$ and $B_z(x)$. Due to the variation of B_y and/or B_z with x, the direction of the magnetic-field vector rotates as we move in the x direction. Such fields are said to be 'sheared'. For sheared fields, the directions of the y and z axes can be chosen so that the field points exactly in the z direction at some selected point, say $x = 0$. The configuration then looks exactly like the one illustrated in Figures 20.1 and 20.2 (with a B_z field added). Thus, in regard to tearing instabilities, our particular example is, in fact, representative of a wider class of sheared-field configurations.

Since these plane slab equilibria are stationary in time and uniform in the y and z directions, linearized perturbations of the equilibria may be Fourier analyzed into normal modes of the form

$$\psi_1(\mathbf{x}, t) = \hat{\psi}_1(x)\exp(\mathrm{i}k_y y + \mathrm{i}k_z z - \mathrm{i}\omega t)$$

where $\psi_1(\mathbf{x}, t)$ is any first-order perturbation quantity. For the particular equilibrium defined by equations (20.1) and (20.2), which has $B_y(x) = 0$ on the surface $x = 0$, the resistive tearing instabilities have $k_z = 0$, i.e. the **k**-vector is exactly perpendicular to **B** at $x = 0$, i.e. $\mathbf{k} \cdot \mathbf{B} = 0$ at the location of the tearing instability. When a B_z field is introduced, so that we have a sheared-field configuration with both $B_y(x)$ and $B_z(x)$, it is clear that all surfaces $x =$ constant are potential locations for tearing instabilities, for we can orient the y and z axes so that the magnetic field lies in the z direction on any particular surface, and we can then choose a **k**-vector in the y direction, subject of course to this being allowed by the boundary conditions. For a plane slab that extends infinitely far in the y and z directions, all k_y and k_z values are allowed: for a slab of finite extent, the allowed values are determined by the boundary conditions, which will then generally limit the surfaces on which tearing instabilities may be located. For the present analysis, we will limit ourselves to the equilibrium of equations (20.1) and (20.2) and perturbations with k_y only, i.e. $k_z = 0$. This simply puts the 'resonant surface' where $\mathbf{k} \cdot \mathbf{B} = 0$ at the location $x = 0$. At this resonant surface, a zeroth-order magnetic field line lies along a line of constant phase in the wave-like perturbation, making it very susceptible to the first-order magnetic perturbation. We will further simplify the notation by dropping the suffix 'y' from k_y, since this is the only non-zero component of the **k**-vector. Thus, for the remainder of this Chapter, the perturbations are assumed to vary as $\exp(iky)$.

20.2 IDEAL MHD STABILITY OF THE CURRENT SLAB

As we saw in our treatment of the Rayleigh–Taylor instability in Chapter 19, some general properties of the magnetic field perturbations can be obtained from the linearized version of the combination of Faraday's law and Ohm's law. First, we consider a perfectly conducting plasma, in which case we obtain

$$
\frac{\partial \mathbf{B}_1}{\partial t} = -\nabla \times \mathbf{E}_1 = \nabla \times (\mathbf{u}_1 \times \mathbf{B}_0)
$$
$$
= (\mathbf{B}_0 \cdot \nabla)\mathbf{u}_1 - (\mathbf{u}_1 \cdot \nabla)\mathbf{B}_0 - \mathbf{B}_0(\nabla \cdot \mathbf{u}_1) \tag{20.3}
$$

noting that the plasma velocity **u** is zero in the equilibrium and has only a perturbed value, denoted by $\mathbf{u}_1$. Unlike the geometry for the Rayleigh–Taylor instability, in the case considered here the field lines become *bent*, i.e. both a first-order B_x component and a first-order perturbed B_y component arise. Accordingly, the x and y components of equation (20.3) provide some non-trivial information, namely

$$
\frac{\partial B_{x1}}{\partial t} = ik B_{y0} u_{x1} \tag{20.4}
$$

and

$$\frac{\partial B_{y1}}{\partial t} = ik B_{y0} u_{y1} - u_{x1} \frac{\partial B_{y0}}{\partial x} - B_{y0}(\nabla \cdot \mathbf{u}_1)$$

$$= -u_{x1} \frac{\partial B_{y0}}{\partial x} - B_{y0} \frac{\partial u_{x1}}{\partial x}$$

$$= -\frac{\partial}{\partial x}(B_{y0} u_{x1}). \tag{20.5}$$

(Equation (20.5) could also have been derived by combining equation (20.4) with the requirement that $\nabla \cdot \dot{\mathbf{B}}_1 = 0$.) For a normal mode with frequency ω, i.e. with perturbation quantities varying as $\exp(-i\omega t)$ such as we are seeking, equation (20.4) can be written simply

$$\omega B_x = -k B_{y0} u_x \tag{20.6}$$

in which, here and henceforth, we drop the suffix '1' from the velocity and field components u_x and B_x, respectively, since these components are zero in the equilibrium. We note, in passing, that equation (20.6) requires that B_x vanish at any point where $B_{y0} = 0$, in particular at $x = 0$ in our example: otherwise, the velocity component u_x would be infinite.

Let us now turn to the linearized first-order perturbed equation of motion, namely

$$\rho_0 \frac{\partial \mathbf{u}_1}{\partial t} = -\nabla p_1 + (\mathbf{j} \times \mathbf{B})_1$$

$$= -\nabla \left(p_1 + \frac{\mathbf{B}_0 \cdot \mathbf{B}_1}{\mu_0} \right) + \frac{1}{\mu_0}[(\mathbf{B}_0 \cdot \nabla)\mathbf{B}_1 + (\mathbf{B}_1 \cdot \nabla)\mathbf{B}_0]. \tag{20.7}$$

We have used $\mathbf{j} = (\nabla \times \mathbf{B})/\mu_0$ and the vector identity for $(\nabla \times \mathbf{B}) \times \mathbf{B}$ (see Appendix D). We have also linearized the magnetic pressure perturbation, writing $(B^2)_1 = 2\mathbf{B}_0 \cdot \mathbf{B}_1$. Both x and y components of this linearized equation of motion provide useful information, namely

$$-i\omega\rho_0 u_x = -\frac{\partial}{\partial x} \left(p_1 + \frac{B_{z0}B_{z1} + B_{y0}B_{y1}}{\mu_0} \right) + \frac{1}{\mu_0} ik B_{y0} B_x \tag{20.8}$$

$$-i\omega\rho_0 u_y = -ik \left(p_1 + \frac{B_{z0}B_{z1} + B_{y0}B_{y1}}{\mu_0} \right) - \frac{1}{\mu_0} \left(B_{y0} \frac{\partial B_x}{\partial x} - B_x \frac{\partial B_{y0}}{\partial x} \right). \tag{20.9}$$

In the second-to-last term on the right-hand side in equation (20.9), we have used $\nabla \cdot \mathbf{B}_1 = 0$ to express B_{y1} in terms of B_x. Just as in our treatment of the Rayleigh–Taylor instability we take note of the fact that, beyond equations (20.8) and (20.9) themselves, we do not have any additional information on either

p_1 or B_{z1}. In principle, we could obtain p_1, for example from an adiabatic equation of state. Normally, we would obtain B_{z1} from the z component of equation (20.3), but this will involve the compressible, i.e. non-divergence-free, part of the plasma fluid velocity, which we expect to be very small. In the approximately incompressible case, B_{z1} is *determined from* either equation (20.8) or equation (20.9); when the value so determined is substituted into the z component of equation (20.3), this will yield a value for the compressible part of the fluid velocity, i.e. for $\nabla \cdot \mathbf{u}_1$, but this is a small quantity that does not enter anywhere else. Physically, the very small B_{z1} produces whatever modification of the almost-uniform magnetic pressure B_z^2 is needed to maintain force balance against small changes in pressure, in approximately incompressible flow. Both the Rayleigh–Taylor (gravitational) instability and the tearing instability are thus essentially independent of plasma pressure. The Rayleigh–Taylor instability is driven by the energy available from the inverted density gradient (relative to the gravitational force), and the tearing instability can be driven purely by the energy available from the sheared magnetic field. We will see, however, that this magnetic energy will become available to the plasma motion only through resistivity.

Just as we did in the case of the Rayleigh–Taylor instability, we can eliminate the two quantities p_1 and B_{z1} by forming the z component of the curl of the equation of motion. Specifically, we take $\partial/\partial x$ of the y component, equation (20.9), and subtract ik times the x component, equation (20.8). This produces

$$-i\omega \left(\frac{\partial}{\partial x}(\rho_0 u_y) - ik\rho_0 u_x \right) = \frac{1}{\mu_0} \left[\frac{\partial}{\partial x}\left(B_x \frac{\partial B_{y0}}{\partial x} - B_{y0}\frac{\partial B_x}{\partial x} \right) + k^2 B_{y0} B_x \right]$$

$$= -\frac{1}{\mu_0} \left\{ \frac{\partial}{\partial x}\left[B_{y0}^2 \frac{\partial}{\partial x}\left(\frac{B_x}{B_{y0}} \right) \right] - k^2 B_{y0} B_x \right\}.$$

(20.10)

At this point, our analysis is still valid for a general equilibrium $B_{y0}(x)$ and is not limited to the equilibrium defined by equation (20.2).

Let us, for the moment, suppose that the plasma motion is exactly incompressible, i.e.

$$0 = \nabla \cdot \mathbf{u}_1 = \frac{\partial u_x}{\partial x} + iku_y.$$

(20.11)

As in the case of the Rayleigh–Taylor instability, this assumption is only approximately valid. Its validity could be verified after we have completed our calculation, in exactly the same way as was done in Chapter 19. Specifically, we could relate $\nabla \cdot \mathbf{u}_1$ to the perturbation B_{z1} produced by compressing the strong magnetic field B_{z0} (see equation (19.20)). We could then relate the force arising from the gradient of the perturbed magnetic pressure $B_{z0}B_{z1}$ to either u_x

or u_y (see, for example, equation (19.22)). Comparing the magnitude of $\nabla \cdot \mathbf{u}_1$ with either of its constituent terms (in this case, $\partial u_x/\partial x$ or iku_y), we would find that $\nabla \cdot \mathbf{u}_1$ is smaller by a factor $\omega^2/k^2 v_A^2$, where v_A is the Alfvén speed, $B_0/(\rho_0\mu_0)^{1/2}$. As in the case of the Rayleigh–Taylor instability, the frequencies (or growth rates) of even the fastest modes that will be found here are much less than kv_A. Hence, again, the compressibility is negligible, and we may to a very good approximation write $\nabla \cdot \mathbf{u}_1 = 0$.

Using equation (20.11) to substitute for u_y in terms of u_x, the left-hand side of equation (20.10) can be expressed entirely in terms of u_x, so that this equation becomes

$$-\frac{\omega\mu_0}{k}\left[\frac{\partial}{\partial x}\left(\rho_0\frac{\partial u_x}{\partial x}\right) - k^2\rho_0 u_x\right] = \frac{\partial}{\partial x}\left[B_{y0}^2\frac{\partial}{\partial x}\left(\frac{B_x}{B_{y0}}\right)\right] - k^2 B_{y0}B_x. \quad (20.12)$$

For perfect conductivity, equation (20.6) is valid and can now be used in the form

$$B_x/B_{y0} = -ku_x/\omega \quad (20.13)$$

to express the right-hand side of equation (20.12) also in terms of u_x. Multiplying through by $-\omega k$ and rearranging terms slightly, equation (20.12) can now be written

$$\frac{\partial}{\partial x}\left((\rho_0\mu_0\omega^2 - k^2 B_{y0}^2)\frac{\partial u_x}{\partial x}\right) - k^2(\rho_0\mu_0\omega^2 - k^2 B_{y0}^2)u_x = 0. \quad (20.14)$$

Equation (20.14) is a homogeneous second-order differential equation for u_x. It describes ideal MHD waves in the configuration being considered. With proper boundary conditions, eigenmode solutions to the equation could be found. However, certain general properties of such waves can be determined by examining the quadratic (in u_x) expression formed by multiplying equation (20.14) by the complex conjugate u_x^* and integrating over all x, i.e. from $-\infty$ to $+\infty$. The result, after integrating by parts and noting that u_x must vanish as $x \to \pm\infty$, is

$$\int_{-\infty}^{\infty}(\rho_0\mu_0\omega^2 - k^2 B_{y0}^2)\left(\left|\frac{\partial u_x}{\partial x}\right|^2 + k^2|u_x|^2\right)dx = 0. \quad (20.15)$$

By examining equation (20.15), it is evident first that ω^2 must be real, so that ω must be either real or pure imaginary. It is further evident that our plasma must be completely stable (under this assumption of perfect conductivity), since an instability must correspond to a pure imaginary value of ω, i.e. $\omega = i\gamma$ for $\gamma > 0$, which would render the left-hand side of equation (20.15) negative-definite, so that it certainly could not be equal to zero.

The stable oscillatory waves that are described by equation (20.14) are the 'shear Alfvén waves' in the low-frequency limit introduced in Chapter 18. We note that their frequencies are typically $\omega \sim k_{\parallel} v_A$, where $k_{\parallel} = \mathbf{k} \cdot \hat{\mathbf{b}} = k_y B_{y0}/B_z$ is the component of the wave vector in the direction of the equilibrium magnetic field. The particular configuration under discussion here, however, has a special property, namely that B_{y0} depends on x. If the value of $\omega(\rho_0\mu_0)^{1/2}$ falls into the range of values assumed by $kB_{y0}(x)$, then equation (20.14) becomes *singular*, in that the coefficient of the second derivative can vanish. Since our main interest here is in instabilities, not stable oscillations, we need not explore this matter further. It is sufficient to note that the spectrum of possible solutions of equation (20.14) contains discrete modes with $\omega > k|B_{y0}|_{max}/(\rho_0\mu_0)^{1/2}$ and a continuum of modes with smaller ω values that are generally subject to strong damping at the location of the singularity due to effects not included in the ideal MHD analysis.

20.3 INCLUSION OF RESISTIVITY: THE TEARING INSTABILITY

Let us now introduce resistivity into the plasma Ohm's law, i.e.

$$\mathbf{E} + \mathbf{u} \times \mathbf{B} = \eta \mathbf{j}. \tag{20.16}$$

Combining this with Faraday's law and linearizing, the magnetic field perturbation is now given by

$$\frac{\partial \mathbf{B}_1}{\partial t} = -\nabla \times \mathbf{E}_1 = \nabla \times (\mathbf{u}_1 \times \mathbf{B}_0) - \eta \nabla \times \mathbf{j}_1 \tag{20.17}$$

where we have taken the resistivity to be uniform. Invoking Ampere's law for $\mathbf{j}_1$, i.e. $\mu_0 \mathbf{j}_1 = (\nabla \times \mathbf{B}_1)$, and making use of the identity $\nabla \times (\nabla \times \mathbf{B}_1) = \nabla(\nabla \cdot \mathbf{B}_1) - \nabla^2 \mathbf{B}_1 = -\nabla^2 \mathbf{B}_1$ (see Appendix D), we obtain

$$\frac{\partial \mathbf{B}_1}{\partial t} = \nabla \times (\mathbf{u}_1 \times \mathbf{B}_0) + \frac{\eta}{\mu_0} \nabla^2 \mathbf{B}_1. \tag{20.18}$$

Using the expansion of the first term on the right-hand side of equation (20.3), the x component of equation (20.18) becomes

$$\omega B_x = -k B_{y0} u_x + \frac{i\eta}{\mu_0} \frac{\partial^2 B_x}{\partial x^2}. \tag{20.19}$$

Here we have approximated $\nabla^2 \approx \partial^2/\partial x^2$ in anticipation of finding that resistivity is important only in a narrow region of x, within which B_x is relatively sharply varying. Equation (20.19) replaces equation (20.6) in the resistive case.

Several important conclusions follow from examination of equation (20.19). First, it is clear that our previous ideal MHD treatment corresponds to the case

$$\omega B_x \gg \frac{\eta}{\mu_0} \frac{\partial^2 B_x}{\partial x^2}. \tag{20.20}$$

For the shear Alfvén waves we have been studying, which generally have quite high frequencies ω compared to resistive diffusion rates, this relation is valid in all but the most resistive plasmas. However, we might legitimately inquire whether other modes of perturbation are possible, which have much lower frequencies or much shorter scale-lengths, such that the two terms in equation (20.20) are comparable.

For such modes, the resistive term in equation (20.19) must be retained. A second important conclusion now follows from equation (20.19): namely, it is no longer necessary for the first-order perturbation B_x to vanish at points where $B_{y0} = 0$, i.e. at $x = 0$ in the particular example shown in Figures 20.1 and 20.2. Physically, relaxing the constraint that $B_x = 0$ wherever $B_{y0} = 0$ allows the plasma much more freedom in finding ways to lower its magnetic energy, corresponding to more possibilities for unstable perturbations. A third conclusion that follows from examination of equation (20.19) is that the resistive term is likely to be most important in a narrow region around the point where $B_{y0} = 0$, i.e. around $x = 0$ in our particular example. We call this the 'resistive layer'. Since $\mathbf{k} \cdot \mathbf{B} = 0$ at $x = 0$, the perturbation can be considered to be 'resonant' at $x = 0$, such that the unperturbed magnetic field lies parallel to wave-fronts on this surface. The non-zero η in the resistive layer then allows the magnetic field lines to *connect across* the resonance, via a finite value of B_x.

Well away from the resistive layer, both to the left and to the right of $x = 0$ in the particular case illustrated in Figure 20.1, we expect the ideal MHD approximation to remain valid. Since the frequencies ω (or, more appropriately, the growth rates γ) are much less than Alfvén-wave frequencies, the perturbations in these ideal MHD regions will be given by equation (20.14) (or, equivalently, equation (20.12)) *but with the inertia terms omitted*. Since it is more convenient to describe the perturbations in the ideal MHD regions in terms of B_x rather than u_x, we prefer to work from equation (20.12), obtaining

$$\frac{\partial}{\partial x}\left[B_{y0}^2 \frac{\partial}{\partial x}\left(\frac{B_x}{B_{y0}} \right) \right] - k^2 B_{y0} B_x = 0. \tag{20.21}$$

This equation describes the perturbations in the 'outer-regions' well to the left and well to the right of the resistive layer around $x = 0$. As $x \to 0$ (either from the left or from the right), taking $B_y(x) \approx B_{y0}' x$, the possible forms for the solution B_x as $x \to 0$ are twofold: either $B_x \propto x$ or $B_x \approx$ constant.

Problem 20.1: Prove the last statement by searching for solutions of equation (20.21) with $B_x \propto x^\beta$ as $x \to 0$. You will find that the first term on the left-hand side of equation (20.21) tends to dominate as $x \to 0$, allowing only solutions with $\beta = 0$ or $\beta = 1$. Why is it safe to assume that there are only these two solutions as $x \to 0$?

It is possible to see, however, that the case $B_x \propto x$ as $x \to 0$, is excluded for solutions that are well behaved as $x \to \pm\infty$ for, if B_x/B_{y0} were finite as $x \to 0$, it would be permissible to multiply equation (20.21) by B_x^*/B_{y0} and integrate from $x = -\infty$ to $x = 0$. If we then integrate the first term by parts, noting that $B_{y0} = 0$ at $x = 0$, we obtain

$$B_x^* B_{y0} \frac{\partial}{\partial x} \left(\frac{B_x}{B_{y0}} \right) \Big|_{-\infty}^{0} - \int_{-\infty}^{0} B_{y0}^2 \left| \frac{\partial}{\partial x} \left(\frac{B_x}{B_{y0}} \right) \right|^2 dx - \int_{-\infty}^{0} k^2 |B_x|^2 dx = 0.$$
(20.22)

Since we want a localized solution in which $B_x \to 0$ as $x \to \infty$ (otherwise there would be infinite magnetic energy $|B_x|^2$, which is not a physically interesting case), the first term on the left-hand side vanishes in the $x \to -\infty$ limit. We then cannot allow $B_x \propto x$ as $x \to 0$, for this would make the first term on the left-hand side vanish in the $x \to 0$ limit also, and we would then have a negative-definite expression on the left, which is required to be zero.

Thus, we conclude that the only allowed solutions of equation (20.21) are such that B_x approaches some non-zero constant as $x \to 0$, either from the left or from the right. Such solutions would not be allowed by the ideal MHD constraint, i.e. equation (20.6), applied *exactly* at the point $x = 0$, *for this constraint requires B_x to be zero*. Such solutions *are* allowed in the resistive case, in which equation (20.19) replaces equation (20.6) in the vicinity of $x = 0$. It is just this non-vanishing of B_x at the point where $B_{y0} = 0$ that characterizes the 'resistive tearing' instability.

It is useful to think of the region around $x = 0$ as forming a 'boundary layer' between the two ideal MHD regions to the left and right of it. Moreover, it is possible to obtain some useful and revealing 'boundary conditions' by integrating various plasma equations over a thin box placed in this boundary layer, as illustrated in Figure 20.3. The box is supposed to have an infinitesimal width in x (but wider than the resistive layer) and a height in y that is finite but much less than the characteristic wavelength of the perturbation; its extent in z is arbitrary, since there are no variations in the z direction. Integrating the equation $\nabla \cdot \mathbf{B}_1 = 0$ over the volume of the box and applying Gauss' theorem, we find that B_x must be continuous across the boundary, i.e.

$$B_x(x \to 0+) = B_x(x \to 0-).$$
(20.23)

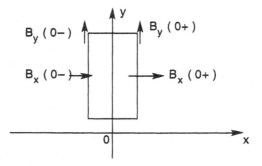

Figure 20.3. Thin box used for obtaining boundary conditions across the resistive layer.

From this we deduce that the value of B_x at each y-value may be *taken to be constant throughout the resistive layer around* $x = 0$. Similarly, integrating $\nabla \times \mathbf{B} = \mu_0 \mathbf{j}$ over the surface of the box in the (x, y)-plane and applying Stokes's theorem for the surface integral of a curl, we find that any discontinuity in B_{y1} must be associated with a first-order 'surface current' J_{z1} flowing in the boundary layer, i.e.

$$B_{y1}(x \to 0+) - B_{y1}(x \to 0-) = \mu_0 J_{z1}. \tag{20.24}$$

(By a 'surface current', we mean here a very large current density j_{z1} concentrated in a very narrow layer of thickness Δx, such that $J_{z1} = j_{z1} \Delta x$ = finite. A highly conducting plasma has the capability to carry such currents; in the limit of resistivity decreasing toward zero, the thickness of the current layer approaches zero, and a true surface current arises.) Equation (20.24) thus indicates that the y component of the field perturbation can be *discontinuous* across the boundary layer. From the divergence-free property of $\mathbf{B}_1$, i.e.

$$\frac{\partial B_x}{\partial x} + ik B_{y1} = 0 \tag{20.25}$$

we note that a discontinuity in B_{y1} implies a discontinuity in $\partial B_x / \partial x$. Thus, although B_x itself is continuous across the boundary layer, its gradient in x is not. Indeed, the quantity

$$\Delta' = \frac{1}{B_x} \left[\frac{\partial B_x}{\partial x} \right]_{x=0} = \frac{1}{B_x} \left(\left. \frac{\partial B_x}{\partial x} \right|_{x=0+} - \left. \frac{\partial B_x}{\partial x} \right|_{x=0-} \right) \tag{20.26}$$

where the notation $[\ ]_{x=0}$ is seen to denote the discontinuous *jump* across the boundary layer at $x = 0$; this is an important quantity, which will turn out to determine the stability of resistive tearing modes.

It is clear that the 'outer-region' solutions will completely determine the quantity Δ'. We could imagine integrating equation (20.21) for B_x in the region well to the left of $x = 0$, applying the appropriate boundary condition (usually $B_x \rightarrow 0$) at $x \rightarrow -\infty$ (or at some intervening boundary, e.g. a conducting wall). Indeed, we could carry out a numerical integration of equation (20.21), beginning at a conducting wall far to the left, where we would set $B_x = 0$ and would choose some arbitrary non-zero value for $\partial B_x / \partial x$, which simply measures the amplitude of our solution for B_x in this region. This solution will give some finite value of B_x at $x = 0$, approaching from the left, and this value provides an alternative measure of the amplitude of our solution. Thus, choosing some arbitrary value for the amplitude B_x at $x = 0$ (noting that the amplitude of a linear perturbation will always be arbitrary, within the confines of the linearized theory), the outer-region solution for B_x is then completely determined for $x < 0$, as is the value of $\partial B_x / \partial x$ at $x = 0-$. Similarly, the outer-region solution for $x > 0$, including the value of $\partial B_x / \partial x$ at $x = 0+$, is completely determined from the boundary condition at $x \rightarrow \infty$ (or at an intervening conducting wall) and the requirement that it have the same amplitude, B_x, at $x = 0$ as has the solution for the left outer-region. It follows that the quantity Δ' is completely determined by the outer-region solutions. Indeed, later in this Chapter, we will calculate Δ' explicitly for our 'plasma current slab' configuration, but first we will analyze the resistive layer in more detail, to determine how it can provide the localized, concentrated currents j_z needed to produce the sharp 'jump' in B_{y1} and in $\partial B_x / \partial x$.

Problem 20.2: Show that the first-order 'surface current density' J_{z1}, i.e. the perturbed volume current density integrated in x across the resistive layer at any point y, is related to the value of B_x at this point y by $\mu_0 J_{z1} = i\Delta' B_x / k$. For the particular choice of phase in which $B_x = \bar{B}_x \sin(ky)$, show that $\mu_0 J_{z1} = (\Delta' \bar{B}_x / k)\cos(ky)$.

20.4 THE RESISTIVE LAYER

It is not sufficient merely to obtain 'boundary conditions' that apply across the resistive layer: it is necessary to resolve the fine-scale structure of this layer in order to determine the growth rate of the resistive tearing mode. Within the layer, we may certainly take $B_{y0} = B'_{y0} x$, and we may also make use of our finding that the perturbed field component B_x is approximately constant throughout the layer; this constant part of B_x will be denoted $\bar{B}_x$.

Equation (20.19) then becomes

$$\omega \bar{B}_x + k B'_{y0} x u_x = \frac{i\eta}{\mu_0} \frac{\partial^2 B_x}{\partial x^2} \tag{20.27}$$

where the term on the right-hand side evidently involves the non-constant part of B_x. Plasma inertia must also be included in the resistive layer, since we will see that the plasma flow velocities tend to peak in this region, implying that the full form of equation (20.12) must be used. However equation (20.12) may be simplified by noting that the x derivatives will tend to dominate over the y derivatives (i.e. the k-factors) in the thin resistive layer. Thus, an approximate form of equation (20.12) will suffice, namely

$$-\omega \rho_0 \mu_0 \frac{\partial^2 u_x}{\partial x^2} = k B'_{y0} \frac{\partial}{\partial x} \left[x^2 \frac{\partial}{\partial x} \left(\frac{B_x}{x} \right) \right]$$

$$= k B'_{y0} \frac{\partial}{\partial x} \left(x \frac{\partial B_x}{\partial x} - B_x \right)$$

$$= k B'_{y0} x \frac{\partial^2 B_x}{\partial x^2}. \tag{20.28}$$

Substituting for $\partial^2 B_x / \partial x^2$ from equation (20.27), this becomes

$$\gamma \eta \rho_0 \frac{\partial^2 u_x}{\partial x^2} = k B'_{y0} x \left(i \gamma \bar{B}_x + k B'_{y0} x u_x \right) \tag{20.29}$$

where we have also written $\omega = i\gamma$ in anticipation of finding the result that the tearing instability is purely growing.

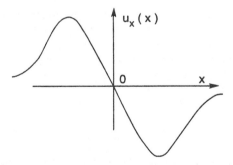

Figure 20.4. Typical form of the function $u_x(x)$ in the resistive layer.

Since $\bar{B}_x$ is constant, equation (20.29) can be solved to find an explicit solution for the x dependence of u_x. Unfortunately, the solution cannot be given in terms of analytic functions but must be evaluated partially numerically. However, it is apparent from equation (20.29) that u_x will decrease steadily away from the resistive layer. Specifically, $u_x \sim -i\gamma \bar{B}_x / k B'_{y0} x \sim 1/x$ as $x \to \infty$ and the term on the left-hand side of equation (20.29) becomes negligible. It is also apparent that the solution u_x will be odd in x; its actual form is sketched in Figure 20.4. This implicitly assumes that the solution of the *inhomogeneous* equation (20.29) is *unique*, i.e. that the homogeneous equation obtained by omitting the term including $x\bar{B}_x$ has no permitted solutions. This latter result can be established easily, by multiplying the homogeneous equation by u_x^* and integrating from $-\infty$ to $+\infty$, thereby obtaining a negative-definite expression that must equal zero for any solution with $u_x \to 0$ as $x \to \infty$. The *characteristic width* of the resistive layer can be determined simply by inspection of equation (20.29). Balancing the term on the left-hand side against the second term on the right-hand side gives a characteristic width

$$x \sim \delta = (\gamma \eta \rho_0)^{1/4}/(kB'_{y0})^{1/2}. \tag{20.30}$$

As we might have expected, the resistive layer becomes thinner as the resistivity η decreases.

To complete the solution and find the growth rate γ, it is necessary to obtain an explicit solution of equation (20.29) in some form. For this purpose, it is convenient to transform to scaled variables X and U, which are defined by

$$X \equiv x/\delta$$
$$U \equiv (\gamma \eta \rho_0)^{1/4}(kB'_{y0})^{1/2}u_x/i\gamma \bar{B}_x. \tag{20.31}$$

In terms of these variables, equation (20.29) becomes

$$\frac{\partial^2 U}{\partial X^2} = X(1 + XU). \tag{20.32}$$

The solution $U(X)$ will be an odd function of X and, as long as $\partial^2 U/\partial X^2$ is well-behaved as $X \to \pm\infty$, $U \to -X^{-1}$ as $X \to \pm\infty$. An explicit solution is obtainable in an integral form, namely

$$U(X) = -\frac{X}{2} \int_0^{\pi/2} \exp\left(-\frac{X^2}{2}\cos\theta\right) \sin^{1/2}\theta d\theta. \tag{20.33}$$

That this is the desired solution can be verified by direct substitution into equation (20.32), after first differentiating equation (20.33) twice to obtain

$$\frac{\partial^2 U}{\partial X^2} = \frac{X}{2} \int_0^{\pi/2} \exp\left(-\frac{X^2}{2}\cos\theta\right) \sin^{1/2}\theta(3\cos\theta - X^2\cos^2\theta)d\theta. \tag{20.34}$$

Using equations (20.33) and (20.34), we then obtain

$$\frac{\partial^2 U}{\partial X^2} - X^2 U = \frac{X}{2} \int_0^{\pi/2} \exp\left(-\frac{X^2}{2}\cos\theta\right) \sin^{1/2}\theta(3\cos\theta + X^2\sin^2\theta)d\theta$$
$$= X \int_0^{\pi/2} \frac{d}{d\theta}\left[\sin^{3/2}\theta\exp\left(-\frac{X^2}{2}\cos\theta\right)\right]d\theta$$
$$= X \tag{20.35}$$

which establishes that equation (20.33) is indeed a solution of equation (20.32). Examination of the asymptotic form of equation (20.33) for large X, where the dominant contribution to the integral arises from values of θ near $\pi/2$, shows that equation (20.33) also has the correct asymptotic form, namely $U \rightarrow -X^{-1}$. This may be seen by changing the integration variable in equation (20.33) from θ to $\varphi = \pi/2 - \theta$, so that the asymptotic form for large X is obtained by approximating the integrand as $\exp(-X^2 \sin\varphi/2) \approx \exp(-X^2\varphi/2)$.

The purpose of analyzing the resistive layer in such detail is to obtain the correct boundary conditions to be applied to the solutions to the left and right of the resistive layer. We have seen in the previous Section that these outer-region solutions are completely defined when the surface current J_{z1} or, equivalently, the jump in B_{y1} or in $\partial B_x/\partial x$, is specified. From our equations for the resistive layer, the jump in $\partial B_x/\partial x$ can easily be obtained, for example by integrating equation (20.27) across the layer:

$$\left[\frac{\partial B_x}{\partial x} \right]_{x=0} = \frac{\mu_0}{i\eta} \int \left(i\gamma \bar{B}_x + k B'_{y0} x u_x \right) \mathrm{d}x. \tag{20.36}$$

Reverting to our scaled variables X and U, and noting that the limits of integration in equation (20.36) may be taken as $\pm\infty$ on the scale of the resistive-layer width, i.e. the scale of X, we obtain

$$\frac{1}{\bar{B}_x} \left[\frac{\partial B_x}{\partial x} \right]_{x=0} = \frac{\gamma^{5/4} \rho_0^{1/4} \mu_0}{\eta^{3/4} (k B'_{y0})^{1/2}} \int_{-\infty}^{\infty} (1 + XU)\mathrm{d}X. \tag{20.37}$$

The integral on the right-hand side of equation (20.37) can be evaluated numerically, using equation (20.33) for $U(X)$. It is also possible to reduce the integral to a particularly simple form using both equation (20.32) and its solution, equation (20.33). To do this, we proceed as follows:

$$\int_{-\infty}^{\infty} (1 + XU)\mathrm{d}X = \int_{-\infty}^{\infty} \frac{1}{X} \frac{\partial^2 U}{\partial X^2} \mathrm{d}X$$

$$= \frac{1}{2} \int_{-\infty}^{\infty} \mathrm{d}X \int_{0}^{\pi/2} \exp(-\tfrac{1}{2}X^2 \cos\theta) \sin^{1/2}\theta (3\cos\theta - X^2\cos^2\theta)\mathrm{d}\theta$$

$$= \frac{1}{2} \int_{0}^{\pi/2} \sin^{1/2}\theta \mathrm{d}\theta \int_{-\infty}^{\infty} \exp(-\tfrac{1}{2}X^2 \cos\theta)(3\cos\theta - X^2\cos^2\theta)\mathrm{d}X$$

$$= \left(\frac{\pi}{2} \right)^{1/2} \int_{0}^{\pi/2} \sin^{1/2}\theta (3\cos^{1/2}\theta - \cos^{1/2}\theta)\mathrm{d}\theta$$

$$= (2\pi)^{1/2} \int_{0}^{\pi/2} \sin^{1/2}\theta \cos^{1/2}\theta \mathrm{d}\theta \approx 2.12 \tag{20.38}$$

where the final integral in equation (20.38) has been evaluated numerically. The left-hand side of equation (20.37) is equated to the quantity Δ' which was introduced in the previous Section and was defined in terms of the outer-region solutions. Equation (20.37) then gives an expression for the growth rate γ, namely

$$\gamma = 0.55\Delta'^{4/5}\eta^{3/5}(kB'_{y0})^{2/5}/\rho_0^{1/5}\mu_0^{4/5}. \tag{20.39}$$

Once the quantity Δ' has been calculated from the properties of the outer solutions, equation (20.39) gives the growth rate of the resistive tearing instability.

Examination of equation (20.39) reveals some important information about the magnitude of the growth rate γ. In many cases of interest, it is appropriate to think of the resistivity η as a small quantity, i.e. the plasma obeys 'ideal magnetohydrodynamics' to a good approximation. The introduction of non-zero resistivity into the equilibrium will produce diffusion of plasma relative to the magnetic field, but only at a very slow rate, proportional to η. The introduction of non-zero resistivity into the stability calculation has, however, produced unstable modes that grow at a much faster rate, proportional to $\eta^{3/5}$.

This argument can be made more quantitative by defining various characteristic times. Let us first introduce a characteristic macroscopic length scale a, e.g. the half-width of the current slab shown in Figure 20.1. One characteristic time is the inverse of the frequency ω_A of a shear Alfvén wave with wave-number k propagating in the y direction, i.e. almost perpendicular to the assumed very strong magnetic field B_z. This shear Alfvén wave has $\omega = k_\parallel v_A = (k_y B_{y0}/B_z)v_A$; evaluating B_{y0} at the edge of the current slab, this time τ_A is defined by

$$\tau_A^{-1} = \omega_A \approx (k_y B'_{y0}a/B_{z0})v_A$$
$$\approx k_y B'_{y0}a/(\rho_0\mu_0)^{1/2}. \tag{20.40}$$

A second characteristic time describes the diffusion of the field B_{y0} into the plasma due to non-zero resistivity; since the 'diffusion coefficient' for this process is η/μ_0 (see, for example, equation (20.18)), this time τ_R is defined by

$$\tau_R \approx a^2\mu_0/\eta. \tag{20.41}$$

Equation (20.39) may be rewritten in terms of τ_A and τ_R, giving

$$\gamma = \frac{0.55(\Delta'a)^{4/5}}{\tau_A^{2/5}\tau_R^{3/5}}. \tag{20.42}$$

Equation (20.42) shows that resistive tearing instabilities grow on time-scales that are *intermediate* between the very short MHD time-scale, τ_A, and the very long resistive time-scale, τ_R. Indeed the relevant time-scale is close to the geometric mean of τ_A and τ_R. Thus, resistive tearing instabilities grow much more slowly than ideal MHD instabilities (e.g. the flute instability, which has characteristic growth time $a/C_s \sim \beta^{-1/2}\tau_A$, i.e. approaching τ_A for finite β values), but much more rapidly than resistive diffusion of the equilibrium configuration. In this discussion, we have implicitly assumed that $\Delta'a$ is a quantity of order unity, which is generally valid, since Δ' is a characteristic of the macroscopic configuration. We will find that this assumption is confirmed, for example, in the case of the current slab analyzed in detail in the next Section.

20.5 THE OUTER MHD REGIONS

Until this point, we have not made use of any specific form for $B_{y0}(x)$ in the outer MHD regions, only that $B_{y0}(x) \approx B'_{y0}x$ in the narrow resistive layer around $x = 0$. Let us now find an explicit solution for the form of the perturbation in the outer-regions for the particular case of the plasma current slab illustrated in Figure 20.1 and specified in equations (20.1) and (20.2). To do this, we must solve equation (20.21) for the particular $B_{y0}(x)$ given in equation (20.2).

First consider the region $x > a$, where $B_y = B'_{yo}a = $ constant. Here, equation (20.21) becomes simply

$$\frac{\partial^2 B_x}{\partial x^2} - k^2 B_x = 0 \qquad (20.43)$$

whose only solution, vanishing as $x \to \infty$, is

$$B_x = C \exp(-kx) \qquad (20.44)$$

where C is an arbitrary constant that measures the amplitude of the perturbation.

Next, consider the region $0 < x < a$, where $B_{y0} = B'_{y0}x$. Here, equation (20.21) takes the form

$$\frac{\partial}{\partial x}\left[x^2 \frac{\partial}{\partial x}\left(\frac{B_x}{x}\right)\right] - k^2 x B_x = 0 \qquad (20.45)$$

but the derivative term can be expanded, i.e.

$$\frac{\partial}{\partial x}\left[x^2 \frac{\partial}{\partial x}\left(\frac{B_x}{x}\right)\right] = \frac{\partial}{\partial x}\left(x \frac{\partial B_x}{\partial x} - B_x\right)$$

$$= x \frac{\partial^2 B_x}{\partial x^2} \qquad (20.46)$$

so that equation (20.45) also becomes simply

$$\frac{\partial^2 B_x}{\partial x^2} - k^2 B_x = 0 \qquad (20.47)$$

whose general solution is

$$B_x = A \exp(kx) + B \exp(-kx) \tag{20.48}$$

where A and B are arbitrary constants.

The solutions in the two regions must be matched at $x = a$. The correct matching conditions are obtained from equation (20.21), which applies throughout the outer region, including both $x < a$ and $x > a$, and they are

$$B_x|_{x=a-} = B_x|_{x=a+}$$

$$\frac{\partial}{\partial x}\left(\frac{B_x}{B_{y0}}\right)\bigg|_{x=a-} = \frac{\partial}{\partial x}\left(\frac{B_x}{B_{y0}}\right)\bigg|_{x=a+} \tag{20.49}$$

the latter following from integrating equation (20.21) across an infinitesimal boundary layer at $x = a$. For the solutions given in equations (20.44) and (20.48), the two conditions expressed in equation (20.49) give

$$A \exp(ka) + B \exp(-ka) = C \exp(-ka)$$

$$A(ka - 1) \exp(ka) - B(ka + 1) \exp(-ka) = -Cka \exp(-ka). \tag{20.50}$$

From these relations, the constants A and B can easily be obtained in terms of C:

$$A = \frac{C}{2ka} \exp(-2ka) \qquad B = \frac{C}{2ka}(2ka - 1). \tag{20.51}$$

This completes the solution for $x > 0$. One arbitrary constant, in this case C, must remain, since the amplitude of a perturbation in linear theory is indeterminate.

Since the form of the equilibrium to the left of $x = 0$ is exactly the same as that to the right of $x = 0$, the solution for $x < 0$ can be obtained by simply substituting $-x$ for x in the solution which we have already found. Specifically, for $-a < x < 0$, the solution is

$$B_x = A \exp(-kx) + B \exp(kx) \tag{20.52}$$

and, for $x < -a$, it is

$$B_x = C \exp(kx) \tag{20.53}$$

with the same values of the constants A, B and C.

It is now possible to calculate the quantity Δ' defined in equation (20.26). Specifically,

$$\Delta' \equiv \frac{1}{B_x}\left[\frac{\partial B_x}{\partial x}\right]_{x=0} = \frac{2k(A - B)}{A + B}. \tag{20.54}$$

Substituting for A and B in terms of C using equation (20.51), we obtain

$$\Delta'a = \frac{2ka[\exp(-2ka) - 2ka + 1]}{\exp(-2ka) + 2ka - 1}. \qquad (20.55)$$

In Figure 20.5, we plot the function $\Delta'a$ versus ka. We see that Δ' is positive for small k (long wavelengths in the y direction) and negative for large k (short wavelengths in the y direction).

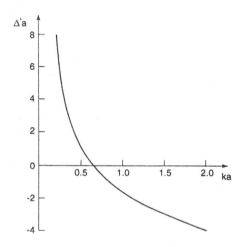

Figure 20.5. The function $\Delta'a$ describing tearing-mode stability plotted against ka.

Since $\Delta' > 0$ is the condition for the resistive tearing mode to be unstable, we have now shown that the 'plasma current slab' equilibrium is, in fact, unstable to all perturbations that are wave-like in the y direction and have sufficiently long wavelength.

As we saw at the beginning of this Chapter, the annihilation of magnetic field, by means of the cancellation of positive and negative B_y components in a small region $|x| < \delta$, is energetically favored, i.e. it lowers the magnetic energy. However, as we have now seen, a magnetic perturbation that is wave-like in the y direction is required to produce the B_x component at $x = 0$ needed for the negative B_y field to connect to, and thereby annihilate, the positive B_y field. This wave-like perturbation necessarily involves bending of the field lines, which requires energy in an amount that increases as the wavelength decreases. Thus it should not be surprising that the resistive tearing mode is unstable only for sufficiently long wavelengths, i.e. wavelengths for which the energy released by field annihilation exceeds that needed for field bending.

We also saw earlier in this Chapter that a general sheared-field plasma slab configuration with both $B_y(x)$ and $B_z(x)$ could be susceptible to resistive tearing instabilities at many locations x, depending on which modes of perturbation are allowed by the boundary conditions. This has an important application to the 'cylindrical tokamak', which is a model configuration with a strong, approximately uniform axial field B_z and a weaker azimuthal field $B_\theta(r)$. The normal modes of perturbation of an infinitely long cylindrical plasma are of the form $\exp(im\theta + ik_z z)$, where m must be an integer but k_z can have any value. However, in the tokamak case, the cylinder is an approximation to a 'straightened out' torus and should therefore be considered to have finite length $2\pi R$, where R is the major (larger) radius of the torus. Moreover, 'periodic boundary conditions' should be applied at the ends of the now-finite-length cylinder, so that we must take $k_z = -n/R$ where n is an integer (the choice of a negative sign being simply for convenience, as we will soon see, since both positive and negative integers are allowed). Such a perturbation can be 'resonant', in the sense that $\mathbf{k} \cdot \mathbf{B} \equiv mB_\theta/r - nB_z/R$ will vanish at a radius r where $q(r) \equiv rB_z/[RB_\theta(r)] = m/n$. This is the equivalent of the resonant surface in our 'slab' calculation at $x = 0$, where $\mathbf{k} \cdot \mathbf{B} \equiv k_y B_{y0} = 0$. For a tokamak with a current distribution $j_z(r)$ that peaks at $r = 0$ and decreases to zero at the plasma edge, $r = a$, the function $q(r)$ will increase monotonically from a minimum value at $r = 0$ to a maximum value at $r = a$. Clearly, infinitely many rational numbers m/n can be 'fitted in' between $q(0)$ and $q(a)$. However, since we have seen that only large wavelengths tend to be unstable to resistive tearing modes, only 'low-order' rationals, i.e. those for which m and n are small integers, are of interest. By far the most unstable mode in a tokamak is that with $m = n = 1$, and the nonlinear evolution of this mode tends to strongly flatten the plasma profiles inside of the resonant surface; however, this mode can arise only when $q(0) < 1$. The mode with $m = 2, n = 1$ is also dangerous, since it can occur whenever $q(0) \approx 1$ and $q(a) > 2$. However, the stability of any particular mode is determined not just by the presence of the associated resonant surface, but also by the form of the plasma current distribution; in many cases, all modes can be stable.

Problem 20.3 Suppose that rigid conducting walls are introduced into our plasma current slab at $x = \pm b$ (with $b > a$). Find the generalization of equation (20.55) for $\Delta'a$ in this case. Do you expect the plasma to be more, or less, stable? Is this expectation confirmed by your expression for $\Delta'a$?

20.6 MAGNETIC ISLANDS

The resistive tearing instability produces a change in the topology of the magnetic field. The magnetic configuration of the plasma current slab before onset of the instability is illustrated in Figure 20.2. The field lines are straight and, assuming that a strong approximately uniform B_z component is added to the B_y component shown in Figure 20.2, lie in flat surfaces parallel to the (y, z) plane. The direction of the B_y component reverses across $x = 0$. After onset of the instability, the magnetic configuration is deformed, and the field lines now lie on modified surfaces, which are still uniform in the z direction (since there is no variation of the perturbation in the z direction) but which intersect the (x, y) plane in curved lines determined by the relations $\mathrm{d}x/\mathrm{d}l = B_x/B$ and $\mathrm{d}y/\mathrm{d}l = B_y/B$. In effect, all of the deformed field lines project in the z direction onto the (x, y) plane to curved lines given by

$$\frac{\mathrm{d}x}{\mathrm{d}y} = \frac{B_x}{B_y}. \tag{20.56}$$

In essence, the configuration illustrated in Figure 20.2 is modified to that given by the solution of equation (20.56).

For small-amplitude perturbations, the B_y component can be approximated by its equilibrium value, $B_y \sim B'_{y0}x$. For a particular choice of phase (in order to deal with real quantities, rather than complex ones such as $\exp(\mathrm{i}ky)$), the B_x component at some particular time t can be written

$$B_x = \bar{B}_x e^{\gamma t} \sin(ky) \tag{20.57}$$

where, as we have seen, the quantity $\bar{B}_x$ can be taken as approximately independent of x within the resistive layer around $x = 0$. Equation (20.56) can then be integrated to give

$$\tfrac{1}{2} B'_{y0} x^2 + \frac{\bar{B}_x}{k} e^{\gamma t} \cos(ky) = \text{constant} \tag{20.58}$$

where different values of the constant give the projections of different field lines onto the (x, y) plane.

The solutions of equation (20.58) can easily be plotted in the (x, y) plane, and a typical example is illustrated in Figure 20.6. At relatively large values of $|x|$, corresponding to large values of the constant in equation (20.58), the field lines are only slightly distorted from the unperturbed configuration shown in Figure 20.2. However, the distortion increases for smaller values of $|x|$, corresponding to smaller values of the constant in equation (20.58), and eventually the field lines become 'closed on themselves'. Inspection of equation (20.58) shows that these 'closed' field lines arise from values of the constant less than $(\bar{B}_x/k)\exp(\gamma t)$, for which only a limited range of y values are possible, since for these values of the constant equation (20.58) does not allow $\cos(ky)$ to reach unity for any real value of x.

The closed field line regions shown in Figure 20.6 are called 'magnetic islands'. When the strong approximately uniform B_z field is taken into account,

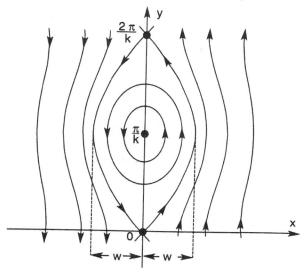

Figure 20.6. Perturbed field line configuration of magnetic islands of half-width w produced by a resistive tearing instability. The pattern is repeated with period $2\pi/k$ in the y direction.

the individual field lines will not actually close on themselves, but will traverse surfaces whose shapes will approximate elliptical cylinders, which are infinitely long in the z direction. In this case, Figure 20.6 depicts the intersections of these surfaces with the (x, y) plane at $z = 0$ or, equivalently, the projection of the field lines onto this plane. A given field line will always remain on the same surface, and its projection onto the (x, y) plane at $z = 0$ will traverse the closed curves shown in Figure 20.6 over and over again as it proceeds further and further in the z direction.

The surface that separates the closed field line surfaces from the open field line surfaces is usually called the 'magnetic separatrix'. The separatrix corresponds to a value of the constant in equation (20.58) exactly equal to $(\bar{B}_x/k)\exp(\gamma t)$. The half-width w of the magnetic island formed by the separatrix, which is of course the largest magnetic island (see Figure 20.6), is simply the value of x given by equation (20.58) for this value of the constant and at $ky = \pi$, namely

$$w = 2(\bar{B}_x/kB'_{y0})^{1/2}\exp(\gamma t/2). \tag{20.59}$$

The half-width of the magnetic island is proportional to the square-root of the field perturbation $\bar{B}_x$, so it increases exponentially in time, as indicated in equation (20.59). In practice, nonlinear effects will limit the growth of magnetic islands when significant modifications are produced in the underlying magnetic configuration on which our stability analysis was based. Such effects begin to appear as soon as the island width becomes comparable to the width of the resistive layer given by equation (20.30), as was shown in a paper by one of the authors of this present book (P H Rutherford 1973 *Phys. Fluids* **16** 1903). When the island grows to a significant fraction of the size of the overall configuration, it can affect the gross current profile, usually acting to reduce the value of $\Delta' a$ and thereby tending to stabilize the tearing mode.

There is clearly a close connection between the magnetic islands and magnetic separatrix obtained here and the islands and separatrices found in the numerical analysis of area-preserving maps presented in connection with particle orbits in Chapter 5. Indeed, the field line equation of motion, equation (20.56), can be represented as a map, where a point is laid down each time a distance $2\pi R$ is traversed in the z direction. The shear in the magnetic field is then equivalent to the sheared particle flow for the problem in Chapter 5, and many of the previous results carry through. The island width at the rational surface, for example, scales in both cases with the square-root of the perturbation strength. Were we to attempt a numerical treatment of the effects of resistive tearing instabilities, we would expect to find, at least in some cases, not only a primary island chain, but also secondary chains of smaller islands, as in Figure 5.2. When the mode amplitude grows so large that secondary islands begin to overlap with the primary island or, in cases where several different modes are unstable, primary island chains begin to overlap with each other, then the magnetic field structure becomes 'stochastic'. When this occurs, an individual field line can find its way completely across the plasma (i.e. in the x direction for the plasma slab configuration considered in this Chapter), if followed a sufficient distance. As a practical consequence, this will generally mean that electron thermal conduction parallel to the magnetic field will rapidly flatten the electron temperature across the stochastic region.

The origin of the name 'tearing mode' is now apparent. The magnetic configuration illustrated in Figure 20.2 'tears' at its weakest points, i.e. along the plane $x = 0$. Provided the conditions for instability are satisfied (i.e. positive Δ'), the plasma current slab will then have a tendency to break up into discrete current 'filaments'.

Problem 20.4 The result of Problem 20.2 implies that the first-order perturbed current density in the z direction is *negative* at the 0-point of the magnetic island, i.e. the point $(0, \pi/k)$ in Figure 20.6, for an unstable mode ($\Delta' > 0$), and *positive* at the X-point of the island, i.e. the point $(0,0)$ in Figure 20.6. (It should be noted that this is a special property of our choice of geometry; the signs are reversed, for example, in a cylindrical tokamak configuration with $dq/dr > 0$.) Verify this for the slab geometry by a different method, as follows. Consider the magnetic flux 'trapped' within the magnetic island. Referring to Figure 20.6, we may view this flux as that of the B_x field crossing the y axis between the X-point and the 0-point; per unit length in the z direction, this flux is

$$\Psi = \int_0^{\pi/k} B_x(0, y)dy.$$

By employing the usual combination of Faraday's law and Ohm's law, show that

$$\frac{d\Psi}{dt} = \eta[j_z(0, 0) - j_z(0, \pi/k)]$$

(Hint: Note that the magnetic field is *exactly* in the z direction at both the 0-point and the X-point, which precludes convection of flux across the boundaries of the surface under consideration.) The trapped flux Ψ must increase as the instability and island-width grow. What does this tell us about the magnitude of the perturbed current density j_z at the island 0-point, versus that at the X-point?

Chapter 21

Drift waves and instabilities*

We have now considered two types of instabilities that can arise in the fluid plasma model: the first, the ideal MHD flute instability (the pressure-driven version of the Rayleigh–Taylor instability), which draws upon the *thermal* energy of the plasma as it expands unstably across a curved (concave toward the plasma) magnetic field, and the second, the resistive tearing instability, which draws upon the energy of the magnetic field in the plasma as it rearranges itself toward a configuration of lower *magnetic* energy. There is yet a third important class of instability of a fluid plasma, the so-called 'drift-wave instability', which requires neither a curved magnetic field nor a magnetic configuration for which lower magnetic-energy states exist. Indeed, drift-wave instabilities occur in the simplest and most 'universal' of configurations, namely a plasma of non-uniform density maintained in equilibrium by a strong and essentially straight magnetic field. Because of the pervasiveness of this situation, instabilities of this type have sometimes been called 'universal instabilities'. Like flute instabilities, drift-wave instabilities draw upon the thermal energy of the plasma as it expands across a magnetic field. Unlike flute instabilities, however, they have finite wavelengths along the field, and the plasma motion is decoupled, to a significant extent, from that of the magnetic field, so as to avoid energetically unfavorable bending of the field lines. Because of the difficulty of drawing upon the thermal energy of expansion in this way, drift-wave instabilities tend to have rather small growth rates—certainly smaller than those characteristic of flute instabilities.

Unlike Rayleigh–Taylor, flute and resistive-tearing instabilities, drift-wave instabilities are not purely growing, but have complex frequencies ω, with the imaginary part, denoted by γ (the growth rate), usually much smaller than the real part. Of course, any such mode of perturbation can be made purely growing by transforming to a moving frame in which the wave is at rest, but in such a frame the plasma itself will acquire a non-zero velocity. Normally, we choose to work in the 'laboratory frame' in which the plasma is assumed to be at rest (more precisely, the mass velocity **u** is taken to be zero) in the unperturbed equilibrium state. In such a frame, the drift-wave instabilities have complex frequencies ω, i.e. they are partly travelling waves and partly growing waves.

Drift waves require non-zero plasma resistivity, or (as we will see in Chapter 26) other forms of dissipation, to be unstable. However, the waves themselves (i.e. without instability) can exist and propagate in any non-uniform plasma. Moreover, as we will see, except at relatively high values of the plasma β (but still $\beta \ll 1$), drift waves do not produce a significant perturbation of the magnetic field. Rather they involve a self-consistent wave-like pattern of density perturbations and flow velocities that propagates partly along and partly across a fixed, approximately uniform, straight magnetic field.

21.1 THE PLANE PLASMA SLAB

We will analyze drift waves in the simplest possible configuration involving a non-uniform plasma, the so-called 'plane plasma slab'. In this configuration, there is a plasma with non-uniform density $n(x)$ and pressure $p(x)$, maintained in equilibrium by a strong magnetic field, B_z. There is no variation of the equilibrium in the y or z directions. The plasma is at rest in the equilibrium configuration, i.e. $\mathbf{u} = 0$, but there is, of course, a non-zero current density $j_y(x)$ needed to provide equilibrium, i.e. to provide a $\mathbf{j} \times \mathbf{B}$ force that balances the pressure gradient ∇p. The magnetic field B_z will be modified (and will acquire a variation with x) as a result of the plasma currents, so that the pressure-balance condition, $p + B_z^2/2\mu_0 = $ constant, is satisfied. However, for low values of β, the non-uniformity of B_z is very small and will be neglected in our analysis. The suffix '0' will be used to denote equilibrium quantities, e.g. $n_0(x)$, $p_0(x)$ and B_{z0}.

The new element in our description of a plasma that is needed to produce drift waves is the full so-called 'generalized' Ohm's law, introduced in equation (8.13), namely

$$\mathbf{E} + \mathbf{u} \times \mathbf{B} = \eta \mathbf{j} + \frac{\mathbf{j} \times \mathbf{B} - \nabla p_e}{ne}. \tag{21.1}$$

Before embarking upon our stability analysis, we must address the question of whether the use of this generalized Ohm's law, rather than the simple version which omits the last two terms on the right-hand side of equation (21.1), has any effect on our description of the *equilibrium* configuration. Clearly, such an effect *does* arise, since satisfying the independent force-balance condition, $\mathbf{j} \times \mathbf{B} = \nabla p$, where $p = p_e + p_i$, will leave an uncanceled term in ∇p_i on the right-hand side of equation (21.1). Thus, it will not be possible to have *both* $\mathbf{u} = 0$ *and* $\mathbf{E} = 0$ in the equilibrium configuration. Physically, we are encountering here the contribution to the fluid velocity from the *ion diamagnetic drift* which we discussed previously in Chapter 7. Specifically, substituting $\mathbf{j} \times \mathbf{B} = \nabla(p_e + p_i)$ on the right-hand side of equation (21.1) and neglecting, for now, the resistivity term, we can solve equation (21.1) for $\mathbf{u}_\perp$, obtaining

$$\mathbf{u}_\perp = \frac{\mathbf{E} \times \mathbf{B}}{B^2} + \frac{\mathbf{B} \times \nabla p_i}{neB^2}. \tag{21.2}$$

Equation (21.2) tells us that the fluid (mass) velocity across the magnetic field is the sum of the $\mathbf{E} \times \mathbf{B}$ drift and the ion diamagnetic velocity, as we would have expected, since the ions make the dominant contribution to the plasma mass. Clearly, in a non-uniform plasma, $\mathbf{u}$ and $\mathbf{E}$ cannot *both* be zero in equilibrium. If we have an equilibrium in which the plasma is at rest, i.e. $\mathbf{u} = 0$, there will necessarily be a non-zero electric field $\mathbf{E}$, and, conversely, if the equilibrium has $\mathbf{E} = 0$, we will need to take into account a non-zero mass-velocity $\mathbf{u}$.

For present purposes, however, we can simplify the analysis by restricting ourselves to the case where the *ion pressure vanishes*, while the electron pressure does not vanish. Physically, this corresponds to a situation where $T_i \ll T_e$, which is a legitimate (and not uncommon) case to consider. Since the equilibrium ion diamagnetic drift is essentially zero, this allows us to assume that $\mathbf{E}_0 = \mathbf{u}_0 = 0$. There would be no *fundamental* difficulty in pursuing the more general case with non-zero ion pressure, for example by keeping a non-zero equilibrium $\mathbf{E}$ field in the stability analysis of a static (i.e. $\mathbf{u} = 0$) equilibrium, but the algebraic complexity would be greater, without adding much more insight into the underlying drift-wave physics.

The plasma is uniform and of infinite extent in the y and z directions. Thus we can assume that perturbations take the form of plane waves in these two directions, so that any perturbation quantity $\psi_1(\mathbf{x}, t)$ can be written

$$\psi_1(\mathbf{x}, t) = \hat{\psi}_1(x)\exp(-i\omega t + ik_y y + ik_z z) \tag{21.3}$$

where $\hat{\psi}_1(x)$ is the amplitude of the wave-like perturbation. Once again, since the equilibrium varies in the x direction, we cannot Fourier decompose into sinusoidal modes in the x direction, but rather must search for eigenfunctions $\hat{\psi}_1(x)$. Our method of analysis will be generally similar to that employed in the derivation of the Rayleigh–Taylor and resistive-tearing instabilities in Chapters 19 and 20, respectively, except that here we have $k_z \neq 0$, implying that the perturbations have a variation *along* the main equilibrium magnetic field. However, we will look for waves satisfying

$$k_z \ll k_y \tag{21.4}$$

and the outcome of our analysis will show that this inequality is valid for a typical drift-wave instability.

For our initial derivation of the drift waves, we will keep the *magnetic* perturbations as well as the electric-field perturbations, but we will then show that, for low-β plasmas, the magnetic perturbations are unimportant relative to the perturbed electric fields $\mathbf{E}$ and the associated $\mathbf{E} \times \mathbf{B}$ flow velocities. If the magnetic perturbations are neglected from the outset, so that the perturbed electric field can be assumed to be derivable from a scalar electric potential, i.e. $\mathbf{E} = -\nabla\phi$, the analysis of drift waves is simplified considerably. We will indeed discuss this 'electrostatic' limit after we have developed the analysis for the more general case. The value of first analyzing the more general 'finite-β' case in some detail is that it demonstrates the connection to the slow shear Alfvén waves discussed in the previous two Chapters (and in Chapter 18), and it shows explicitly how the new drift-wave branch of the spectrum arises at frequencies much lower than all Alfvén wave frequencies, i.e. $\omega \ll k_z v_A \ll k_y v_A$.

21.2 THE PERTURBED EQUATION OF MOTION IN THE INCOMPRESSIBLE CASE

We begin with the perturbed equation of motion

$$\rho_0 \frac{\partial \mathbf{u}_1}{\partial t} = -\nabla p_1 + (\mathbf{j} \times \mathbf{B})_1$$
$$= -\nabla\left(p_1 + \frac{\mathbf{B}_0 \cdot \mathbf{B}_1}{\mu_0}\right) + \frac{1}{\mu_0}[(\mathbf{B} \cdot \nabla)\mathbf{B}]_1 \qquad (21.5)$$

where, as usual, we use the suffix '1' to denote perturbed quantities. Noting that the equilibrium magnetic field is entirely in the z direction, the two components of equation (21.5) perpendicular to this equilibrium field are

$$-i\omega\rho_0 u_x = -\frac{\partial}{\partial x}\left(p_1 + \frac{B_{z0}B_{z1}}{\mu_0}\right) + \frac{ik_z}{\mu_0}B_{z0}B_x \qquad (21.6)$$

$$-i\omega\rho_0 u_y = -ik_y\left(p_1 + \frac{B_{z0}B_{z1}}{\mu_0}\right) + \frac{ik_z}{\mu_0}B_{z0}B_y. \qquad (21.7)$$

Here, and henceforth in this Chapter, we omit the suffix '1' from perturbed quantities whose equilibrium values are zero, e.g. u_x, u_y, B_x and B_y. In deriving equations (21.6) and (21.7), we have noted that $\mathbf{B}_0$ has only a component in the z direction, so $(\mathbf{B}_1 \cdot \nabla)\mathbf{B}_0$ does not contribute anything to the x and y components of equation (21.5).

We now argue that the term in B_{z1} in equations (21.6) and (21.7) contributes significantly to the right-hand side of these equations, i.e. to the force arising from the gradient of the magnetic pressure, even for B_{z1} values that are so small that they do not make a significant contribution to the divergence of the magnetic field. Using equation (21.7) for our estimates, we see that the contribution from B_{z1} to the perturbed magnetic-pressure gradient is comparable to the contribution from B_y if

$$B_{z1} \sim (k_z/k_y)B_y. \tag{21.8}$$

The condition that the perturbed magnetic field be divergence-free is

$$\frac{\partial B_x}{\partial x} + ik_y B_y + ik_z B_{z1} = 0 \tag{21.9}$$

and we see immediately that the contribution from B_{z1} is negligible compared with that from B_y if $k_z \ll k_y$, as we have assumed. Thus, the divergence-free magnetic field condition becomes essentially

$$\frac{\partial B_x}{\partial x} + ik_y B_y = 0 \tag{21.10}$$

the same as for the Rayleigh–Taylor (flute) and resistive-tearing instabilities, both of which had $k_z = 0$.

We will also see below that B_{z1} values which contribute significantly to the magnetic-pressure gradients in equations (21.6) and (21.7) are much smaller than those which would arise if we allowed the main magnetic field to be compressed significantly. Thus, as in the case of the Rayleigh–Taylor (flute) and resistive-tearing instabilities, we want to look for solutions with the property that the flow $\mathbf{u}$ is such that the B_z field is not compressed. The consequence of these approximations is that the perturbed magnetic-field component B_{z1} will play no role in determining the plasma flows and density perturbation, and it will not, finally, appear anywhere else in our analysis, except in equations (21.6) and (21.7).

Accordingly, it is convenient to use our now-familiar technique for eliminating B_{z1} from equations (21.6) and (21.7), namely taking the x derivative of equation (21.7) and subtracting ik_y times equation (21.6). We obtain

$$-i\omega\left(\frac{\partial}{\partial x}(\rho_0 u_y) - ik_y \rho_0 u_x\right) = \frac{ik_z B_{z0}}{\mu_0}\left(\frac{\partial B_y}{\partial x} - ik_y B_x\right)$$

$$= -\frac{k_z B_{z0}}{\mu_0 k_y}\left(\frac{\partial^2 B_x}{\partial x^2} - k_y^2 B_x\right) \tag{21.11}$$

where we have used $\partial/\partial x$ of equation (21.10) to obtain the second form of the right-hand side.

Any flow $\mathbf{u}$ that arises will be associated with an electric field $\mathbf{E}_\perp \approx \mathbf{u} \times \mathbf{B}$ and will result in compression of the magnetic field B_z, described by

$$\frac{\partial B_z}{\partial t} \approx [\boldsymbol{\nabla} \times (\mathbf{u} \times \mathbf{B})]_z \tag{21.12}$$

the approximate equality indicating that some smaller terms in Ohm's law are being neglected. Equation (21.12), to first order, gives

$$-i\omega B_{z1} \approx -B_{z0}\left(\frac{\partial u_x}{\partial x} + ik_y u_y\right). \tag{21.13}$$

Unless the right-hand side of equation (21.13) effectively vanishes, there would arise from compression of the B_z field a perturbation of magnitude given roughly by $B_{z1} \sim B_{z0}k_y u_y/\omega$. If we were to substitute this into equation (21.7), we would find the ratio of the inertia term on the left to the term in B_{z1} on the right to be $\omega^2/k_y^2 v_A^2$, where v_A is the Alfvén speed, $B/(\rho_0\mu_0)^{1/2}$. Similarly, if we used the term u_x in equation (21.13) to eliminate B_{z1}, then we would substitute this into equation (21.6) and would find the ratio of the inertia term on the left to the term in B_{z1} on the right to be $\omega^2/k_x^2 v_A^2$, where $k_x \sim \partial/\partial x$. Since we will find that drift waves are generally characterized by $k_x \sim k_y$, these two estimates are similar. However, we want to look for frequencies *much smaller* than $k_y v_A$ (at most of order $k_z v_A$, where $k_z \ll k_y \sim k_x$), so we cannot allow the large B_{z1} values that would arise if this degree of compression of the B_z field were to occur. Thus, we can write

$$\frac{\partial u_x}{\partial x} + ik_y u_y = 0. \tag{21.14}$$

We note that this is *not*, in this case, the condition for *exactly* incompressible fluid flow, which would involve an additional term $ik_z u_z$ on the left-hand side of equation (21.14). Indeed, it is only the flow perpendicular to the magnetic field that is required to be incompressible; an arbitrary flow *along* the field can be added without contributing anything to the compression of the magnetic field. Nonetheless, for most cases of interest, including drift waves, both k_z and u_z are relatively small, so that a term $ik_z u_z$, even if added to the left-hand side of equation (21.14), would make little difference.

Our argument for incompressibility, which has been invoked for the Rayleigh–Taylor (or flute) instability, the resistive-tearing instability and now for the drift wave, can be expressed in terms of the various types of Alfvén waves discussed in Chapter 18. Essentially, these three instabilities all arise in the *linearly polarized shear Alfvén wave branch* of the low-frequency 'spectrum', rather than in the *magnetosonic wave branch*. The physical reason for this is that these shear Alfvén waves do not require the large amount of energy that would be needed to compress the magnetic field, with the result that they are most easily driven unstable by relatively weak sources of free energy. Since perpendicular compression is not involved, the shear Alfvén waves can also have much smaller frequencies, in the case $k_z \ll k_y$. For the case of drift waves, for which we will derive a dispersion relation that displays explicitly the coupling to the shear Alfvén waves, we will find frequencies in the range

$\omega < k_z v_A$ (often $\omega \ll k_z v_A$), to be compared with the much larger frequencies, $\omega \sim k_y v_A$, characteristic of the magnetosonic waves.

Using the incompressibility condition, equation (21.14), to substitute for u_y in terms of u_z on the left-hand side, equation (21.11) becomes

$$\frac{\omega \rho_0}{k_y}\left(\frac{\partial^2 u_x}{\partial x^2} - k_y^2 u_x\right) = -\frac{k_z B_{z0}}{\mu_0 k_y}\left(\frac{\partial^2 B_x}{\partial x^2} - k_y^2 B_x\right) \qquad (21.15)$$

where, on the left-hand side, we have made the simplifying assumption that ρ_0 is not strongly varying with x on the scale of distances over which the perturbations vary significantly. Basically, we are assuming here that the effective wavelength of the perturbation in the x direction is much shorter than the scale-length of the equilibrium density variation.

In cases such as this, where the wavelength of a perturbation is much shorter than the scale-length over which an equilibrium varies, we can use the 'WKB approximation', introduced in Chapter 15. The perturbation will adopt an approximately wave-like form, although the local wave-number k_x will adjust itself gradually to local conditions. For any general perturbed quantity $\psi_1(x)$, the WKB approximation is adopted by writing

$$\psi_1(x) = \hat{\psi}_1 \exp\left(i\int^x k_x dx\right) \qquad (21.16)$$

where the amplitude $\hat{\psi}_1$ and the effective wave-number k_x are both *slowly varying* functions of x, i.e. they vary on the scale of the equilibrium variation. A full application of the WKB approximation allows actual eigenfunctions to be obtained, i.e. forms for $\hat{\psi}_1(x)$ as well as for $k_x(x)$, but for present purposes it is sufficient simply to introduce a wave-number k_x, as in equation (21.16), implying that the perturbation is wave-like in x. (Effectively, the WKB approximation generates eigenfunctions by approximating to successive orders in an expansion in $(k_x L_n)^{-1}$, where L_n is the typical scale-length of the density non-uniformity; equation (21.16) represents the lowest-order eigenfunction.) When x derivatives are taken, we may simply use the rule $\partial/\partial x \to ik_x$, just as if the perturbation were *exactly* of plane-wave form.

Applying this technique to equation (21.15), we obtain

$$-\frac{\omega \rho_0}{k_y} k_\perp^2 u_x = \frac{k_z B_{z0}}{\mu_0 k_y} k_\perp^2 B_x \qquad (21.17)$$

where $k_\perp^2 = k_x^2 + k_y^2$. Equation (21.17) may be rewritten

$$\omega u_x = -k_z v_A^2 B_x / B_{z0}. \qquad (21.18)$$

Equation (21.18) is as much information as we can obtain from the perpendicular components of the perturbed equation of motion, because we have now reduced the independent variables to two, namely u_x and B_x, which will be related to each other also through Ohm's law.

21.3   THE PERTURBED GENERALIZED OHM'S LAW

We turn next to the generalized Ohm's law for the first-order perturbed quantities, namely

$$\mathbf{E}_1 + \mathbf{u}_1 \times \mathbf{B}_0 = \eta \mathbf{j}_1 + \frac{1}{ne}(\mathbf{j} \times \mathbf{B} - \nabla p_e)_1 \tag{21.19}$$

which, when coupled with Faraday's law, i.e.

$$\frac{\partial \mathbf{B}_1}{\partial t} = -\nabla \times \mathbf{E}_1 \tag{21.20}$$

must yield another relation between B_x and u_x to combine with equation (21.18). Substituting equation (21.19) into equation (21.20) and employing our usual expansion of $\nabla \times (\mathbf{u}_1 \times \mathbf{B}_0)$ (see, for example, equation (19.6)), we obtain

$$\frac{\partial \mathbf{B}_1}{\partial t} = (\mathbf{B}_0 \cdot \nabla)\mathbf{u}_1 - (\mathbf{u}_1 \cdot \nabla)\mathbf{B}_0 - \mathbf{B}_0(\nabla \cdot \mathbf{u}_1) - \nabla \times \left(\eta \mathbf{j}_1 + \frac{1}{ne}(\mathbf{j} \times \mathbf{B} - \nabla p_e)_1 \right). \tag{21.21}$$

Examination of the size of the various terms in the generalized Ohm's law shows that the additional terms on the right-hand side of equation (21.19), i.e. the last two terms, are of much more importance in the component *parallel* to the magnetic field than they are in the components *perpendicular* to the field. To see this, we simply note that the equation of motion tells us that

$$(\mathbf{j} \times \mathbf{B} - \nabla p_e)_1 \approx \rho_0 \frac{\partial \mathbf{u}_1}{\partial t} = -i\omega \rho_0 \mathbf{u}_1 \tag{21.22}$$

and so the ratio of the magnitude of the last two terms on the right-hand side in the perpendicular components of equation (21.19) to the magnitude of the second term on the left-hand side is of order $\omega \rho_0 |\mathbf{u}_1| / ne |\mathbf{u}_1| B \approx \omega M / eB \approx \omega / \omega_{ci}$, where ω_{ci} is the Larmor frequency of the ions. For waves, with $\omega \ll \omega_{ci}$, these additional terms on the right-hand side in the perpendicular components of equation (21.19) are unimportant and may be neglected. However, the new terms must be retained in the *parallel* component of the generalized Ohm's law, which becomes

$$E_{\parallel} = \eta j_{\parallel} - \frac{1}{ne} \nabla_{\parallel} p_e. \tag{21.23}$$

Noting that the equilibrium magnetic field is entirely in the z direction, equation (21.23) to first order in the perturbations can be written

$$E_z = \eta j_z - \frac{1}{ne} \left(ik_z p_{e1} + \frac{B_x}{B_{z0}} \frac{dp_{e0}}{dx} \right) \tag{21.24}$$

where, in accordance with our usual convention, we have dropped the suffix '1' from the perturbed quantities E_z and j_z, whose equilibrium values are zero. Note the appearance of the third term on the right-hand side of equation (21.24), which arises from observing that the operator $\nabla_\parallel$ means $(\hat{\mathbf{b}} \cdot \nabla)$, where $\hat{\mathbf{b}}$ is the unit vector in the direction of **B**, so that

$$(\nabla_\parallel p_e)_1 = [(\hat{\mathbf{b}} \cdot \nabla) p_e]_1 = \hat{\mathbf{b}}_0 \cdot \nabla p_{e1} + \hat{\mathbf{b}}_1 \cdot \nabla p_{e0}$$
$$= ik_z p_{e1} + \frac{B_x}{B_{z0}} \frac{dp_{e0}}{dx}. \tag{21.25}$$

(Strictly, we should note that j_y is non-zero in the equilibrium, and hence will require a small but non-zero $-u_{x0}B_{z0} = \eta j_{y0}$. As we saw in Chapter 12, this u_{x0} is the fluid velocity due to collisional diffusion. In the perturbed form of equation (21.23), there will be an additional term $\eta B_y j_{y0}/B_{z0}$ on the right-hand side. This extra term is very small, since the resistivity η is generally very small; comparing it with the last term on the right-hand side of equation (21.24), we find it to be of relative order ν_{ei}/ω_{ce}, where we have written η in term of ν_{ei} and assumed $B_x \sim B_y$.)

Using equation (21.24) for the parallel component of the generalized Ohm's law, but assuming that $\mathbf{E}_\perp = -\mathbf{u} \times \mathbf{B}$ is a satisfactory approximation for the perpendicular components, so that the vector inside the curl operator in the last term in equation (21.21) retains only its component parallel to **B**, i.e. $[\eta j_\parallel - (\nabla_\parallel p_e)/ne]_1 \hat{\mathbf{b}}$, the x and y components of equation (21.21) can be written

$$-i\omega B_x = ik_z B_{z0} u_x - ik_y \left[\eta j_z - \frac{1}{ne} \left(ik_z p_{e1} + \frac{B_x}{B_{z0}} \frac{dp_{e0}}{dx} \right) \right]$$
$$-i\omega B_y = ik_z B_{z0} u_y + \frac{\partial}{\partial x} \left[\eta j_z - \frac{1}{ne} \left(ik_z p_{e1} + \frac{B_x}{B_{z0}} \frac{dp_{e0}}{dx} \right) \right] \tag{21.26}$$

although the second of these equations is redundant once equations (21.10) and (21.14) have been established, and so it is not used further in our analysis. We now use Ampere's law, $\nabla \times \mathbf{B} = \mu_0 \mathbf{j}$, together with equation (21.10) to express B_y in terms of B_x, to obtain an expression for j_z in terms of B_x:

$$j_z = \frac{1}{\mu_0} \left(\frac{\partial B_y}{\partial x} - ik_y B_x \right)$$
$$= \frac{i}{\mu_0 k_y} \left(\frac{\partial^2 B_x}{\partial x^2} - k_y^2 B_x \right)$$
$$\approx -\frac{ik_\perp^2}{\mu_0 k_y} B_x \tag{21.27}$$

with $k_\perp^2 = k_x^2 + k_y^2$ and where the WKB approximation has been invoked in the final step. From equation (21.26), we now obtain

$$\omega B_x + k_z B_{z0} u_x = -\frac{i\eta}{\mu_0} k_\perp^2 B_x - \frac{k_y}{ne}\left(ik_z p_{e1} + \frac{B_x}{B_{z0}}\frac{dp_{e0}}{dx}\right). \tag{21.28}$$

The electron pressure perturbation p_{e1} still needs to be eliminated in favor of B_x and u_x. It will be determined by an equation of state, which relates p_{e1} to the density perturbation n_{e1}, which in turn will be determined from the perturbed continuity equation. Physically, the most appropriate assumption will be that the electrons are isothermal, which is equivalent to assuming that the electron thermal conductivity is sufficiently large to maintain a uniform temperature T_e *along* the magnetic field, i.e.

$$\mathbf{B} \cdot \nabla T_e = 0. \tag{21.29}$$

Allowing for the possibility of a temperature gradient *across* the field in the equilibrium, i.e. $T_{e0} = T_{e0}(x)$, the perturbed form of equation (21.29) is

$$ik_z T_{e1} + \frac{B_x}{B_{z0}}\frac{dT_{e0}}{dx} = 0. \tag{21.30}$$

Using $p_{e1} = T_{e0}n_{e1} + n_{e0}T_{e1}$, it follows that the term in parenthesis on the right-hand side of equation (21.28) is given by

$$\begin{aligned} ik_z p_{e1} + \frac{B_x}{B_{z0}}\frac{dp_{e0}}{dx} &= ik_z T_{e0}n_{e1} + \frac{B_x}{B_{z0}}\left(\frac{dp_{e0}}{dx} - n_{e0}\frac{dT_{e0}}{dx}\right) \\ &= T_{e0}\left(ik_z n_{e1} + \frac{B_x}{B_{z0}}\frac{dn_{e0}}{dx}\right). \end{aligned} \tag{21.31}$$

Equation (21.31) may be substituted into equation (21.28), which has the effect of eliminating the pressure perturbation p_{e1} in favor of the density perturbation n_{e1}.

The continuity equation to first order in the perturbations, i.e.

$$\frac{\partial n_{e1}}{\partial t} + \mathbf{u}_\perp \cdot \nabla n_{e0} + \nabla_\parallel(n_{e0}u_\parallel) = 0 \tag{21.32}$$

where we have used $\nabla \cdot \mathbf{u}_\perp = 0$, can be written

$$-i\omega n_{e1} + u_x \frac{dn_{e0}}{dx} + ik_z n_{e0}u_z = 0. \tag{21.33}$$

The perturbed velocity parallel to the equilibrium magnetic field, u_z, must be obtained from the parallel component of the equation of motion. Although we have already made use of the perpendicular components of the equation of motion, we have not yet used the parallel component, which is

$$\rho_0 \frac{\partial u_\parallel}{\partial t} = -\nabla_\parallel p_e. \tag{21.34}$$

To first order, the perturbed form of equation (21.34) becomes

$$
\begin{aligned}
-i\omega\rho_0 u_z &= -ik_z p_{e1} - \frac{B_x}{B_{z0}} \frac{dp_{e0}}{dx} \\
&= -T_{e0}\left(ik_z n_{e1} + \frac{B_x}{B_{z0}} \frac{dn_{e0}}{dx}\right)
\end{aligned}
\tag{21.35}
$$

where we have again used equation (21.31). We can substitute equation (21.35) into equation (21.33), to obtain

$$-i\left(\omega - \frac{k_z^2 T_{e0}}{\omega M}\right) n_{e1} + u_x \frac{dn_{e0}}{dx} + \frac{k_z T_{e0}}{\omega M} \frac{B_x}{B_{z0}} \frac{dn_{e0}}{dx} = 0. \tag{21.36}$$

We have now obtained an expression for the density perturbation, and hence also the electron pressure perturbation, in terms of u_x and B_x.

We now substitute equation (21.31) into equation (21.28) and then substitute for n_{e1} from equation (21.36). This involves a significant amount of straightforward manipulation, which proceeds most easily by first noting that equation (21.36) can be rewritten

$$ik_z n_{e1} + \frac{B_x}{B_{z0}} \frac{dn_{e0}}{dx} = \frac{\omega}{B_{z0}} \frac{dn_{e0}}{dx} \frac{\omega B_x + k_z B_{z0} u_x}{\omega^2 - k_z^2 C_s^2} \tag{21.37}$$

where $C_s = (T_e/M)^{1/2}$ is the plasma sound speed (i.e. the ion thermal speed evaluated with the electron temperature). Using this in equation (21.31), which is then substituted into equation (21.28), we obtain

$$(\omega B_x + k_z B_{z0} u_x)\left(1 - \frac{k_y v_{de}}{\omega - k_z^2 C_s^2/\omega}\right) = -\frac{i\eta}{\mu_0} k_\perp^2 B_x. \tag{21.38}$$

Here

$$v_{de} = -\frac{T_{e0}}{n_{e0}e B_{z0}} \frac{dn_{e0}}{dx} \tag{21.39}$$

is very similar in form to the electron diamagnetic drift velocity (see Chapter 7), the minus sign coming from the electron's charge, $-e$. (Note that v_{de} is not *exactly* the electron diamagnetic drift velocity, as defined in Chapter 7, in which dp_{e0}/dx would appear, rather than $T_e(dn_{e0}/dx)$. Thus, v_{de} differs from the diamagnetic drift velocity if there is a temperature gradient across the magnetic field. In magnitude and sign, however, the two velocities are, of course, generally similar.)

21.4   THE DISPERSION RELATION FOR DRIFT WAVES

By combining equation (21.38) with the relation between u_x and B_x obtained from the perpendicular components of the equation of motion, i.e. equation (21.18), we obtain the dispersion relation for the waves under investigation, namely

$$\left(\omega - \frac{k_z^2 v_A^2}{\omega}\right)\left(1 - \frac{k_y v_{de}}{\omega - k_z^2 C_s^2/\omega}\right) = -\frac{i\eta}{\mu_0}k_\perp^2. \tag{21.40}$$

In the limit of zero resistivity, we see that there are two distinct branches of the dispersion relation. One branch has

$$\omega = k_z v_A \tag{21.41}$$

and clearly corresponds to the shear Alfvén wave. The second branch has a dispersion relation

$$\omega - k_y v_{de} - \frac{k_z^2 C_s^2}{\omega} = 0 \tag{21.42}$$

and corresponds to the 'drift waves'. In a uniform plasma, for which $v_{de} = 0$, this is just the ion sound wave encountered in Chapter 16, with $k\lambda_D \ll 1$. Since equation (21.42) is quadratic in ω, for given values of k_y and k_z there are two branches of the drift wave, i.e. two possible values of ω, as shown in Figure 21.1. The branch for which ω has the same sign as $k_y v_{de}$ (upper curve in Figure 21.1) is usually called the 'electron drift wave'; the other branch (lower curve in Figure 21.1) is usually called the 'ion branch' of the drift wave, although, for reasons that will soon be apparent, this branch is of less interest. In the limit in which $k_z C_s \ll k_y v_{de}$, the electron drift wave has the frequency

$$\omega \approx k_y v_{de}. \tag{21.43}$$

(The ion branch of the drift wave as shown in Figure 21.1 violates the convention introduced in Chapter 15 that real frequencies ω are taken to be positive. If we are interested in this branch, we can satisfy the convention by simply reversing the sign of k_y. Physically, the ion branch of the drift wave propagates in the direction opposite to that of the electron diamagnetic drift.)

Problem 21.1: By solving the quadratic equation, equation (21.42), for ω exactly, draw a more accurate version of Figure 21.1, plotting the dimensionless frequency $\omega/k_y v_{de}$ versus the dimensionless quantity $k_z C_s/k_y v_{de}$.

Equation (21.40) indicates that the effects of non-zero resistivity are to couple the shear Alfvén and drift-wave branches of the spectrum together and to add an imaginary part (either a growth rate or a damping decrement, depending on sign) to the frequencies of each of the branches.

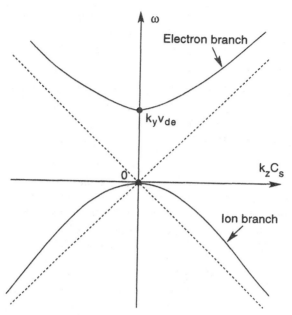

Figure 21.1. Electron and ion branches of the drift-wave dispersion relation. Both branches approach asymptotes $\omega = \pm k_z C_s$.

In order to proceed further, we must consider the typical magnitudes of the various frequencies appearing in equation (21.40). First, we note that $C_s \equiv (T_e/M)^{1/2}$ and $v_A \equiv B/(\mu_0 n M)^{1/2}$, so that

$$C_s/v_A = (\mu_0 n T_e)^{1/2}/B \approx (\beta/2)^{1/2} \qquad (21.44)$$

indicating that the sound-wave frequency, $k_z C_s$, is very much smaller than the shear Alfvén wave frequency in all cases where the plasma β value is very small.

Second, we note that $v_{de} = T_e/eBL_n$, where $L_n = n/(dn/dx)$, the scale length of the density non-uniformity, so that

$$v_{de}/C_s = (MT_e)^{1/2}/eBL_n \approx r_{Ls}/L_n \qquad (21.45)$$

where $r_{Ls} = (MT_e)^{1/2}/eB = C_s/\omega_{ci}$, the average Larmor radius of the ions evaluated as if the ions had the *electron* temperature. The ion Larmor radius

r_{Li} in a magnetized plasma is usually very much smaller than any macroscopic scale-length. Furthermore, although our treatment of drift waves has assumed, for simplicity of analysis, that $T_i \ll T_e$, the disparity in the two temperatures is not usually sufficient to make r_{Ls} more than a few times, at most, larger than r_{Li}. Thus, in many cases of interest, we can assume that $v_{\text{de}} \ll C_s$. It follows that the ratio of the two frequencies appearing in the drift-wave dispersion relation, equation (21.42), namely $k_y v_{\text{de}}/k_z C_s$, is very small, unless

$$k_z \ll k_y \qquad (21.46)$$

or, more specifically, $k_z/k_y \sim r_{\text{Ls}}/L_n$ for the two frequencies $k_y v_{\text{de}}$ and $k_z C_s$ to be comparable.

Since a finite number of wavelengths $\lambda_y = 2\pi/k_y$ and $\lambda_z = 2\pi/k_z$ must 'fit' into the plasma in the y and z directions, respectively, it follows that our 'plane plasma slab' must be much more extended in the z direction than in the y direction, by roughly the ratio L_n/r_{Ls}, for the drift wave to be clearly distinguishable from the ion sound wave. If the plasma slab is infinite in both y and z directions, as it strictly is within our model, then all k_y and k_z values are allowed but, as we will see, the most unstable perturbations will be much more extended in the z direction. The infinite plasma slab will be a good representation of a finite-size plasma, provided the wavelengths in both y and z directions are much shorter than the y and z dimensions of the finite plasma, respectively.

To retain both branches of the drift waves shown in Figure 21.1, we take $k_y v_{\text{de}} \sim k_z C_s$, in which case the typical ordering of the frequencies in equation (21.40) is

$$k_y v_{\text{de}} \sim k_z C_s \ll k_z v_{\text{A}} \qquad (21.47)$$

the inequality following from $\beta \ll 1$.

In this case, even with resistivity included, equation (21.40) divides into a higher-frequency branch, the shear Alfvén wave, with

$$\omega - \frac{k_z^2 v_{\text{A}}^2}{\omega} = -\frac{i\eta}{\mu_0} k_\perp^2 \qquad (21.48)$$

and a lower-frequency branch, the drift wave, with

$$\omega - k_y v_{\text{de}} - \frac{k_z^2 C_s^2}{\omega} = \frac{i\eta k_\perp^2}{\mu_0} \frac{\omega^2 - k_z^2 C_s^2}{k_z^2 v_{\text{A}}^2}. \qquad (21.49)$$

This separation into two branches of the dispersion relation (21.40) can be derived by first looking for high-frequency solutions, $\omega \sim k_z v_{\text{A}}$, for which the inequality given in equation (21.47) implies that the second of the two factors in parentheses on the left in equation (21.40) is approximately unity, thereby yielding equation (21.48). Next, looking for low-frequency solutions,

$\omega \sim k_y v_{de} \sim k_z C_s$, the same inequality, i.e. equation (21.47), implies that the first of the two factors in parentheses on the left-hand side of equation (21.40) is simply $-k_z^2 v_A^2/\omega$, thereby yielding equation (21.49). The fundamental assumption that permits this division into two distinct branches of the dispersion relation is that $\beta \ll 1$, which produces a wide separation between the lower frequency drift waves with $\omega \sim k_y v_{de} \sim k_z C_s$, and the higher frequency shear Alfvén waves with $\omega \sim k_z v_A$.

We examine the effect of resistivity first on the shear Alfvén waves. Neglecting the imaginary term from resistivity in equation (21.48), we have at lowest-order the familiar solution $\omega \sim \pm k_z v_A$. Treating the imaginary term on the right-hand side of equation (21.48) as a small correction and allowing ω to acquire a correspondingly small imaginary part, $\omega \to \omega + i\gamma$ (where ω and γ are now both assumed real, with $\gamma/\omega \ll 1$), the imaginary part of the left-hand side of equation (21.48) is simply $\gamma + (k_z^2 v_A^2/\omega^2)\gamma \approx 2\gamma$, which yields $\gamma \approx -\eta k_\perp^2/2\mu_0$, indicating that the shear Alfvén waves are damped by resistivity (negative γ). The damping decrement is essentially the rate of resistive diffusion of magnetic field over a distance of order a perpendicular wavelength—a physically intuitive result unrelated to the present topic of drift-wave physics.

Carrying out a similar analysis of equation (21.49), we find a lowest-order dispersion relation for ω that is the same as equation (21.42) whose solutions are shown in Figure 21.1. Then, letting $\omega \to \omega + i\gamma$ and equating the imaginary part of order γ on the left-hand side of equation (21.49), which is $\gamma + (k_z^2 C_s^2/\omega^2)\gamma$, to the imaginary expression on the right-hand side, in which only the real part of the frequency ω need be used, we obtain

$$\gamma = \frac{\eta k_\perp^2}{\mu_0} \frac{\omega^2(\omega^2 - k_z^2 C_s^2)}{k_z^2 v_A^2(\omega^2 + k_z^2 C_s^2)}. \tag{21.50}$$

Equation (21.50) shows that the drift wave is unstable whenever $|\omega| > |k_z C_s|$. Referring to Figure 21.1, we see that the electron drift wave (the upper curve in Figure 21.1) is always unstable (positive γ), although the growth rate will diminish rapidly as ω approaches the asymptote $k_z C_s$, whereas the ion branch of the drift wave (the lower curve in Figure 21.1) is always damped. The electron drift wave destabilized by resistivity is usually called the 'resistive drift instability'.

In the simple case where $k_y v_{de} \gg k_z C_s$, the frequency and growth rate of the resistive drift wave instability are given by

$$\omega = k_y v_{de} \qquad \gamma = \eta k_\perp^2 k_y^2 v_{de}^2/\mu_0 k_z^2 v_A^2 \approx \nu_{ei} k_\perp^2 r_{Ls}^2 k_y^2 v_{de}^2/k_z^2 v_{t,e}^2 \tag{21.51}$$

where, in the second form of the expression for the growth rate γ, we have substituted $\eta \approx \nu_{ei} m/ne^2$, where ν_{ei} is the electron–ion collision frequency, and

$v_{t,e} = (T_e/m)^{1/2}$ is the electron thermal velocity. The growth rates of resistive drift instabilities tend to be quite small. Specifically, since $k_y v_{de} \ll k_z v_A$, the first expression for γ in equation (21.51) shows that the growth rate must be very small compared with the rate of resistive diffusion of magnetic field over a distance of order a perpendicular wavelength, i.e. $\eta k_\perp^2/\mu_0$. For perpendicular wavelengths much longer than the ion Larmor radius (evaluated with the electron temperature), i.e. $k_\perp r_{Ls} \lesssim 1$, and for $k_y v_{de} \lesssim k_z C_s \ll k_y v_{t,e}$, the second expression for γ in equation (21.51) shows that the growth rate must also be very much less than the electron–ion collision frequency ν_{ei}. On the other hand, since $\gamma \propto k_\perp^2 k_y^2/k_z^2$, the growth rate increases rapidly as the perpendicular wavelength decreases or as the parallel wavelength increases. Thus, for very short perpendicular wavelengths (down to some limit of order the ion Larmor radius, below which our analysis would not be valid) and for very long parallel wavelengths, the growth rates of resistive drift instabilities can be appreciable. Since the parallel wavelength is limited only by the length of the plasma slab in the z direction, drift-wave instabilities tend to be most serious for plasmas that are very extended along a straight, unidirectional magnetic field. Not surprisingly, drift waves are quite strongly affected by the introduction of magnetic shear, i.e. an equilibrium component $B_{y0}(x)$, as was discussed in the context of resistive tearing instabilities in Chapter 20.

Problem 21.2: Using the same dimensionless quantities for the two axes, add the shear Alfvén wave, whose dispersion relation is given by equation (21.41), to the figure drawn in Problem 21.1. To do this, you need to choose a specific value of β in order to relate C_s to v_A using equation (21.44): take $\beta = 0.02$. Using equation (21.40), indicate which branches of the dispersion relation in the upper (electron) half of your figure become unstable when a small amount of resistivity η is added. By what factor must our 'plane plasma slab' be more extended in the z direction than in the y direction to allow waves with $\omega \sim k_y v_{de} \sim k_z v_A$: give your answer in terms of the quantities r_{Ls}/L_n and β.

Problem 21.3: Examine analytically the region where the two branches of the dispersion relation in the upper half of the figure which you have produced in Problem 21.2 appear to cross each other, i.e. the region $\omega \approx k_y v_{de} \approx k_z v_A$. For the purpose of this analytic calculation, you may assume $\beta \to 0$, i.e. $C_s/v_A \to 0$. By choosing some particular k_z value in this region, for example that given exactly by $k_z v_A = k_y v_{de}$, show from equation (21.40) that there is an instability with a growth rate that scales like $\eta^{1/2}$, rather than like η, for small values of the resistivity. (Hint: You will find it useful to note that the frequency is given approximately by $\omega \approx k_y v_{de} = k_z v_A$, so that equation (21.40) may then be used only to calculate the small complex correction to this frequency.) This more-rapidly growing instability arises from a coupling between the drift wave and the shear Alfvén wave.

21.5 'ELECTROSTATIC' DRIFT WAVES

The astute reader may suspect that the limit $\omega \ll k_z v_A$, in which the lower-frequency drift wave separates from the shear Alfvén wave in the dispersion relation equation (21.40), corresponds to the case where the *magnetic* perturbations play essentially no role in the dynamics. In this sense, the drift wave is sometimes called 'electrostatic'.

We can see this by noting that our analysis of the perturbed generalized Ohm's law, with the added assumption that the perturbed electric field is constrained so as to produce negligible magnetic perturbations, is essentially sufficient by itself to produce the drift-wave dispersion relation: comparing equation (21.38) with equation (21.40), we see that the shear Alfvén wave branch of the dispersion relation arises from retaining the term ωB_x in the first factor on the left-hand side of equation (21.38). This, in turn, arises from retaining the $\dot{B}$ term in the perturbed Ampere's law, i.e. the term on the left-hand side of equation (21.26). Neglecting these terms is equivalent to looking for modes in which the perturbed **E** fields adjust themselves so as to avoid producing significant magnetic perturbations. This will necessarily involve a non-zero perturbed $E_\parallel$ as well as $\mathbf{E}_\perp$, but the generalized Ohm's law allows this perturbed $E_\parallel$ to be balanced by the parallel perturbed electron pressure gradient. If we neglect the term ωB_x in the first factor of the left-hand side of equation (21.38), but keep all of the other terms, using equation (21.18) to provide another relation between u_x and B_x, we obtain the drift-wave branch of the dispersion relation, i.e. equation (21.49).

The derivation of the drift-wave dispersion relation is simplified considerably if we make this 'electrostatic' assumption from the outset. Specifically, the 'electrostatic'' approximation amounts to assuming that the components of the perturbed electric field, $\mathbf{E}_1$, are related to each other by the requirement that $\nabla \times \mathbf{E}_1 = 0$, which implies that the perturbed electric field can be written as the gradient of a scalar potential ϕ, i.e.

$$\mathbf{E} = -\nabla \phi \qquad (21.52)$$

where we have dropped the subscript '1', since both **E** and ϕ are zero in the equilibrium.

As we have seen, the generalized Ohm's law for the perturbed quantities, i.e. equation (21.19), divides into components perpendicular to the magnetic field, for which the approximation

$$\mathbf{u}_\perp \approx \mathbf{E} \times \mathbf{B}/B^2 \tag{21.53}$$

will suffice, and a component parallel to the magnetic field, in which all of the terms must be retained, i.e.

$$E_\parallel = \eta j_\parallel - \frac{1}{ne} \nabla_\parallel p_e. \tag{21.54}$$

Noting that the equilibrium magnetic field is in the z direction and that the perturbed magnetic field is to be neglected, equation (21.54) to first order in the perturbations can be written

$$E_z = \eta j_z - \frac{\mathrm{i} k_z p_{e1}}{ne}. \tag{21.55}$$

In the electrostatic approximation, equation (21.53) tells us that

$$u_x = E_y/B_{z0} = -\mathrm{i} k_y \phi / B_{z0} \tag{21.56}$$

so that

$$E_z = -\mathrm{i} k_z \phi = k_z B_{z0} u_x / k_y \tag{21.57}$$

in which case equation (21.55) becomes

$$k_z B_{z0} u_x = k_y \left(\eta j_z - \frac{\mathrm{i} k_z p_{e1}}{ne} \right) = k_y \left(\eta j_z - \frac{\mathrm{i} k_z T_{e0}}{ne} n_{e1} \right). \tag{21.58}$$

In the second form of equation (21.58), we have again made the assumption that the electron temperature must remain uniform along the (now straight and unperturbed) magnetic field.

To obtain the density perturbation, n_{e1}, in terms of u_z, we proceed in much the same way as before, i.e. we combine the continuity equation

$$-\mathrm{i}\omega n_{e1} + u_x \frac{\mathrm{d} n_{e0}}{\mathrm{d} x} + \mathrm{i} k_z n_{e0} u_z = 0 \tag{21.59}$$

with the parallel component of the equation of motion

$$-\mathrm{i}\omega \rho_0 u_z = -\mathrm{i} k_z T_{e0} n_{e1} \tag{21.60}$$

(see equations (21.33) and (21.35)). We substitute for u_z from equation (21.60) into equation (21.59), thereby obtaining n_{e1} in terms of u_x, which is then substituted into equation (21.58). This gives

$$\left(1 - \frac{k_y v_{de}}{\omega - k_z^2 C_s^2/\omega} \right) u_x = \frac{k_y \eta}{k_z B_{z0}} j_z. \tag{21.61}$$

It remains only to relate the perturbed current density j_z to the mass velocity u_x by the equation of motion. Our procedure here is somewhat different from before, in that we do not want to express the forces arising from current-density perturbations, such as j_z, in terms of the perturbed magnetic fields, as was done in equations (21.5)–(21.11), because these perturbed magnetic fields are neglected, and so are not being otherwise calculated. Rather, we want to deal with the current-density perturbations directly. The x and y components of the perturbed equation of motion, equation (21.5), can be written

$$-i\omega\rho_0 u_x = -\frac{\partial p_1}{\partial x} + j_{y1}B_{z0}$$
$$-i\omega\rho_0 u_y = -ik_y p_1 - j_{x1}B_{z0} \tag{21.62}$$

noting that terms such as $j_z B_y$ and $j_z B_x$ will be second order in the perturbations and may therefore be omitted. Taking $\partial/\partial x$ of the second of these and subtracting ik_y times the first, thereby eliminating the pressure perturbation p_1 (a familiar procedure), we obtain

$$-i\omega\left(\frac{\partial}{\partial x}(\rho_0 u_y) - ik_y \rho_0 u_x\right) = -B_{z0}\left(\frac{\partial j_{x1}}{\partial x} + ik_y j_{y1}\right)$$
$$= ik_z B_{z0} j_z \tag{21.63}$$

where, in the second step, we have made use of the divergence-free property of the perturbed current density. Invoking the incompressibility of $\mathbf{u}_\perp$, i.e. equation (21.14), and using the WKB approximation to express $\partial/\partial x$ as $-ik_x$, equation (21.63) gives

$$j_z = \frac{i\omega\rho_0}{k_y k_z B_{z0}} k_\perp^2 u_x \tag{21.64}$$

where $k_\perp^2 = k_x^2 + k_y^2$. Substituting this into equation (21.61) gives a final dispersion relation

$$\omega - k_y v_{de} - \frac{k_z^2 C_s^2}{\omega} = \frac{i\eta k_\perp^2}{\mu_0}\frac{\omega^2 - k_z^2 C_s^2}{k_z^2 v_A^2} \tag{21.65}$$

exactly the same as equation (21.49). In the case where $k_y v_{de} \ll k_z C_s$, the frequency of the drift wave becomes simply $\omega \approx k_y v_{de}$ and its growth rate is given in equation (21.51).

We conclude that magnetic-field perturbations play no essential role in the dynamics of the low-β drift wave. Rather, the drift wave is produced by a perturbed electric field, whose perpendicular components give rise to perpendicular plasma flows, and whose parallel component is force-balanced self-consistently by the perturbed electron pressure gradient along the magnetic field. Without resistivity, equation (21.54) tells us that the peaks in the electron pressure (or equivalently, the electron density) along the magnetic field coincide exactly with the peaks in the electric potential ϕ. Indeed, assuming as before that the electron temperature remains uniform along the magnetic field, equation (21.54) (without the resistivity term) has the familiar exact nonlinear solution $n_e \propto \exp(e\phi/T_{e0})$, which reflects the tendency of the electrons to adopt a Boltzmann distribution along the magnetic field. In the drift wave, without resistivity, the electron density perturbation will be exactly *in phase* with electric potential perturbations. Introduction of non-zero resistivity produces a small phase shift between the density and potential perturbations. It is this phase shift that allows the drift-wave flow pattern to extract energy from the thermal energy available in the pressure gradient of the electrons to provide for unstable growth of the wave energy.

The analysis of drift waves presented in this Chapter has made several simplifying assumptions, in particular that the equilibrium magnetic field is straight and essentially uniform and that the ions are essentially 'cold', i.e. $T_i \ll T_e$. The introduction of non-zero ion temperature, i.e. $T_i \sim T_e$, would have the predictable effect of bringing the ion diamagnetic drift into the theory, in addition to the electron diamagnetic drift. However, this would not introduce any qualitative change in the stability properties of the drift wave, at least not for the 'electron branch'. The frequency of the 'ion branch' of the drift wave would be modified and, if additional dissipative effects are included, this branch can sometimes be destabilized, but we defer this topic until we are able to treat drift waves from a 'kinetic' viewpoint (see Chapter 26). Modifications to the equilibrium geometry of greatest impact are those that eliminate very small values of the wave-vector parallel to the magnetic field, namely k_z in our case of a straight, uniform field. Finite-length limitations, or the periodic boundary conditions that would be appropriate for a toroidal plasma, rather than an infinitely long plasma slab, are examples where lower limits are imposed on k_z. If the magnetic field is slightly sheared, i.e. a component $B_y(x)$ is added to the larger B_z component (see Chapter 20), then the effective parallel component of the wave-vector becomes $k_\parallel = \mathbf{k} \cdot \hat{\mathbf{B}} \approx k_z + k_y B_y(x)/B_z$, which assumes a range of values as a function of x depending on the width of the mode in the x direction. All 'finite-length' and 'shear' effects tend to be stabilizing, but a detailed analysis of these effects is outside the scope of this book.

Of perhaps more fundamental concern is the validity of the fluid model itself, with its implied assumption that the electrons remain Maxwellian, with a temperature that remains uniform along the magnetic field. We have seen in Chapter 12 that the electron thermal diffusivity along a magnetic field is a quantity of order $v_{t,e}^2/\nu_{ei}$. For the electron temperature to remain essentially uniform along the magnetic field in the presence of a drift wave with frequency ω and wave-number k_z along the field requires that $\omega \ll k_z^2 v_{t,e}^2/\nu_{ei}$. Thus, *the electron collision frequency cannot become arbitrarily large* without violating our assumption of isothermal electrons and requiring a more complete fluid model including parallel temperature gradients. Moreover, inspection of the second form of the growth rate γ given in equation (21.51) shows that for $\omega \sim k_y v_{de}$ the growth rate is then limited to values satisfying $\gamma/\omega \ll k_\perp^2 r_{Ls}^2$. Again, we see that drift-wave growth rates are appreciable only for perpendicular wavelengths that do not exceed by much the ion Larmor radius, although it should be noted that, because of our assumption that $T_i \ll T_e$, our analysis has not implied an expansion in $k_\perp r_{Ls}$. The validity of the fluid model also requires that *the electron collision frequency not be too small*. Specifically, for collisions to maintain a *Maxwellian* distribution along the magnetic field, the mean-free path must be shorter than the parallel wavelength, which requires $k_z v_{t,e} \ll \nu_{ei}$. If this latter requirement is not satisfied, a 'kinetic' version of the 'electron branch' of the drift wave must be found, which is discussed in Chapter 26.

There is a vast literature on drift waves in non-uniform plasmas. An account of the early work in the field is to be found in an article by N A Krall (1968, in *Advances in Plasma Physics 1*, edited by A Simon and W B Thompson New York: Interscience), which discusses the 'kinetic' versions of the drift wave, to be introduced in Chapter 26, as well as the fluid versions which have been described in the present Chapter.

UNIT 6

KINETIC THEORY OF PLASMAS

Fluid (e.g. magnetohydrodynamic) models provide by far the simplest descriptions of plasmas and are sufficiently accurate to describe the majority of macroscopic (i.e. large-scale) plasma phenomena. We recall, however, that our derivation of a closed set of fluid equations for a plasma in Chapter 6 depended on an *ad hoc* assumption regarding the plasma pressure tensor, namely that it is isotropic with the scalar pressure p obeying an equation of state, such as the adiabatic 'gas law'. (We also considered a 'double adiabatic' case with separate pressures perpendicular and parallel to a magnetic field.) There are some plasma phenomena, however, for which these simple forms for the plasma pressure tensor are inapplicable, because the fluid treatment is intrinsically inadequate. For these phenomena, which will occur mainly in low-collisionality plasmas, we need to work explicitly with the *velocity distribution function* $f(\mathbf{x}, \mathbf{v}, t)$ itself for each of the species of particles in the plasma; such a treatment is called a 'kinetic theory'.

In this Unit, we will first derive the fundamental equation for $f(\mathbf{x}, \mathbf{v}, t)$, i.e. the Vlasov equation, which will then be used to analyze from the viewpoint of kinetic theory the simplest type of wave-like disturbance in a plasma, namely electron plasma (Langmuir) waves. When this problem is analyzed correctly, as was first done by Landau, the waves are found to exhibit damping in time, even in fully collisionless situations. Ion acoustic waves are subject to a similar kind of damping. We will also find that certain velocity distributions that depart significantly from the Maxwellian can be subject to new types of 'velocity-space' instabilities. Finally, the kinetic formulation is generalized to low-frequency phenomena in non-uniform plasmas, yielding 'collisionless' versions of the drift wave.

Chapter 22

The Vlasov equation

The fluid approximation is sufficiently accurate to describe the majority of macroscopic (i.e. large-scale) plasma phenomena that are typically encountered, such as the instabilities discussed in the previous Unit. We have also seen that the fluid model is sufficient for providing a good description of important types of wave-like behavior that are possible in a plasma. There are some phenomena, however, for which a fluid treatment is inadequate. For these, we need to work with the *velocity distribution function* $f(\mathbf{x}, \mathbf{v}, t)$, introduced in Chapter 1, for each of the species of particles in the plasma: such a treatment is called a 'kinetic theory'.

22.1 THE NEED FOR A KINETIC THEORY

In fluid theory the relevant dependent variables, such as density, fluid velocity and pressure, are functions of $\mathbf{x}$ and t only. This is possible because the velocity distribution of each species about some mean velocity is implicitly assumed to be Maxwellian (see Figure 22.1(a)) everywhere—uniquely specified by only two parameters, namely the density and temperature. In the hydrodynamics of ordinary fluids and gases, interparticle collisions are usually sufficiently frequent to maintain Maxwellian distributions of particles everywhere in the fluid. In high-temperature plasmas, however, interparticle collisions are relatively infrequent, and deviations from local thermodynamic equilibrium can be maintained for long times. For example, velocity distributions of the type shown in Figure 22.1(b) can often be created in a plasma, as well as, in the three-dimensional case, anisotropic distributions in which the 'temperatures' are different for different velocity-vector directions, e.g. parallel and perpendicular to a magnetic field.

Since collisions are so infrequent in high-temperature plasmas, one might well wonder why a kinetic theory is not needed for *all* plasma problems. Why

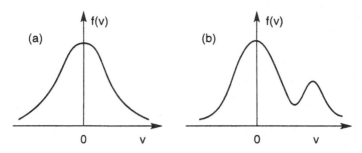

Figure 22.1. Examples of (a) Maxwellian and (b) non-Maxwellian one-dimensional velocity distributions. The distribution (b) has a 'beam' of suprathermal particles superimposed on an approximately Maxwellian background distribution.

does a fluid theory work at all? The reason is that a strong magnetic field can play the role of collisions in maintaining approximately Maxwellian distributions and in providing the 'localizing influence' that is the essential ingredient in a fluid theory. For plasma phenomena that are 'slow' and 'large scale' in relation to particle gyration, in the sense that their typical time-scale is long compared with the Larmor period and their typical spatial length-scale large compared with the Larmor radius, all particles remain close to their initial field lines. An initially Maxwellian distribution of such particles would *remain* approximately Maxwellian. For this reason, the *two-dimensional flow* of a plasma *perpendicular* to a strong magnetic field can often be treated by magnetohydrodynamics, even when collisions are very infrequent. For an example of a 'two-dimensional flow' exactly perpendicular to the magnetic field, we could cite the Rayleigh–Taylor (flute) instability discussed in Chapter 19. Often the plasma flow perpendicular to the magnetic field can be treated by magnetohydrodynamics even for phenomena that are not *exactly* two-dimensional, because commonly the length-scales are such that $L_\parallel \gg L_\perp$: the tearing and drift instabilities discussed in Chapters 20 and 21 fall into this category.

For flow *along* the magnetic field, however, the fluid theory will be valid only if collisions are frequent enough (specifically if the mean-free path is much shorter than some characteristic distance along the field). In the case where there are no collisions, the individual particles making up the plasma will freely stream for large distances along the field. To treat such problems, we need a kinetic theory, in which individual particle velocities are taken into account. Such a theory will also be needed to treat problems involving flow *across* a magnetic field in the case where the magnetic field is very weak, in the sense that the gyration period and gyration radius are *not* small compared with characteristic time-scales and length-scales of the flow.

In summary, therefore, kinetic theory is needed to treat (i) problems involving flow along a magnetic field (or in the absence of a magnetic field) in the case of long mean-free path ($\lambda_{mfp} \gtrsim L_{\parallel}$, where $L_{\parallel}$ is the scale-length of the gradients along the field), and (ii) problems of high-frequency ($\omega \gtrsim \omega_c$) and/or short-wavelength ($k_{\perp} r_L \gtrsim 1$) flow across a magnetic field.

Before beginning our development of the kinetic theory of plasmas, however, we must first establish some important properties of the particle distribution function.

22.2 THE PARTICLE DISTRIBUTION FUNCTION

The basic element in the kinetic description of a plasma is the distribution function $f(\mathbf{x}, \mathbf{v}, t)$ that describes how particles are distributed in both physical space and velocity space.

Consider a plasma as a collection of N charged particles, each with its own position and velocity vectors $\mathbf{x}_i$ and $\mathbf{v}_i$. If the forces $\mathbf{F}_i$ on the various particles are given, the position and velocity vectors will evolve according to

$$d\mathbf{x}_i/dt = \mathbf{v}_i$$
$$d\mathbf{v}_i/dt = \mathbf{a}_i = \mathbf{F}_i/m. \tag{22.1}$$

The forces $\mathbf{F}_i$ will, in general, be composed of both a macroscopic, or slowly varying part, together with a microscopic, or rapidly varying part, due to short-range interparticle forces, i.e. collisions. The *macroscopic* part will be approximately the same for all particles with about the same $\mathbf{x}_i$ and $\mathbf{v}_i$. Our fundamental assumption here is that the macroscopic forces are dominant over the microscopic, i.e. collisional, forces. Accordingly, rather than treat the entire array of $6N$ equations represented by equation (22.1), we simplify the problem by a statistical treatment based on the assumption that we need not distinguish between particles that have about the same velocity and are located at about the same place.

Specifically, we can average over distances that are large compared with the interparticle spacing, $n^{-1/3}$, but small compared with the Debye length, λ_D, which characterizes the minimum length-scale for the so-called 'collective' plasma phenomena we will treat. (Remember that a key definition of a plasma is that $n\lambda_D^3 \gg 1$, so these two distances are very far apart.) Averaging over distances that are only a small fraction of the Debye length can effectively eliminate binary collisions, despite the fact that the Coulomb logarithm includes collisions out to impact parameters of order λ_D. For example, if we average all the electric fields over a distance of order one-tenth of the Debye length, we would be effectively reducing the Coulomb logarithm, $\ln \Lambda$ (a quantity typically in the range 16–20), by only $\ln 10 \sim 2.3$. Since the kinetic theory is generally applied to plasma phenomena for which binary collisions are relatively unimportant (indeed the Vlasov equation which we are about to derive describes the case where binary collisions are entirely neglected), a possible error of this magnitude in estimating the Coulomb logarithm will not matter. More important, 'collective' plasma phenomena, i.e. phenomena arising from electric fields averaged in this way, do occur for characteristic length-scales as small as a Debye length, or even somewhat smaller. It is important to note that our averaging process preserves the capability to describe such phenomena.

As was done in Chapter 1, we can define a *distribution function* $f(\mathbf{x}, \mathbf{v}, t)$ that represents the number density of particles found 'near' a point in the six-dimensional space $(\mathbf{x}, \mathbf{v})$. Specifically, the number of particles located within a volume element d^3x of physical space and having velocities lying within a volume element d^3v of velocity space is defined to be $f(\mathbf{x}, \mathbf{v}, t)\mathrm{d}^3v\mathrm{d}^3x$. The six-dimensional space whose volume element is $\mathrm{d}^3x\mathrm{d}^3v$ is called 'phase space'.

The number density of particles in physical space is given simply by

$$n(\mathbf{x}, t) = \int f(\mathbf{x}, \mathbf{v}, t)\mathrm{d}^3v. \tag{22.2}$$

The mean (fluid) velocity of the particles is given by

$$n\mathbf{u} = \int \mathbf{v} f(\mathbf{x}, \mathbf{v}, t)\mathrm{d}^3v. \tag{22.3}$$

A scalar pressure can be defined by

$$p(\mathbf{x}, t) = \frac{m}{3} \int v^2 f(\mathbf{x}, \mathbf{v}, t)\mathrm{d}^3v \tag{22.4}$$

although it must be noted that f may not be isotropic in velocity space, in which case the concept of a scalar pressure may be inappropriate. Indeed, we recall that, as early as Chapter 1, we introduced the idea of different pressures parallel and perpendicular to the magnetic field, and a general pressure *tensor* was defined in Chapter 6. In equations (22.2)–(22.4), the integrals go from $-\infty$ to $+\infty$ for each of the three velocity components, v_x, v_y and v_z.

As discussed in Chapter 1, in thermal equilibrium, i.e. after many interparticle collisions have occurred, particle distribution functions will always relax toward the (three-dimensional) Maxwellian velocity distribution:

$$f_{\mathrm{M}}(\mathbf{v}) = n \left(\frac{m}{2\pi T} \right)^{3/2} \exp\left(-\frac{mv^2}{2T} \right) \tag{22.5}$$

where the density n and temperature T will, in general, both be functions of $\mathbf{x}$ and t. The Maxwellian distribution is isotropic, and the mean square velocity is the same in any direction, namely

$$\langle v_x^2 \rangle = \langle v_y^2 \rangle = \langle v_z^2 \rangle = \frac{1}{n} \int \langle v_x^2 \rangle f_{\mathrm{M}}(\mathbf{x}, \mathbf{v}, t)\mathrm{d}^3v = \frac{T}{m}. \tag{22.6}$$

The pressure is then given by $p = nT$, as usual. We can generalize the Maxwellian distribution to allow a non-zero mean velocity $\mathbf{u}$, in which case v^2 should be replaced by $|\mathbf{v} - \mathbf{u}|^2$ in the expression for $f_M(\mathbf{v})$; for present purposes, however, it will generally be sufficient to consider Maxwellian distributions with zero mean velocity.

Often we are interested in the one-dimensional velocity distribution, obtained by integrating over the other two components of velocity. For example, the distribution of velocities v_x is given by

$$F(v_x) = \int_{-\infty}^{\infty} \int_{-\infty}^{\infty} f(v_x, v_y, v_z) dv_y dv_z. \tag{22.7}$$

For a Maxwellian distribution

$$F_M(v_x) = n \left(\frac{m}{2\pi T}\right)^{1/2} \exp\left(-\frac{mv_x^2}{2T}\right). \tag{22.8}$$

Since the three-dimensional distribution $f_M(v_x, v_y, v_z)$ is isotropic, it is sometimes more convenient to work in spherical velocity coordinates. The volume element in spherical velocity coordinates (v, θ, ϕ) is given by $d^3v = v^2 \sin\theta d\phi d\theta dv$, where v takes on all values from 0 to ∞, θ all values from 0 to π, and ϕ all values from 0 to 2π. Since f_M is independent of θ and ϕ, we can integrate over these two coordinates from $\phi = 0$ to 2π and from $\theta = 0$ to π, respectively, to obtain $\int \sin\theta d\phi d\theta = 4\pi$. Having integrated over θ and ϕ in this way, the volume element d^3v becomes simply $4\pi v^2 dv$, which is of course simply the volume of a thin spherical shell in velocity space. We can now define a distribution $g_M(v)$ which represents the number of particles per unit volume and per unit *magnitude* v of the three-dimensional velocity-vector $\mathbf{v}$ (with v going from 0 to ∞), namely

$$g_M(v) = 4\pi n \left(\frac{m}{2\pi T}\right)^{3/2} v^2 \exp\left(-\frac{mv^2}{2T}\right). \tag{22.9}$$

In a similar way, a distribution $g(v)$ could be defined for any $f(\mathbf{v})$ that is isotropic in velocity space. In such cases, the number density is given by

$$n(\mathbf{x}, t) = \int_0^{\infty} g(v) dv \tag{22.10}$$

and the scalar pressure is given by

$$p(\mathbf{x}, t) = \frac{m}{3} \int_0^{\infty} v^2 g(v) dv. \tag{22.11}$$

22.3 THE BOLTZMANN–VLASOV EQUATION

We want now to obtain an equation for the evolution of the particle distribution function, $f(\mathbf{x}, \mathbf{v}, t)$. We can do this rather simply by invoking conservation of particle number, as we follow a group of particles through phase space. Consider a small region of phase space, representing the position and velocity coordinates of a group of particles. For simplicity of portrayal, we might suppose that phase space contained only one dimension of physical space, x, and one dimension of velocity, v_x, in which case our group of particles might occupy the small region A shown in Figure 22.2. After a certain time interval, the particles occupying the region A in Figure 22.2 will have moved to the region B.

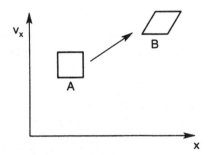

Figure 22.2. One-dimensional phase space (x, v_x) in which a group of particles occupying the region A move, after a certain time interval, to the region B.

Problem 22.1 Explain why the square shape of A changes to the parallelogram shape of B.

The points on the surface of any volume in phase space move according to the rules

$$\frac{\mathrm{d}\mathbf{x}}{\mathrm{d}t} = \mathbf{v} \qquad \frac{\mathrm{d}\mathbf{v}}{\mathrm{d}t} = \frac{\mathbf{F}}{m} \qquad (22.12)$$

where $\mathbf{F}$ is the external force. The number of particles N in a volume of phase space is given by

$$N = \int f(x, v, t)\mathrm{d}^3 v \mathrm{d}^3 x. \qquad (22.13)$$

Conservation of the number of particles demands that the total time derivative of N must vanish, where the 'total' time derivative means that we allow the boundary surface to move with the particles that lie on it:

$$0 = \frac{\mathrm{d}N}{\mathrm{d}t} = \int \frac{\partial f}{\partial t}\mathrm{d}^3 v \mathrm{d}^3 x + \int f\mathbf{U} \cdot \mathrm{d}\mathbf{S}. \qquad (22.14)$$

Here, the second term on the far right-hand side arises from the additional volume captured or lost by the moving surface. It is important to note that the 'velocity vector' **U** and the 'surface area' d**S** in this equation are both *six-vectors* in (**x**, **v**) space: the six components of **U** are $(\dot{\mathbf{x}}, \dot{\mathbf{v}}) = (\mathbf{v}, \mathbf{F}/m)$. By using the divergence theorem in six-dimensional space, we can rewrite the conservation equation

$$0 = \frac{dN}{dt} = \int \left(\frac{\partial f}{\partial t} + \boldsymbol{\nabla} \cdot (f\mathbf{U}) \right) d^3 v d^3 x. \tag{22.15}$$

Here $\boldsymbol{\nabla}$ denotes a six-component divergence operator, whose components are $(\boldsymbol{\nabla}_x, \boldsymbol{\nabla}_v)$. Since equation (22.15) must hold for every volume element in phase space, we must have

$$\frac{\partial f}{\partial t} + \boldsymbol{\nabla} \cdot (f\mathbf{U}) = 0 \tag{22.16}$$

where, again, both $\boldsymbol{\nabla}$ and **U** are six-vectors.

In terms of ordinary three-vectors, where we write simply $\boldsymbol{\nabla}$ for $\boldsymbol{\nabla}_x$ and $\partial/\partial \mathbf{v}$ for $\boldsymbol{\nabla}_v$, equation (22.16) may be written, for certain forms of **F**, as

$$\frac{\partial f}{\partial t} + \mathbf{v} \cdot \boldsymbol{\nabla} f + \frac{\mathbf{F}}{m} \cdot \frac{\partial f}{\partial \mathbf{v}} = 0. \tag{22.17}$$

In obtaining this form, we have assumed

$$\boldsymbol{\nabla} \cdot \mathbf{U} = \boldsymbol{\nabla}_x \cdot \mathbf{v} + \boldsymbol{\nabla}_v \cdot \frac{\mathbf{F}}{m} = 0. \tag{22.18}$$

In equation (22.18), the first term on the right-hand side vanishes since **v** is not a function of **x**: indeed **x** and **v** are independent coordinates in our six-dimensional space. The second term vanishes provided the force **F** is not a function of **v** (as will be true, for example, for electric and gravitational forces).

Before proceeding further, we must stop and note that the *Lorentz* force *is* a function of **v**, namely

$$\mathbf{F} = q\mathbf{v} \times \mathbf{B}. \tag{22.19}$$

However, evaluating the velocity-space divergence of this force, component by component, we obtain

$$\boldsymbol{\nabla}_v \cdot (\mathbf{v} \times \mathbf{B}) = \frac{\partial}{\partial v_x}(v_y B_z - v_z B_y) + \ldots = 0 \tag{22.20}$$

so that our result, i.e. equation (22.17), remains valid also for the particular **v**-dependence of the Lorentz force.

We may now give our final result for the case of a plasma whose particles are acted upon by electric and magnetic forces, namely

$$\mathbf{F} = q(\mathbf{E} + \mathbf{v} \times \mathbf{B}). \tag{22.21}$$

For this case, the equation for the evolution of the particle distribution function **f** becomes

$$\frac{\partial f}{\partial t} + \mathbf{v} \cdot \nabla f + \frac{q}{m}(\mathbf{E} + \mathbf{v} \times \mathbf{B}) \cdot \frac{\partial f}{\partial \mathbf{v}} = 0. \tag{22.22}$$

We note that this equation could be thought of as $Df/Dt = 0$, where D/Dt means the total derivative following the particle along its trajectory in six-dimensional phase space.

When collisions are important, an additional term $(\partial f/\partial t)_{\text{coll}}$ must be included on the right-hand side of equation (22.22) to describe the effect of short-range interparticle forces, especially binary collisions whose effect was excluded in our initial 'averaging' of electric fields over some distance intermediate between the interparticle spacing and the Debye length. This additional term describes a local evolution of the velocity-space distribution at each point in physical space; it has no direct explicit effect on the distribution function elsewhere in physical space. For small-angle-scattering Coulomb collisions, an approximate form $(\partial f/\partial t)_{\text{coll}}$ is the Fokker–Planck form derived in Chapter 13. When collisions are fully included, equation (22.22) is usually called the 'Boltzmann equation', after Boltzmann who was the first to obtain an expression for $(\partial f/\partial t)_{\text{coll}}$ for the case of short-range interparticle forces. The form of Boltzmann's collision term is unlike that of the Fokker–Planck equation in that it is appropriate mainly for situations where large-angle scattering is dominant. However, the Boltzmann collision term can accommodate an arbitrary dependence of the cross section, σ, on the impact parameter, b, and on the velocities of the colliding particles. When the Coulomb cross section is used and only the (dominant) contributions from small-angle scattering are retained, the Boltzmann form of the collision term reduces to the Fokker–Planck form. For more on this topic, the reader is referred to the monograph by D C Montgomery and D A Tidman (1964 *Plasma Kinetic Theory* New York: McGraw-Hill).

When collisions of all kinds are neglected, the equation is usually called the 'Vlasov equation', after A A Vlasov who was the first to formulate the 'collisionless' equation in the form of equation (22.22) (1938 *Zh. Eksp. Teor. Fiz.* **8** 291 (in Russian)).

22.4 THE VLASOV–MAXWELL EQUATIONS

We have now derived the Vlasov equation, (22.22), for the evolution of the particle distribution function $f(\mathbf{x}, \mathbf{v}, t)$ in a collisionless plasma. Typically, the electric and magnetic fields **E** and **B** making up the force $q(\mathbf{E} + \mathbf{v} \times \mathbf{B})$ are partly due to externally applied fields and partly due to internally generated fields. In order to have a closed set of equations, we must find some way of deriving the 'internally generated' parts of the electric and magnetic fields from the distribution function that describes the plasma particles themselves.

The internally generated force on a charged particle in a plasma can be divided roughly into two parts; one part is the average force due to a large number of relatively distant particles, and the other part is the force due to the near-neighbor particles, i.e. collisions. In a collisionless plasma, the former greatly exceeds the latter. Since 'collisions' in a plasma have been shown to include all Coulomb interactions at impact parameters up to a distance that greatly exceeds the interparticle spacing $n^{-1/3}$ and approaches the Debye length, a 'collisionless' plasma is one in which the forces due to more distant particles (at distances of order the Debye length and larger) are dominant. Moreover, the average force due to *distant* particles does not depend on the exact location of these particles, but only on the density of such particles averaged over a small region with $L \gg n^{-1/3}$ at each distant location. In other words, it depends only on the distribution function $f(\mathbf{x}, \mathbf{v}, t)$ itself. This average force due to distant plasma particles can be combined with any externally applied force. In this spirit, the electric and magnetic fields that make up the average force in the Vlasov equation are to be *calculated self-consistently from the Maxwell equations*:

$$\nabla \cdot (\epsilon_0 \mathbf{E}) = \sigma$$
$$\nabla \times \mathbf{B} = \mu_0 \mathbf{j} + \frac{1}{c^2} \frac{\partial \mathbf{E}}{\partial t} \qquad (22.23)$$

which have been written in terms of the electric and magnetic fields $\mathbf{E}$ and $\mathbf{B}$ that appear in the force law, eliminating $\mathbf{D}(= \epsilon_0 \mathbf{E})$ and $\mathbf{H}(= \mathbf{B}/\mu_0)$ by using the free-space permittivity, ϵ_0, and permeability, μ_0, respectively. For completeness, we repeat the other two Maxwell equations:

$$\nabla \cdot \mathbf{B} = 0$$
$$\nabla \times \mathbf{E} = -\frac{\partial \mathbf{B}}{\partial t}. \qquad (22.24)$$

In equation (22.23), *the charge density and current density are to be obtained at each point in space from the appropriate integrals of the distribution function itself:*

$$\sigma = \sum q \int f \mathrm{d}^3 v$$
$$\mathbf{j} = \sum q \int \mathbf{v} f \mathrm{d}^3 v \qquad (22.25)$$

where the summation is over the species of particles present in the plasma.

Superficially, the Vlasov–Maxwell equations, (22.22)–(22.24), *resemble* the Liouville equation for an ensemble of particles moving in externally generated $\mathbf{E}$ and $\mathbf{B}$ fields with fixed charge and current densities. The Liouville equation, by itself, contains no terms that describe interactions among particles of the ensemble. However, the Vlasov–Maxwell equations are *conceptually distinct* from the Liouville equation, because a part (the main part in a collisionless plasma) of the interaction with all the other particles is retained by including 'smeared-out' charge and current densities σ and $\mathbf{j}$ derived from the distribution function itself (see equation (22.25)) in calculating the electric and magnetic fields. Although information is lost in treating the other particles in this 'smeared-out' way, the Vlasov–Maxwell equations *do* contain the main part of the interparticle interaction in low-collisionality plasmas, and they provide the most realistic description of such a plasma that is analytically tractable.

The addition of a Fokker–Planck collision term to the right-hand side of the Vlasov equation, i.e. to equation (22.22), provides an even superior description, in that this equation applies also to collisional plasmas and thus includes a very wide range of plasma-physical effects. However, analytic tractability is sacrificed substantially if the full Fokker–Planck term is used. It is sometimes possible to employ much simplified forms of the collision term in kinetic descriptions of complex phenomena such as plasma waves, where collision effects are often subdominant. For example, the expression $(\partial f/\partial t)_{\text{coll}} = -\nu(f - f_{\text{max}})$ is often used, where ν represents a typical collision frequency, and where f_{max} is a Maxwellian distribution with the same number density and same energy density (i.e. temperature) as the distribution f (and sometimes in the case of like-particle collisions which conserve momentum, the same mass velocity). This simplified collisional model does not describe correctly the physical effects of collisional velocity-space scattering, but it does describe relaxation toward an appropriate Maxwellian distribution and it is, at least in some sense, *linear* in f.

For the applications of kinetic models of plasmas that take up the remainder of this book, we will limit ourselves to low-collisionality plasmas for which equation (22.22) provides an adequate description, *without* the addition of a Fokker–Planck collision term to the right-hand side.

Chapter 23

Kinetic effects on plasma waves: Vlasov's treatment

The Vlasov–Maxwell equations can be used to determine how the various types of plasma waves that were discussed in Chapters 15–18 are affected by the presence of a *distribution* of particle velocities. Even in the case where the velocity distribution in the absence of waves is Maxwellian, significant 'kinetic effects' can enter due to the presence of particles streaming along at speeds which are comparable to the wave phase velocity. The velocity distribution may, instead, be quite strongly non-Maxwellian, in which case new kinds of plasma waves can arise, including some that can be unstable, i.e. grow exponentially in amplitude in time.

As our first illustration of the use of the Vlasov equation for treating plasma waves, we will derive the dispersion relation for electron plasma (Langmuir) waves. We have seen in Chapter 16 that electron plasma waves with $k = 0$ have a frequency

$$\omega = \omega_p = (ne^2/\epsilon_0 m)^{1/2} \tag{23.1}$$

but that 'thermal effects' (treated in Chapter 16 via a fluid model) modify this dispersion relation for electron plasma waves with non-zero wave-number k.

Plasma waves are high-frequency oscillations in which electrons move back and forth relative to fixed ions, creating an oscillating space charge that is self-consistent with the oscillating electric field driving the electron motion. The oscillations take place either *along* a magnetic field, or *in the absence* of any significant magnetic field; in both cases, the magnetic field plays no role. The only Maxwell equation that enters into the theory of these waves is the Poisson equation, which relates the charge density to the electric field. Thus, the equations to be used for treating plasma waves are the Vlasov equation for the distribution function f (including an electric field E) and the Poisson equation in which the charge density is expressed in terms of the appropriate integral of the distribution function f over velocity space.

23.1 THE LINEARIZED VLASOV EQUATION

We will assume here, as in Chapter 16, that the plasma waves are of very small amplitude, i.e. they represent only a small perturbation away from an initial equilibrium. Moreover, since the wavelength of plasma waves is very small, sometimes only a few times the Debye length, the background plasma equilibrium will be assumed to be spatially uniform over such short distances. Thus, in the equilibrium that exists before the perturbation is applied, the distribution function f_0 can be considered to be a function of $\mathbf{v}$ only, and not of $\mathbf{x}$. The electron and ion equilibrium distribution functions f_0 must be chosen such that the electron and ion number densities are equal, so that they correspond to the physical case in which the plasma is charge-neutral. There will then be no electric field in the equilibrium state: the electric field will arise only when the perturbation is applied.

We will further assume (see Chapter 16) that the plasma wave being considered takes the form of a plane wave travelling in the x direction, with the electric field having an x component only, given by

$$E(x, t) = \hat{E}\exp(-i\omega t + ikx) \tag{23.2}$$

where $\hat{E}$ represents the wave amplitude.

Since the v_y and v_z velocity components are not affected by the electric field, we may integrate the three-dimensional electron velocity distribution $f(\mathbf{v})$ over v_y and v_z and work with the resulting one-dimensional distribution, which in Chapter 22 we denoted $F(v_x)$ but which here we will simply denote $f(v_x)$ to retain the familiar appearance of the Vlasov equation. We may also drop the subscript x from v_x, since the problem has in effect become purely one-dimensional. This one-dimensional electron distribution, now denoted simply $f(v)$, must satisfy a Vlasov equation which is the same as would apply if the problem had been strictly one-dimensional from the outset, namely

$$\frac{\partial f}{\partial t} + v\frac{\partial f}{\partial x} - \frac{e}{m}E\frac{\partial f}{\partial v} = 0 \tag{23.3}$$

where we have set $q = -e$ for the electron charge. Moreover, for *small-amplitude* waves, the oscillating E field is small and leads to a small perturbation $f_1(x, v, t)$ (in effect, the first-order term in an expansion of the exact distribution function f in powers of E) away from the initial, spatially uniform velocity distribution $f_0(v)$, so that, if we write

$$f(x, v, t) = f_0(v) + f_1(x, v, t) \tag{23.4}$$

we can assume that f_1 is small compared with f_0. Simple estimation of the magnitude of the third term in equation (23.3) compared with the first term shows that the expansion is in the dimensionless parameter $eE/(\omega m v)$; for this parameter to be small for a typical particle requires that the acceleration in a wave period produces only a small change in velocity relative to v_t, the thermal velocity. The 'linearized Vlasov equation' becomes

$$\frac{\partial f_1}{\partial t} + v\frac{\partial f_1}{\partial x} - \frac{e}{m}E\frac{\partial f_0}{\partial v} = 0 \qquad (23.5)$$

where we have neglected second-order terms that involve products of first-order quantities f_1 and E.

For electron plasma waves, the only relevant Maxwell equation is the Poisson equation

$$\epsilon_0 \nabla \cdot \mathbf{E} = \sigma = -e\int f_1 \mathrm{d}^3 v \qquad (23.6)$$

where we have taken account of the fact that only the electrons, and not the ions, contribute significantly to the oscillating charge density. For our one-dimensional problem, in which our f_1 has *already* been integrated with respect to the other two velocity coordinates, we can write this Poisson equation

$$\epsilon_0\frac{\partial E}{\partial x} = -e\int_{-\infty}^{\infty} f_1 \mathrm{d}v. \qquad (23.7)$$

The task is now to solve simultaneously equations (23.5) and (23.7).

23.2 VLASOV'S SOLUTION

To Vlasov, who was the first to attempt this problem, it seemed reasonable to suppose that f_1 also has a wave-like form in both space and time, just like equation (23.2) for $E(x, t)$, namely

$$f_1(x, v, t) = \hat{f}_1(v)\exp(-\mathrm{i}\omega t + \mathrm{i}kx). \qquad (23.8)$$

The linearized Vlasov equation, equation (23.5), can then be written

$$-\mathrm{i}(\omega - kv)\hat{f}_1 = \frac{e}{m}\hat{E}\frac{\partial f_0}{\partial v} \qquad (23.9)$$

which can be 'solved' to give

$$\hat{f}_1 = \frac{\mathrm{i}e\hat{E}}{m}\frac{\partial f_0/\partial v}{\omega - kv}. \qquad (23.10)$$

This 'solution' for $\hat{f}_1$ may be substituted into the Poisson equation, equation (23.7), which now takes the form

$$
\begin{aligned}
ik\epsilon_0 \hat{E} &= -e \int_{-\infty}^{\infty} \hat{f}_1 \mathrm{d}v \\
&= -\frac{ie^2 \hat{E}}{m} \int_{-\infty}^{\infty} \frac{\partial f_0/\partial v}{\omega - kv} \mathrm{d}v.
\end{aligned}
\tag{23.11}
$$

After dividing through by $ik\epsilon_0 \hat{E}$, noting that $\hat{E}$ cannot be zero by hypothesis (or else there would be no perturbation to study) and collecting both terms on the left-hand side, we obtain the relation

$$
D(k, \omega) \equiv 1 + \frac{e^2}{mk\epsilon_0} \int_{-\infty}^{\infty} \frac{\partial f_0/\partial v}{\omega - kv} \mathrm{d}v = 0.
\tag{23.12}
$$

The function $D(k, \omega)$ is often called the 'plasma dispersion function', and the equation $D(k, \omega) = 0$ defines the *dispersion relation*, since this equation can, in principle, be solved to produce a relation of the type $\omega = \omega(k)$. The function $D(k, \omega)$, which was first obtained by A A Vlasov (1945 *J. Phys. USSR* **9** 25), is also sometimes called the 'plasma dielectric function', because the oscillating charge density σ can be viewed as *internal* to the plasma and absorbed into a frequency and wavelength-dependent kinetic 'dielectric constant', which would in this case be just $\epsilon_0 D(k, \omega)$, so that Poisson's equation becomes $\nabla \cdot \mathbf{D} = 0$ where $\mathbf{D} = \epsilon_0 D(k, \omega)\mathbf{E}$.

It should be emphasized that the dispersion function given in equation (23.12) applies only to the case of electron plasma waves in an unmagnetized plasma. It corresponds to the high-frequency kinetic generalization of the 'electrostatic' term in the cold-plasma dielectric tensor introduced in Chapter 18, i.e. the term $P\hat{z}\hat{z}$ in the dielectric tensor for $\theta = 0$ given in equation (18.16). A more complicated dispersion, or 'dielectric', function (in the form of a tensor) must be derived to describe the full range of plasma waves in a 'hot', kinetic plasma. This would correspond to the kinetic generalization of the full dielectric tensor. This kinetic dielectric tensor gives rise to a new class of waves in a magnetized plasma in the vicinity of *harmonics* of the ion and electron cyclotron frequencies, the so-called 'Bernstein waves'. The full range of waves in a magnetized plasma described by kinetic theory is discussed in specialized texts, for example T H Stix (1992 *Waves in Plasmas* New York: American Institute of Physics), but is beyond the scope of the present book. In this Chapter and the following two Chapters, we will limit ourselves to the case of electrostatic waves in an unmagnetized plasma (or with **k** and **E** vectors directed *along* the magnetic field, so that the Lorentz force plays no role). The first fully kinetic treatment of waves in a magnetized plasma, including perpendicular wavelengths of order the Larmor radius and frequencies of order the cyclotron frequency and harmonics thereof, was given in a paper by I B Bernstein (1958 *Phys. Rev.* **109** 10).

In *principle*, we have now solved our problem of high-frequency electrostatic waves in an unmagnetized plasma from a kinetic viewpoint. Given some initial velocity distribution f_0 and some wave-number k, we can carry out the integration over v in equation (23.12) to obtain an explicit form for the dispersion function $D(k, \omega)$. We can then find the frequency ω by solving the dispersion relation $D(k, \omega) = 0$ for any given k value. In practice, this is difficult to do, because the integral over v can rarely be done analytically.

23.3 THERMAL EFFECTS ON ELECTRON PLASMA WAVES

An approximate solution for electron plasma waves can be obtained by assuming that, for almost all particle velocities v, the relation $\omega \gg kv$ will be a good approximation. Since this states that the phase velocity of the wave is much larger than a typical particle velocity, we expect it to correspond to the adiabatic approximation in a fluid treatment. We may then expand the integrand in equation (23.12):

$$\frac{1}{\omega - kv} = \frac{1}{\omega} + \frac{kv}{\omega^2} + \frac{k^2 v^2}{\omega^3} + \frac{k^3 v^3}{\omega^4} + \dots . \qquad (23.13)$$

For a Maxwellian distribution f_0, i.e. the distribution given in equation (22.8), the integrals over the one-dimensional velocity v can then be carried out explicitly, noting that

$$\int_{-\infty}^{\infty} \frac{\partial f_0}{\partial v} dv = 0 \qquad \int_{-\infty}^{\infty} \frac{\partial f_0}{\partial v} v \, dv = -n$$

$$\int_{-\infty}^{\infty} \frac{\partial f_0}{\partial v} v^2 dv = 0 \qquad \int_{-\infty}^{\infty} \frac{\partial f_0}{\partial v} v^3 dv = -3n v_t^2 \qquad (23.14)$$

where $v_t = (T/m)^{1/2}$. Going this far but no further, the dispersion relation, equation (23.12), becomes

$$D(k, \omega) \equiv 1 - \frac{\omega_p^2}{\omega^2} \left(1 + \frac{3k^2 v_t^2}{\omega^2} \right) = 0 \qquad (23.15)$$

where $\omega_p = (ne^2/m\epsilon_0)^{1/2}$. We may solve equation (23.15) by successive approximations, assuming $\omega^2 \gg k^2 v_t^2$. First, we neglect completely the term in $k^2 v_t^2/\omega^2 \ll 1$, thereby obtaining the zeroth-order approximation to the solution, namely just $\omega^2 = \omega_p^2$. Next we retain the correction term in $k^2 v_t^2/\omega^2$ but, to obtain a solution that is correct to first order in this small parameter, it suffices to evaluate the correction term using the zeroth-order solution for ω^2, namely $\omega^2 = \omega_p^2$. Thus, correct to first order, the term in parenthesis in equation (23.15) can be written $1 + 3k^2 v_t^2/\omega_p^2$, which makes equation (23.15) trivial to solve for ω^2. Thus, for an approximate solution of equation (23.15) correct to first order in $k^2 v_t^2/\omega_p^2$, we obtain

$$\omega^2 \approx \omega_p^2 + 3k^2 v_t^2 \qquad (23.16)$$

which is exactly the dispersion relation obtained in the fluid treatment of Chapter 16 for the adiabatic case.

Problem 23.1: In the same limit, $k^2 v_t^2/\omega_p^2 \ll 1$, find the other solution of equation (23.15) treated as a quadratic equation for ω^2, and explain why this solution is unphysical.

Thus 'thermal effects' lead to a modification of the simple dispersion relation, $\omega = \omega_p$, for electron plasma waves. However, remembering that $v_t/\omega_p = \lambda_D$, the Debye length, we see that, as in the fluid model, the thermal corrections are small for wavelengths much longer than the Debye length. Moreover, we must be wary of using equation (23.16) for wavelengths as short as a Debye length (although, as we saw in Chapter 22, the underlying Vlasov–Maxwell equations are valid for length-scales as short as this), because the assumption on which our expansion was based, namely $k^2 v_t^2/\omega_p^2 \ll 1$, would be violated. Nonetheless, even where the thermal effects represent small corrections to the dispersion relation, as we saw in Chapter 16, these effects provide plasma waves with features that would otherwise be absent; for example, there is now a non-zero group velocity $d\omega/dk$, with the result that energy may be propagated by the waves from one part of the plasma to another.

We should beware, however, that equation (23.16) does not account *fully* for thermal effects on electron plasma waves, since it has resulted from an *approximate* evaluation of the integral appearing in the dispersion relation, equation (23.12). We will return to this topic in Chapter 24.

23.4 THE TWO-STREAM INSTABILITY

For initial velocity distributions f_0 that depart substantially from Maxwellian, it is possible for electron plasma waves to become *unstable*. For example, suppose that the electron velocity distribution consists of two identical but oppositely directed streams with velocities $\pm v_0$. For simplicity, let us assume that each stream is 'cold', i.e. the particles in each stream have negligible thermal spread about their streaming velocity. Using δ-functions as a way of expressing one-dimensional velocity distributions without any thermal spread (e.g. a one-dimensional Maxwellian with zero temperature is the δ-function $n\delta(v)$), the distribution function for two such electron streams is

$$f_0(v) = \tfrac{1}{2}n[\delta(v - v_0) + \delta(v + v_0)] \tag{23.17}$$

where n is the total density of electrons, i.e. in both streams together.

We substitute this distribution f_0 into the Vlasov dispersion relation, equation (23.12). At first sight, the velocity integral might appear intractable, since the integrand involves a derivative of a δ-function. However, integrals of this type may readily be evaluated using integration by parts, as follows:

$$
\begin{aligned}
\int_{-\infty}^{\infty} \frac{\partial f_0/\partial v}{\omega - kv}\,dv &= -\int_{-\infty}^{\infty} f_0 \frac{\partial}{\partial v}\left(\frac{1}{\omega - kv}\right)dv + \left[\frac{f_0}{\omega - kv}\right]_{-\infty}^{\infty} \\
&= -k\int_{-\infty}^{\infty} \frac{f_0}{(\omega - kv)^2}\,dv \\
&= -\frac{kn}{2}\left(\frac{1}{(\omega - kv_0)^2} + \frac{1}{(\omega + kv_0)^2}\right)
\end{aligned}
\tag{23.18}
$$

the last form for the particular f_0 given in equation (23.17). In this way, the dispersion relation for this case becomes

$$D(k, \omega) \equiv 1 - \frac{1}{2}\left(\frac{\omega_p^2}{(\omega - kv_0)^2} + \frac{\omega_p^2}{(\omega + kv_0)^2}\right) = 0 \tag{23.19}$$

where, as usual, $\omega_p^2 = ne^2/m\epsilon_0$.

The function $D(k, \omega)$ plotted against ω is shown in Figure 23.1 with either Case A or Case B possible in the region $-kv_0 < \omega < kv_0$.

Since the quartic equation for ω represented by $D(k, \omega) = 0$ must always have four roots, real or complex, Case B of Figure 23.1 must have two complex roots for ω, since there are clearly only two real roots corresponding to the two crossings of the real axis. Since the complex roots ω of a polynomial with real coefficients must be complex conjugates of one another, one of them must have a positive imaginary part, corresponding to exponential growth in time, i.e. *instability*. The condition for instability (i.e. Case B rather than Case A) can be expressed as $D(k, 0) < 0$, which becomes $k^2v_0^2 < \omega_p^2$. Thus, the condition for instability is satisfied for all sufficiently long wavelengths. This is known as the 'two-stream instability'.

The two-stream instability prevents two oppositely directed uniform beams of electrons from passing through each other, even if the electrons are neutralized by a uniform background of ions. The instability produces strong spatial inhomogeneities, in which the electrons become 'bunched' together, ultimately

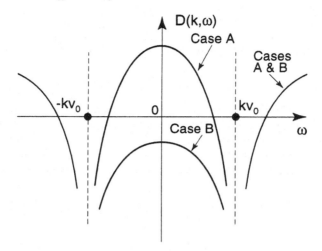

Figure 23.1. The dispersion function $D(k, \omega)$ for the two-stream instability plotted against ω for the Cases A (four real roots ω of $D(k, \omega) = 0$) and B (two real roots ω of $D(k, \omega) = 0$).

allowing the energy of the electron beams to be significantly dissipated into plasma waves. Other versions of the two-stream instability arise, some of which are discussed in Problems 23.2 and 23.5.

Problem 23.2: A uniform plasma with fixed ions has a density n of 'cold' bulk electrons at rest, together with a 'beam' of electrons with density n_b and streaming velocity u. The density of electrons in the beam is much less than the density of bulk electrons, i.e. $n_b/n = \epsilon \ll 1$. Neglecting thermal effects (i.e. assuming that the background plasma has zero temperature and that the beam has negligible velocity 'spread' about the streaming speed u), show that the dispersion relation for this version of the two-stream instability is

$$D(k, \omega) \equiv 1 - \frac{\omega_p^2}{\omega^2} - \frac{\epsilon \omega_p^2}{(\omega - ku)^2} = 0.$$

By sketching $D(k, \omega)$ as a function of ω show that, in the limit of infinitesimally small but non-zero ϵ, instability occurs for $ku \lesssim \omega_p$, approximately. Considering the case where ku is exactly equal to ω_p, and guessing that the (complex) frequency ω will be very close to ku and ω_p, i.e. $\omega = \omega_p + \Delta\omega = ku + \Delta\omega$ where $\Delta\omega$ is small, show that the growth rate has a magnitude $\gamma \sim \epsilon^{1/3}\omega_p$.

23.5 ION ACOUSTIC WAVES

When both ions and electrons are allowed to oscillate in the wave-field **E**, a new type of 'electrostatic' wave (i.e. having its wave vector **k** parallel to **E**) can arise, called an 'ion acoustic wave'. This wave was discussed in the fluid approximation in Chapter 16.

To obtain the dispersion relation for ion acoustic waves, it is necessary to include perturbed distribution functions f_1 for both ions and electrons, and to sum the two contributions to the charge-density perturbations in the Poisson equation. When this is done, a generalization of the dispersion relation given in equation (23.12) is obtained, namely

$$D(k, \omega) \equiv 1 + \sum \frac{e^2}{mk\epsilon_0} \int_{-\infty}^{\infty} \frac{\partial f_0/\partial v}{\omega - kv} \mathrm{d}v = 0. \tag{23.20}$$

Problem 23.3: If ion motions are included, derive the result quoted in equation (23.20) for the Vlasov plasma dispersion function, where the summation Σ is over species, i.e. ions and electrons.

As we have seen, electron plasma waves have phase velocities ω/k that are much larger than the electron thermal speed $v_{t,e}$, and therefore very much larger than the ion thermal speed $v_{t,i}$. In this case, inclusion of the ions would not significantly change the dispersion relation.

To find ion acoustic waves, however, we must look for waves with phase velocities that are *intermediate* between the thermal velocities of electrons and ions, i.e.

$$kv_{t,i} \ll \omega \ll kv_{t,e}. \tag{23.21}$$

The approximation for the ions is the same as the one used for electrons in our treatment of electron plasma waves. The new approximation for electrons is, however, quite different from the one used previously, and a new method of evaluating the electron contribution to the dispersion relation is needed.

For the ions, we may expand the factor in the integrand in equation (23.20):

$$\frac{1}{\omega - kv} \approx \frac{1}{\omega} + \frac{kv}{\omega^2} \tag{23.22}$$

which is similar to the expansion already given in equation (23.13), except that here only the first two terms are retained, which is equivalent to neglecting ion thermal motion. We obtain

$$\int_{-\infty}^{\infty} \frac{\partial f_0/\partial v}{\omega - kv} \mathrm{d}v \approx -\frac{nk}{\omega^2}. \tag{23.23}$$

For the electrons, we must expand in the opposite limit, i.e.

$$\frac{1}{\omega - kv} \approx -\frac{1}{kv} \tag{23.24}$$

to obtain

$$\int_{-\infty}^{\infty} \frac{\partial f_0/\partial v}{\omega - kv} dv \approx \frac{n}{kv_{t,e}^2}. \tag{23.25}$$

Equation (23.25) was obtained by evaluating $\partial f_0/\partial v$ for a Maxwellian electron distribution, namely

$$\frac{\partial f_0}{\partial v} = -\frac{v f_0}{v_{t,e}^2} \tag{23.26}$$

and noting that the factor v cancels between the numerator and the approximate form of the denominator. The dispersion function (23.20) becomes

$$D(k, \omega) \equiv 1 + \frac{\omega_p^2}{k^2 v_{t,e}^2} - \frac{\Omega_p^2}{\omega^2}. \tag{23.27}$$

Here, ω_p and Ω_p denote electron and ion plasma frequencies, respectively:

$$\omega_p^2 = \frac{ne^2}{m\epsilon_0} \qquad \Omega_p^2 = \frac{ne^2}{M\epsilon_0} \tag{23.28}$$

and $v_{t,e}$ is the electron thermal velocity, $(T_e/m)^{1/2}$. For wavelengths much longer than the Debye length, i.e. $k\lambda_D \equiv kv_{t,e}/\omega_p \ll 1$, the first term on the right-hand side of equation (23.27), i.e. the constant unity, can be neglected compared with the other terms, in which case the dispersion relation $D(k, \omega) = 0$, gives

$$\omega = kC_s \tag{23.29}$$

where $C_s = v_{t,e}(\Omega_p/\omega_p) = v_{t,e}(m/M)^{1/2} = (T_e/M)^{1/2}$. The phase velocity of these waves, C_s, is called the 'sound' or 'ion acoustic' speed. As we have seen before, it is the appropriate phase velocity for ion acoustic waves when ion thermal effects are neglected; it is the thermal velocity of ions evaluated with the *electron* temperature.

In Chapter 16, we obtained a slightly more general result for the dispersion relation of ion acoustic waves in the limit $k\lambda_D \ll 1$, which was still of the form $\omega = kC_s$, but with a sound speed $C_s = [(T_e + \gamma_i T_i)/M]^{1/2}$. This same result (with $\gamma_i = 3$, see Problem 23.4) could have been obtained here by retaining two more terms in the expansion, equation (23.22), used in evaluating the ion integral in the dispersion function, as was done for the electrons in the previous Section, i.e. equation (23.13). It follows that equation (23.29) is a good approximation only in the case $T_e \gg T_i$. However, recalling that our expansion of the ion integral is based on the assumption that $\omega \gg k v_{t,i}$, we see that the analysis leading to the dispersion relation, equation (23.29), is valid only if $C_s \gg v_{t,i}$, which *itself* requires $T_e \gg T_i$. The extra contribution to C_s arising from including the finite T_i terms in the ion integral can be retained as a small correction in the limit $T_i \ll T_e$, but this correction would be large only in cases where the approximations used to derive it are not valid. The fluid model allows ion acoustic waves with *some* value of γ_i for all ratios T_i/T_e. We will see in the next Chapter, however, that the kinetic treatment introduces an important, qualitative change in the dispersion relation for these waves in the case $T_i \sim T_e$.

Problem 23.4: Carry out the calculation referred to above, by using the full expansion, equation (23.13), rather than just equation (23.22) in evaluating the ion integral in the ion acoustic wave dispersion function. You should still assume $T_i \ll T_e$, but should retain first-order corrections in T_i in your dispersion relation for the case $k\lambda_D \ll 1$. Show that the resulting dispersion relation is the same as equation (23.29), except that first-order corrections in T_i modify the sound speed to $C_s = [(T_e + 3T_i)/M]^{1/2}$. By analogy with isothermal and adiabatic fluids, explain why there are different coefficients of T_e and T_i in the sound speed C_s.

Problem 23.5: A uniform plasma has ions that are initially at rest, but its electrons are streaming through the ions with velocity u. Again neglecting both ion and electron thermal effects (i.e. the ions are 'cold', and the electrons have negligible velocity 'spread' about the streaming speed u), show that the dispersion relation for electrostatic oscillations involving both ions and electrons is

$$D(k, \omega) \equiv 1 - \frac{m}{M}\frac{\omega_p^2}{\omega^2} - \frac{\omega_p^2}{(\omega - ku)^2} = 0.$$

Show that, for any streaming speed u, the plasma is always unstable to modes with sufficiently long wavelengths, i.e. sufficiently small values of k. By analogy with Problem 23.2, show that the typical growth rate γ has magnitude $\gamma \sim (m/M)^{1/3}\omega_p$.

23.6 INADEQUACIES IN VLASOV'S TREATMENT OF THERMAL EFFECTS ON PLASMA WAVES

Despite the apparent success of Vlasov's treatment in reproducing the dispersion relation describing the effect of thermal motions on electron plasma waves, namely equation (23.16), as well as in yielding dispersion relations for other types of plasma waves such as two-stream instabilities and ion acoustic waves, there are serious inadequacies in this method of solving the Vlasov–Poisson equations.

Using the Vlasov equation to find f_1, which was then substituted into the Poisson equation, we obtained a dispersion relation for electron plasma waves including thermal effects, namely equation (23.12), from which an approximate description of thermal corrections to the dispersion relation for plasma waves was obtained by expanding the integrand for $\omega \gg k v_{t,e}$, taking a Maxwellian for f_0. The problem with this solution is that the integral in equation (23.12) is *singular* at $v = \omega/k$, and we have developed no prescription for how to treat this singularity. In the case of ion acoustic waves, singular integrals of this type arise in *both* electron and ion contributions (see equation (23.20)) and, in this case, the electron integral was expanded in the opposite limit, i.e. $\omega \ll k v_{t,e}$, while the ions were treated in the limit $\omega \gg k v_{t,i}$, without addressing the problem of the singularity in either case.

For electron plasma waves, we have found the electrons to behave as an adiabatic fluid at this level of approximation and, for ion acoustic waves with $T_i \ll T_e$, we have found that the electrons behave as an isothermal fluid whereas the ions behave as an adiabatic fluid. We could proceed to carry our expansions to higher order, thereby uncovering additional physics, but it is more fundamental at this point to address the problem of how to treat the singularity in the integrals at $v = \omega/k$, because behind this mathematical problem lies the new physics of strong wave–particle interactions.

Vlasov argued that imaginary contributions to $D(k, \omega)$ cannot be allowed and concluded that the *principal values* of the singular integrals should be taken. The principal value of an integral of this type is defined as

$$\text{Pr} \int_{-\infty}^{\infty} = \lim_{\epsilon \to 0} \left\{ \int_{-\infty}^{\omega/k-\epsilon} + \int_{\omega/k+\epsilon}^{\infty} \right\}. \tag{23.30}$$

The principal value of a singular integral avoids the singularity by stopping infinitesimally short of it on the left, and starting again at an exactly equal distance on the right. This definition eliminates any possible imaginary contributions from integrating *around* the singularity in the complex plane. A proper treatment must, however, find some way to *formulate the problem physically so as to avoid the singularity from the beginning*. We cannot simply make the *ad hoc* assumption that the principal value must be used because we do not like the answer which we would otherwise obtain. The correct treatment was first given by L Landau (1946 *J. Phys. USSR* **10** 25), who found Vlasov's prescription for treating the singularity to be incorrect. Landau's treatment is the topic of the next Chapter.

Chapter 24

Kinetic effects on plasma waves: Landau's treatment

Landau used the method of Laplace transformation to obtain the full correct solution for the effects of a distribution of particle velocities on electron plasma waves, thereby correcting the treatment of Vlasov. His result extends the linear kinetic theory of small-amplitude perturbations to include the effects of particles traveling close to the wave's phase velocity, and so 'resonating' with the wave. Before describing Landau's treatment, we briefly review the mathematics of Laplace transforms and their inversion.

24.1 LAPLACE TRANSFORMATION

Laplace transformation is a well-developed mathematical technique for solving linear differential equations formulated as initial value problems. The technique may be summarized briefly as follows. To determine the time dependence of a function $f(t)$ that is determined by a linear differential equation, we first define a 'Laplace transform':

$$\tilde{f}(s) = \int_0^\infty f(t) e^{-st} dt \tag{24.1}$$

which is defined only for complex s with $\mathrm{Re}(s)$ positive and sufficiently large, i.e. $\mathrm{Re}(s) > s_0$, so that the integral converges at $t \to \infty$.

We then solve for $\tilde{f}(s)$ instead of $f(t)$ by performing a Laplace transform on each term of the equation. Since the time derivative $\dot{f} = \mathrm{d}f/\mathrm{d}t$ will appear in the differential equation for $f(t)$, it is useful to have a rule for the Laplace transform of a time derivative. This is given by

$$\tilde{\dot{f}} = s\tilde{f}(s) - f(0) \tag{24.2}$$

409

where $f(0)$ is the value of $f(t)$ at $t = 0$. In this way, the initial condition is introduced explicitly into the solution for the Laplace transform. This rule may be proved easily using integration by parts. If a second derivative with respect to time should appear, the above rule may be applied twice, which will introduce a further initial condition, namely the value of $\dot{f}(0)$. Thus, Laplace transformation is an appropriate technique for solving differential equations describing *initial value problems*. The technique transforms the problem from a differential equation for $f(t)$ to an algebraic equation for $\tilde{f}(s)$.

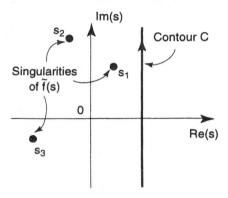

Figure 24.1. The Laplace inversion contour C for the general case where the transform $\tilde{f}(s)$ has singularities in both the right half-plane and the left half-plane.

Having obtained $\tilde{f}(s)$, we must *invert* the transformation to find $f(t)$. The appropriate inversion formula is

$$f(t) = \frac{1}{2\pi i} \int_C e^{st} \tilde{f}(s)\,ds \tag{24.3}$$

where C is a contour in the complex plane that runs from $-i\infty + s_0$ to $+i\infty + s_0$ sufficiently far to the right of the imaginary axis (i.e. sufficiently large positive s_0) to ensure that all singularities of $\tilde{f}(s)$ lie to the left of the contour, as shown in Figure 24.1. Singularities in the right half-plane correspond to exponentially growing terms in $f(t)$ which cause the integral in equation (24.1) defining $\tilde{f}(s)$ to diverge unless Re(s) is sufficiently large (see Problem 24.1). Often such terms are absent, in which case the choice of a contour C running just to the right of the imaginary axis in the Laplace inversion formula is satisfactory. As a general approach, this integral is then evaluated by 'closing the contour to the left', i.e. adding the semi-circle at Re$(s) = -\infty$ (which contributes nothing for $t > 0$) and noting that the integral around such a closed contour must equal $2\pi i$ times the sum of the residues of all singularities within the closed contour. In some cases, however, there may be an infinite number of singularities encountered as

$\text{Re}(s) \rightarrow -\infty$, in which case this approach will not yield a solution in closed form.

Problem 24.1: As an exercise in Laplace transforms, find the transform of the function

$$f(t) = \Sigma_i a_i \exp(-s_i t).$$

Carry out the inversion of $\tilde{f}(s)$ explicitly, using the Laplace inversion formula, to obtain $f(t)$ again.

The reader interested in pursuing further the theory of Laplace transforms is referred to J Matthews and R L Walker (1970 *Mathematical Methods of Physics* 2nd edn, Menlo Park, CA: Benjamin/Cummings).

24.2 LANDAU'S SOLUTION

The problem with our previous solution of the Vlasov equation for electron plasma waves lies in the singularity at $v = \omega/k$ that arises when we assume normal mode solutions for $E(x, t)$ and $f_1(x, v, t)$, i.e. solutions that vary like $\exp(ikx - i\omega t)$. In particular, our assumption of Fourier normal modes in time effectively supposes that the wave can have existed for all time and will continue to exist to $t = \infty$ with a sinusoidal time dependence, i.e. $\exp(-i\omega t)$. Possibly, then, it is the assumption of normal mode solutions—especially for $f_1(x, v, t)$— that is invalid. In other words, there may simply not exist such a solution. Instead of assuming that the solutions are *necessarily* of this form, let us try to solve a problem that *must* be well-posed, namely the *initial-value* problem. Specifically, we create some initial disturbance at $t = 0$, which we assume to be wave-like in x space, and we follow its subsequent development without assuming that it must behave as $\exp(-i\omega t)$. We will find that this new approach provides a prescription for how to treat the singularity at $v = \omega/k$ that arose in the integrals obtained in Vlasov's normal mode method. In some cases, we will find that normal modes varying exactly as $\exp(-i\omega t)$ *do* exist, but in most cases they do not, and a sinusoidal variation with t represents at best an approximation to a more complicated time dependence.

The solution must remain wave-like in physical space, since first-order perturbations of a spatially uniform equilibrium can be Fourier analyzed into independent Fourier components. Thus we can still assume

$$E(x, t) = \hat{E}(t)\exp(ikx)$$
$$f_1(x, v, t) = \hat{f}_1(v, t)\exp(ikx) \tag{24.4}$$

but the time-dependence is, for now, not specified. (We henceforth drop the 'hats' from the quantities $\hat{E}(t)$ and $\hat{f}_1(v, t)$ without creating any ambiguity. As was discussed in Chapter 15, the quantities $E(t)$ and $f_1(v, t)$ are complex, but here only as associated with *spatial* phase differences.) We can, of course, specify the initial perturbations $E(0)$ and $f_1(v, 0)$ at will, provided that the charge density given by $f_1(v, 0)$ is self-consistent with $E(0)$ in the Poisson equation.

The linearized Vlasov equation for electrons, Fourier analyzed in space but not in time, is

$$\frac{\partial f_1}{\partial t} + ikvf_1 - \frac{e}{m}E\frac{\partial f_0}{\partial v} = 0. \tag{24.5}$$

In principle, we could integrate f_1 forward in time to obtain a complete solution. In this sense, the problem is well-posed.

Since it is a standard mathematical technique for solving initial value problems of this kind, the method of Laplace transformation is very appropriate here. The Laplace transform of $f_1(v, t)$ is

$$\tilde{f}_1(v, s) = \int_0^\infty f_1(v, t)e^{-st}dt. \tag{24.6}$$

Taking the Laplace transform of the linearized Vlasov equation, and using the rule for the Laplace transform of a time derivative, we obtain

$$(s + ikv)\tilde{f}_1(v, s) - \frac{e}{m}\tilde{E}(s)\frac{\partial f_0}{\partial v} = f_1(v, 0) \tag{24.7}$$

where $\tilde{E}(s)$ is the Laplace transform of $E(t)$. We solve equation (24.7) for $\tilde{f}_1(v, s)$ and substitute it into the Laplace transform of the Poisson equation

$$ik\epsilon_0\tilde{E}(s) = -e\int_{-\infty}^\infty \tilde{f}_1(v, s)dv \tag{24.8}$$

to obtain, after some straightforward algebra,

$$D(k, s)\tilde{E}(s) = \frac{ie}{k\epsilon_0}\int_{-\infty}^\infty \frac{f_1(v, 0)}{s + ikv}dv \tag{24.9}$$

where

$$D(k, s) \equiv 1 - \frac{ie^2}{mk\epsilon_0}\int_{-\infty}^\infty \frac{\partial f_0/\partial v}{s + ikv}dv. \tag{24.10}$$

In principle, given some initial perturbation $f_1(v, 0)$, equations (24.9) and (24.10) give the complete solution for the time evolution of the electric field. Then equation (24.7) gives the solution for the perturbed velocity distribution. Of course, an explicit expression for $E(t)$ can only be obtained by substituting

$\tilde{E}(s)$ into the formula for inverting a Laplace transform. Although this cannot, in many cases, be done analytically, we can draw some conclusions about the behavior of $E(t)$ from the general properties of $D(k, s)$.

We note first that the expression for $D(k, s)$ is very reminiscent of the Vlasov 'dispersion function' $D(k, \omega)$ given in equation (23.12), which was obtained by Vlasov by assuming normal modes that are oscillating in time, with frequency ω. To avoid confusion, for the purposes of the present discussion we re-label the Vlasov dispersion function as $D_V(k, \omega)$. We then see that the two functions $D(k, s)$ and $D_V(k, \omega)$ are exactly the same if we substitute $s \to -i\omega$ or $\omega \to is$. However, since the Laplace transform is defined only for $Re(s) > 0$, and is only needed in this region to apply the inversion formula, the singularity that appeared in the definition of $D_V(k, \omega)$ (for real frequency ω) is absent from the definition of $D(k, s)$.

To obtain $E(t)$, we must invert $\tilde{E}(s)$ using the Laplace inversion formula. The Laplace inversion formula is

$$E(t) = \frac{1}{2\pi i} \int_C \tilde{E}(s) e^{st} ds \qquad (24.11)$$

where the contour C runs from $-i\infty + s_0$ to $+i\infty + s_0$, at some real distance $s_0 > 0$ to the right of the imaginary axis, such that all singularities are located to the left of s_0. This requirement can be understood by noting that the Laplace transform of $E(t)$, namely

$$\tilde{E}(s) = \int_0^\infty E(t) e^{-st} dt \qquad (24.12)$$

is defined only if $Re(s)$ is sufficiently positive to overcome any exponentially growing terms in $E(t)$. Thus, the integration contour C in the Laplace inversion formula must lie sufficiently to the right in the complex s-plane and, in particular, to the right of all singularities of $\tilde{E}(s)$. If we denote the singularities of $\tilde{E}(s)$ by $s_1, s_2, \ldots$, where s_1 is the singularity furthest to the right, s_2 is next, etc, the contour C must be of the type shown in Figure 24.1, and $s_0 > Re(s_1)$.

In order to find the dominant behavior of $E(t)$ as $t \to \infty$, we would like to move the contour C, in an appropriate fashion, as far as possible toward the left half of the s-plane, ultimately closing it by the infinite semi-circle at $Re(s) = -\infty$. Since the functions in the integral will all be analytic, or if defined only in certain regions of the s-plane can be extended into other regions by analytic continuation, except for singularities which we specifically identify, then the contour C can indeed by moved around in this way without changing the result for $E(t)$, provided no singularity is crossed. The reason for seeking to move C as far as possible to the left can be seen by examining equation (24.11). The first 'obstacle', i.e. singularity, that will be encountered as we do this will give the dominant contribution as $t \to \infty$, with contributions from smaller values of Re(s) being subdominant. Once we have accounted for the contributions from all singularities, we have left only the contribution from the semicircle at infinity in the left half-plane, i.e. at Re(s) $= -\infty$, which vanishes for all positive t. Accordingly, we imagine moving C to the left by analytic continuation of $\tilde{E}(s)$ (if necessary) until we encounter the first singularity of $\tilde{E}(s)$, which we have called s_1. This first singularity may lie in the right half of the s-plane (as shown in Figure 24.1), or it may lie in the left half-plane.

Before continuing, we must investigate the possible origins of singularities in $\tilde{E}(s)$. These may arise (i) from singularities on the right-hand side of equation (24.9), i.e. singularities of the numerator $N(k, s)$ in the expression for $\tilde{E}(s)$

$$D(k, s)\tilde{E}(s) = N(k, s) \equiv \frac{ie}{k\epsilon_0} \int \frac{f_1(v, 0)}{s + ikv} dv \qquad (24.13)$$

or (ii) from zeros of the denominator, $D(k, s)$. Singularities of origin (i) cannot arise, because the integral in equation (24.13) defines (for smooth enough initial perturbations $f_1(v, 0)$) an entire function (finite everywhere) for Re(s) > 0, which may be analytically continued into the region Re(s) < 0, where it must remain finite at least for some finite distance into the left half-plane. Hence, *all singularities of $\tilde{E}(s)$, with the possible exception of those finitely into the left half-plane, must arise from zeros of $D(k, s)$.* Singularities of $\tilde{E}(s)$ that arise from singularities of $N(k, s)$ in the left half-plane describe the damping-out of peculiar features of the initial velocity-distribution perturbation $f_1(v, 0)$ and are of little interest to us here. Singularities that arise from zeros of $D(k, s)$, on the other hand, describe the collective oscillations of the plasma, in this case the electron plasma waves. Let us consider three different cases, distinguished from each other by the location of these zeros.

24.2.1   Case 1: First zero of $D(k, s)$ has Re(s) > 0

As we try to move our Laplace inversion contour C toward the left, suppose that we first encounter a singularity of $\tilde{E}(s)$ (a zero of $D(k, s)$) at $s = s_1$ in the right half-plane, i.e. with Re(s_1) > 0. We can move our contour past s_1, provided we include in $E(t)$ a term arising from the residue at the pole at $s = s_1$. Specifically, we can write

$$E(t) = \text{Res}(s_1)e^{s_1 t} + \frac{1}{2\pi i} \int_{C'} \tilde{E}(s)e^{st} ds \qquad (24.14)$$

where Res(s_1) denotes the residue at s_1, and the contour C' is now to the left of s_1, as shown in Figure 24.2. As $t \to \infty$, the residue term dominates, giving

$$E(t) \to \text{Res}(s_1)e^{s_1 t}. \qquad (24.15)$$

Since $\text{Re}(s_1) > 0$, this dominant term represents an *instability*, i.e. an electric field that grows exponentially in time. Of course, in many cases, certainly including the case of a Maxwellian f_0, there is no such zero s_1, and no such instability.

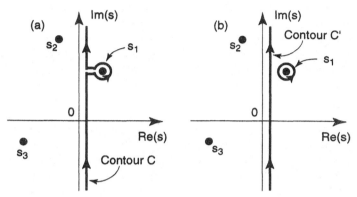

Figure 24.2. The Laplace inversion contour C of Figure 24.1 is moved as far as possible to the left, as shown in (a), and the contour is then 'snapped' into two pieces, one of which is the contour C' and the other is a contour that encircles the singularity at s_1, as shown in (b). The contributions from the two horizontal lines connecting the vertical contour to the circle around the singularity in (a) cancel each other in the limit that these lines touch, and so leave no net contribution after the contour is 'snapped' into two pieces.

We have seen in Chapter 23, however, that there are certain distributions f_0 (in particular, 'two-stream' distributions) that *do* lead to instability. For such cases, we note that for $\text{Re}(s) > 0$, a zero s_1 of our function $D(k, s)$ corresponds *exactly* to a zero ω_1 of the Vlasov dispersion function $D_V(k, \omega)$, obtained by setting $s_1 = -i\omega_1$. Since $\text{Im}(\omega_1) > 0$, there is *no singularity* in the integral over v in the Vlasov dispersion function $D_V(k, \omega)$. In this case, a problem does not arise with Vlasov's normal mode solution, in which the perturbation quantities were assumed to vary as $\exp(-i\omega t)$: the problem is resolved because ω is now *complex*. Our conclusion from the Vlasov analysis for this case is as follows: an instability arising from a zero of the Vlasov dispersion function $D_V(k, \omega)$ with $\text{Im}(\omega) > 0$ *is* a pure normal mode; i.e. it has a single (complex) frequency ω. For this case, Vlasov's treatment turns out to have been valid.

Our conclusion from the Landau analysis of the initial-value problem for this case is that, if such a zero of the function $D(k, s)$ given in equation (24.10) with $\text{Re}(s) > 0$ exists, say at $s = s_1$, inversion of the Laplace transform shows that the dominant term in the solution for $E(t)$ as $t \to \infty$ will be an exponentially growing term, i.e. an instability. There can be various types of 'subdominant' terms, including other exponentially growing terms with smaller growth rates (which could be found by looking for the next zero, s_2, etc.) and terms that describe oscillatory and damped terms in $E(t)$.

Although Landau's Laplace transform approach can, in principle, yield the full time-dependent solution of the initial-value problem, it does not add much to Vlasov's solution for this particular case. If there is a zero of $D(k, s)$ in the right half-plane, both approaches indicate that there will be an instability, i.e. an exponentially growing term in $E(t)$, which will represent the dominant time dependence as $t \to \infty$.

24.2.2   Case 2: All zeros of $D(k, s)$ have $\text{Re}(s) < 0$

If $D(k, s)$ has no zeros in the right half-plane, we may move the contour C leftward until it lies along the imaginary axis. Provided there are no zeros on the imaginary axis itself, the contour C may be moved into the left half-plane, now running from $-i\infty - \delta$ to $+ i\infty - \delta$. Just as before, the dominant term in $E(t)$ as $t \to \infty$ is still the contribution from the residue at the first pole s_1 that is encountered, i.e.

$$E(t) = \text{Res}(s_1)e^{s_1 t} + \frac{1}{2\pi i} \int_{C''} \tilde{E}(s)e^{st} ds \qquad (24.16)$$

where the contour C'' is now even further to the left, as shown in Figure 24.3. However, the dominant term now describes a perturbation that *decays in time*, i.e. $\text{Re}(s_1) < 0$. This is the case that occurs for a Maxwellian f_0; the damping is called 'Landau damping'. (There is, of course, an intermediate case where $D(k, s)$ has its first zero exactly on the imaginary axis. This results in an oscillation that is neither growing nor damped.)

In moving the contour C from the right half of the s-plane to the left half, the integrand—in particular, the function $D(k, s)$—must be defined by *analytic continuation*, starting from the right half-plane, $\text{Re}(s) > 0$. To ensure the proper analytic continuation of the function

$$D(k, s) \equiv 1 - \frac{ie^2}{mk\epsilon_0} \int_{-\infty}^{\infty} \frac{\partial f_0 / \partial v}{s + ikv} dv \qquad (24.17)$$

the *contour of integration in the v-plane must be deformed so that the singularity at $v = is/k$ always lies on the same side of the v-contour*. For values of s with a positive real part, the original contour is by definition correct, and as s is moved toward the imaginary axis, the contour still does not need to be deformed in the case $\text{Re}(s) = \text{Im}(\omega) > 0$ (instability; case (a) in Figure 24.4). However, if s moves to the left of the imaginary axis, the contour in v-space must be deformed to ensure smooth analytic continuation of $D(k, s)$ as a function of s. Thus, the

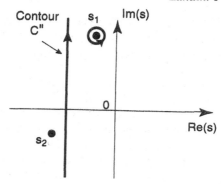

Figure 24.3. The Laplace inversion contour for the case where there are no singularities in the right half-plane. The contour C of Figure 24.1 has been 'snapped' into two pieces, one of which is the contour C'' that is now well into the left half-plane and the other is a contour that encircles the first singularity s_1 encountered in the left half-plane.

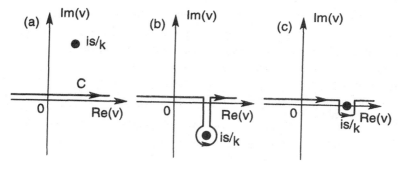

Figure 24.4. The contour of integration is the v-plane for the cases: (a) instability, i.e. there is a zero of the dispersion function with $\text{Re}(s) > 0$, corresponding to $\text{Im}(\omega) > 0$; (b) strong damping, i.e. all zeros of the dispersion function have $\text{Re}(s) < 0$; and (c) weak damping, i.e. there is a zero of the dispersion function with $\text{Re}(s) \approx 0$, corresponding to $\text{Im}(\omega) \approx 0$.

contour *does* need to be deformed in the case $\text{Re}(s) = \text{Im}(\omega) < 0$ (damping; case (b) in Figure 24.4).

It follows that a damped mode ($\text{Im}(\omega) < 0$) is *not* a solution of the *Vlasov* dispersion relation, but rather of a modified relation in which the contour of integration in the v-plane is deformed as described above. It therefore follows that this damped solution is *not a normal mode*, i.e. it does *not* correspond to a solution in which all perturbation quantities have only a single Fourier component in time, i.e. vary exactly as $\exp(-i\omega t)$. For if this solution *were* a normal mode, it would satisfy the Vlasov dispersion relation as originally derived, i.e. with the v-integration running simply along the real axis, rather than along a deformed contour. True normal modes could be produced only by choosing very special non-physical initial perturbations $f_1(v, 0)$. For most physical cases, the Landau mode described by the first term on the right-hand side in equation (24.16) represents the longest-lived term in a solution with a very complicated (but always decaying) time dependence.

Our conclusion from the Landau analysis for this case is that all terms in $E(t)$ damp out as $t \to \infty$. In general, we cannot identify any dominant 'mode' in this case, nor can we easily describe in detail the time-dependence of $E(t)$, except to say that $E(t) \to 0$ as $t \to \infty$. However, to the extent that s_1, the first zero of $D(k, s)$ lies *just* to the left of the imaginary axis, there *is* an identifiable dominant term as $t \to \infty$ which has only *weak* damping. This is a case of great physical interest and is considered next.

24.2.3   Case 3: First zero of $D(k, s)$ lies just to the left of the imaginary axis

This is really a sub-category of Case 2, but it is of particular interest, since it describes *weakly damped* oscillations, which occur commonly in plasmas. We suppose that the first zero of $D(k, s)$ lies just to the left of a point $s = -i\omega$ on the imaginary axis. In this case, the contour of integration in the v-plane must be deformed to include a path passing below the singularity at $v = \omega/k$, as shown in Figure 24.4(c).

Using $s = -i\omega$ in order to return to more familiar notation, the dispersion function for ω exactly real becomes

$$D \equiv 1 + \frac{e^2}{mk\epsilon_0} \left(\mathrm{Pr} \int_{-\infty}^{\infty} \frac{\partial f_0/\partial v}{\omega - kv} \mathrm{d}v - \frac{\pi i}{k} \frac{\partial f_0}{\partial v}\Big|_{v=\omega/k} \right) \qquad (24.18)$$

where Pr denotes the principal value of the integral, defined in Chapter 23. The imaginary term comes from going 180° around the pole and is πi times the residue. (Note that this expression can be regarded as providing a general prescription for resolving the singularity originally noted by Vlasov in the integral in $D_V(k, \omega)$.)

As in our previous treatment of thermal effects on plasma waves (Chapter 23), we can expand the principal-value integral in the limit $\omega \gg kv_t$, to obtain

$$D \approx 1 - \frac{\omega_p^2}{\omega^2} + \dots . \qquad (24.19)$$

Thermal corrections of the type discussed in the previous Chapter (and from the fluid treatment in Chapter 16) would be obtained by retaining the next non-vanishing term in this expansion. For a Maxwellian f_0, i.e.

$$f_0 = \frac{n}{(2\pi)^{1/2} v_t} \exp\left(-\frac{v^2}{2v_t^2}\right) \tag{24.20}$$

where $v_t = (T/m)^{1/2}$, equation (24.18) indicates that we must also include an imaginary term arising from the pole, giving altogether

$$D \equiv 1 - \frac{\omega_p^2}{\omega^2} + i\left(\frac{\pi}{2}\right)^{1/2} \frac{\omega_p^2 \omega}{k^3 v_t^3} \exp\left(-\frac{\omega^2}{2k^2 v_t^2}\right). \tag{24.21}$$

Since this is explicitly an analytic function of ω, it can be employed for values of ω slightly away from the real axis. Treating the last term as a small correction, we can solve the dispersion relation, $D = 0$, corresponding to the Landau pole we have been treating, by iteration, obtaining

$$\omega = \omega_p - \frac{i}{2}\left(\frac{\pi}{2}\right)^{1/2} \frac{\omega_p^4}{k^3 v_t^3} \exp\left(-\frac{\omega_p^2}{2k^2 v_t^2}\right). \tag{24.22}$$

This is our final expression for the frequency ($\approx \omega_p$) and Landau-damping decrement γ of electron plasma waves. We see that plasma waves are, in fact, always slightly damped. Remembering that $v_t/\omega_p = \lambda_D$, the Debye length, we see that the damping is exponentially small for long wavelengths ($k\lambda_D \ll 1$), but is large ($\gamma \sim \omega_p$) for wavelengths of order the Debye length. The small Landau damping for $k\lambda_D \ll 1$ can be interpreted as due to the fact that there are very few particles (i.e. very small f_0 and $\partial f_0/\partial v$) at $v = \omega/k \approx \omega_p/k \approx v_t/k\lambda_D \gg v_t$.

However, the physical interest in Landau damping of plasma waves does not lie in the numerical magnitude of the damping decrement, which is usually small. Rather, it lies in the surprising discovery of *wave-damping in an entirely collisionless system*. This might seem to violate our sense that there are no strictly dissipative terms in the Vlasov–Maxwell equations. Moreover, the phenomenon of Landau damping appears in many other contexts in plasma physics—indeed, whenever there are particles whose velocity is approximately in resonance with the phase velocity ω/k of some type of plasma wave.

Problem 24.2: For certain simple equilibrium distributions other than the Maxwellian, the Landau damping can be calculated explicitly. For the distribution

$$f_0(v) = \frac{n}{\pi}\frac{a}{v^2 + a^2}$$

use contour integration to evaluate the dispersion function $D(k, s)$ explicitly, and show that plasma oscillations damp as $\exp(-kat)$. (Hint:

the explicit evaluation of $D(k, s)$ uses contour integration in the v-plane, summing the contributions from poles within a closed contour; you should take care to choose the most convenient way to close the v-plane contour.) Qualitatively, why is the Landau damping larger for this case than for the Maxwellian distribution?

24.3 PHYSICAL MEANING OF LANDAU DAMPING

Physically, it is clear that Landau damping is associated with those particles in the distribution that have a velocity nearly equal to the phase velocity of the wave, ω/k, since the contribution to the dispersion function, equation (24.18), that gives rise to Landau damping is the term in $(\partial f_0/\partial v)|_{\omega/k}$. These may be called 'resonant particles'. Resonant particles travel along at almost the same speed as the wave and tend to see a relatively static electric field, rather than a rapidly fluctuating one. They can, therefore, exchange energy very effectively with the wave.

The electrons with $v \approx \omega/k$, which are nearly resonant with the plasma wave in the Landau problem, are analogous to the resonant particles in the mapping problem of Chapter 5. They see an essentially steady electric field, which can be positive or negative depending on their phase relative to the wave. Thus, some nearly resonant particles are accelerated by the wave, while others are decelerated. A resonant individual particle has an equal chance of being accelerated or decelerated, after averaging over all possible phases. Thus the population of particles that was originally moving slightly faster than ω/k is *mixed* with the population that was moving slightly slower.

However, a Maxwellian distribution has *more slower electrons than faster ones*. Consequently, there are more particles being accelerated on average by this mixing process than being decelerated. Since this results in a net transfer of energy from the wave to the particles, the wave is damped.

As particles with velocities near the phase velocity ω/k are speeded-up or slowed down in this way by the wave, the distribution $f(v)$ (averaged over wave phase) tends to be 'flattened' in this region. Effectively, there arises a wave-induced diffusion in velocity space, concentrated in the region around the phase velocity ω/k. The new, modified distribution function contains the same number of particles, but it has gained a little energy at the expense of the wave. Strictly, this flattening of the distribution function is a *nonlinear* effect, because it is quadratic in the amplitude of the perturbation. For infinitesimal perturbations, the flattening would be imperceptible, but it is sufficient to account for the loss of wave energy, which is also quadratic in the perturbation amplitude. For larger wave amplitudes, such as those arising from unstable modes of perturbation, wave-induced velocity diffusion can often be the dominant nonlinear effect, as in the 'quasi-linear theory' discussed in the next Chapter.

If the amplitude of the perturbation is large, another specifically nonlinear effect can arise, namely the 'trapping' of particles at locations of minimal potential energy in the wave. This is analogous to the formation of islands in the mapping problem of Chapter 5. For the electrons, these trapping locations will be at the maxima of the electric potential. Trapping of electrons in a plasma wave will 'compete' with Landau damping, since once the electrons become trapped they can no longer take any more energy from the wave. For small-amplitude waves, for which the linear treatment is valid, trapping does not play a significant role. However, the trapping phenomenon will be discussed in the context of plasma-wave *instabilities*, for which larger field amplitudes can be expected to occur, in the next Chapter.

It is important to point out that, unless collisions or some equivalent dissipative effects (such as orbit stochasticity, see Chapter 25) are introduced, Landau damping is not really a dissipative or irreversible process. Indeed, the 'information' that was present in the initial perturbation is retained in a time- and space-dependent 'microstructure' of the velocity distribution function, even after the electric field has damped out almost to zero. Landau damping has been demonstrated in the laboratory by J H Malmberg and C B Wharton (1966 *Phys. Rev. Lett.* **17** 175). Moreover, the reconstruction of a damped electric field perturbation from information contained entirely in the perturbed distribution function has also been demonstrated in an experiment on 'plasma echoes' by J H Malmberg, C B Wharton, R W Gould and T M O'Neill (1968 *Phys. Fluids* **11** 1147) and independently by A Y Wong and D R Baker (1969 *Phys. Rev.* **188** 326) using ion acoustic waves.

24.4 THE NYQUIST DIAGRAM*

It is possible to give a formal proof that there are no instabilities, i.e. no zeros of the dispersion relation $D(k, s) = 0$ with $\mathrm{Re}(s) > 0$, in the case of a spatially homogeneous plasma with Maxwellian distribution f_0. Physically, of course, this result is obvious, since otherwise thermodynamics would be contradicted, in that a tendency would exist for the Maxwellian distribution to evolve toward some other distribution. Nonetheless, it is valuable to develop a technique for proving this rigorously, since the same technique can often be useful in searching for possible instabilities for non-Maxwellian distributions f_0 (see Problem 24.3).

The technique, which is derived from a powerful electrical engineering technique due to Nyquist, is to consider a closed semicircular contour (with the semicircular part at infinity) that encloses the entire right half of the s-plane, as shown in Figure 24.5(a). As the complex s value traverses this contour in the anti-clockwise sense, so that the area $\mathrm{Re}(s) > 0$ lies always on our *left*

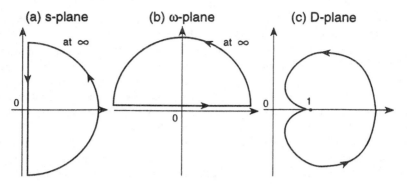

Figure 24.5. The Nyquist diagram for a Maxwellian distribution f_0. The analytic function $D(k, s)$ transforms the contour in the s-plane (a), or equivalently the contour in the ω-plane (b) where $s = -i\omega$, into the contour in the D-plane (c).

as we trace out the contour, the value of the function $D(k, s)$ will trace out a corresponding closed contour in the complex D-plane. Assuming that this contour in the D-plane is simply connected (i.e. does not cross over itself), then the area enclosed by tracing out the D-plane contour, specifically the area formed to the *left* of the contour, corresponds to the area $\text{Re}(s) > 0$ in the s-plane. This mapping of regions to the left of a contour in one complex plane onto regions to the left of the corresponding contour in the other complex plane follows from the analytic nature of the function which defines the mapping, in this case $D(k, s)$ as a function of the complex variable s in the half-plane $\text{Re}(s) > 0$. (See, for example, the Chapters on conformal mappings in R V Churchill (1960 *Complex Variables and Applications* 2nd edn, New York: McGraw-Hill) or in E G Phillips (1957 *Functions of a Complex Variable* 8th edn, Edinburgh: Oliver and Boyd).) If this area in the D-plane contains the point $D = 0$, then there must be a solution of the dispersion relation, $D(k, s) = 0$, with $\text{Re}(s) > 0$, i.e. an instability. On the other hand, if the area in the D-plane does *not* contain the point $D = 0$, then there will be no instability. The case where the contour in the D-plane is not simply connected represents a straightforward generalization. The area in the D-plane that corresponds to the area $\text{Re}(s) > 0$ in the s-plane will always be the area that lies on our left as we trace out the D-plane contour, when this contour is traced out by an anticlockwise traversal of the s-plane contour. Sometimes, of course, this area in the D-plane will extend to infinity. At other times, there will be certain areas of the D-plane that are encircled *twice* by the D-plane contour (on its left); within such areas, there will be *two* values of s with $\text{Re}(s) > 0$ for every D value. The s-plane contour and its corresponding D-plane contour is called the 'Nyquist diagram'. The contour in the D-plane is sometimes called the 'Nyquist contour'.

The Nyquist-diagram technique can be illustrated by applying it to the case of the dispersion relation $D(k, s) = 0$ for a Maxwellian f_0. It is convenient, however, to transform again from the variable s to a variable ω, defined by

$$s = -i\omega. \tag{24.23}$$

The semi-circular contour in the s-plane shown in Figure 24.5(a) becomes the contour in the ω-plane shown in Figure 24.5(b). Just as the straight part of the contour in Figure 24.5(a) lies infinitesimally to the right of the imaginary-s axis, so the corresponding part of the contour in Figure 24.5(b) lies infinitesimally above the real-ω axis. The corresponding contour in the D-plane is obtained by allowing ω to trace out this contour, obtaining D values from the expression

$$D \equiv 1 + \frac{e^2}{mk\epsilon_0} \int_{-\infty}^{\infty} \frac{\partial f_0/\partial v}{\omega - kv} dv \approx 1 - \frac{\omega_p^2}{\omega^2} \tag{24.24}$$

for ω on the semicircle at infinity, and

$$D \equiv 1 + \frac{e^2}{mk\epsilon_0} \left(\text{Pr} \int_{-\infty}^{\infty} \frac{\partial f_0/\partial v}{\omega - kv} dv - \frac{\pi i}{k} \frac{\partial f_0}{\partial v} \Big|_{v=\omega/k} \right) \tag{24.25}$$

for ω on the straight part of the contour along the real axis. For a Maxwellian f_0, it follows from equation (24.25) that

$$\text{Im}(D) = \left(\frac{\pi}{2} \right)^{1/2} \frac{\omega_p^2 \omega}{k^3 v_t^3} \exp \left(-\frac{\omega^2}{2k^2 v_t^2} \right). \tag{24.26}$$

The contour in the D-plane can now easily be traced out. The entire semicircle at infinity in the ω-plane transforms into the point $D = 1$. The contour in the D-plane can cross the real axis again only once, namely at the point corresponding to $\omega = 0$. Moreover, for $\text{Re}(\omega) < 0$, we have $\text{Im}(D) < 0$, and for $\text{Re}(\omega) > 0$, we have $\text{Im}(D) > 0$. Finally, at the point corresponding to $\omega = 0$, we have

$$\text{Re}(D) = 1 - \frac{e^2}{mk^2\epsilon_0} \int_{-\infty}^{\infty} \frac{\partial f_0/\partial v}{v} dv$$

$$= 1 + \frac{e^2}{mk^2\epsilon_0 v_t^2} \int_{-\infty}^{\infty} f_0 dv$$

$$= 1 + \frac{\omega_p^2}{k^2 v_t^2} \tag{24.27}$$

which exceeds unity, implying that the crossing of the real D-axis corresponding to the point $\omega = 0$ lies to the right of the crossing corresponding to ω values on the semicircle at infinity. It follows that the D-plane contour must have the general shape shown in Figure 24.5(c). Tracing out this contour in the sense that corresponds to an anticlockwise traversal of the ω-plane contour, we see that the region of the D-plane to the left of the contour is the area *inside* the contour *which excludes the point $D = 0$*. Thus, there can be no solution of $D(k, s) = 0$ with $\text{Re}(s) > 0$, i.e. no instability.

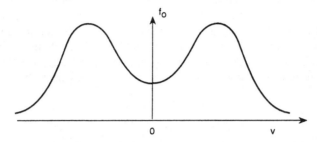

Figure 24.6. A symmetric 'double-humped' electron distribution function corresponding to equal and opposite streams, each of which is given a spread in velocities (see Problem 24.4).

To change this conclusion, for example for distributions f_0 other than Maxwellian, it is generally necessary for the Nyquist contour to have multiple crossings of the real axis in the D-plane. From equation (24.25), it is evident that this can occur only if there are multiple velocities v where $\partial f_0 / \partial v = 0$. For example, 'double-humped' distributions of the type shown in Figure 22.1(b) would have three velocities where $\partial f_0 / \partial v = 0$, and their Nyquist contours would have three crossings of the real axis in the D-plane. Since an extreme form of the double-humped distribution could be the two-stream distribution of equation (23.17), the instabilities arising with such double humped distributions are all essentially versions of the two-stream instability. An example of the use of the Nyquist diagram to determine the stability properties of a symmetrical double-humped distribution, shown in Figure 24.6, is given in Problem 24.3.

Problem 24.3: Use the Nyquist diagram technique to search for possible instabilities of electron plasma waves in the case of a *symmetric* 'double-humped' electron distribution function f_0, such as that shown in Figure 24.6. Show that the condition for instability is that

$$\frac{e^2}{mk^2\epsilon_0} \int_{-\infty}^{\infty} \frac{f_0(v) - f_0(0)}{v^2} \, dv > 1.$$

(Hint: It is not necessary to know the exact shape of the distribution function f_0, only that it is symmetric in v, which implies that a minimum occurs at $v = 0$. This is sufficient to be able to sketch the Nyquist diagram. However, it will be necessary to locate approximately the points where the Nyquist contour crosses the real axis. To do this, you will need to develop estimates for the principal-value term in equation (24.25). The symmetry of $f_0(v)$ will help in this regard.)

By examining the instability condition that you have derived, do you conclude that *all* distributions of the type shown in Figure 24.6 will be unstable, or only those in which the 'double-humpedness' is sufficiently pronounced?

24.5   ION ACOUSTIC WAVES: ION LANDAU DAMPING

Electrons are not the only particles that can be in resonance with waves in a plasma. If the wave has a small enough phase velocity to match the thermal velocity of ions, then strong *ion* Landau damping can occur. For example, 'ion acoustic waves' have phase velocities of order the sound speed, defined here as $[(T_e + 3T_i)/M]^{1/2}$, and should be strongly affected by ion Landau damping if $T_i \sim T_e$.

It is trivial to generalize our derivation of the Landau form for the plasma dispersion function to the case where both species (electrons and ions) participate in the oscillation, just as was done in Chapter 23 for the Vlasov treatment. Referring to equations (23.20) and (24.18), we obtain

$$D(k, \omega) \equiv 1 + \Sigma \frac{e^2}{mk\epsilon_0} \left(\mathrm{Pr} \int_{-\infty}^{\infty} \frac{\partial f_0/\partial v}{\omega - kv} \mathrm{d}v - \frac{\pi \mathrm{i}}{k} \frac{\partial f_0}{\partial v}\bigg|_{v=\omega/k} \right) \qquad (24.28)$$

where the summation is over species. (Our dispersion function is, of course, still limited to the case of electrostatic oscillations propagating along, or in the absence of, a magnetic field.)

As in the treatment of ion acoustic waves given in Chapter 23, we assume that

$$\omega \ll kv_{t,e} \qquad \omega \gg kv_{t,i}. \qquad (24.29)$$

Approximate expressions for the two principal-value integrals in equation (24.28) have been given in equations (23.23) and (23.25), assuming the relationships given in equation (24.29). (The approximations employed in Chapter 23 were, in effect, evaluating the integrals as principal values, since they did not include any contributions from the singularity in the integrand.)

Reproducing these previous results, we have for the electrons

$$\mathrm{Pr} \int_{-\infty}^{\infty} \frac{\partial f_0/\partial v}{\omega - kv} \mathrm{d}v \approx \frac{n}{kv_t^2} \qquad (24.30)$$

and, for the ions, neglecting finite ion temperature effects (i.e. assuming $T_i \ll T_e$):

$$\text{Pr} \int_{-\infty}^{\infty} \frac{\partial f_0/\partial v}{\omega - kv} \, dv \approx -\frac{nk}{\omega^2}. \tag{24.31}$$

Keeping the imaginary (Landau damping) terms from both electrons and ions, with $\partial f_0/\partial v|_{v=\omega/k}$ evaluated in the appropriate limits, we obtain

$$D(k, \omega) \equiv 1 + \frac{\omega_p^2}{k^2 v_{t,e}^2} - \frac{\Omega_p^2}{\omega^2}$$
$$+ i \left(\frac{\pi}{2}\right)^{1/2} \left[\frac{\omega_p^2 \omega}{k^3 v_{t,e}^3} + \frac{\Omega_p^2 \omega}{k^3 v_{t,i}^3} \exp\left(-\frac{\omega^2}{2k^2 v_{t,i}^2}\right) \right]. \tag{24.32}$$

This dispersion relation is identical with equation (23.27) except for the imaginary (Landau damping) terms which now appear.

For wavelengths much longer than the Debye length ($k\lambda_D = kv_{t,e}/\omega_p \ll 1$), the dispersion relation $D(k, \omega) = 0$ gives $\omega \approx kC_s - i\gamma$, where $C_s = (T_e/M)^{1/2}$ is the sound speed (the thermal speed of ions at the electron temperature) for the case where $T_i \ll T_e$, and γ is a damping rate given by

$$\gamma = \frac{1}{2} \left(\frac{\pi}{2}\right)^{1/2} \left[\frac{\omega^2}{kv_{t,e}} + \frac{T_e}{T_i} \frac{\omega^2}{kv_{t,i}} \exp\left(-\frac{\omega^2}{2k^2 v_{t,i}^2}\right) \right]$$
$$= \frac{1}{2} \left(\frac{\pi}{2}\right)^{1/2} kC_s \left[\left(\frac{m}{M}\right)^{1/2} + \left(\frac{T_e}{T_i}\right)^{3/2} \exp\left(-\frac{T_e}{2T_i}\right) \right]. \tag{24.33}$$

The Landau damping from electrons is always small, of order $(m/M)^{1/2}$. The Landau damping from ions is small only if $T_e \gg T_i$, in which case it is exponentially small. We conclude that undamped (or weakly damped) ion acoustic waves occur only in the case $T_e \gg T_i$; otherwise, they are subject to strong ion Landau damping.

For $T_e \gg T_i$, ion Landau damping of acoustic waves is small for the usual reason: the phase velocity ω/k is much larger than the ion thermal velocity, so there are very few particles in the resonance region and the slope $\partial f_0/\partial v$ is very small. On the other hand, *electron* Landau damping of acoustic waves is small for a quite different reason: the phase velocity ω/k is much *smaller* than the electron thermal velocity, so the resonance falls into the low-velocity region of $f_0(v)$ where the slope $\partial f_0/\partial v$ is also very small. The situation is illustrated in Figure 24.7.

We saw in Chapter 16 and again in Chapter 23 (see Problem 23.3) that, when finite-T_i corrections are retained in the dispersion function, for $k\lambda_D \ll 1$ the dispersion relation for ion acoustic waves remains $\omega \approx kC_s$, but the sound

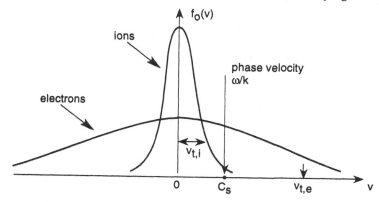

Figure 24.7. Ion and electron distribution functions and the phase velocity ω/k of the ion acoustic wave in the case where $T_e \gg T_i$, so that $C_s \gg v_{t,i}$.

speed is modified to $C_s = [(T_e + 3T_i)/M]^{1/2}$, although this result was limited still to the case $T_i \ll T_e$ in the kinetic treatment of Chapter 23. If, nonetheless, we use this result to obtain an order-of-magnitude estimate for the ion Landau damping in the case $T_i \approx T_e$, by substituting $\omega/k = C_s \sim 2(T/M)^{1/2}$ into the second term on the right-hand side on the first line of equation (24.33), we obtain $\gamma/\omega \sim 0.2$. Such a large value of the damping decrement, γ, indicates that the ion acoustic wave is essentially non-existent in such a plasma.

Chapter 25

Velocity-space instabilities and nonlinear theory

For initial velocity distributions that depart substantially from Maxwellian, we have already seen that it is possible for electron plasma waves to become unstable. The two-stream instability, discussed in Chapter 23, provides a simple example of this kind of 'velocity-space instability'—an instability that can arise in a non-Maxwellian but homogeneous plasma.

More generally, many different types of velocity-space instability are possible in a plasma, and these can arise in various modes of oscillation, not only in the electron plasma waves. Moreover, in some cases, a velocity-space instability can arise in situations where the unperturbed distribution function $f_0(v)$ departs only *slightly* from the Maxwellian. We will discuss two examples of velocity-space instabilities of this type, the first of which arises in the electron plasma waves, and the second in the ion acoustic waves.

25.1 'INVERSE LANDAU DAMPING' OF ELECTRON PLASMA WAVES

The physical picture of Landau damping presented in Chapter 24 gives an immediate indication as to the type of velocity-space distribution function that will tend to destabilize the electron plasma wave. Specifically, if the unperturbed distribution function $f_0(v)$ contains more fast particles than slow particles in the vicinity of $v = \omega/k$, Landau damping is 'inverse', and the electron plasma waves can become unstable. If we suppose that the distribution $f_0(v)$ is *approximately* Maxwellian, except in some small region of suprathermal velocities, then we can assume a Maxwellian in calculating the principal-value integral in equation (24.18), but not in calculating the contribution from the pole at $v = \omega/k$. Keeping only the lowest-order contribution to the principal-value integral for $\omega \gg k v_t$, we obtain the dispersion function

$$D(k, \omega) \equiv 1 - \frac{\omega_p^2}{\omega^2} - \frac{\pi i e^2}{mk^2\epsilon_0} \frac{\partial f_0}{\partial v}\bigg|_{v=\omega/k}. \tag{25.1}$$

Treating the imaginary term as a small correction, the solution of the dispersion relation $D(k, \omega) = 0$ is given by

$$\begin{aligned}
\omega &= \omega_p + \frac{\pi i e^2 \omega_p}{2mk^2\epsilon_0} \frac{\partial f_0}{\partial v}\bigg|_{v=\omega/k} \\
&= \omega_p + \frac{\pi i \omega_p^3}{2nk^2} \frac{\partial f_0}{\partial v}\bigg|_{v=\omega/k}.
\end{aligned} \tag{25.2}$$

We see that the plasma wave becomes *unstable* if the distribution f_0 is 'double-humped' in some region of relatively large v, i.e. if $\partial f_0/\partial v > 0$ at $v = \omega/k$. Such a distribution is shown in Figure 25.1. For obvious reasons, this is sometimes called the 'bump-on-the-tail' distribution. Distributions of this kind are found in laboratory plasmas, such as those first studied by Langmuir, where energetic 'primary' electrons provide the power for plasma ionization. Such distributions are also common in magnetospheric plasmas, when energetic electrons precipitate to lower altitudes due to geomagnetic disturbances.

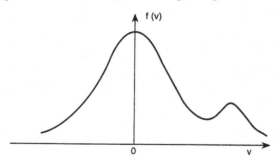

Figure 25.1. A 'double-humped' electron distribution function.

Waves with phase velocities ω/k in the region of the positive slope of $f_0(v)$ will be unstable, gaining energy at the expense of the resonant particles. Although the values of ω are all close to the plasma frequency ω_p, there is almost complete freedom in the choice of k value. The only restriction is that k values with $k\lambda_D \gtrsim 1$ are excluded, since these would introduce large thermal corrections to the dispersion relation. The effect of this restriction is to exclude phase velocities in the thermal range, i.e. $\omega/k \lesssim v_t$. Thus, with this one restriction, we can always find a phase velocity ω/k that lies in the region of positive slope of $f_0(v)$, implying that there is always *some* mode that is unstable.

Clearly, this type of instability is a generalization of the two-stream instability discussed in Chapter 23 (see also Problem 23.2). The distribution $f_0(v)$ shown in Figure 25.1 is the 'finite temperature' generalization of the distribution considered in Problem 23.2, in which each of the two streams is given a finite thermal spread. We see at once how much thermal spread is needed to stabilize the mode: the mode is stabilized only when the thermal spreads are sufficient to remove the minimum in $f_0(v)$, so that the distribution now becomes 'single-humped'. The 'bump-on-the-tail' distribution is different, in this respect, from the symmetric double-humped distribution, shown in Figure 24.6 and analyzed in Problem 24.3. In the latter case, a degree of 'double humpedness' can be tolerated before instability arises. A region with $\partial f_0/\partial v > 0$ (for positive v, the opposite inequality for negative v) is necessary, but not sufficient, for instability: there must also be a mode with phase velocity, ω/k, in this region of reversed slope.

25.2 QUASI-LINEAR THEORY OF UNSTABLE ELECTRON PLASMA WAVES*

Thus far, our discussion of plasma instabilities has been limited to the linear regime, i.e. to perturbations that have very small amplitude. The theory of linear instabilities treats the perturbations as infinitesimal, thereby giving rise to homogeneous linear equations that are relatively amenable to mathematical analysis. In particular, if the unperturbed equilibrium is essentially uniform in space (at least on a scale-length much greater than that of the perturbations), then the eigenfunctions of linear perturbations will be sinusoidal in space, i.e. they will vary as $\exp(i\mathbf{k}\cdot\mathbf{x})$. Each such perturbation (i.e. each value of the wave-vector $\mathbf{k}$) will have a temporal behavior like $\exp(-i\omega_k t)$, where the frequency ω_k is in general complex (real for an oscillatory mode, imaginary for a purely damped or growing mode). In linear theory, we can consider one Fourier component, i.e. one value of the wave-vector $\mathbf{k}$, at a time: there will be no 'interference' between different Fourier components.

For oscillatory modes (and even more so for damped modes), the linear approximation is often quite adequate, since the amplitude of the perturbation does not increase beyond its initial value. For unstable modes, however, the linear approximation must eventually break down, because it predicts that the amplitude of the mode increases exponentially in time without limit. Clearly, there must arise some *nonlinear effects* that limit the amplitude of the perturbation, or that change the 'equilibrium' in such a way that the mode is no longer unstable.

Nonlinear effects in plasma physics can generally be divided into two categories: *wave–particle interactions* and *wave–wave interactions*. For certain types of plasma instabilities, for example the unstable electron plasma waves that arise with 'bump-on-the-tail' distributions, the wave-particle interactions are sufficient to stabilize the waves at relatively small amplitude. In other cases, the waves grow to such large amplitude that interactions between the different waves are important in determining the final wave spectrum. Wave–wave interactions are beyond the scope of this book, but the interested reader is referred to the monographs by R Z Sagdeev and A A Galeev (1969 *Nonlinear Plasma Theory* edited by T M O'Neill and D L Book, New York: Benjamin) and by R C Davidson (1971 *Methods in Nonlinear Plasma Theory* New York: Academic).

Our discussion of the physical meaning of Landau damping has shown that even a small-amplitude wave can interact strongly with particles that are almost resonant with it, tending to mix those particles that are travelling slightly slower than the wave with those particles that are travelling slightly faster than the wave. The result of this will be a flattening of the distribution function in the region of the phase velocity ω/k. A larger-amplitude wave existing for a long enough time can also 'trap' a significant number of particles in its potential troughs, analogous to the island-confined trajectories of the Chirikov–Taylor map in Chapter 5. These are examples of the nonlinear effects of *wave–particle interactions*.

The 'quasi-linear theory' of unstable electron plasma waves provides a technique for describing the effects of wave–particle interactions. Quasi-linear theory assumes that the amplitudes of the various modes that are excited are still small enough that the structure, frequency and instantaneous growth rates of the modes are all adequately described by the linear theory. Thus, even in the small-amplitude limit, quasi-linear theory allows us to understand and quantify how waves and particles exchange energy, and how the flattening of the distribution function leads ultimately to the 'saturation' of the instability, i.e. the cessation of growth of the amplitude of the perturbation electric field. It is a basic assumption of the quasi-linear theory presented here that this process of saturation of the instability arises before the waves have sufficient amplitude that wave–wave interactions become important; otherwise the theory must be extended to include such effects.

As we have seen, the linear theory of electron plasma waves proceeds from the Vlasov equation for the one-dimensional electron distribution f:

$$\frac{\partial f}{\partial t} + v\frac{\partial f}{\partial x} - \frac{e}{m}E\frac{\partial f}{\partial v} = 0 \qquad (25.3)$$

which is linearized by writing

$$f(x, v, t) = f_0(v) + f_1(v)\exp(-i\omega t + ikx)$$
$$E(x, t) = E\exp(-i\omega t + ikx) \qquad (25.4)$$

where the perturbations f_1 and E are assumed small enough that second-order terms involving products of f_1 and E may be neglected. With these approximations, the linearized Vlasov equation becomes

$$-\mathrm{i}(\omega - kv)f_1 = \frac{e}{m} E \frac{\partial f_0}{\partial v}. \tag{25.5}$$

For the present treatment of an *unstable* mode, it is not necessary to introduce the more cumbersome Laplace-transform analysis of Landau, since we found that an unstable mode (complex frequency ω having a positive imaginary part) is always an exact normal mode. For a weakly unstable (i.e. almost oscillatory) mode, we saw in Chapter 24 that the singularity at $v = \omega/k$ should be resolved by noting that ω has a small positive imaginary part, i.e. $\omega \rightarrow \omega + \mathrm{i}\gamma$.

For a specific example of unstable electron plasma waves, we will consider the case of a 'bump-on-the-tail' electron distribution, as shown in Figure 25.1. We have already found the frequency and growth rate of the unstable mode that arises in this situation, and these are given in equation (25.2).

The essence of *quasi-linear theory* is to suppose that the distribution f_0 does not merely describe some initial state, but also describes a slowly evolving 'background' distribution that is changing due to the effects of the unstable waves themselves. With this in mind, we adopt a formal definition of f_0 as the spatially averaged (i.e. averaged over many wavelengths) part of the complete distribution function $f(x, v, t)$. We also assume that a continuous spectrum of waves with different k values is excited, as would be the case for the instability we are considering here. (The opposite case—where there is only one wave excited—is discussed later in this Chapter.)

Accordingly, we generalize equation (25.4) to allow perturbations containing waves with different k values:

$$
\begin{aligned}
f(x, v, t) &= f_0(v) + f_1(x, v, t) \\
f_1(x, v, t) &= \sum_k f_{1k}(v)\exp(-\mathrm{i}\omega_k t + \mathrm{i}kx) \\
E(x, t) &= \sum_k E_k \exp(-\mathrm{i}\omega_k t + \mathrm{i}kx).
\end{aligned}
\tag{25.6}
$$

We continue to limit ourselves to the 'one-dimensional' case which, in effect, assumes that the **k**-vectors of all excited waves are in the same direction, in this case the x direction, so that we may integrate out the other two velocity components and work with the one-dimensional Vlasov equation, equation (25.3), with one velocity component, in this case v_x. It is also important to remind ourselves of our convention (see Chapter 15) for describing waves with 'exponential notation' in expressions such as those in equation (25.6), which is that the *measurable* physical quantity is the *real part* of the right-hand side;

for example, if there is only one k value with a real E_k, then equation (25.6) is understood to mean $E(x, t) = E_k \cos(\omega_k t - kx)$. We have also adopted the convention (without any loss of generality) that all frequencies ω_k are positive (or, if complex, that their real parts are positive), noting of course that we must then include both positive and negative k values in order to allow both leftward and rightward propagating waves. The summations in equation (25.6) are then over whatever waves are present, perhaps some with positive and some with negative k values (if there are symmetric 'bumps-on-the-tail' of $f_0(v)$); each physically distinct wave corresponds to one k value and therefore one term in the summation over k. It is important to understand that we have chosen *not* to introduce an $E_{-k} = E_k^*$ for each k value just to make the summations in equation (25.6) real, since this would require two terms in the summation over k to describe each physically distinct wave and would also require negative frequencies (since $\omega_{-k} = -\omega_k$ when $\pm k$ values are used to describe the same wave). This is, however, an alternative convention that is often used.

We obtain an equation for the *slow* evolution in time of $f_0(v)$ by spatially averaging equation (25.3) over many wavelengths, yielding

$$\frac{\partial f_0}{\partial t} = \frac{e}{m} \left\langle E \frac{\partial f_1}{\partial v} \right\rangle \tag{25.7}$$

noting that the term in $\partial f_0 / \partial v$ must vanish, since all contributions to $E(x, t)$ are oscillating in time. Moreover, only terms in E and f_1 with the same k value survive in the time averaging, since all other terms will be oscillating at some frequency $\omega_k \pm \omega_{k'}$, which will be a finite frequency in all cases except where the negative sign is chosen and $k' = \pm k$. (This assumes that the waves have at least *some* dispersion, which is almost always the case, even for electron plasma waves.) Recalling (from Chapter 15) that in our convention the time average of two first-order quantities A_1 and B_1 written in 'exponential notation' is $\text{Re}(A_1^* B_1 / 2)$, we find that the spatially averaged part of the distribution function evolves according to the equation

$$\frac{\partial f_0}{\partial t} = \frac{e}{2m} \text{Re} \left(\sum_k E_k^* \frac{\partial f_{1k}}{\partial v} \right). \tag{25.8}$$

Since the right-hand side of equation (25.8) is second order in the perturbations, the evolution of f_0 is very slow. Thus, at any given time, we may treat f_0 as essentially constant for the purpose of describing the time dependence of the waves themselves. In particular, we may substitute for f_{1k} from the linearized Vlasov equation, i.e. equation (25.5), for a mode with complex frequency ω_k,

$$f_{1k} = \frac{ie E_k}{m} \frac{\partial f_0 / \partial v}{\omega_k - kv} \tag{25.9}$$

to obtain

$$\frac{\partial f_0}{\partial t} = -\frac{e^2}{2m^2}\frac{\partial}{\partial v}\left[\text{Im}\left(\sum_k |E_k|^2\frac{1}{\omega_k - kv}\right)\frac{\partial f_0}{\partial v}\right] \qquad (25.10)$$

where we have written $E_k^* E_k = |E_k|^2$. If we divide the complex ω_k into a real part ω_k (frequency) and an imaginary part γ_k (growth rate), i.e. $\omega_k \to \omega_k + i\gamma_k$, and take the imaginary part as indicated in equation (25.10), we obtain

$$\frac{\partial f_0}{\partial t} = \frac{e^2}{2m^2}\frac{\partial}{\partial v}\left(\sum_k |E_k|^2\frac{\gamma_k}{(\omega_k - kv)^2 + \gamma_k^2}\frac{\partial f_0}{\partial v}\right). \qquad (25.11)$$

Our use here of the linearized perturbation f_{1k} is equivalent to the use of zeroth-order trajectories in calculating the first-order effect of the waves on the particle distribution function. It assumes implicitly that the distribution function does not develop significant structure in phase space beyond that involved in the linear calculation of f_{1k}. In the nonlinear case, the trajectories are of course strongly modified for special groups of particles, in particular those with velocities close to the phase velocity of the unstable waves. Indeed, nonlinearly, a 'microstructure' in the phase-space, i.e. (x, v), plot of the particle trajectories can arise, similar to the 'islands' of the mapping problem discussed in Chapter 5, making the quasi-linear approach invalid. When a sufficiently broad spectrum of waves is present, however, the microstructure in the particle orbits is destroyed, just as the 'islands' in the mapping were destroyed by 'stochastic overlap'. Interparticle collisions, even if relatively infrequent overall, can also serve to destroy the phase-space microstructure in the particle orbits. In these cases, the use of zeroth-order trajectories in calculating f_{1k} is valid; the opposite case, where quasi-linear theory would be inappropriate, is considered later in this Chapter in the context of trapping in a single wave.

Often the spectrum of waves is sufficiently dense that the summation over discrete k values in equation (25.11) can be replaced by an integral over a continuous variable k. The validity of this assumption must be verified in the particular case under consideration. For the 'bump-on-the-tail' distribution discussed in the previous Section, the phase velocities of unstable waves lie in some band of width Δv about a mean phase velocity $\omega/k = v_0$, where v_0 is in the region of the inverted slope of the equilibrium distribution $f_0(v)$, and Δv is the width of this region of inverted slope. Since the frequencies are all approximately ω_p, the wave-numbers of unstable modes form a band of width $\Delta k \approx k_0(\Delta v/v_0)$ about some mean value $k_0 \approx \omega_p/v_0 = v_{t,e}/v_0\lambda_D$. If the overall size of the plasma in the x direction is L and periodic boundary conditions are applied at the plasma extremities, the wave-numbers allowed by the boundary conditions are $k = 2\pi n/L$ for all integers n. The number $N_{\Delta k}$ of modes (i.e. the number of integers n) which have k values that 'fit' within the prescribed range Δk is

$$N_{\Delta k} = \left(\frac{L}{2\pi \lambda_D} \right) \left(\frac{v_{t,e}}{v_0} \right) \left(\frac{\Delta v}{v_0} \right). \tag{25.12}$$

The factors $v_{t,e}/v_0$ and $\Delta v/v_0$ are both moderately small quantities; for the distribution depicted in Figure 25.1, they are about 0.3 and 0.1, respectively. However, the first factor on the right-hand side of equation (25.12) is an exceedingly large quantity for a plasma, by definition, usually 10^4 or larger. (Even if we divide our band of unstable k values, width Δk, into a large number of narrower 'sub-bands', within which the modes have not only the same frequency ω_k but also the same growth rate γ_k (because the slope of f_0 is essentially the same within the corresponding sub-band of phase velocities), the number of modes within each sub-band will still be large.) Thus, at least for this particular case, it is legitimate to consider the wave-number k as a continuous variable. It then becomes useful to define a measure of the amplitude of the field perturbation in a continuous spectrum. The appropriate quantity for our present needs is the *energy density* residing in the field perturbations within each band of k values, or more precisely within each differential dk of the continuous variable k.

The energy density of an electric field has been found (see Chapter 8, Problem 8.2) to be $\epsilon_0 E^2/2$. For the fluctuating electric field given by equation (25.6), remembering our result regarding the average of the product of two first-order wave quantities expressed in 'exponential notation', we have a spatially averaged energy density

$$W_E = \frac{\epsilon_0}{2} \left\langle \left(\mathrm{Re} \sum_k E_k \exp(-i\omega_k t + ikx) \right)^2 \right\rangle$$
$$= \frac{\epsilon_0}{4} \mathrm{Re} \sum_k E_k^* E_k = \frac{\epsilon_0}{4} \sum_k |E_k|^2. \tag{25.13}$$

We now introduce a quantity $\mathcal{E}(k)$, which is the *density in k-space* of the average energy per unit volume of the electric-field perturbations. Specifically, we define $\mathcal{E}(k)dk = (\epsilon_0/4) \sum_k^{k+dk} |E_k|^2$, where $\sum_k^{k+dk}$ means the sum over all k values lying within the infinitesimal dk. With this definition, we may replace summations over all k values, e.g. the summation appearing in equation (25.11), by an integral over the continuous variable k, provided we substitute

$$\frac{1}{2} \sum_k |E_k|^2 \longrightarrow \frac{2}{\epsilon_0} \int_{-\infty}^{\infty} \mathcal{E}(k)dk. \tag{25.14}$$

The quantity $\mathcal{E}(k)$ is usually called the 'spectral energy density' of the electric field for the particular wave and k value. With our conventions that the real part is to be understood in expressions such as equation (25.6) and that all

ω_k are positive, so that positive and negative k values therefore correspond to rightward and leftward propagating waves, respectively, the spectral energy density $\mathcal{E}(k)$ is *not* (necessarily) symmetric in k, as it is with some alternative conventions often used. In our case, $\mathcal{E}(k)$ for positive k is the spectral energy density in rightward propagating waves, which, for example, will be relatively large for phase velocities ω/k in the region of inverted slope of the 'bump-on-the-tail' distribution shown in Figure 25.1, while $\mathcal{E}(k)$ for negative k is the spectral energy density of leftward propagating waves, which will be negligible in the case of the (positive-v) 'bump-on-the-tail' distribution of Figure 25.1. The concept of spectral energy density can easily be generalized to three dimensions, in which case the electric field energy per unit volume $\epsilon_0 \langle |\mathbf{E}(x, t)|^2 \rangle / 2$ becomes $\int \mathcal{E}(\mathbf{k}) d^3 k$. By analogy with the one-dimensional case, the spectral energy density is defined by the relation $\mathcal{E}(\mathbf{k}) d^3 k = (\epsilon_0/4) \sum_k^{k+dk} |E_k|^2$, where $\sum_k^{k+dk}$ means the sum over all $\mathbf{k}$ values lying within an infinitesimal volume $d^3 k$, which is the volume of a cuboid whose sides are dk_x, dk_y and dk_z.

When we use the procedure of equation (25.14) to transform the right-hand side of equation (25.11) to an integral over k, we may also make use of the fact that the growth rates γ_k are very small, so that the resonance term in equation (25.11) can be approximated by a δ-function, i.e.

$$\frac{\gamma_k}{(\omega_k - kv)^2 + \gamma_k^2} \approx \pi \delta(\omega_k - kv). \tag{25.15}$$

Problem 25.1: Verify equation (25.15) by the following procedure. First sketch the left-hand side as a function of ω_k for fixed kv and for successively smaller values of γ_k, tending toward zero. Then, by integrating the left-hand side over ω_k, show that the area under each of the curves you have sketched is π. (Hint: the integral is best evaluated by substituting $\omega_k - kv = \gamma_k \tan\theta$.)

In this case, we can write equation (25.11) as

$$\frac{\partial f_0}{\partial t} = \frac{2\pi e^2}{\epsilon_0 m^2} \frac{\partial}{\partial v} \left[\left(\int_{-\infty}^{\infty} dk \mathcal{E}(k) \delta(\omega_k - kv) \right) \frac{\partial f_0}{\partial v} \right]. \tag{25.16}$$

This is the 'quasi-linear' equation for the evolution of f_0.

The equation has the form of a *diffusion equation in velocity space*, i.e. an equation of the form

$$\frac{\partial f_0}{\partial t} = \frac{\partial}{\partial v} \left(D(v) \frac{\partial f_0}{\partial v} \right). \tag{25.17}$$

Moreover, the diffusion coefficient

$$
\begin{aligned}
D(v) &= \frac{2\pi e^2}{\epsilon_0 m^2} \int_{-\infty}^{\infty} \mathcal{E}(k)\delta(\omega_k - kv)\mathrm{d}k \\
&= \frac{2\pi e^2}{\epsilon_0 m^2 v}\mathcal{E}(\omega/v)
\end{aligned}
\tag{25.18}
$$

is non-zero only in regions of velocity v corresponding to phase velocities ω/k of excited waves. This corresponds nicely to our physical picture of Landau damping (or the inverse process leading to wave growth) arising from velocity-space diffusion in the region of wave–particle resonance. (Note that, for plasma waves, the frequencies ω_k are all approximately the same, i.e. $\omega_k \approx \omega \approx \omega_\mathrm{p}$. This lack of a significant dependence on k was necessary to allow the simple evaluation of the δ-function integral in equation (25.18).) For the case of our 'bump-on-the-tail' electron distribution, phase velocities of unstably excited waves occur only in regions where the slope of f_0 is inverted, i.e. where $\partial f_0/\partial v > 0$. We suppose, as implied by Fig. 25.1, that this occurs for positive values of v, and this has already been assumed in the second form of equation (25.18). Unless there is some external 'driver' maintaining the inverted slope on the distribution function (such as an energetic electron beam injected continuously into the plasma), we can now see at once that the final state will be a distribution function that has become 'flattened' in the offending region. The original distribution shown by the full curve in Figure 25.2(a) is replaced by the flattened distribution shown by the broken curve in Figure 25.2(a). We see that the time evolution of the quasi-linear process has led to the distribution function being modified between the velocities v_1 and v_2, i.e. a slightly broader region than that of the original inverted slope.

When the distribution function flattens, the growth of the unstable mode is arrested. We can, in fact, calculate the final spectral energy density of the electric field at each k value, by noting that $\mathcal{E}(k)$ is proportional to $|E_k|^2$ and so will increase exponentially with an instantaneous growth rate of $2\gamma_k$:

$$
\begin{aligned}
\frac{\mathrm{d}\mathcal{E}(k)}{\mathrm{d}t} &= 2\gamma_k \mathcal{E}(k) \\
&= \frac{\pi e^2 \omega}{mk^2 \epsilon_0}\mathcal{E}(k)\left.\frac{\partial f_0}{\partial v}\right|_{v=\omega/k}
\end{aligned}
\tag{25.19}
$$

where we have used the expression for γ_k given in equation (25.2). Combining this with the equation

$$
\frac{\partial f_0}{\partial t} = \frac{\partial}{\partial v}\left(\frac{2\pi e^2}{\epsilon_0 m^2 v}\mathcal{E}(\omega/v)\frac{\partial f_0}{\partial v}\right)
\tag{25.20}
$$

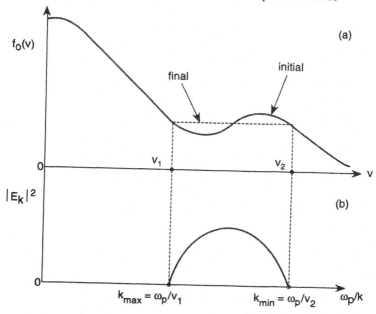

Figure 25.2. Results of quasi-linear evolution of the instability: (a) initial and final distribution functions; (b) final electric-field perturbation spectrum. During this evolution the spectrum spreads out somewhat, so as to extend throughout the range k_{min} to k_{max}, corresponding to phase velocities from v_1 to v_2, which is seen to be somewhat broader than the range of velocities for which the initial distribution has an inverted slope.

we obtain

$$\frac{\partial f_0}{\partial t} = \frac{\partial}{\partial v}\left(\frac{2\omega}{mv^3}\frac{d\mathcal{E}(\omega/v)}{dt}\right). \tag{25.21}$$

Integrating this equation with respect to both v and t, we see that the final spectral energy density of field perturbations (for negligible initial amplitude) is given by

$$\mathcal{E}(k) = \frac{m\omega^2}{2k^3}\int_{v_1}^{v=\omega/k}\left[f_0(v,\infty) - f_0(v,0)\right]dv \tag{25.22}$$

where $f_0(v,\infty)$ and $f_0(v,0)$ denote the final and initial distribution functions, respectively. The lower limit of integration, v_1, is the left-hand end of the flattened region, as indicated in Figure 25.2(a). The integral in equation (25.22) is always positive for ω/k within the flattened region; at the right-hand end, where $v = v_2$, the spectral energy density $\mathcal{E}(k)$ at $k = \omega/v_2$ has fallen to zero again, because of conservation of the total number of particles, i.e.

$$\int_{v_1}^{v_2} f_0(v, \infty)\mathrm{d}v = \int_{v_1}^{v_2} f_0(v, 0)\mathrm{d}v. \tag{25.23}$$

Our result for the final spectral energy density of field perturbations is illustrated in Figure 25.2(b).

The quasi-linear theory of electron plasma waves was formulated first by W E Drummond and D Pines (1962 *Plasma Physics and Controlled Nuclear Fusion Research, Nuclear Fusion, 1962 Supplement, Part 3* p 1049) and independently by A A Vedenov, E P Velikhov and R Z Sagdeev (1961 *Nuclear Fusion* 1 82 (in Russian)).

25.3 MOMENTUM AND ENERGY CONSERVATION IN QUASI-LINEAR THEORY*

We have already used the fact that the total number of particles is conserved in quasi-linear theory. Indeed, this follows immediately from the 'diffusion-equation' form of equations (25.11) or (25.17), since these equations can be integrated over all velocities to show that the total number of particles is conserved, provided only that $\partial f_0/\partial v$ vanishes at $v \to \pm\infty$. In quasi-linear saturation of the 'bump-on-the-tail' instability, particles with velocities near the phase velocity are simply redistributed (on balance toward lower velocities) as shown in Figure 25.2.

Since, in this particular case, there has clearly been a loss of momentum of these 'resonant' particles, it is of interest to explore how overall momentum is conserved in quasi-linear theory. We return to the more exact form of the quasi-linear diffusion equation for f_0, namely equation (25.11), multiply by mv and integrate over all v, integrating by parts on the right-hand side, to obtain

$$\frac{\mathrm{d}}{\mathrm{d}t} \int_{-\infty}^{\infty} mv f_0 \mathrm{d}v = -\frac{e^2}{2m} \sum_k |E_k|^2 \int_{-\infty}^{\infty} \frac{\gamma_k}{(\omega_k - kv)^2 + \gamma_k^2} \frac{\partial f_0}{\partial v} \mathrm{d}v \tag{25.24}$$

where the initially complex frequency ω_k has here been written explicitly in terms of a real frequency and a growth rate, $\omega_k \to \omega_k + \mathrm{i}\gamma_k$. The δ-function approximation given in equation (25.15) can be used in the resonant region of velocities, whereas the integrand in equation (25.24) can be expanded for $\omega_k \gg kv \gg \gamma_k$ in the non-resonant thermal region of velocities. Keeping terms of order kv/ω_k, but not terms of order γ_k^2/ω_k^2 in the non-resonant region, and integrating by parts in v to evaluate the integral, equation (25.24) becomes

$$\frac{\mathrm{d}}{\mathrm{d}t} \int_{-\infty}^{\infty} mv f_0 \mathrm{d}v = -\frac{e^2}{2m} \sum_k |E_k|^2 \left(\frac{2\gamma_k nk}{\omega_k^3} - \frac{\pi}{k} \frac{\partial f_0}{\partial v} \Big|_{v=\omega/k} \right)$$

$$= 0 \tag{25.25}$$

where in the last step we have substituted for γ_k from equation (25.2), and have also written $\omega_k \approx \omega_p$. Despite this last approximation, momentum conservation is of course *exact*, as can be seen by noting that the integral on the right-hand side of equation (25.24) vanishes by virtue of the vanishing of the imaginary part of the *exact* Vlasov dispersion relation.

Problem 25.2: Verify this last statement by taking the imaginary part of the Vlasov dispersion relation, i.e. equation (23.12), for an unstable electron plasma wave with frequency ω_k and growth rate γ_k. Remember that the Vlasov dispersion relation describes *unstable* waves without need for the Landau theory.

Equation (25.25) indicates that the loss of momentum of the resonant particles is balanced by a small gain in the momentum of all the other particles. Purely electrostatic field perturbations themselves carry no momentum.

Next we consider conservation of energy. In the particular case shown in Figure 25.2, the resonant particles have clearly lost energy, which must be accounted for. Again, beginning with equation (25.11), we now multiply by $mv^2/2$ and integrate over all v, integrating by parts on the right-hand side, to obtain

$$\frac{d}{dt}\int_{-\infty}^{\infty}\frac{mv^2}{2}f_0\,dv = -\frac{e^2}{2m}\sum_k|E_k|^2\int_{-\infty}^{\infty}\frac{v\gamma_k}{(\omega_k-kv)^2+\gamma_k^2}\frac{\partial f_0}{\partial v}\,dv. \quad (25.26)$$

As before, the δ-function approximation given in equation (25.15) can be used in the resonant region of velocities, whereas the integral can be expanded for $\omega_k \gg kv \gg \gamma_k$ in the non-resonant regions. Keeping only the zeroth-order term in the expansion in the non-resonant regions and integrating again by parts in v, equation (25.26) becomes

$$\begin{aligned}
\frac{d}{dt}\int_{-\infty}^{\infty}\frac{mv^2}{2}f_0\,dv &= \frac{e^2}{2m}\sum_k|E_k|^2\left(\frac{\gamma_k n}{\omega_k^2}-\frac{\pi\omega_k}{k^2}\left.\frac{\partial f_0}{\partial v}\right|_{v=\omega/k}\right)\\
&= -\frac{ne^2}{2m}\sum_k\frac{\gamma_k}{\omega_k^2}|E_k|^2\\
&= -\frac{\epsilon_0}{2}\sum_k\gamma_k|E_k|^2.
\end{aligned} \quad (25.27)$$

Here, in the next-to-last step, we have substituted for γ_k from equation (25.2), and in the last step we have written $\omega_k \approx \omega_p$. (Energy conservation is of course *exact*; only our particular expressions for the loss of energy from resonant particles and its gain by non-resonant particles are approximate.) Energy resides also in the electric field perturbations themselves, and the electric field energy at each k value, namely $\epsilon_0 |E_k|^2/4$ per unit volume, is growing exponentially with growth rate $2\gamma_k$. Thus, the rate of change of the total energy per unit volume in the electric field is

$$\frac{\mathrm{d}W_E}{\mathrm{d}t} = \frac{\epsilon_0}{2} \sum_k \gamma_k |E_k|^2. \tag{25.28}$$

Adding equations (25.27) and (25.28), we obtain the energy conservation relation

$$\frac{\mathrm{d}}{\mathrm{d}t} \left(\int_{-\infty}^{\infty} \frac{mv^2}{2} f_0 \mathrm{d}v + \int_{-\infty}^{\infty} \mathcal{E}(k)\mathrm{d}k \right) = 0 \tag{25.29}$$

where we have gone to the continuous-k case, in which $W_E = \int_{-\infty}^{\infty} \mathcal{E}(k)\mathrm{d}k$.

The energy lost from the resonant particles goes partly into electric field energy and partly into kinetic energy of the non-resonant particles oscillating in the wave. Sometimes these last two are combined and called the 'wave energy'. With this terminology, the growth in wave energy is exactly balanced by the loss of energy in resonant particles.

Problem 25.3: For electron plasma waves with $\omega \approx \omega_\mathrm{p}$, such as those destabilized by a bump-on-the-tail distribution, show that the wave energy is divided *equally* between electric field energy and kinetic energy of the oscillating non-resonant electrons. By examining the first form of the right-hand side of equation (25.27), confirm that half the loss of energy from resonant particles is balanced by a gain in kinetic energy of non-resonant particles, the other half going into electric field energy.

25.4   ELECTRON TRAPPING IN A SINGLE WAVE*

Our discussion of the quasi-linear theory of plasma waves destabilized by a bump-on-the-tail electron distribution has noted that the number of unstable k values is usually very large, so much so that the spectrum becomes sufficiently dense for stochastic destruction of microstructures in the phase space of particle orbits to occur, giving rise to a continuous region of velocity-space diffusion. Nonetheless it is of interest to examine the opposite situation, where dominantly only one k value is excited. Let us choose the phase so that the electric field of this single wave may be written

$$E(x, t) = \bar{E}\sin(kx - \omega t). \tag{25.30}$$

The motion of an electron in this wave is given by

$$m\frac{\mathrm{d}v}{\mathrm{d}t} = -e\bar{E}\sin(kx - \omega t) \tag{25.31}$$

and if the (assumed real) amplitude, $\bar{E}$, of the electric field is small, so that we may integrate equation (25.31) along the unperturbed orbit, $\mathrm{d}x/\mathrm{d}t = v_0$, we obtain

$$v = v_0 - \frac{e\bar{E}}{m}\frac{\cos(kx - \omega t)}{\omega - kv_0} \tag{25.32}$$

to first order in $\bar{E}$. If we plot v as a function of $(x - \omega t/k)$ for a variety of values of v_0, as is done in Figure 25.3, we obtain a pictorial depiction of the 'trajectories' $v(x, t)$ of all electrons relative to a frame moving at a velocity ω/k. Ignoring for the moment the 'islands' at the center of Figure 25.3 (which are not derived from equation (25.32)), we see that electrons with $v_0 \gg \omega/k$ move always rightward, although their velocities have maxima at $x - \omega t/k = 0, \pm 2\pi, \pm 4\pi$, etc, whereas electrons with $v_0 \ll \omega/k$ move always leftward, their velocities having minima at these same values of $x - \omega t/k$. (Figure 25.3 is drawn for the case where $\bar{E}$ and k are both positive.) However, if v_0 is very close to ω/k, it is no longer permissible to integrate equation (25.31) along the unperturbed orbits. Specifically, equation (25.32) fails (because the first-order term is as large as the zeroth-order term) when

$$\left(v_0 - \frac{\omega}{k}\right)^2 \sim \frac{e\bar{E}}{mk} \tag{25.33}$$

i.e. for electrons whose kinetic energy in the frame moving at the phase velocity is of order their potential energy in the electric field perturbation.

Problem 25.4: Identify the equivalent of equation (25.33) in the mapping problem discussed in Chapter 5.

The *exact* trajectories of electrons in a single electrostatic plasma wave can be obtained by first deriving an applicable constant of the motion, which is essentially the sum of the kinetic energy of the electron in a frame moving at the phase velocity and the potential energy of the electron in the electric field. We define an electric potential

$$\phi(x, t) = \bar{\phi}\cos(kx - \omega t) \tag{25.34}$$

so that

$$E = -\partial\phi/\partial x \qquad \bar{\phi} = \bar{E}/k. \tag{25.35}$$

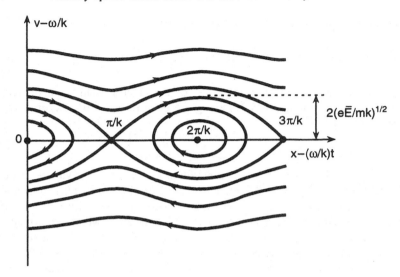

Figure 25.3. Phase-space trajectories of electrons in a frame moving at the phase velocity, ω/k, of a plasma wave with $E = \bar{E}\sin(kx - \omega t)$.

Multiplying the electron's equation of motion, equation (25.31), by $(v - \omega/k)$, we obtain

$$\frac{\mathrm{d}}{\mathrm{d}t}\left[\frac{m}{2}\left(v - \frac{\omega}{k}\right)^2\right] = -e\bar{\phi}(kv - \omega)\sin(kx - \omega t)$$

$$= e\left(\frac{\partial}{\partial t} + v\frac{\partial}{\partial x}\right)\bar{\phi}\cos(kx - \omega t)$$

$$= e\frac{\mathrm{d}}{\mathrm{d}t}\left[\bar{\phi}\cos(kx - \omega t)\right] \qquad (25.36)$$

so that

$$\frac{m}{2}\left(v - \frac{\omega}{k}\right)^2 - e\bar{\phi}\cos(kx - \omega t) = \text{constant}. \qquad (25.37)$$

(Equation (25.37) is equivalent to equation (5.22) in the mapping problem of Chapter 5.)

The trajectories shown in Figure 25.3 are the plots of $v - \omega/k$ versus $x - \omega t/k$ for various values of the constant on the right-hand side of equation (25.37). Values of this constant greater than $e\bar{\phi}$ or less than $-e\bar{\phi}$ give the 'open' trajectories along which an electron's velocity remains of one sign. Values of the constant in the range $(-e\bar{\phi}, e\bar{\phi})$ give the 'closed' trajectories, or 'islands', corresponding to electrons that are 'trapped' near locations of maximum electric potential (minimum electron potential energy).

Figure 25.3 can be viewed as a continuous, rather than discrete, 'area-preserving map' of one-dimensional phase space (x, v) from the equilibrium configuration to the perturbed configuration. (We know that the area of phase space occupied by any given set of particles is preserved.) We have encountered discrete mappings (and the associated 'island chains', leading eventually to the onset of stochasticity) in our analysis of non-J-conserving particle orbits (see Chapter 5) and other continuous structures of this kind in our treatment of the reconnection of magnetic flux by a resistive tearing instability (see Chapter 20). These previous discussions allow us to draw some immediate qualitative conclusions about the relationship between quasi-linear theory and the trapping of electrons in a single nonlinear wave.

For the case of a single wave, particle trapping is the long-time consequence of the nonlinear wave–particle interaction. At early times, when the wave is still of very small amplitude, near-resonant particles are speeded-up and slowed-down by their interaction with the wave and, if the distribution function $f_0(v)$ has an inverted slope at the phase-velocity ω/k, these interactions cause diffusion in velocity space that on balance feeds energy into the wave, so that it grows exponentially in time. The distribution function $f_0(v)$, averaged in space and time, then tends to flatten in the vicinity of the phase velocity. In some cases, this may be sufficient to stabilize the wave growth. However, as the wave grows, it traps an increasing number of particles, which can no longer contribute any more energy to increasing the wave amplitude; eventually, a saturated state may be reached in which further growth of the wave amplitude, accompanied by trapping of more particles, is energetically disfavored. In the case of a *stable* wave, which must be excited initially by some external disturbance of finite amplitude, particle trapping will 'compete' against Landau damping. If the initial amplitude is sufficiently large that significant particle trapping occurs before the damping has proceeded very far, then a saturated state can arise in which Landau damping is effectively 'turned off', since the oscillating motion no longer produces further flattening of the distribution function. In this case, Landau damping will prevail only for times up to the 'bounce' time of a particle trapped in the wave unless, as is very often the case, interparticle collisions or stochastic overlaps arising from a spectrum of wave effectively destroy the phase-space microstructure. The effect of particle trapping on Landau damping was first analyzed in a paper by T M O'Neill (1965 *Phys. Fluids* **8** 2255).

Now suppose that additional discrete waves with slightly different k values and phase velocities ω/k are unstable. As single waves, these will give rise to additional island chains centered at different values of $v - \omega/k$ in the particle trajectories shown in Figure 25.3. Indeed, Figure 25.3 would begin to look like the non-J-conserving particle-orbit maps depicted in Figure 5.2. When these different island chains begin to overlap, the trajectories become stochastic. In the context of the present discussion of electron trapping in a nonlinear plasma wave, this means that velocity-space diffusion, leading to flattening of the distribution function $f_0(v)$, will extend over the entire range of v corresponding to phase velocities of unstable waves—exactly as predicted by quasi-linear theory. Similarly, in the stable case, if a wave-packet with a range of k values is excited initially, rather than a pure single wave, particle trapping can be inhibited by island overlap, in which case Landau damping will continue essentially indefinitely. As we have seen, in many cases of interest, such as the electron plasma wave destabilized by a bump-on-the-tail distribution, there are *very many* unstable modes excited, with a large number of different k values. In such cases, quasi-linear theory generally gives a good description of the nonlinear behavior. (By contrast, we saw in Chapter 20 that only *very few* (typically not more than one, or at most two) resistive tearing modes can be unstable at the same time in the standard tokamak configuration. Individual chains of 'magnetic islands' are a fairly common occurrence in tokamaks, if such modes arise at all. However, when these island chains *do* overlap, the magnetic fields become stochastic, and severe plasma losses can occur.)

25.5   ION ACOUSTIC WAVE INSTABILITIES

The kinetic theory of ion acoustic waves was presented in Chapter 24. For Maxwellian distributions, we found that these waves have phase velocity $\omega/k \approx C_s \equiv (T_e/M)^{1/2}$ and are subject to only weak Landau damping if $T_i \ll T_e$.

It is interesting to consider whether ion acoustic waves can be *destabilized*. By analogy with the case of electron plasma waves, which are destabilized by creating a 'double-humped' or 'bump-on-the-tail' electron distribution $f_0(v)$ with a region of positive slope $\partial f_0/\partial v$, we might expect acoustic waves to be destabilized by a region of positive slope $\partial f_0/\partial v$ on *either* the electron or ion distribution function in the region of the phase velocity ω/k.

One important case in which this can occur is where the electrons are carrying a non-zero current, i.e. where the electron distribution f_{0e} is 'shifted' relative to the ion distribution f_{0i}. This would correspond to the case of a plasma carrying an electrical current and could be produced, for example, by an equilibrium electric field. Although the exact electron distribution function in this case would be that given by solving the Fokker–Planck equation, for a simple example we might suppose that the electron distribution function is still Maxwellian, but about some *non-zero mean velocity u*. Specifically, taking the non-zero mean velocity to be in the x direction and integrating out the unimportant v_y and v_z components of the velocity, the distribution function for electrons becomes

$$f_{0,e} = n \left(\frac{m}{2\pi T}\right)^{1/2} \exp\left(-\frac{m(v_x - u)^2}{2T}\right). \tag{25.38}$$

It is clear from Figure 25.4 that a potential for instability exists if the mean velocity u exceeds the sound speed $C_s = (T_e/M)^{1/2}$.

The analysis is very similar to that given in the Section on ion acoustic waves in Chapter 24. Indeed, to obtain the dispersion function $D(k, \omega)$, it is only necessary to make the substitution $\omega \rightarrow \omega - ku$ in the electron contribution given in equation (24.32). In this way, we obtain

$$D(k, \omega) \equiv 1 + \frac{\omega_p^2}{k^2 v_{t,e}^2} - \frac{\Omega_p^2}{\omega^2}$$
$$+ i \left(\frac{\pi}{2}\right)^{1/2} \left[\frac{\omega_p^2(\omega - ku)}{k^3 v_{t,e}^3} + \frac{\Omega_p^2 \omega}{k^3 v_{t,i}^3} \exp\left(-\frac{\omega^2}{2k^2 v_{t,i}^2}\right)\right]. \tag{25.39}$$

Problem 25.5: Verify that equation (25.39) is indeed the correct dispersion function for ion acoustic waves for the shifted Maxwellian electron distribution given in equation (25.38), with $u \ll v_{t,e}$.

As before, for wavelengths longer than the Debye length, the solution is $\omega = kC_s + i\gamma$, where C_s is still the sound speed, but where γ is now an *instability growth rate* given by

$$\gamma = \frac{1}{2}\left(\frac{\pi}{2}\right)^{1/2}\left[\frac{\omega(ku - \omega)}{kv_{t,e}} - \frac{T_e}{T_i}\frac{\omega^2}{kv_{t,i}}\exp\left(-\frac{\omega^2}{2k^2 v_{t,i}^2}\right)\right]$$
$$= \frac{1}{2}\left(\frac{\pi}{2}\right)^{1/2}kC_s\left[\left(\frac{m}{M}\right)^{1/2}\left(\frac{u}{C_s} - 1\right) - \left(\frac{T_e}{T_i}\right)^{3/2}\exp\left(-\frac{T_e}{2T_i}\right)\right]. \tag{25.40}$$

The first (electron) term in equation (25.40) is destabilizing if $u > C_s$. Whether or not the wave is actually unstable is a competition between the destabilizing electron term, which is small of order $(m/M)^{1/2}$, and the stabilizing ion term, which is small if $T_i \ll T_e$.

The overall conclusion is that, if we try to drive a current in a plasma with $T_i \ll T_e$ (a common case), ion acoustic wave instabilities may set in when the electron streaming speed u exceeds the ion sound speed $(T_e/M)^{1/2}$. Clearly, this constitutes a much lower instability threshold than for two-stream instabilities

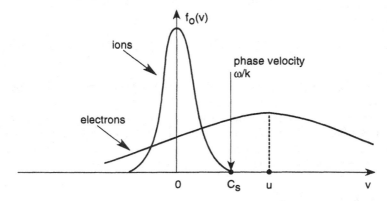

Figure 25.4. Ion and electron distribution functions and the phase velocity ω/k of the ion acoustic wave for the case where the electron distribution has a streaming speed u (cf Figure 24.7). As in Figure 24.7, the case $T_e \gg T_i$ so that $C_s \gg v_{t,i}$, is shown.

of electron plasma waves (see, for example, Problem 24.3), which require the electron streaming speed to exceed the *electron* thermal speed $(T_e/m)^{1/2}$. Whether or not ion acoustic wave instabilities impede current flow has to do with the nonlinear effects of the waves. Do they create additional 'resistivity', or do they merely redistribute the electron velocities to create a stable, flattened distribution function f_{0e} in the low-velocity region where wave resonances occur? The quasi-linear theory, discussed earlier in this Chapter for the case of electron plasma waves, suggests strongly that the latter possibility will occur, and this is indeed the case: the electron distribution f_{0e} shown in Figure 25.4 becomes flattened in the region $0 < v < u$, while still having a non-zero mean velocity, i.e. a non-zero net current. To the extent that electron–electron collisions are important, however, there will be a tendency to restore the shifted-Maxwellian distribution. A competition will arise between these collisional effects and the quasi-linear effects of weakly unstable ion acoustic waves, the former tending to maintain an electron distribution with a positive slope, the latter attempting to flatten the electron distribution in the region $0 < v < u$.

Chapter 26

The drift-kinetic equation and kinetic drift waves*

In Chapter 22, we introduced the Vlasov equation, which describes the evolution in time of the particle distribution function, $f(\mathbf{x}, \mathbf{v}, t)$, in six-dimensional phase space $(\mathbf{x}, \mathbf{v})$. In principle, the Vlasov equation may be used to generate, from some 'given' electric and magnetic fields, charge and current densities which may then modify the electric and magnetic fields through Maxwell's equations. Although, in the general case, the Vlasov–Maxwell description of a plasma would be highly intractable mathematically (and even computationally), we succeeded in Chapters 23–25 in solving the relevant equations in a few interesting cases, specifically for 'one-dimensional' linearized wave-like perturbations in which the wave-vector $\mathbf{k}$ has a single component directed along (or in the absence of) a magnetic field. We also limited ourselves to electrostatic waves, in which there are no significant magnetic perturbations, so that the Maxwell equations reduce, essentially, to the Poisson equation. The plasma equilibria were assumed to be spatially uniform, at least on the scale of the perturbation wavelength.

The same 'kinetic' approach can be extended to more realistic situations involving spatially non-uniform plasmas, perturbations with wave-vectors $\mathbf{k}$ having all three components, and magnetic perturbations. For example, in uniform magnetized plasmas, the full 'hot plasma' dispersion relation can be derived using a kinetic approach (see T H Stix (1992 *Waves in Plasmas* New York: American Institute of Physics)). In at least one case, namely where the perturbations have frequencies much smaller than the frequency of Larmor gyration and wavelengths much larger than the Larmor radius, non-uniform plasmas also become reasonably tractable. In this latter case, we can replace the particle distribution function, $f(\mathbf{x}, \mathbf{v}, t)$, by a guiding-center distribution $f_{gc}(\mathbf{x}_{gc}, v_\perp, v_\parallel, t)$, where $\mathbf{x}_{gc}$ is the position of the guiding-center and $v_\perp$ and

$v_\parallel$ are velocity coordinates representing, respectively, the velocity perpendicular to the magnetic field (i.e. the speed with which the circular Larmor orbit is traversed) and the velocity parallel to the magnetic field (i.e. the speed at which the guiding center moves along the magnetic field). The motion of the guiding centers *across* the magnetic field can then be described in terms of the guiding-center drifts derived in Chapters 2–4. The 'kinetic' (i.e. Vlasov-like) equation for f_{gc} which can be formulated in this way is usually called the 'drift-kinetic equation'. The drift-kinetic equation can be constructed for quite complex magnetic configurations, in which case the ∇B and curvature drifts must certainly appear in the guiding-center motion, but we will limit ourselves here to a simple geometry, namely the 'plane plasma slab' introduced in Chapter 21, in which the magnetic field is essentially straight and uniform. We will also limit ourselves to 'electrostatic' perturbations in which the equilibrium magnetic field is undisturbed.

26.1 THE 'LOW-β' PLANE PLASMA SLAB

We consider an equilibrium in the form of a 'plane plasma slab' of the type that was used in our analysis of drift waves in Chapter 21. Indeed, the main application of our present analysis will be to look for kinetic modifications of the same drift waves, as well as new types of drift wave that arise only in a kinetic treatment. There is a strong, approximately uniform, straight magnetic field, taken to be in the z direction. The equilibrium plasma is non-uniform in one direction, taken to be the x direction, and it is assumed to be of infinite extent in the other two directions, i.e. the y and z directions. In the fluid treatment of Chapter 21, we described this non-uniformity by means of x-dependent equilibrium densities and temperatures $n_0(x)$ and $T_0(x)$, respectively. In the present 'kinetic' analysis, we must further describe the equilibrium by means of an x-dependent guiding-center distribution $f_0(x, v_\perp, v_z)$, noting that, for a fixed magnetic field in the z direction, the parallel velocity $v_\parallel$ becomes simply v_z. This distribution may or may not be Maxwellian in regard to its velocity-space dependences, but the Maxwellian case is obviously of particular interest, and so we will assume it here, i.e.

$$f_0(x, v) = n_0(x) \left(\frac{m}{2\pi T_0(x)} \right)^{3/2} \exp\left(-\frac{mv^2}{2T_0(x)} \right) \qquad (26.1)$$

where, of course, $v = (v_\perp^2 + v_z^2)^{1/2}$, the magnitude of the total velocity. (It is important to remind ourselves that all distribution functions employed in this Chapter, including the equilibrium distribution $f_0(x, v_\perp, v_z)$, are *guiding-center distributions*, not particle distributions, and so strictly the equilibrium distribution should be written $f_{gc}(x_{gc}, v_\perp, v_z)$. However, for simplicity of notation, we will omit the subscript 'gc' throughout this Chapter. Note that, in this case, the particle distribution function is *not* symmetric in v_y, reflecting the non-zero mean drift (diamagnetic drift) in the y direction, while the guiding-center distribution *is* symmetric in v_y. Particle and guiding-center distributions with these properties have already been encountered in our discussion of the diamagnetic drift in Chapter 7.)

We will consider wave-like perturbations of this equilibrium in which components of the k-vector are allowed in the y and z directions, so that any perturbation quantity ψ can be written

$$\psi(\mathbf{x}, t) = \hat{\psi}(x)\exp(-i\omega t + ik_y y + ik_z z) \tag{26.2}$$

where $\hat{\psi}$ denotes the amplitude of the perturbed quantity ψ.

We will limit ourselves, however, to the 'electrostatic' approximation, in which perturbations of the magnetic field may be neglected. In some general sense, this approximation will be valid for instabilities in low-β plasmas, in which the available plasma thermal energy is insufficient to disturb significantly the magnetic field, which has much larger energy. More specifically, as we have seen in Chapter 21, the electrostatic approximation applies to low-frequency waves and instabilities, such as drift waves and ion acoustic waves, whose frequencies ($k_y v_{de}$ and $k_z C_s$, respectively) are much less than the frequency of the shear Alfvén wave ($\sim k_z v_A$). The electrostatic approximation implies that the perturbed electric field is derivable from a scalar potential ϕ, i.e.

$$\mathbf{E} = -\nabla\phi. \tag{26.3}$$

26.2 DERIVATION OF THE DRIFT-KINETIC EQUATION

Let us begin by deriving the so-called 'drift-kinetic equation' for the electron *guiding-center distribution* $f_e(\mathbf{x}, v_\perp, v_z, t)$. The reason for beginning with the *electrons* is that the assumptions underlying the guiding-center description, namely that the Larmor gyration frequency is very large and the Larmor radius very small compared with macroscopic time- and length-scales, respectively, are better satisfied for electrons than for ions. In particular, we will find it necessary to include some second-order corrections to the first-order guiding-center drifts, e.g. the polarization drift, in treating the ions.

To derive the drift-kinetic equation, we follow the approach used in Chapter 22 to obtain the Vlasov equation. The total number of electron guiding-centers in a volume V of six-dimensional phase space is given by

$$N_e = \int f_e d^3 v d^3 x = \int f_e dV. \tag{26.4}$$

(In a sense, for guiding-centers, phase space has only *five* dimensions, since the guiding-center velocities are completely defined by the two 'cylindrical' velocity coordinates $v_\perp$ and v_z (equivalent to r and z in standard cylindrical coordinates in physical space), so that $d^3v = 2\pi v_\perp dv_\perp dv_z$. In the ensuing analysis, the volume element d^3v in velocity space should be understood to mean the thin annular ring of volume $2\pi v_\perp dv_\perp dv_z$.) Conservation of the number of guiding centers demands that the total time derivative of N must vanish, i.e.

$$0 = \frac{dN_e}{dt} = \int \frac{\partial f_e}{\partial t} dV + \int f \mathbf{U} \cdot d\mathbf{S} \tag{26.5}$$

where $\mathbf{U}$ is a 'six-dimensional' velocity, components $(\dot{\mathbf{x}}, \dot{v})$, which describes the motion of the phase-space surface that bounds the volume V. Applying the divergence theorem to equation (26.5), we obtain

$$0 = \int \left(\frac{\partial f_e}{\partial t} + \nabla \cdot (f_e \mathbf{U}) \right) dV \tag{26.6}$$

and since this must apply to every volume element dV, we can write simply

$$
\begin{aligned}
0 &= \frac{\partial f_e}{\partial t} + \nabla \cdot (f_e \mathbf{U}) \\
&= \frac{\partial f_e}{\partial t} + \nabla_{\mathbf{x}} \cdot (\dot{\mathbf{x}} f_e) + \nabla_{\mathbf{v}} \cdot (\dot{\mathbf{v}} f_e) \\
&= \frac{\partial f_e}{\partial t} + \nabla \cdot (\dot{\mathbf{x}} f_e) + \frac{1}{v_\perp} \frac{\partial}{\partial v_\perp} (v_\perp \dot{v}_\perp f_e) + \frac{\partial}{\partial v_z} (\dot{v}_z f_e).
\end{aligned} \tag{26.7}
$$

Here, in the second form, we have expressed the six-dimensional divergence in terms of its three-dimensional components and, in the third form, we have noted that the velocity-space coordinates applicable to the guiding-center description are the 'cylindrical' coordinates $v_\perp$ and v_z.

In the low-β 'plane plasma slab' equilibrium, equation (26.7) simplifies dramatically. The magnetic field B_z is straight and essentially uniform; the only guiding-center drift that enters is the $\mathbf{E} \times \mathbf{B}$ drift. (The polarization drift is small for electrons.) Thus the guiding-center motion is described by

$$\dot{\mathbf{x}} = \mathbf{v}_E + v_z \hat{\mathbf{z}} \tag{26.8}$$

where $\mathbf{v}_E = \mathbf{E} \times \mathbf{B}/B^2$, and $\hat{\mathbf{z}}$ is a unit vector in the z direction. Moreover, since in the electrostatic limit the $\mathbf{E}$ field is derivable from a scalar potential, as given by equation (26.3), and B_z is essentially uniform, the $\mathbf{E} \times \mathbf{B}$ drift produces incompressible flow, i.e.

$$
\begin{aligned}
\nabla \cdot \mathbf{v}_E = \nabla_\perp \cdot \mathbf{v}_E &= \frac{\partial}{\partial x} \left(\frac{E_y}{B_z} \right) - \frac{\partial}{\partial y} \left(\frac{E_x}{B_z} \right) \\
&= -\frac{1}{B_z} \frac{\partial^2 \phi}{\partial x \partial y} + \frac{1}{B_z} \frac{\partial^2 \phi}{\partial x \partial y} = 0.
\end{aligned} \tag{26.9}
$$

Moreover, the constancy of the magnetic moment $mv_\perp^2/2B$ in a uniform field B_z implies that

$$\dot{v}_\perp = 0. \tag{26.10}$$

Finally, the acceleration of electron guiding-centers along the magnetic field will be determined by the electric field, i.e.

$$\dot{v}_z = -\frac{e}{m}E_z. \tag{26.11}$$

Substituting equations (26.8), (26.9), (26.10) and (26.11) into equation (26.7), we obtain our simplified form for the drift-kinetic equation:

$$\frac{\partial f_e}{\partial t} + \frac{\mathbf{E} \times \mathbf{B}}{B^2} \cdot \boldsymbol{\nabla}_\perp f_e + v_z \frac{\partial f_e}{\partial z} - \frac{e}{m}E_z\frac{\partial f_e}{\partial v_z} = 0. \tag{26.12}$$

Here we have identified the local electron density with the density of guiding centers because of the very small electron Larmor radius. Once this equation is solved, the electron density n_e can be obtained by integrating over all velocities, i.e.

$$n_e = \int f_e d^3v = 2\pi \int f_e v_\perp dv_\perp dv_z. \tag{26.13}$$

We could write down a similar drift-kinetic equation for the *ions*. Indeed, the equation would be exactly the same as equation (26.12), with $e \to -e$ and $m \to M$. However, it is usually necessary to include *second-order* guiding-center drifts for the ions, i.e. the polarization drift which has been neglected in equation (26.12), and it is sometimes necessary to include *other second-order* effects, such as the corrections to the $\mathbf{E} \times \mathbf{B}$ drift that arise from taking the 'finite size' of the ion Larmor orbit into account. For present purposes, however, it is possible to circumvent these difficulties by supposing that the ions are 'cold', i.e. by limiting ourselves to the case where $T_i \ll T_e$. (We had a similar limitation on our fluid analysis of drift waves in Chapter 21.) Physically, the assumption $T_i \ll T_e$ suppresses effects due to the finite size of the ion Larmor orbits, while retaining effects due to the plasma dielectric constant, $\varepsilon_\perp$, which arises from the ion polarization drift but does not require finite ion temperature. Furthermore, if the ion thermal velocities are negligible, then a kinetic description involving the distribution function f_i is unnecessary. Rather, it is sufficient to treat the ions as a 'cold fluid', obeying a continuity equation

$$\frac{\partial n_i}{\partial t} + \boldsymbol{\nabla} \cdot (n_i\mathbf{u}) = 0 \tag{26.14}$$

in which the velocity perpendicular to the magnetic field, $\mathbf{u}_\perp$, is simply the sum of the $\mathbf{E} \times \mathbf{B}$ and polarization drifts

$$\mathbf{u}_\perp = \frac{\mathbf{E} \times \mathbf{B}}{B^2} + \frac{M\dot{\mathbf{E}}_\perp}{eB^2}. \tag{26.15}$$

The first part of this velocity, i.e. the $\mathbf{E} \times \mathbf{B}$ drift, will be divergence-free in the 'electrostatic' case, as indicated in equation (26.9), but the second part, i.e. the polarization drift, will have a non-zero divergence.

The velocity along the field, u_z, is determined from the accelerating parallel electric field

$$M\frac{du_z}{dt} = eE_z. \tag{26.16}$$

Equations (26.14)–(26.16) will suffice to determine the ion density n_i in terms of the electric field $\mathbf{E}$. At this point, then, we have equations for obtaining the electron and ion densities in terms of a single scalar variable, depending only on the spatial coordinates, namely the electric potential ϕ from which $\mathbf{E}$ is derived, as given by equation (26.3).

When the electron and ion densities, n_e and n_i, have both been obtained, they are to be substituted into the Poisson equation

$$\epsilon_0 \nabla^2 \phi = -e(n_i - n_e) \tag{26.17}$$

from which the self-consistent electric potential ϕ can, in principle, be determined. When $k\lambda_D \ll 1$ and $\omega \ll \omega_p$, as is usually the case for phenomena described by the drift-kinetic equation (given its requirements on ω and k for applicability), it is a satisfactory approximation to replace the Poisson equation by the quasi-neutrality condition, namely

$$n_i \approx n_e. \tag{26.18}$$

Even this approximation, however, leaves a highly nonlinear equation to be solved for ϕ. For mathematical tractability, we limit ourselves here to a *linearized* treatment of small amplitude perturbations, for which the electric field $\mathbf{E}$ and the perturbations that it produces in the distribution function are both assumed to be infinitesimally small.

26.3 'COLLISIONLESS' DRIFT WAVES

As an example of the use of the electron drift-kinetic equation to treat small-amplitude waves (and instabilities), we will consider the so-called *'collisionless'* *drift waves*. These are, in a sense, the 'kinetic versions' of the resistive drift waves and instabilities (in the 'electrostatic' approximation) discussed in Chapter 21. Here, kinetic effects, rather than resistivity, provide the dissipation needed to release the energy that is available to make drift waves unstable.

We have already described the equilibrium configuration to be considered, namely the 'plane plasma slab', with an assumed Maxwellian equilibrium electron distribution fraction f_{e0}, as given in equation (26.1). For an initial case to consider, we will suppose that there is a density gradient, but *no* temperature gradient, i.e. $dT_{e0}/dx = 0$. Thus, differentiating equation (26.1) with respect to x, we have

$$\frac{\partial f_{e0}}{\partial x} = \frac{f_{e0}}{n_{e0}} \frac{dn_{e0}}{dx}. \tag{26.19}$$

There is no electric field in the equilibrium.

We now proceed to *linearize* the electron drift-kinetic equation, equation (26.12), about this equilibrium, denoting the perturbation in the distribution function f_{e1}. (The suffix '1' can be dropped from the electric field **E**, which is zero in the equilibrium.) All perturbation quantities, including both f_{e1} and **E**, can be expressed in the wave-like form given in equation (26.2). The linearized drift-kinetic equation then becomes

$$-\mathrm{i}(\omega - k_z v_z)f_{e1} + \frac{E_y}{B_{z0}} \frac{\partial f_{e0}}{\partial x} - \frac{e}{m} E_z \frac{\partial f_{e0}}{\partial v_z} = 0. \tag{26.20}$$

Writing $E_y = -\mathrm{i}k_y\phi$ and $E_z = -\mathrm{i}k_z\phi$, using equation (26.19) for $\partial f_{e0}/\partial x$, and noting that $\partial f_{e0}/\partial v_z = -(v_z/v_{t,e}^2)f_{e0}$, where $v_{t,e} \equiv (T_{e0}/m)^{1/2}$ is the electron thermal velocity, equation (26.20) can be solved for f_{e1}. We obtain

$$\begin{aligned}
f_{e1} &= \frac{k_y v_{de} - k_z v_z}{\omega - k_z v_z} \frac{e\phi f_{e0}}{T_{e0}} \\
&= \frac{e\phi f_{e0}}{T_{e0}} \left(1 - \frac{\omega - k_y v_{de}}{\omega - k_z v_z}\right)
\end{aligned} \tag{26.21}$$

where we have again defined an electron 'diamagnetic drift', v_{de}, given by

$$v_{de} = -\frac{T_{e0}}{n_{e0}eB_{z0}} \frac{dn_{e0}}{dx} \tag{26.22}$$

as in Chapter 21.

The electron density perturbation is obtained by integrating equation (26.21) over all velocities, giving

$$n_{e1} = \frac{n_{e0}e\phi}{T_{e0}} - \frac{e\phi}{T_{e0}}(\omega - k_y v_{de}) \int \frac{f_{e0}d^3v}{\omega - k_z v_z}. \tag{26.23}$$

The first term on the right-hand side of equation (26.23) reflects the tendency of the electrons to relax toward a Boltzmann distribution, $n_e \approx n_{e0} \exp(e\phi/T_{e0})$, along the magnetic field. In the second term on the right-hand side, the integrations over the perpendicular velocity components are trivial, i.e.

$$\int \frac{f_{e0}d^3v}{\omega - k_z v_z} = \int_{-\infty}^{\infty} \frac{F_{e0}(v_z)dv_z}{\omega - k_z v_z} \tag{26.24}$$

where $F_{e0}(v_z)$ is now the 'one-dimensional' Maxwellian distribution

$$F_{e0}(v_z) = n_{e0} \left(\frac{m}{2\pi T_{e0}} \right)^{1/2} \exp \left(-\frac{mv_z^2}{2T_{e0}} \right). \tag{26.25}$$

The most interesting case to consider is where

$$\omega \ll k_z v_{t,e} \tag{26.26}$$

which minimizes electron Landau damping of ion acoustic waves (see Chapter 24) and implies that the typical (i.e. thermal) electron streams along the magnetic field with a speed much faster than the phase velocity with which the wave itself moves along the field. Since, as we have already seen in Chapter 21, the phase velocity of drift waves along the magnetic field tends to be of order the ion acoustic, or 'sound', speed $C_s \approx (T_e/M)^{1/2}$, this assumption that the electrons stream at a much faster speed will generally be valid, since $v_{t,e}/C_s \approx (M/m)^{1/2} \gg 1$. In this case, assuming $\omega \sim k_y v_{de}$, a rough estimation of the magnitude of the second term on the right-hand side of equation (26.23) shows it to be much smaller than the first term on the right-hand side, by a factor of about $\omega/k_z v_{t,e}$. Examining equation (26.23), however, we observe that the integral in the second term on the right-hand side is *singular*, the integrand becoming infinite at $v_z = \omega/k_z$. Fortunately, the analysis of Landau has provided us with a prescription of how to treat singular integrals such as this. Assuming that we will indeed find a wave with an approximately real frequency ω, we should evaluate the singular integral *as if ω had a small positive imaginary part*. (If, in fact, we find an *instability*, i.e. an ω value that actually *does* have a positive imaginary part, the difficulty would not have arisen in the first place. Indeed, we saw in Chapter 24 that Landau's detailed analysis of time behavior

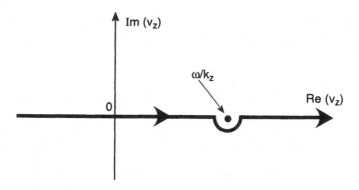

Figure 26.1. Contour of integration in the v_z-plane for evaluating the integral in equation (26.24).

is unnecessary in the case of an unstable mode, which can be a single pure eigenmode with complex frequency ω with a positive imaginary part.)

Thus, according to Landau's prescription, the integration over v_z in equation (26.24) should be taken along a contour in the v_z-plane that lies essentially along the real axis, but is deformed slightly so that it passes *below* the pole at $v_z = \omega/k_z$, as shown in Figure 26.1. (This assumes positive k_z; the contour passes above the pole in the case of negative k_z. Both cases correspond to $\mathrm{Im}(\omega) > 0$.) In the limit $\omega \ll k_z v_{t,e}$, the contribution from the small semicircular path around the pole actually gives the *dominant* contribution to the integral, i.e. larger than the contribution from the rest of the real axis. The contribution from the pole (i.e. $\pi\mathrm{i}$ times the residue) is evaluated as follows:

$$
\mathrm{Res} \int_{-\infty}^{\infty} \frac{F_{e0}(v_z)\mathrm{d}v_z}{\omega - k_z v_z} = \frac{-\pi\mathrm{i}}{|k_z|} F_{e0}\left(\frac{\omega}{k_z}\right)
$$

$$
= -\mathrm{i}\left(\frac{\pi}{2}\right)^{1/2} \frac{n_{e0}}{|k_z|v_{t,e}} \exp\left(-\frac{\omega^2}{2k_z^2 v_{t,e}^2}\right)
$$

$$
\approx -\mathrm{i}\left(\frac{\pi}{2}\right)^{1/2} \frac{n_{e0}}{|k_z|v_{t,e}}. \tag{26.27}
$$

The contribution from the rest of the real axis (i.e. the principal value of the integral) may be estimated as follows:

$$
\mathrm{Pr} \int_{-\infty}^{\infty} \frac{F_{e0}(v_z)\mathrm{d}v_z}{\omega - k_z v_z} = \mathrm{Pr} \int_{0}^{\infty} \frac{2\omega F_{e0}(v_z)}{\omega^2 - k_z^2 v_z^2}\mathrm{d}v_z \sim \mathrm{O}\left(\frac{n_{e0}\omega}{k_z^2 v_{t,e}^2}\right) \tag{26.28}
$$

where, in the first step here, we have used the fact that $F_e(v_z)$ is symmetric in v_z, so that the contributions from positive and negative values of v_z may be combined. The contribution from the pole, given in equation (26.27), is seen to be larger than the principal-value contribution, given in equation (26.28), by a factor of about $k_z v_{t,e}/\omega$. Moreover, when inserted into equation (26.23), the contribution given in equation (26.28) is seen to be much smaller, by a factor of about $\omega^2/k_z^2 v_{t,e}^2$, than the first term on the right in equation (26.23). Accordingly, we neglect the principal-value contribution given in equation (26.28).

Substituting equation (26.27) into equation (26.24), which is then used in equation (26.23), the electron density perturbation becomes

$$
n_{e1} = \frac{n_{e0}e\phi}{T_{e0}}\left[1 + \mathrm{i}\left(\frac{\pi}{2}\right)^{1/2} \frac{\omega - k_y v_{de}}{|k_z|v_{t,e}}\right]. \tag{26.29}
$$

Let us now calculate the *ion* density perturbation, using linearized versions of equations (26.14)–(26.16). The linearization of equation (26.14) in the presence of a density gradient, using equation (26.15) for the perpendicular velocity component $\mathbf{u}_\perp$, gives

$$-i\omega n_{i1} + \frac{E_y}{B_{z0}}\frac{dn_{i0}}{dx} - i\omega n_{i0}\nabla_\perp \cdot \left(\frac{ME_\perp}{eB_{z0}^2}\right) + ik_z n_{i0}u_z = 0. \qquad (26.30)$$

Here, in the term arising from the polarization drift, we have assumed that the gradient scale-length of perturbation quantities is much shorter than the scale-length of the equilibrium density variation, so that n_{i0} may be taken outside of the divergence operator. (If we were to compare the orders-of-magnitude of the second and third terms in equation (26.30) for a drift wave with frequency $\omega \sim k_y v_{de}$, where v_{de} is the electron diamagnetic drift speed, we would find that the third term is smaller by a factor of order $k_\perp^2 C_s^2/\omega_{ci}^2$, which is formally of second order the ratio of the ion Larmor radius to the perpendicular wavelength, since C_s/ω_{ci} is of order the ion Larmor radius, although evaluated at the electron temperature. Indeed a term arising from the polarization drift should be expected to be of this order. However, we retain this term in order to treat short perpendicular wavelengths, as will be seen below.) Equation (26.16) gives simply

$$-i\omega M u_z = eE_z. \qquad (26.31)$$

Using $E_y = -ik_y\phi$ and $E_z = -ik_z\phi$ to express the components of the electric field perturbation in terms of the first-order scalar potential ϕ, we can combine equations (26.30) and (26.31) to obtain a final result for the ion density perturbation, namely

$$\begin{aligned}
n_{i1} &= -\frac{k_y\phi}{\omega B_{z0}}\frac{dn_{i0}}{dx} + \frac{n_{i0}ek_z^2\phi}{M\omega^2} - \frac{n_{i0}k_\perp^2 M\phi}{eB_{z0}^2} \\
&= \frac{n_{i0}e\phi}{T_{e0}}\left(-\frac{k_y T_{e0}}{n_{i0}eB_{z0}\omega}\frac{dn_{i0}}{dx} + \frac{k_z^2 T_{e0}}{M\omega^2} - \frac{k_\perp^2 M T_{e0}}{e^2 B_{z0}^2}\right) \\
&= \frac{n_{i0}e\phi}{T_{e0}}\left(\frac{k_y v_{de}}{\omega} + \frac{k_z^2 C_s^2}{\omega^2} - k_\perp^2 r_{Ls}^2\right). \qquad (26.32)
\end{aligned}$$

Here, we have introduced the equilibrium electron temperature, T_{e0}, in order to express our result in terms of familiar quantities such as the electron diamagnetic drift v_{de}, given in equation (26.22), the sound speed $C_s = (T_{e0}/M)^{1/2}$, and the ion Larmor radius evaluated with the electron temperature, $r_{Ls} = C_s/\omega_{ci} = (MT_{e0})^{1/2}/eB_{z0}$. We have also made use of the charge neutrality of the equilibrium, i.e. $n_{i0} = n_{e0}$. In the last term on the right-hand side of equation (26.32), which arises from the divergence of the polarization drift, we have written

$$\nabla_\perp^2 = -k_\perp^2 = -\left(k_x^2 + k_y^2\right) \qquad (26.33)$$

invoking the 'WKB approximation' in which the perturbation is assumed to vary more rapidly in the x direction than does the equilibrium, so that any perturbation quantity $\psi(x)$ can be approximated as having the wave-like form, $\hat{\psi} \exp(i \int^x k_x dx)$.

The term in $k_\perp^2 r_{Ls}^2$ in equation (26.32) arises from the polarization drift and is formally of second order in the Larmor radius, in the sense noted above. However, the term becomes of order unity for short perpendicular wavelengths, i.e. those of order r_{Ls}. Since r_{Ls} is defined using the *electron* temperature, even the case $k_\perp r_{Ls} \sim 1$ does not violate our assumption that $k_\perp r_{Li} \ll 1$, since we have from the outset chosen to limit our analysis to the case $T_i \ll T_e$. Although our restricted analysis will prove sufficient to identify most of the important classes of drift waves, the theory becomes more complicated in the case where $k_\perp r_{Li}$ is not so restricted, and in particular in the case $k_\perp r_{Li} \sim 1$ (see, for example, N A Krall and A W Trivelpiece (1986 *Principles of Plasma Physics* San Francisco, CA: San Francisco Press) and T H Stix (1992 *Waves in Plasmas* New York: American Institute of Physics)).

Employing the quasi-neutrality approximation, $n_{e1} = n_{i1}$, and using equations (26.29) and (26.32) for n_{e1} and n_{i1}, respectively, we obtain our final

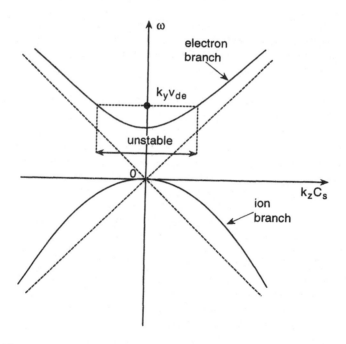

Figure 26.2. Electron and ion branches of the collisionless drift wave. The electron branch is unstable if $\omega < k_y v_{de}$, i.e. in the region shown. Both branches approach asymptotes $\omega = \pm k_z C_s$.

dispersion relation for collisionless drift waves with $T_i \ll T_e$, namely

$$\omega(1 + k_\perp^2 r_{Ls}^2) - k_y v_{de} - \frac{k_z^2 C_s^2}{\omega} = -i \left(\frac{\pi}{2}\right)^{1/2} \frac{\omega(\omega - k_y v_{de})}{|k_z| v_{t,e}}. \tag{26.34}$$

Let us examine this dispersion relation in some detail. First, neglecting the imaginary term on the right-hand side and solving the resulting quadratic equation for ω, we find that there are two 'branches' of the dispersion relation, an 'electron branch' on which the frequency ω has the same sign as $k_y v_{de}$, and an 'ion branch' on which the frequency has the opposite sign. The two branches of the dispersion relation are shown in Figure 26.2, where we have taken $k_y v_{de}$ to be positive, so that the frequency ω on the electron branch is also positive. (As was noted in the discussion of resistive drift waves in Chapter 21, the ion branch of the drift wave shown in Figure 26.2 violates the convention introduced in Chapter 15 that real frequencies ω are taken to be positive. If we are interested in this branch, we could satisfy the convention by simply reversing the sign of k_y.)

Problem 26.1: Return to the figure that you drew in Problem 21.2, and add the plot of $\omega/(k_y v_{de})$ versus $k_z C_s/(k_y v_{de})$ obtained by setting the right-hand side of equation (26.34) to zero, for the case $k_\perp r_{Ls} = 0.3$.

In the limit where $k_\perp r_{Ls} \ll 1$ and $k_z C_s \ll k_y v_{de}$, the electron branch has simply

$$\omega \approx k_y v_{de} \tag{26.35}$$

and it is this mode that is usually called the 'drift wave', or sometimes the 'electron drift wave'. Including the other two terms on the left-hand side of equation (26.34) as small corrections, we obtain a more accurate dispersion relation for the electron drift wave, namely

$$\omega = k_y v_{de}(1 - k_\perp^2 r_{Ls}^2) + \frac{k_z^2 C_s^2}{k_y v_{de}}. \tag{26.36}$$

Because of the factor $(\omega - k_y v_{de})$ appearing in the imaginary term on the right-hand side of equation (26.34), these correction terms are needed to provide a perturbative estimate for the non-zero imaginary part of the complex frequency ω. Letting $\omega \to \omega + i\gamma$ and equating the imaginary parts on the two sides of equation (26.34), assuming that $k_\perp r_{Ls}$ and $k_z C_s/k_y v_{de}$ are small but non-zero quantities, we obtain

$$\gamma = \left(\frac{\pi}{2}\right)^{1/2} \frac{k_y^2 v_{de}^2}{|k_z| v_{t,e}} \left(k_\perp^2 r_{Ls}^2 - \frac{k_z^2 C_s^2}{k_y^2 v_{de}^2}\right). \tag{26.37}$$

We see that the 'collisionless' electron drift wave is unstable only if $k_\perp^2 r_{Ls}^2 > k_z^2 C_s^2 / k_y^2 v_{de}^2$, implying ω values that are less than $k_y v_{de}$ (see equation (26.36)). The region of instability is indicated on Figure 26.2. Since $v_{de} \ll C_s$, we see that instability tends to arise only in situations where $k_z \ll k_y$ and for perpendicular wavelengths that are at most a few times longer than the ion Larmor radius evaluated at the electron temperature. In the limit $k_z \to 0$, modes with smaller values of $k_\perp$ will be unstable, but at some point our assumption that the perpendicular wavelength is much smaller than the scale-length of the equilibrium density gradient will be violated; the limit $k_z \to 0$ also implies an extremely large plasma scale-length along the magnetic field.

The physical process that gives rise to instability is clearly *inverse Landau damping* by electrons, i.e. the electrons in resonance with the parallel phase velocity ω/k_z become destabilizing. We have already encountered this type of instability mechanism in Chapter 25, but only in cases where velocity distributions departed significantly from the Maxwellian (e.g. the 'bump-on-the-tail' distribution). Here, we have found a case of inverse Landau damping for a Maxwellian distribution, but where there is a non-zero density gradient. Examination of equation (26.34) shows that the essential feature that has allowed the interaction between the wave and resonant electrons to be destabilizing is that the wave frequency ω is slightly *less than the diamagnetic frequency*, $k_y v_{de}$.

From an energy viewpoint, the resonant interaction between the wave and the low-v_z electrons allows the release of some of the energy that is available through expansion of the spatially non-uniform plasma in the x direction. The low-v_z electrons drift outwards and inwards in the first-order fluctuating electric drifts, E_y/B_z; by analogy with the cases considered in Chapter 25, the principal nonlinear effect of this type of drift wave will be a flattening of the density gradient in *physical* space, rather than velocity-space flattening. That this flattening of the density gradient releases net energy to the wave can be seen as follows. Suppose that ω, k_y and v_{de} are all positive, implying that $dn_{e0}/dx < 0$ (see equation (26.22)); if k_z is also positive, the wave will be resonant with positive-v_z electrons. The motion of the resonant electrons in the fluctuating electric fields is given by $v_x = E_y/B$ and $dv_z/dt = -eE_z/m$. Since $E_y/E_z = k_y/k_z > 0$, we see that v_x and dv_z/dt are exactly 180° out-of-phase, i.e. dv_z/dt is negative when v_x is positive and *vice versa*. Although resonant electrons drift in both the positive-x and negative-x directions in the fluctuating field, there will be a preponderance of electrons drifting in the positive-x direction because $dn_{e0}/dx < 0$. Thus, on balance, the resonant electrons will *lose* parallel energy (i.e. the net dv_z/dt is negative for positive-v_z electrons) as a result of the flattening of the density gradient. This is the energy that becomes available to drive the unstable wave.

It is interesting to compare this collisionless drift instability with the resistive drift instability derived in Chapter 21. In the electrostatic limit, the resistive drift instability was found to have frequency and growth rate given by equation (21.51). We see that the frequencies of the two modes are essentially the same, namely $\omega \approx k_y v_{de}$, and the growth rates (where the first term in the parenthesis in equation (26.37) is assumed to dominate over the second term) are of similar form, with the growth rate of the resistive mode being larger by a factor $\nu_{ei}/|k_z|v_{t,e}$. Since $v_{t,e}/\nu_{ei}$ is the electron collisional mean-free path, we see that the growth rate of the resistive mode is the larger if the mean-free path is shorter than the parallel wavelength. This is what might have been expected, since in this case the electron motion along the magnetic field in response to the perturbed electric field will be disrupted by collisions before the electron can remain in resonance with the wave for a full wavelength.

The other branch of the dispersion relation, equation (26.34), i.e. the 'ion branch' shown in Figure 26.2, has a frequency ω opposite in sign to $k_y v_{de}$. Again letting $\omega \to \omega + i\gamma$ and equating the imaginary parts on the two sides of equation (26.34), we see immediately that this branch has $\gamma < 0$, i.e. a damping decrement. Within the limitations of the present analysis (especially the assumption of cold ions), this branch is stable and is thus of less interest.

Problem 26.2 Find the correction to the drift wave dispersion relation, i.e. equation (26.34), that arises when the Poisson equation is used rather than the quasi-neutrality approximation. In what regimes of plasma parameters would this correction be important?

26.4   EFFECT OF AN ELECTRON TEMPERATURE GRADIENT

The preceding analysis was limited to the case where there is a density gradient, but no electron temperature gradient. It is interesting to generalize our result to the case where the equilibrium electron distribution function is still Maxwellian, i.e. of the form given in equation (26.1), but where *both* the density, $n_{e0}(x)$, and the temperature, $T_{e0}(x)$, have significant variation in the x direction.

In this case, equation (26.19) must be replaced by a more complicated expression

$$\frac{\partial f_{e0}}{\partial x} = \frac{f_{e0}}{n_{e0}}\frac{dn_{e0}}{dx} - \frac{f_{e0}}{T_{e0}}\frac{dT_{e0}}{dx}\left(\frac{3}{2} - \frac{v^2}{2v_{t,e}^2}\right)$$

$$= \frac{f_{e0}}{n_{e0}}\frac{dn_{e0}}{dx}\left[1 - \eta_e\left(\frac{3}{2} - \frac{v^2}{2v_{t,e}^2}\right)\right] \qquad (26.38)$$

where

$$\eta_e \equiv \frac{1}{T_{e0}} \frac{dT_{e0}}{dx} \left(\frac{1}{n_{e0}} \frac{dn_{e0}}{dx} \right)^{-1} = \frac{\nabla(\ln T_{e0})}{\nabla(\ln n_{e0})}. \tag{26.39}$$

(Here, ∇ simply means d/dx.) Equation (26.38) is obtained by differentiating equation (26.1) with respect to x, which requires that *both* appearances of $T_{e0}(x)$ in equation (26.1) be included in the differentiation.

The parameter η_e is a dimensionless quantity, typically of order unity, and it is the ratio of the scale-length of the density variation to the scale-length of the temperature variation. When equation (26.38) is used in equation (26.20), the resulting expression for f_{e1} given in equation (26.21) must be modified, becoming

$$f_{e1} = \frac{e\phi f_{e0}}{T_{e0}} - \frac{1}{\omega - k_z v_z} \left\{ \omega - k_y v_{de} \left[1 - \eta_e \left(\frac{3}{2} - \frac{v^2}{2v_{t,e}^2} \right) \right] \right\} \frac{e\phi f_{e0}}{T_{e0}}. \tag{26.40}$$

The electron density perturbation is now

$$n_{e1} = \frac{n_{e0}e\phi}{T_{e0}} - \frac{e\phi}{T_{e0}} \int \frac{f_{e0} d^3 v}{\omega - k_z v_z} \left\{ \omega - k_y v_{de} \left[1 - \eta_e \left(\frac{3}{2} - \frac{v^2}{2v_{t,e}^2} \right) \right] \right\}. \tag{26.41}$$

The velocity-space integral in equation (26.41) is more complicated than that in equation (26.23), in view of the additional velocity-dependence in the term involving η_e. Nonetheless, the integrations over the perpendicular velocity components can still be carried out, noting that $v^2 = v_\perp^2 + v_z^2$ and that the average of $v_\perp^2/2$ for a Maxwellian distribution is simply v_t^2. We obtain

$$n_{e1} = \frac{n_{e0}e\phi}{T_{e0}} - \frac{e\phi}{T_{e0}} \int_{-\infty}^{\infty} \frac{F_{e0}(v_z) dv_z}{\omega - k_z v_z} \left\{ \omega - k_y v_{de} \left[1 - \eta_e \left(\frac{1}{2} - \frac{v_z^2}{2v_{t,e}^2} \right) \right] \right\} \tag{26.42}$$

where $F_{e0}(v_z)$ is the 'one-dimensional' Maxwellian distribution given in equation (26.25).

As before, we limit ourselves to the case $\omega \ll k_z v_{t,e}$, where the dominant contribution to the integral in equation (26.42) comes from the pole at $v_z = \omega/k_z$. Deforming the contour in the v_z-phase so that it passes just below this pole, as shown in Figure 25.1, and evaluating the contribution from the pole as π times the residue, we obtain

$$n_{e1} = \frac{n_{e0}e\phi}{T_{e0}} \left\{ 1 + i \left(\frac{\pi}{2} \right)^{1/2} \frac{\exp(-\omega^2/2k_z^2 v_{t,e}^2)}{|k_z| v_{t,e}} \right.$$
$$\times \left\{ \omega - k_y v_{de} \left[1 - \eta_e \left(\frac{1}{2} - \frac{\omega^2}{2k_z^2 v_{t,e}^2} \right) \right] \right\} \Bigg\}$$
$$\approx \frac{n_{e0}e\phi}{T_{e0}} \left[1 + i \left(\frac{\pi}{2} \right)^{1/2} \frac{\omega - k_y v_{de}(1 - \eta_e/2)}{|k_z| v_{t,e}} \right] \tag{26.43}$$

where, in the second form, we have used $\omega \ll k_z v_{t,e}$. Equation (26.43) replaces equation (26.29). We see that the only effect of the electron temperature gradient is to modify the small imaginary term.

The ion density perturbation, n_{i1}, is unaltered by the introduction of an electron temperature gradient, and is still given by equation (26.32). Thus, the dispersion relation obtained by setting $n_{e1} = n_{i1}$ is the same as equation (26.34), except for the modification of the imaginary term on the right-hand side:

$$\omega(1 + k_\perp^2 r_{Ls}^2) - k_y v_{de} - \frac{k_z^2 C_s^2}{\omega} = -i\left(\frac{\pi}{2}\right)^{1/2} \frac{\omega[\omega - k_y v_{de}(1 - \eta_e/2)]}{|k_z|v_{t,e}}. \quad (26.44)$$

The additional term in η_e on the right-hand side of equation (26.44) produces a qualitative change in the stability properties of the electron branch of the drift wave for, considering the simplest limit where $k_\perp r_{Ls} \ll 1$ and $k_z C_s \ll k_y v_{de}$, the real part of the frequency is, as usual,

$$\omega \approx k_y v_{de} \quad (26.45)$$

but the imaginary part no longer vanishes when this lowest-order frequency is used on the right-hand side of equation (26.44), giving a 'growth rate'

$$\gamma \approx -\left(\frac{\pi}{2}\right)^{1/2} \frac{k_y^2 v_{de}^2}{|k_z|v_{t,e}} \eta_e. \quad (26.46)$$

For most cases, where the density and temperature gradients are in the same direction, so that

$$\eta_e \equiv \frac{\nabla(\ln T_e)}{\nabla(\ln n_e)} > 0 \quad (26.47)$$

the value of γ will be negative, implying that the effect of a temperature gradient is to damp, rather than destabilize, the electron drift wave. Indeed, adding the damping decrement given in equation (26.46) to the growth rate given in equation (26.37), we see that the temperature gradient will stabilize all drift waves with $k_\perp^2 r_{Ls}^2 < \eta_e$, i.e. all except the shortest wavelength modes (assuming $\eta_e \sim 1$). From an energy viewpoint, it is perhaps somewhat surprising that the effect of the temperature gradient is stabilizing in this case, since the temperature gradient adds another source of energy available through expansion of the plasma. However, drift waves resonate only with low-v_z electrons; in this region of velocity space, the spatial gradient of the 'one-dimensional' Maxwellian distribution function (at a fixed value of v_z, much less than the thermal velocity) has *opposite* contributions from density and temperature gradients.

On the other hand, situations could occur where the density and temperature gradients are *oppositely* directed, i.e.

$$\eta_e \equiv \frac{\nabla(\ln T_e)}{\nabla(\ln n_e)} < 0. \quad (26.48)$$

In such cases, equation (26.46) shows that the electron drift wave is quite strongly *destabilized*. Presumably, the effect of this instability in its nonlinear regime is to produce some kind of 'turbulent convection' of particles and heat that tends to reduce, or even eliminate, the oppositely directed density and temperature gradients.

26.5 EFFECT OF AN ELECTRON CURRENT

Let us now consider another case of interest, namely where the Maxwellian electron distribution is given a non-zero mean velocity, or 'streaming speed', u_{e0}, along the magnetic field. In this case, the equilibrium distribution function, replacing equation (26.1), takes the form

$$f_0(x, v_\perp, v_z) = n_0(x) \left(\frac{m}{2\pi T_0} \right)^{3/2} \exp \left(-\frac{mv_\perp^2}{2T_0} - \frac{m(v_z - u_{e0})^2}{2T_0} \right). \quad (26.49)$$

To simplify the analysis, we will limit ourselves to the case where there is no temperature gradient, i.e. T_0 = constant. Although the 'shifted Maxwellian' electron distribution function given in equation (26.49) is representative of the case where the plasma carries a net electrical current, i.e. the electrons have a non-zero average streaming speed relative to the ions, it must be noted that if this current is produced by a driving electric field, the electron distribution will have a somewhat different form, to be obtained by balancing the electric force against the collisional friction with the ions (see Chapter 13). Nonetheless, the analysis using equation (26.49) is qualitatively representative of the real situation. (see Problem 26.3).

Going back to the linearized drift-kinetic equation for electrons, equation (26.20), we see that we must now write

$$\partial f_{e0}/\partial v_z = -[(v_z - u_{e0})/v_{t,e}^2] f_{e0}$$

so that equation (26.21) is replaced by

$$\begin{aligned} f_{e1} &= \frac{k_y v_{de} - k_z(v_z - u_{e0})}{\omega - k_z v_z} \frac{e\phi f_{e0}}{T_{e0}} \\ &= \frac{e\phi f_{e0}}{T_{e0}} - \frac{\omega - k_y v_{de} - k_z u_{e0}}{\omega - k_z v_z} \frac{e\phi f_{e0}}{T_{e0}}. \end{aligned} \quad (26.50)$$

Proceeding as before, we find that the electron density perturbation becomes

$$r_{e1} = \frac{n_{e0}e\phi}{T_{e0}} - \frac{e\phi}{T_{e0}} (\omega - k_y v_{de} - k_z u_{e0}) \int_{-\infty}^{\infty} \frac{F_{e0}(v_z)}{\omega - k_z v_z} dv_z \quad (26.51)$$

where $F_{e0}(v_z)$ is again the 'one-dimensional' electron distribution, which is now

$$F_{e0}(v_z) = n_{e0} \left(\frac{m}{2\pi T_{e0}}\right)^{1/2} \exp\left(-\frac{m(v_z - u_{e0})^2}{2T_{e0}}\right). \qquad (26.52)$$

As before, we consider only the case $\omega \ll k_z v_{t,e}$, and we further assume that $u_{e0} \ll v_{t,e}$, i.e. that the electron streaming speed is small compared with the electron thermal velocity (an assumption that is not very restrictive for many realistic conditions). The dominant contribution to the integral in equation (26.51) again comes from the pole at $v_z = \omega/k_z$, which may, as before, be evaluated as πi times the residue. We obtain

$$\begin{aligned}
n_{e1} &= \frac{n_{e0}e\phi}{T_{e0}}\left[1 + i\left(\frac{\pi}{2}\right)^{1/2}\frac{\omega - k_y v_{de} - k_z u_{e0}}{|k_z|v_{t,e}}\exp\left(-\frac{(\omega - k_z u_{e0})^2}{2k_z^2 v_{t,e}^2}\right)\right] \\
&\approx \frac{n_{e0}e\phi}{T_{e0}}\left[1 + i\left(\frac{\pi}{2}\right)^{1/2}\frac{\omega - k_y v_{de} - k_z u_{e0}}{|k_z|v_{t,e}}\right]. \qquad (26.53)
\end{aligned}$$

Equation (26.53) replaces equation (26.29). We see that the only effect of the electron streaming speed u_{e0} is to modify the small imaginary term.

The ion density perturbation, n_{i1}, is still given by equation (26.32), and so the dispersion relation obtained by setting $n_{e1} = n_{i1}$ is the same as equation (26.34), except for modification of the imaginary term on the right-hand side:

$$\omega(1 + k_\perp^2 r_{Ls}^2) - k_y v_{de} - \frac{k_z^2 C_s^2}{\omega} = -i\left(\frac{\pi}{2}\right)^{1/2}\frac{\omega(\omega - k_y v_{de} - k_z u_{e0})}{|k_z|v_{t,e}}. \qquad (26.54)$$

As in the case of a temperature gradient, the modification of the imaginary term on the right-hand side produces a qualitative change in the stability properties of the electron drift wave. Again considering the simplest limit where $k_\perp r_{Ls} \ll 1$, the real part of the frequency of the electron branch is given by

$$\omega \approx k_y v_{de} + \frac{k_z^2 C_s^2}{k_y v_{de}} \qquad (26.55)$$

where we have also assumed that $k_z C_s \ll \omega \approx k_y v_{de}$, but have kept the first-order correction in the small quantity $k_z^2 C_s^2/\omega^2$, while neglecting the correction of order $k_\perp^2 r_{Ls}^2$. The imaginary part of equation (26.54) then gives a growth rate

$$\gamma \approx \left(\frac{\pi}{2}\right)^{1/2}\frac{k_y v_{de}k_z u_{e0} - k_z^2 C_s^2}{|k_z|v_{t,e}}. \qquad (26.56)$$

We see from equation (26.56) that there is a range of k_z values for which $\gamma > 0$, i.e. for which the wave is unstable. (It might seem that we have implicitly assumed that $u_{e0} > 0$, i.e. that the electrons are streaming in the positive direction along the magnetic field. However this is not so, since the same instability would arise for $u_{e0} < 0$, but would then have a negative k_z value. More generally, if we also abandon the convention that $k_y v_{de}$ is positive, instability will arise in cases where $k_z u_{e0}$ and $k_y v_{de}$ have the same sign.)

Problem 26.3 Expand the distribution function given in equation (26.49) as a power series in u_{e0} keeping two terms in the expansion, i.e. the zeroth-order and first-order terms in u_{e0}. For what range of values of u_{e0} will this be valid? Next write down the distribution function that arises from solving the Fokker–Planck equation in the Lorentz-gas approximation when a driving electric field is present, i.e. equations (13.15) and (13.21). In equation (13.21) for f_{e1}, use equation (13.22) to substitute for E_z in terms of j_z and then write $j_z = -neu_{e0}$. Now compare these two distributions, both of which have a Maxwellian zeroth-order term and a correction of first-order in u_{e0}. How are these distributions similar? How are they different? In which case will the current-driven drift waves be more unstable, i.e. which has the larger value of $\partial F_{e0}/\partial v_z$ at some fixed value $v_z \ll v_{t,e}$?

It is appropriate at this point to ask whether we have found anything different from the ion acoustic wave instability driven by a non-zero electron streaming speed, which was derived in Chapter 25. The ion acoustic wave was found to be unstable only in cases where the electron streaming speed exceeded the ion thermal velocity. By contrast, within the limitations of the present analysis, there is no such 'threshold' value of the electron streaming speed which must be exceeded for the drift wave to be destabilized. Indeed, provided that we are allowed to have any k_z value, then we can certainly choose a value sufficiently small that $\omega \approx k_y v_{de} \gg k_z C_s \gtrsim k_z v_{t,i}$. Here we have introduced the ion thermal velocity, $v_{t,i} = (T_{i0}/M)^{1/2}$, which is less than, or comparable to, the sound speed, $C_s = (T_{e0}/M)^{1/2}$, in all cases where $T_{i0} \lesssim T_{e0}$. For these k_z values, the approximations that led to equation (26.56) are valid, as also is the neglect of ion Landau damping, which was initially a consequence of our assumption that $T_{i0} \ll T_{e0}$ but is now seen to be valid more generally. For any non-zero electron steaming speed u_{e0}, we can certainly choose k_z values that are within the range of validity of equation (26.56) and that are also small enough for the right-hand side of equation (26.56) to be positive. Thus, within the limitations of the present analysis, the *electron drift wave is unstable for any non-zero streaming speed, however small*. Nonetheless, the inclusion of an electron temperature gradient (with $\eta_e > 0$) introduces a stabilizing effect that modifies this conclusion, as would the introduction of a finite parallel length, by setting a lower limit on the k_z values.

Problem 26.4 Consider the electron drift wave in the case where there is *both* an electron temperature gradient, described by an η_e value, *and* a non-zero electron streaming speed, u_{e0}. Assuming $k_\perp^2 r_{Ls}^2 \ll 1$, show that the growth rate of the electron drift wave is

$$\gamma \approx \left(\frac{\pi}{2}\right)^{1/2} \frac{k_y v_{de} k_z u_{e0} - k_z^2 C_s^2 - k_y^2 v_{de}^2 \eta_e}{|k_z| v_{t,e}}.$$

For positive values of η_e, show that the plasma is completely stable, in the sense that there are no values of k_y and k_z for which $\gamma > 0$, if

$$\eta_e > 0.25 u_{e0}^2 / C_s^2.$$

It should be noted, finally, that the dispersion relation for the ion acoustic wave can, of course, be obtained from equation (26.54) by going to the limit $\omega \sim k_z C_s \sim k_z u_{e0} \gg k_y v_{de}$. In this limit, the diamagnetic drift frequency $k_y v_{de}$ disappears, and we are left with two waves with $\omega \approx \pm k_z C_s$, one of which is destabilized if $|u_{e0}| > C_s$. This is, of course, exactly the same result as was obtained in Chapter 25. For frequencies much larger than the diamagnetic drift frequency, the non-uniformity of the plasma does not enter in any significant way. It should not be surprising that such cases can be treated adequately with the Vlasov equation for a uniform plasma and do not require the drift-kinetic formulation.

26.6   THE 'ION TEMPERATURE GRADIENT' INSTABILITY

For the electron drift waves that have been considered so far in this Chapter, it has been sufficient to treat the ions as a 'cold fluid': the drift-kinetic equation was needed only for the electrons. This was justified by limiting ourselves to the case $T_i \ll T_e$, which implies both that the ion diamagnetic drift speed is much less than the electron diamagnetic drift speed and that the ion thermal velocity, $v_{t,i}$, is much less than the ion-sound speed, C_s. It is interesting to consider whether there are different types of drift waves (especially unstable waves) that arise from Landau-like resonances between the wave and thermal motion of the ions. In such cases, we will need to use the drift-kinetic equation for *ions*:

$$\frac{\partial f_i}{\partial t} + \frac{\mathbf{E} \times \mathbf{B}}{B^2} \cdot \nabla_\perp f_i + v_z \frac{\partial f_i}{\partial z} + \frac{e}{M} E_z \frac{\partial f_i}{\partial v_z} = 0. \qquad (26.57)$$

The use of the drift-kinetic equation in this form for the ions effectively neglects the ion polarization drift (of second order in the ion Larmor radius), as well as corrections of order $k_\perp^2 r_{\mathrm{Li}}^2$ to the ion $\mathbf{E} \times \mathbf{B}$ drift. However, the instability that we will find in this Section can arise even in the case $k_\perp r_{\mathrm{Li}} \ll 1$, so it is sufficient for present purposes to carry out the analysis neglecting all effects of order $k_\perp^2 r_{\mathrm{Li}}^2$.

Let us consider the case where the equilibrium distribution function for ions is Maxwellian, i.e. as given by equation (26.1), and where both density and temperature gradients are included, so that both n_{i0} and T_{i0} are functions of x. We consider electrostatic perturbations of this equilibrium by introducing a small wave-like electric field, described by an electric potential ϕ, exactly as before, and we linearize equation (26.57) to obtain the perturbed distribution function, f_{i1}, from which the ion density perturbation, n_{i1}, is obtained by integrating over all velocities. The analysis is exactly analogous to that carried out already for electrons in equations (26.20)–(26.24) and (with the inclusion of a temperature gradient) in equations (26.38)–(26.42). Indeed, the result given in equation (26.42) can be taken over in its entirety (with straightforward modifications for changing from electrons to ions), giving

$$n_{i1} = -\frac{n_{i0}e\phi}{T_{i0}} + \frac{e\phi}{T_{i0}} \int_{-\infty}^{\infty} \frac{F_{i0}(v_z)\mathrm{d}v_z}{\omega - k_z v_z} \left\{ \omega - k_y v_{\mathrm{di}} \left[1 - \eta_i \left(\frac{1}{2} - \frac{v_z^2}{2v_{t,i}^2} \right) \right] \right\}.$$

(26.58)

Here, we have defined an ion diamagnetic drift

$$v_{\mathrm{di}} = \frac{T_{i0}}{n_{i0}e B_{z0}} \frac{\mathrm{d}n_{i0}}{\mathrm{d}x}$$

(26.59)

and a dimensionless measure of the ion temperature gradient

$$\eta_i = \frac{1}{T_{i0}} \frac{\mathrm{d}T_{i0}}{\mathrm{d}x} \left(\frac{1}{n_{i0}} \frac{\mathrm{d}n_{i0}}{\mathrm{d}x} \right)^{-1} = \frac{\nabla(\ln T_{i0})}{\nabla(\ln n_{i0})}.$$

(26.60)

We have also defined the 'one-dimensional' Maxwellian distribution

$$F_{i0}(v_z) = n_{i0} \left(\frac{M}{2\pi T_{i0}} \right)^{1/2} \exp\left(-\frac{M v_z^2}{2 T_{i0}} \right).$$

(26.61)

Since we are interested in effects associated with strong resonant interactions between a wave and the ions, we must consider the case where $\omega \sim k_z v_{t,i}$, so that the integral in equation (26.58) cannot be evaluated by any simple expansion of the integrand.

For the electron density perturbation, we can use the results already obtained, as given for example in equations (26.29) or (26.43). These expressions have been obtained under the assumption that $\omega \ll k_z v_{t,e}$, but they have retained a small imaginary term of order $\omega/|k_z|v_{t,e}$. For the case $\omega \sim k_z v_{t,i}$, this assumption will be well satisfied, and the small imaginary term will be of order $(m/M)^{1/2}$. In fact, we will neglect this small imaginary term now altogether, since we will find, at least in some cases, unstable ion drift waves with growth rates that are as large as their frequencies. For such modes, terms of order $(m/M)^{1/2}$ will produce only very small corrections. Thus, neglecting the small imaginary terms in equations (26.29) and (26.43), we have simply

$$n_{e1} = \frac{n_{e0}e\phi}{T_{e0}}. \tag{26.62}$$

Physically, this describes a situation in which the electrons have relaxed completely to a Boltzmann distribution, $n_e \approx n_{e0}\exp(e\phi/T_{e0})$, along the magnetic field.

The dispersion relation is obtained, as usual, by setting $n_{e1} = n_{i1}$. Using equations (26.58) and (26.62), we obtain the dispersion relation

$$1 + \frac{T_{i0}}{T_{e0}} = D(\omega) \tag{26.63}$$

where

$$D(\omega) \equiv \frac{1}{n_{i0}} \int_{-\infty}^{\infty} \frac{F_{i0}(v_z)dv_z}{\omega - k_z v_z} \left\{ \omega - k_y v_{di} \left[1 - \frac{\eta_i}{2} \left(1 - \frac{v_z^2}{v_{t,i}^2} \right) \right] \right\} \tag{26.64}$$

Our task now is to solve this dispersion relation for ω, without making any *a priori* assumptions about the relative magnitude of ω and $k_z v_{t,i}$.

The Nyquist diagram technique, introduced in Chapter 24, allows us to determine whether the dispersion relation given by equations (26.63) and (26.64) has any solutions corresponding to unstable modes, i.e. solutions with $\text{Im}(\omega) > 0$. To apply this technique, we must let ω trace out a closed contour in the complex ω-plane that is composed of the real axis, going from $-\infty$ to $+\infty$, together with a semicircle at infinity in the upper half-plane, traced out in an anti-clockwise sense. A contour of this type, which encloses on its left the entire region with $\text{Im}(\omega) > 0$, is shown in Figure 24.5(b). In evaluating a singular integral, such as the one in equation (26.64), the ω-plane contour should be taken to lie *just above* the real axis, rather than exactly on it, so that ω can be considered to have an infinitesimal *positive* imaginary part. As ω traces out this closed contour, the function $D(\omega)$ given in equation (26.64) will trace out some closed contour in the complex D-plane, which we have called the 'Nyquist contour'. If the point $D = 1 + T_{i0}/T_{e0}$ falls in a region encircled by, and lying to the left of, this contour, then the dispersion relation must have a root with $\text{Im}(\omega) > 0$, i.e. an instability. In the contrary case where the area enclosed on the left of the Nyquist contour does not contain the point $D = 1 + T_{i0}/T_{e0}$, there can be no unstable modes.

To apply this technique in the present case, we first evaluate $D(\omega)$ for very large values of $|\omega|$, i.e. for ω lying at either of the two extremities of the real axis or on the semicircle at infinity in the upper half-plane. Expanding

$$\frac{1}{\omega - k_z v_z} \approx 1 + \frac{k_z v_z}{\omega} + \dots \qquad (26.65)$$

and keeping only these two terms (but noting that the second term makes no contribution because it is odd in v_z), we obtain

$$D(\omega) \approx 1 - \frac{k_y v_{di}}{\omega}. \qquad (26.66)$$

Thus the entire semicircle at infinity in the ω-plane maps onto the point $D = 1$ in the D plane. To be specific on how the Nyquist contour passes through the point $D = 1$, we must make some choice for the sign of $k_y v_{di}$. Since the electron and ion diamagnetic velocities are of opposite sign, for consistency with Figure 26.2 we choose $k_y v_{di} < 0$, noting that such a choice can be made without loss of generality, since the dispersion relation is invariant under simultaneous sign changes for $k_y v_{di}$, k_z and $\text{Re}(\omega)$. (Of course, making the choice $k_y v_{di} < 0$ means that we must allow both positive and negative values for the solutions $\text{Re}(\omega)$ of the dispersion relation, thereby abandoning our usual convention that $\text{Re}(\omega) > 0$. In any case, the Nyquist diagram technique for obtaining information on the roots of a dispersion relation requires that both positive and negative values of $\text{Re}(\omega)$ be considered. If we chose to adhere to our convention that $\text{Re}(\omega) > 0$ always, then we would construct the Nyquist diagram in terms of the variable $\text{Re}(\omega)/k_y$, rather than ω, and this variable can take on both positive and negative values depending on the sign of k_y. There would, of course, be no difference in the physical results obtained. We have chosen here to describe the Nyquist technique in its standard form, in which $\text{Re}(\omega)$ takes on both positive and negative values.) For real values of ω that are large and positive, equation (26.66) shows that the value of $D(\omega)$ slightly exceeds unity, whereas for large and negative ω values, $D(\omega)$ is slightly less than unity. Thus, as ω passes anti-clockwise around the semicircle at infinity (i.e. going from $+\infty$ to $-\infty$), the Nyquist contour passes *leftward* through the point $D = 1$ (i.e. going from D values just greater than, to values just less than, unity). The cases shown in Figure 26.3 all have this property.

Next we evaluate $D(\omega)$ on (or just above) the real axis in the ω-plane. The integral in equation (26.64) must be evaluated as the principal-value integral together with πi times the residue at the singularity $v_z = \omega/k_z$ (for the case

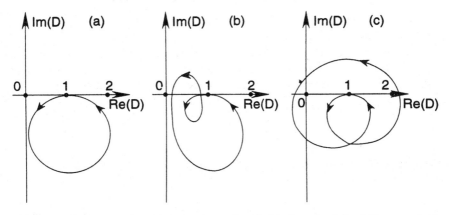

Figure 26.3. The Nyquist D-plane contour for the ion branch of the drift wave, in the cases (a) $\Lambda(2 - \eta_i) > 1$, (b) $0 < \Lambda(2 - \eta_i) < 1$ and (c) $\eta_i > 2$.

$k_z > 0$, and $-\pi i$ times the residue for the case $k_z < 0$). We obtain

$$D(\omega) = \frac{1}{n_{i0}} \mathrm{Pr} \int_{-\infty}^{\infty} \frac{F_{i0}(v_z)\mathrm{d}v_z}{\omega - k_z v_z} \left\{ \omega - k_y v_{di} \left[1 - \frac{\eta_i}{2} \left(1 - \frac{v_z^2}{v_{t,i}^2} \right) \right] \right\}$$
$$- i \frac{(\pi/2)^{1/2}}{|k_z| v_{t,i}} \left\{ \omega - k_y v_{di} \left[1 - \frac{\eta_i}{2} \left(1 - \frac{\omega^2}{k_z^2 v_{t,i}^2} \right) \right] \right\}$$
$$\times \exp\left(-\frac{\omega^2}{2k_z^2 v_{t,i}^2} \right). \tag{26.67}$$

For large $|\omega|$ (i.e. for ω values approaching either $-\infty$ or $+\infty$), the imaginary part of $D(\omega)$ is extremely small, and its sign is the same as the sign of the product $\eta_i k_y v_{di}$ (determined by the sign of the term in ω^2 in the imaginary part of D). We will limit our analysis to the case $\eta_i > 0$, and we recall that we chose $k_y v_{di} < 0$ for consistency with Figure 26.2. Thus, $\mathrm{Im}(D) < 0$ for large $|\omega|$, which tells us that the Nyquist contour in the vicinity of the point $D = 1$ must lie just *below* the real axis, again as shown in all three cases of Figure 26.3.

The Nyquist contour in the D plane can cross the real axis only if there are values of ω for which the imaginary part of $D(\omega)$ vanishes. From equation (26.67), we see that this will occur only if there are real roots ω of the quadratic equation

$$\frac{\eta_i \omega^2}{2k_z^2 v_{t,i}^2} - \frac{\omega}{k_y v_{di}} + 1 - \frac{\eta_i}{2} = 0. \tag{26.68}$$

Using the usual formula for the roots of a quadratic equation, the roots of equation (26.68) are given by

$$\frac{\omega}{k_y v_{di}} = \frac{1}{\Lambda}\left\{1 \pm [1 - \Lambda(2 - \eta_i)]^{1/2}\right\} \tag{26.69}$$

where

$$\Lambda = \eta_i k_y^2 v_{di}^2 / k_z^2 v_{t,i}^2. \tag{26.70}$$

In order to determine the shape of the Nyquist curve, it is now important to determine the value of $\text{Re}(D)$ where the Nyquist contour crosses the real axis in the D plane. (Of course, if there are no real roots of equation (26.68), the contour does not cross the real axis anywhere.) This can be done by substituting into the principal-value integral in equation (26.67) the value of ω at which the imaginary part vanishes. The simplest approach algebraically is to rearrange equation (26.68) as an equation for ω in terms of ω^2, and substitute this form into the principal-value integral in equation (26.67), noting the various cancellations which then occur. We obtain

$$
\begin{aligned}
\text{Re}(D) &= \frac{1}{n_{i0}}\text{Pr}\int_{-\infty}^{\infty}\frac{F_{i0}(v_z)\mathrm{d}v_z}{\omega - k_z v_z}\frac{\eta_i k_y v_{di}}{2}\left(\frac{\omega^2}{k_z^2 v_{t,i}^2} - \frac{v_z^2}{v_{t,i}^2}\right)\\
&= \frac{1}{n_{i0}}\frac{\eta_i k_y v_{di}}{2 k_z^2 v_{t,i}^2}\text{Pr}\int_{-\infty}^{\infty}F_{i0}(v_z)(\omega + k_z v_z)\mathrm{d}v_z\\
&= \frac{\omega \eta_i k_y v_{di}}{2 k_z^2 v_{t,i}^2} = \frac{\Lambda}{2}\frac{\omega}{k_y v_{di}}\\
&= \tfrac{1}{2}\left\{1 \pm [1 - \Lambda(2 - \eta_i)]^{1/2}\right\}
\end{aligned} \tag{26.71}
$$

where, in the final form, we have substituted for ω from equation (26.69). By comparing equation (26.71) with equation (26.69), we see that the values of $\text{Re}(D)$ at crossings of the real-D axis are closely related to the corresponding values of ω at these points.

There are three cases to be considered. The first is where $\Lambda(2 - \eta_i) > 1$, in which case equation (26.69) shows that there are no real roots of the quadratic equation, equation (26.68), and therefore no crossings of the real-D axis by the Nyquist contour: the Nyquist contour must be of the form illustrated in Figure 26.3(a). Clearly, the area to the left of the Nyquist contour does not contain the point $D = 1 + T_{i0}/T_{e0}$, which is the dispersion relation, equation (26.63), and so there can be no unstable mode. The second case is where $0 < \Lambda(2 - \eta_i) < 1$, in which case there are two real roots of the quadratic equation, equation (26.68), given in equation (26.69), both of which have $\omega/k_y v_{di} > 0$. However, equation (26.71) shows that the values of $\text{Re}(D)$ at these ω values are both less than unity, with the more negative root ω corresponding to the larger value of $\text{Re}(D)$, remembering that we have chosen

$k_y v_{di} < 0$. For this case, the Nyquist contour must cross the real axis twice to the left of the point $D = 1$ and must be of the form illustrated in Figure 26.3(b). Again we see that the area encircled to the left of the Nyquist contour does not contain the point $D = 1 + T_{i0}/T_{e0}$, and so again there can be no unstable mode.

The third case is where $\eta_i > 2$, in which case equation (26.69) shows that there are two real roots with opposite signs for $\omega/k_y v_{di}$. Equation (26.71) shows that the root with a positive value of $\omega/k_y v_{di}$ (negative ω value) has $\mathrm{Re}(D) > 1$, whereas the root with a negative value of $\omega/k_y v_{di}$ (positive ω value) has $\mathrm{Re}(D) < 0$. For this case, the Nyquist contour (as ω travels along the real axis from $-\infty$ to $+\infty$) crosses the real axis in the D-plane first to the right of the point $D = 1$ and subsequently crosses it again to the left of the origin: the contour must be of the form illustrated in Figure 26.3(c). This contour encloses to its left an area that includes the point $D = 1 + T_{i0}/T_{e0}$ if the first (i.e. rightmost) crossing of the real axis occurs to the right of this point. Using equation (26.71) for the value of D at each crossing of the real axis, this occurs when

$$\tfrac{1}{2} \left\{ 1 + [1 - \Lambda(2 - \eta_i)]^{1/2} \right\} > 1 + \frac{T_{i0}}{T_{e0}} \tag{26.72}$$

which is therefore the condition that an unstable mode exists. By some straightforward manipulation, inequality (26.72) can be expressed as a condition on η_i, namely

$$\eta_i > 2 + \frac{4}{\Lambda} \frac{T_{i0}}{T_{e0}} \left(1 + \frac{T_{i0}}{T_{e0}} \right). \tag{26.73}$$

For a plane plasma slab that is infinite in both y and z directions, it is possible to choose the wave-vector components k_y and k_z at will, so that the parameter Λ given in equation (26.70) takes on all values. In particular, choosing sufficiently small values of the ratio k_z/k_y that the parameter Λ greatly exceeds unity, the condition for instability will approach a limiting case

$$\eta_i > 2. \tag{26.74}$$

However, since the diamagnetic drift speed is generally much less than the ion thermal speed, specifically $v_{di}/v_{t,i} \sim r_{Li}/L_\perp \ll 1$, where r_{Li} is the ion Larmor radius and $L_\perp$ is the scale-length of the density gradient perpendicular to the magnetic field, the ratio k_z/k_y must be exceedingly small to give $\Lambda \gg 1$. In cases where arbitrarily small k_z values are not allowed, such as a torus which is approximated as a finite-length cylinder with 'periodic boundary conditions', the condition for instability could be significantly more demanding than that given by equation (26.74). The addition of 'magnetic shear', i.e. where a small field $B_y(x)$ is added to the main field B_z, also serves effectively to impose a lower limit on the component of the k-vector along the magnetic field. In this case also, stability is improved. On the other had, inclusion of shorter wavelength modes, specifically these with $k_\perp r_{Li} \sim 1$, is found to lower the instability threshold for η_i to values close to unity. The effect of ∇B and curvature drifts in geometries other than the plane slab is also found to be destabilizing.

This 'ion temperature gradient' instability poses a significant threat to confinement in high-temperature fusion plasmas—for which the present 'collisionless' approximation is applicable. As we saw in Chapter 10, neutral atoms will not penetrate far into a fusion-reactor plasma. There will thus be no source of deuterium–tritium 'fuel', except at the very edge of the plasma. Thus, unless turbulent flows drive net inward convection, the process of turbulent internal diffusion will establish an equilibrium density that is approximately uniform over almost all of the plasma, falling to zero only in a narrow edge layer. The density gradient will therefore be very small in the main part of the plasma (with gradient scale-length much greater than the plasma linear dimension). Thermal conduction will carry the heat that is generated from the charged particles produced by the fusion reactions from the central part of the plasma (where the temperature will be highest) to the edge of the plasma (where the temperature will be lowest). Thus the temperature gradient will be substantial in the main part of the plasma (with gradient scale-length of order the plasma linear dimension). It follows from these general considerations that η_i values may be quite large in the main part of a fusion-reactor plasma, implying that ion temperature gradient instabilities might arise and cause turbulence and perhaps a highly enhanced rate of heat conduction. Indeed this has been a longstanding concern in fusion research. Fortunately, however, effects not considered in the present simple analysis tend to stabilize the ion temperature gradient mode. Moreover, even when the instability does arise, the enhanced thermal conduction caused by it may not exceed the 'anomalous' transport produced by a variety of drift-wave-like and other small-scale instabilities and turbulent processes— all of which are predicted to still allow an acceptable overall level of plasma confinement for fusion power production.

There is an extensive literature on low-frequency drift waves and other related small-scale instabilities in magnetically confined plasmas. A review article, which describes all of the basic modes and their linear growth rates and gives simple estimates of the level of turbulent transport to be expected from them, has been written by two of the most prolific contributors to this field, B B Kadomtsev and O P Pogutse (1972 *Review of Plasma Physics 5* edited by M A Leontovich, pp 249–400, New York: Consultants Bureau). More recently, this field has been developed to include linear calculations in very realistic plasma geometries, as distinct from the 'plane plasma slab' considered here, and to nonlinear calculations of the plasma turbulence produced by drift-wave-like instabilities. As might be expected, both of these developments involve extensive numerical computation.

Over the years, many authors have attempted to explain experimental results on anomalous transport in tokamaks using theoretical modes of drift wave turbulence, generally with only limited success. However, it is encouraging to note that, as the field becomes more sophisticated in its use of realistic geometries and advanced computational techniques, the level of agreement with experimental data appears to be improving markedly. The development of computational techniques for following the kinetics of gyrating ions into regimes of nonlinear perturbations has been responsible for some of the most notable recent advances.

Appendix A

Physical quantities and their SI units

Quantity	Symbol	SI unit Name	SI unit Abbrev.	Conversion formula to Gaussian units
Length	L, a, r, R	meter	m	$1 \text{ m} = 10^2 \text{ cm}$
Time	t	second	s	
Velocity	$\mathbf{u}, \mathbf{v}$	meter per second	m s^{-1}	$1 \text{ m s}^{-1} = 10^2 \text{ cm s}^{-1}$
Mass	m, M	kilogram	kg	$1 \text{ kg} = 10^3 \text{ g}$
Mass density	ρ	kilogram per cubic meter	kg m^{-3}	1 kg m^{-3} $= 10^{-3} \text{ g cm}^{-3}$
Force	$\mathbf{F}$	newton	N	$1 \text{ N} = 10^5 \text{ dyne}$
Energy	W	joule	J	$1 \text{ J} = 10^7 \text{ erg}$
Power	P	watt (J s^{-1})	W	$1 \text{ W} = 10^7 \text{ erg s}^{-1}$
Pressure	p	pascal	Pa	$1 \text{ Pa} = 10 \text{ dyne cm}^{-2}$
Temperature	T	kelvin	K	$1 \text{ eV} = 1.16 \times 10^4 \text{ K}$
Charge	q, e	coulomb	C	$1 \text{ C} = 3 \times 10^9 \text{ esu}$
Charge density	σ	coulomb per cubic meter	C m^{-3}	1 C m^{-3} $= 3 \times 10^3 \text{ esu cm}^{-3}$
Surface charge density	σ_s	coulomb per square meter	C m^{-2}	1 C m^{-2} $= 3 \times 10^5 \text{ esu cm}^{-2}$
Current	I	ampere (C s^{-1})	A	$1 \text{ A} = 3 \times 10^9 \text{ esu}$
Current density	$\mathbf{j}$	ampere per square meter	A m^{-2}	1 A m^{-2} $= 3 \times 10^5 \text{ esu cm}^{-2}$
Electric field	$\mathbf{E}$	volt per meter	V m^{-1}	$1 \text{ V m}^{-1} = 10^{-4}/3 \text{ esu}$
Electric potential	V, ϕ	volt	V	$1 \text{ V} = 10^{-2}/3 \text{ esu}$
Magnetic field	$\mathbf{B}$	tesla	T	$1 \text{ T} = 10^4 \text{ gauss}$
Magnetic flux	Φ	weber (T m^2)	Wb	$1 \text{ Wb} = 10^8 \text{ maxwell}$
Electric resistance	R	ohm	Ω	$1 \ \Omega = (10^{-11}/9) \text{ s cm}^{-1}$
Resistivity	η	ohm-meter	Ω m	$1 \ \Omega \text{ m} = (10^{-9}/9) \text{ s}$

Appendix B

Equations in the SI system

Maxwell's equations (SI units):

$$\nabla \cdot \mathbf{B} = 0$$

$$\nabla \cdot (\epsilon_0 \mathbf{E}) = \sigma \qquad \text{(Poisson's equation)}$$

$$\nabla \times \mathbf{E} = -\partial \mathbf{B}/\partial t \qquad \text{(Faraday's law)}$$

$$\nabla \times \mathbf{B} = \mu_0 \mathbf{j} + (1/c^2)\partial \mathbf{E}/\partial t \qquad \text{(Ampere's law)}$$

Lorentz force on charge q (SI units):

$$\mathbf{F} = q(\mathbf{E} + \mathbf{v} \times \mathbf{B})$$

Appendix C

Physical constants

Physical constant	Symbol	Value in SI units
Elementary charge	e	1.60×10^{-19} C
Electron mass	m	9.11×10^{-31} kg
Proton mass	M	1.67×10^{-27} kg
Boltzmann constant*	k	1.38×10^{-23} J K^{-1}
		1.60×10^{-19} J eV^{-1}
Speed of light in vacuum	c	3.00×10^8 m s^{-1}
Planck constant $(h/2\pi)$	$\hbar$	1.05×10^{-34} J s
Permittivity of free space	ϵ_0	8.85×10^{-12} C m^{-1} V^{-1}
Permeability of free space	μ_0	$4\pi \times 10^{-7} = 1.26 \times 10^{-6}$ T m A^{-1}

* Throughout this book, for simplicity of notation, the plasma temperature T is always in 'energy units', i.e. joules, so that the Boltzmann constant k never appears. The two values of k given here will allow a temperature in kelvin (K) or electron-volts (eV) to be converted into joules (J). Note that the quantities T/e and W/e, where W is an energy, have units of volts. Thus, for example, the value of T/e for a 10 eV temperature is 10 V.

Appendix D

Useful vector formulae

D.1 VECTOR IDENTITIES

$$\mathbf{A} \cdot (\mathbf{B} \times \mathbf{C}) = (\mathbf{A} \times \mathbf{B}) \cdot \mathbf{C}$$
$$\mathbf{A} \times (\mathbf{B} \times \mathbf{C}) = (\mathbf{A} \cdot \mathbf{C})\mathbf{B} - (\mathbf{A} \cdot \mathbf{B})\mathbf{C}$$
$$\nabla \cdot (\psi\mathbf{A}) = \psi(\nabla \cdot \mathbf{A}) + \mathbf{A} \cdot \nabla\psi$$
$$\nabla \times (\psi\mathbf{A}) = \psi(\nabla \times \mathbf{A}) + \nabla\psi \times \mathbf{A}$$
$$\nabla \cdot (\mathbf{A} \times \mathbf{B}) = \mathbf{B} \cdot \nabla \times \mathbf{A} - \mathbf{A} \cdot \nabla \times \mathbf{B}$$
$$\nabla \times (\mathbf{A} \times \mathbf{B}) = \mathbf{A}(\nabla \cdot \mathbf{B}) - \mathbf{B}(\nabla \cdot \mathbf{A}) + (\mathbf{B} \cdot \nabla)\mathbf{A} - (\mathbf{A} \cdot \nabla)\mathbf{B}$$
$$\mathbf{A} \times (\nabla \times \mathbf{B}) = (\nabla\mathbf{B}) \cdot \mathbf{A} - (\mathbf{A} \cdot \nabla)\mathbf{B}$$
$$\nabla \times (\nabla \times \mathbf{A}) = \nabla(\nabla \cdot \mathbf{A}) - \nabla^2\mathbf{A}$$

D.2 MATRIX NOTATION

Note that we employ the Einstein convention, in which repeated suffices are to be summed over the values 1, 2, 3.

D.2.1 Kronecker deltas

$$\delta_{ij} \equiv \begin{cases} 1 & i = j \\ 0 & i \neq j \end{cases}$$

D.2.2 Levi-Civita symbols

$$\epsilon_{ijk} \equiv \begin{cases} 1 & i \neq j \neq k \text{ cyclic permutation of } 1, 2, 3 \\ -1 & i \neq j \neq k \text{ anti} - \text{cyclic permutation of } 1, 2, 3 \\ 0 & i = j \text{ or } j = k \text{ or } i = k \end{cases}$$

$$\epsilon_{ijk}\epsilon_{ilm} = \delta_{jl}\delta_{km} - \delta_{jm}\delta_{kl}$$

$$\mathbf{A} \cdot \mathbf{B} = A_i B_i$$

$$(\mathbf{A} \times \mathbf{B})_i = \epsilon_{ijk} A_j B_k$$

$$(\nabla\psi)_i = \frac{\partial\psi}{\partial x_i}$$

$$\nabla \cdot \mathbf{A} = \frac{\partial A_i}{\partial x_i}$$

$$(\nabla \times \mathbf{A})_i = \epsilon_{ijk}\frac{\partial A_k}{\partial x_j}$$

$$(\nabla\mathbf{B})_{ij} = \frac{\partial B_j}{\partial x_i}$$

$$\mathbf{A} \cdot \nabla\psi = A_i\frac{\partial\psi}{\partial x_i}$$

$$(\mathbf{A} \cdot \nabla\mathbf{B})_i = A_j\frac{\partial B_i}{\partial x_j}$$

$$(\mathbf{AB})_{ij} = A_i B_j$$

Matrix notation with the Einstein convention can be used to derive all of the vector identities given in Section D.1 of this Appendix. For example, we derive the expression for $\nabla \times (\mathbf{A} \times \mathbf{B})$ by proceeding as follows.

$$
\begin{aligned}
[\nabla \times (\mathbf{A} \times \mathbf{B})]_i &= \epsilon_{ijk}\frac{\partial}{\partial x_j}(\epsilon_{klm} A_l B_m) \\
&= \epsilon_{ijk}\epsilon_{klm}\left(A_l\frac{\partial B_m}{\partial x_j} + B_m\frac{\partial A_l}{\partial x_j}\right) \\
&= \epsilon_{kij}\epsilon_{klm}\left(A_l\frac{\partial B_m}{\partial x_j} + B_m\frac{\partial A_l}{\partial x_j}\right) \\
&= (\delta_{il}\delta_{jm} - \delta_{im}\delta_{jl})\left(A_l\frac{\partial B_m}{\partial x_j} + B_m\frac{\partial A_l}{\partial x_j}\right) \\
&= A_i\frac{\partial B_j}{\partial x_j} - A_j\frac{\partial B_i}{\partial x_j} + B_j\frac{\partial A_i}{\partial x_j} - B_i\frac{\partial A_j}{\partial x_j} \\
&= A_i(\nabla \cdot \mathbf{B}) - (\mathbf{A} \cdot \nabla)B_i + (\mathbf{B} \cdot \nabla)A_i - B_i(\nabla \cdot \mathbf{A}) \\
&= [\mathbf{A}(\nabla \cdot \mathbf{B}) - (\mathbf{A} \cdot \nabla)\mathbf{B} + (\mathbf{B} \cdot \nabla)\mathbf{A} - \mathbf{B}(\nabla \cdot \mathbf{A})]_i
\end{aligned}
$$

which, after rearranging terms on the right-hand side, is the desired expression.

Appendix E

Differential operators in cartesian and curvilinear coordinates

E.1 CARTESIAN COORDINATES (x, y, z)

Gradient:

$$\nabla \psi = \left(\frac{\partial \psi}{\partial x}, \frac{\partial \psi}{\partial y}, \frac{\partial \psi}{\partial z} \right)$$

Divergence:

$$\nabla \cdot \mathbf{A} = \frac{\partial A_x}{\partial x} + \frac{\partial A_y}{\partial y} + \frac{\partial A_z}{\partial z}$$

Curl:

$$\nabla \times \mathbf{A} = \left(\frac{\partial A_z}{\partial y} - \frac{\partial A_y}{\partial z}, \frac{\partial A_x}{\partial z} - \frac{\partial A_z}{\partial x}, \frac{\partial A_y}{\partial x} - \frac{\partial A_x}{\partial y} \right)$$

Laplacian:

$$\nabla^2 \psi = \frac{\partial^2 \psi}{\partial x^2} + \frac{\partial^2 \psi}{\partial y^2} + \frac{\partial^2 \psi}{\partial z^2}$$

Laplacian of a vector:

$$\nabla^2 \mathbf{A} = (\nabla^2 A_x, \nabla^2 A_y, \nabla^2 A_z)$$

Divergence of a tensor:

$$(\nabla \cdot P)_x = \frac{\partial P_{xx}}{\partial x} + \frac{\partial P_{yx}}{\partial y} + \frac{\partial P_{zx}}{\partial z}$$

$$(\nabla \cdot P)_y = \frac{\partial P_{xy}}{\partial x} + \frac{\partial P_{yy}}{\partial y} + \frac{\partial P_{zy}}{\partial z}$$

$$(\nabla \cdot P)_z = \frac{\partial P_{xz}}{\partial x} + \frac{\partial P_{yz}}{\partial y} + \frac{\partial P_{zz}}{\partial z}$$

E.2   CYLINDRICAL COORDINATES (r, θ, z)

Gradient:

$$\nabla \psi = \left(\frac{\partial \psi}{\partial r}, \frac{\partial \psi}{r \partial \theta}, \frac{\partial \psi}{\partial z} \right)$$

Divergence:

$$\nabla \cdot \mathbf{A} = \frac{1}{r} \frac{\partial}{\partial r}(r A_r) + \frac{\partial A_\theta}{r \partial \theta} + \frac{\partial A_z}{\partial z}$$

Curl:

$$\nabla \times \mathbf{A} = \left(\frac{\partial A_z}{r \partial \theta} - \frac{\partial A_\theta}{\partial z}, \frac{\partial A_r}{\partial z} - \frac{\partial A_z}{\partial r}, \frac{\partial (r A_\theta)}{r \partial r} - \frac{\partial A_r}{r \partial \theta} \right)$$

Laplacian:

$$\nabla^2 \psi = \frac{1}{r} \frac{\partial}{\partial r}\left(r \frac{\partial \psi}{\partial r} \right) + \frac{1}{r^2} \frac{\partial^2 \psi}{\partial \theta^2} + \frac{\partial^2 \psi}{\partial z^2}$$

Laplacian of a vector:

$$\nabla^2 \mathbf{A} = \left(\nabla^2 A_r - \frac{2}{r^2} \frac{\partial A_\theta}{\partial \theta} - \frac{A_r}{r^2}, \nabla^2 A_\theta + \frac{2}{r^2} \frac{\partial A_r}{\partial \theta} - \frac{A_\theta}{r^2}, \nabla^2 A_z \right)$$

Divergence of a tensor:

$$(\nabla \cdot P)_r = \frac{1}{r} \frac{\partial}{\partial r}(r P_{rr}) + \frac{1}{r} \frac{\partial P_{\theta r}}{\partial \theta} + \frac{\partial P_{zr}}{\partial z} - \frac{P_{\theta\theta}}{r}$$

$$(\nabla \cdot P)_\theta = \frac{1}{r} \frac{\partial}{\partial r}(r P_{r\theta}) + \frac{1}{r} \frac{\partial P_{\theta\theta}}{\partial \theta} + \frac{\partial P_{z\theta}}{\partial z} - \frac{P_{\theta r}}{r}$$

$$(\nabla \cdot P)_z = \frac{1}{r} \frac{\partial}{\partial r}(r P_{rz}) + \frac{1}{r} \frac{\partial P_{\theta z}}{\partial \theta} + \frac{\partial P_{zz}}{\partial z}$$

E.3   SPHERICAL COORDINATES (r, θ, ϕ)

Gradient:

$$\nabla \psi = \left(\frac{\partial \psi}{\partial r}, \frac{\partial \psi}{r \partial \theta}, \frac{1}{r \sin \theta} \frac{\partial \psi}{\partial \phi} \right)$$

Divergence:

$$\nabla \cdot \mathbf{A} = \frac{1}{r^2} \frac{\partial}{\partial r} (r^2 A_r) + \frac{1}{r\sin\theta} \frac{\partial}{\partial \theta} (\sin\theta A_\theta) + \frac{1}{r\sin\theta} \frac{\partial A_\phi}{\partial \phi}$$

Curl:

$$\nabla \times \mathbf{A} = \left(\frac{1}{r\sin\theta} \frac{\partial}{\partial \theta} (\sin\theta A_\phi) - \frac{1}{r\sin\theta} \frac{\partial A_\theta}{\partial \phi}, \frac{1}{r\sin\theta} \frac{\partial A_r}{\partial \phi} - \frac{1}{r} \frac{\partial}{\partial r} (r A_\phi), \right.$$
$$\left. \frac{1}{r} \frac{\partial}{\partial r} (r A_\theta) - \frac{1}{r} \frac{\partial A_r}{\partial \theta} \right)$$

Laplacian:

$$\nabla^2 \psi = \frac{1}{r^2} \frac{\partial}{\partial r} \left(r^2 \frac{\partial \psi}{\partial r} \right) + \frac{1}{r^2\sin\theta} \frac{\partial}{\partial \theta} \left(\sin\theta \frac{\partial \psi}{\partial \theta} \right) + \frac{1}{r^2\sin^2\theta} \frac{\partial^2 \psi}{\partial \phi^2}$$

Appendix F

Suggestions for further reading

There are many textbooks on plasma physics that go into greater detail and cover somewhat more advanced material than has been possible in the present text. In particular, students at the graduate level specializing in plasma physics have found N A Krall and A W Trivelpiece (1963) *Principles of Plasma Physics* (New York: McGraw-Hill, reprinted 1986 by San Francisco Press), particularly useful. Similar material, with a stronger emphasis on fusion applications, can be found in K Miyamoto (1989) *Plasma Physics for Nuclear Fusion* (Cambridge, MA: MIT Press). A recent graduate-level text, which provides a good introduction to astrophysical, geophysical as well as fusion plasmas is P A Sturrock (1994) *Plasma Physics* (Cambridge: Cambridge University Press). Textbooks that take a kinetic, or statistical, approach to the formulation of basic plasma theory include S Ichimaru (1973) *Basic Principles of Plasma Physics* (Reading, MA: Benjamin/Cummings), D R Nicholson (1983) *Introduction to Plasma Theory* (New York: Wiley) and K Nishikawa and M Wakatani (1990) *Plasma Physics* (Berlin: Springer). A recent treatment that emphasizes the theoretical foundations of the subject from a fusion perspective can be found in R D Hazeltine and J D Meiss (1992) *Plasma Confinement* (New York: Addison-Wesley). A more advanced text that focuses on developing and applying the magnetohydrodynamic model is J P Freidberg (1987) *Ideal Magnetohydrodynamics* (New York: Plenum Press).

The topic of waves in plasmas, including also instabilities such as drift waves etc, is treated extensively in T H Stix (1992) *Waves in Plasmas* (New York: American Institute of Physics). Students interested in the experimental techniques used for measuring plasma quantities in laboratory and fusion plasmas are referred to I H Hutchinson (1987) *Principles of Plasma Diagnostics* (Cambridge: Cambridge University Press).

Those interested primarily in astrophysical and solar plasmas should first develop an overall understanding of modern astrophysics, for example by studying F H Shu (1991, 1992) *The Physics of Astrophysics* Volumes I and II (Mill Valley, CA: University Science Books). They are then encouraged to read D B Melrose (1980) *Plasma Astrophysics* Volumes 1 and 2 (New York: Gordon and Breach) and H K Moffatt (1978) *Magnetic Field Generation in Electrically Conducting Fluids* (Cambridge: Cambridge University Press). Geophysical and space plasmas are described in G K Parks (1991) *Physics of Space Plasmas* (New York: Addison-Wesley).

A series of articles outlining the fundamentals of magnetically confined fusion plasmas and the status (circa 1980) of fusion experiments can be found in *Fusion* Volume I, Parts A and B, ed E Teller (1981, New York: Academic). Fusion reactors from a more engineering perspective are described in R A Gross (1984) *Fusion Energy* (New York: Wiley) and in W M Stacey (1984) *Fusion* (New York: Wiley).

Those who are interested in pursuing further the theory of plasma confinement and stability in tokamak configurations are referred to J Wesson (1989) *Tokamaks* (Oxford: Clarendon Press), to R B White (1989) *Theory of Tokamak Plasmas* (Amsterdam: North-Holland), to B B Kadomtsev (1992) *Tokamak Plasma: A Complex Physical System* (Bristol: Institute of Physics Publishing) and to D Biskamp (1993) *Nonlinear Magnetohydrodynamics* (Cambridge: Cambridge University Press).

Index